PROBLEMS & SOLUTIONS OF

Network and Systems

PROBLEMS & SOLUTIONS OF

Network and Systems

A.K. JAIRATH
M.E., F.I.E., F.I.E.T.E., H.D.S.E.

CBS

CBS PUBLISHERS & DISTRIBUTORS PVT. LTD.
New Delhi • Bangalore • Pune • Cochin • Chennai (India)

This book is dedicated to my wife Geeta and son Varun who are my source of inspiration and encouragement.

ISBN : 81-239-0949-7

First Edition : 2003
Reprint : 2009, 2010

Published by Satish Kumar Jain and produced by V.K. Jain for
CBS Publishers & Distributors Pvt. Ltd.,
CBS Plaza, 4819/XI Prahlad Street, 24 Ansari Road, Daryaganj,
New Delhi - 110002, India. • Website: www.cbspd.com
e-mail: delhi@cbspd.com, cbspubs@vsnl.com, cbspubs@airtelmail.in
Ph.: 23289259, 23266861, 23266867 • Fax: 011-23243014

Branches:

- ***Bangalore:*** Seema House, 2975, 17th Cross, K.R. Road, Bansankari 2nd Stage, Bangalore - 560070 Ph.: 26771678/79 Fax: 080-26771680 • e-mail: bangalore@cbspd.com
- ***Pune:*** Bhuruk Prestige, Sr. No. 52/12/2+1+3/2, Narhe, Haveli (Near Katraj-Dehu Road by Pass), Pune-411051 Ph.: +91-20-32404169 • Fax: 020-24464059 e-mail: pune@cbspd.com
- ***Cochin:*** 36/14, Kalluvilakam, Lissie Hospital Road, Cochin - 682018, Kerala • e-mail: cochin@cbspd.com Ph.: 0484-4059061-65 • Fax: 0484-4059065
- ***Chennai:*** 20, West Park Road, Shenoy Nagar, Chennai - 600030 e-mail: chennai@cbspd.com Ph.: 044-26260666-26202620 Fax: 044-45530020

Printed at :
J.S. Offset Printers, Delhi

Preface

It gives me immense pleasure in bringing out this book. This book is designed as a compliment to be read alongwith basic text books on network analysis. Solution of important problems have been compiled topicwise under nine different chapters for easier understanding of the theory. The problems solved inside have appeared in various university examinations which have not been mentioned below the problems, so that the book does not get generalised for a particular university.

Effort has been made to keep mistakes to the minimum. Even then, if the students are able to locate any technical or printing mistakes, the same may be brought to the notice of the author for rectification in the next edition.

A.K. JAIRATH
ME, F.I.E., F.I.E.T.E, HDSE

Contents

1

Graph Theory

Problem No. 1.1 For the given graph shown in Fig. 1.1, write the tie-set matrix

Solution Tie-set matrix gives the relation between branch currents and link currents. Link current is circulating current in a closed path. Let us redraw Fig. 1.1 by marking branches a, b, c, d, e and f, link current as i_a, i_b and i_c and branch currents as j_a, j_b, j_c, j_d, j_e and j_f as shown in Fig. 1.2.

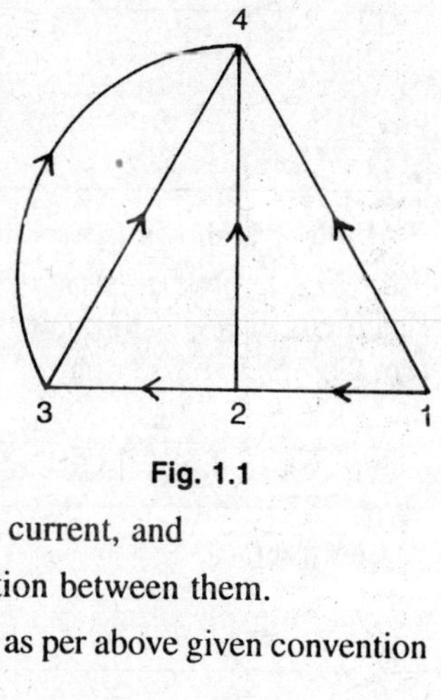

Fig. 1.1

Let us now write down the tie-set matrix by using the following convention:

(a) Enter +1, if branch current and link current coincide

(b) Enter −1, if branch current opposes link current, and

(c) Enter zero (0) when there exists no relation between them.

Form a table which when filled with entries as per above given convention will be called tie-set matrix.

Table 1.1

Link Current	Branch No					
	a	b	c	d	e	f
i_a	+1				+1	
i_b		+1		+1		
i_c			+1			

A part of the tie-set matrix, linking link current and branch current is shown in Table 1.1. In this entry of +1 has been made at those places where the link current and circulating current coincide i.e. they flow in same direction. Now let us make the entry −1 at these places where link current and branch current do not coincide i.e. they oppose each other.

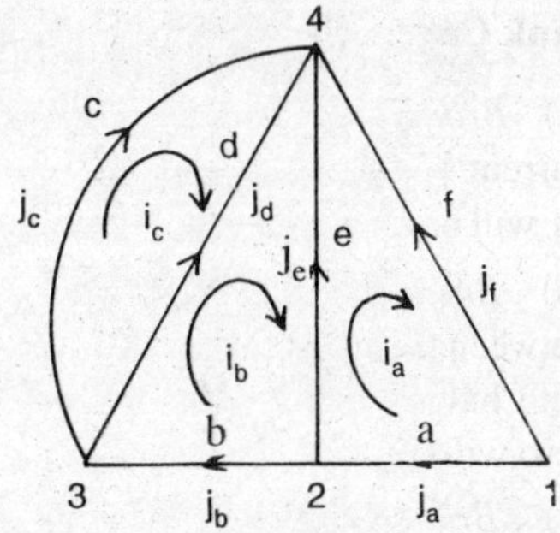

Fig. 1.2

Table 1.2

Link Current	Branch No					
	a	b	c	d	e	f
i_a	+1				+1	−1
i_b		+1		+1	−1	
i_c			+1	−1		

Table 1.2 shows that branch currents j_d, j_e and j_f oppose flow of link current i_c, i_b and i_a respectively. There exists no relation between other branch currents and link current and hence these places are entered with a zero. Complete tie-set matrix is shown in Table 1.3.

Table 1.3

Link Current	Branch No					
	a	b	c	d	e	f
i_a	+1	0	0	0	+1	−1
i_b	0	+1	0	+1	−1	0
i_c	0	0	+1	−1	0	0

Let us see the change in tie-set matrix if the graph is modified (Fig. 1.3). Let the

(a) Link currents be i_a, i_b and i_c

(b) Branches be a, b, c, d, e and f

(c) Branch currents be j_a, j_b, j_c, j_d, j_e and j_f.

Link Current 'i_a'

(a) *Branch 'a'*: Link current i_a and branch current j_a are in same direction and hence +1 will be entered against them in tie-matrix.

(b) *Branch 'b'*: i_a and j_b have no relation between them as i_a is not flowing in a closed path having branch 'b' and hence entry of zero will be made against them in tie-matrix.

(c) *Branch 'c'*: i_a is flowing in opposite direction to j_c. Therefore, entry of −1 will be made against them in tie-matrix.

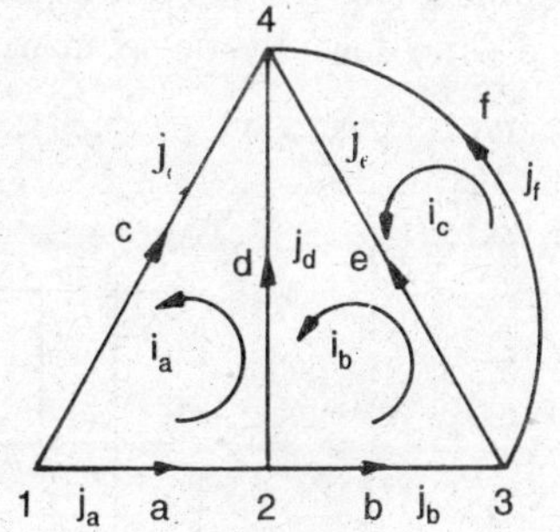

Fig. 1.3

(d) *Branch 'd'*: i_a and j_d are flowing in same direction and hence entry of +1 will be made in tie-matrix

(c) *Branch 'e'*: There exists no relation between i_a and j_e as i_a is not circulating in closed path formed by branch 'e' and hence entry of zero (0) will be made against them in tie-matrix.

(f) *Branch 'f'*: There exists no relation between i_a and j_f as i_a is not circulating in closed path formed by branch 'f' and hence entry of zero (0) will be made against them in tie-matrix.

From the above explained deductions let us complete tie-matrix (Table 1.4) relating link current i_a with all the branch currents.

Table 1.4

Link Current	Branch No					
	a	b	c	d	e	f
i_a	+1	0	−1	+1	0	0

In similar fashion relations are formulated linking link current i_b and i_c with branch currents respectively. Complete tie-matrix is tabulated in Table 1.5.

Table 1.5

Link Current	Branch No					
	a	b	c	d	e	f
i_a	+1	0	−1	+1	0	0
i_b	0	+1	0	−1	+1	0
i_c	0	0	0	0	−1	+1

Problem No. 1.2 For a resistive network shown in Fig. 1.4, draw its graph and write down the tie-set matrix and write down the KVL equations

(Circuits and Theory, 1993)

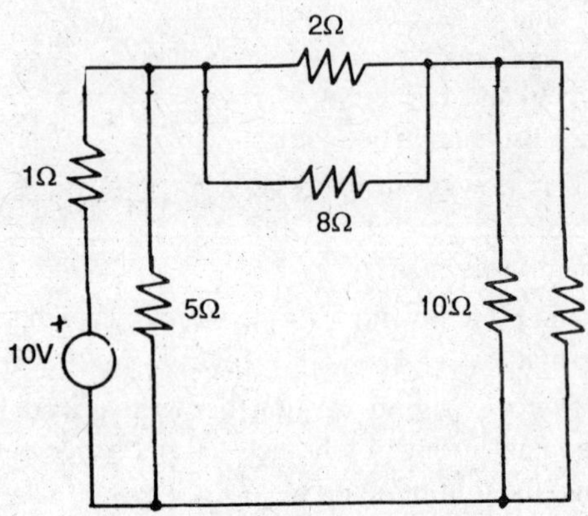

Fig. 1.4

Solution. Let us redraw the circuit as shown in Fig. 1.5.

Directed graph of the resistive network is shown in Fig. 1.6 and is marked with branch numbers, branch currents and link currents.

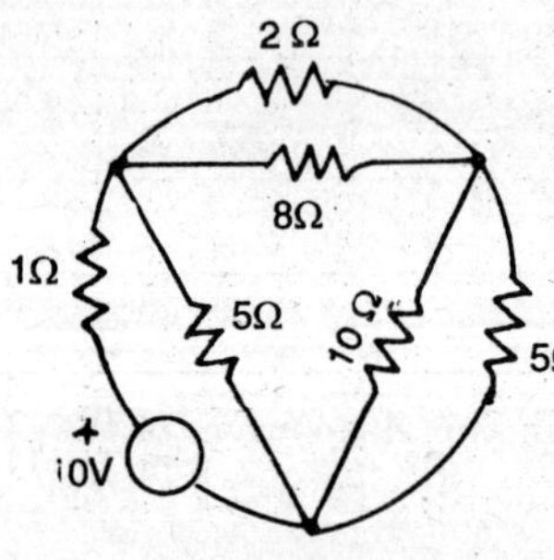

Fig. 1.5

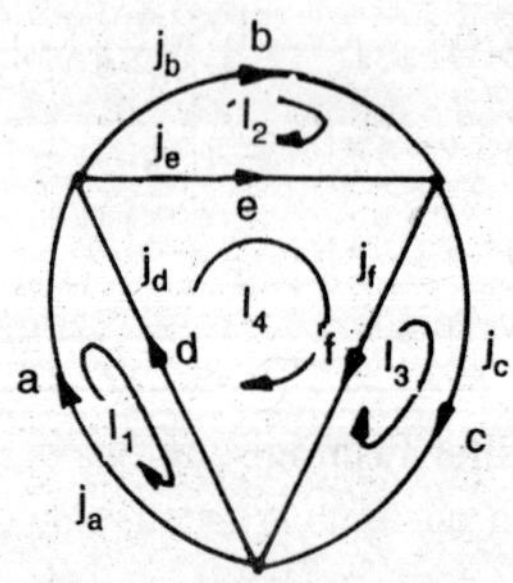

Fig. 1.6

Let

(a) Link currents be I_1, I_2, I_3 and I_4;

(b) Branch Nos be a, b, c, d, e and f and

(c) Branch currents be j_a, j_b, j_c, j_d, j_e and i_f respectively.

Based on the logical explanation given in Problem No. 1.1, tie-set matrix is shown in Table 1.6.

Table 1.6

Link Current	Branch No					
	a	b	c	d	e	f
I_1	+1	0	0	−1	0	0
I_2	0	+1	0	0	−1	0
I_3	0	0	+1	0	0	−1
I_4	0	0	0	+1	+1	+1

The columns of the tie-matrix give us

$$\left.\begin{aligned} &\text{(a)}\quad j_a = I_1, \\ &\text{(b)}\quad j_b = I_2, \\ &\text{(c)}\quad j_c = I_3, \\ &\text{(d)}\quad j_d = I_4 - I_1, \\ &\text{(e)}\quad j_e = I_4 - I_2 \text{ and} \end{aligned}\right\} \quad \text{...(1)}$$

(f) $j_f = I_4 - I_3$

The rows of the tie-set matrix give us

$$\left.\begin{aligned} &\text{(a)}\quad v_a - v_d = 0, \\ &\text{(b)}\quad v_b - v_e = 0, \\ &\text{(c)}\quad v_c - v_f = 0 \text{ and} \end{aligned}\right\} \quad \text{...(2)}$$

(d) $v_d + v_e + v_f = 0.$

Now, relating branch currents and branch voltages

$$\left.\begin{aligned} v_a &= r_1 j_a - 10 = 1 \times j_a - 10 = j_a - 10 \\ v_b &= r_2 j_b \quad = 2 \times j_b \quad = 2 j_b \\ v_c &= r_3 j_c \quad = 5 \times j_c \quad = 5 j_c \\ v_d &= r_4 j_d \quad = 5 \times j_d \quad = 5 j_d \\ v_e &= r_5 j_e \quad = 8 \times j_e \quad = 8 j_e \\ v_f &= r_6 j_f \quad = 10 \times j_f \quad = 10 j_f \end{aligned}\right\} \quad \text{...(3)}$$

Substituting the values of voltages so obtained in equation (3) in equation (2), we get

$$j_a - 10 - 5j_d = 0;$$
$$2j_b - 8j_e = 0;$$
$$5j_c - 10j_f = 0; \text{ and}$$
$$5j_d + 8j_e + 10j_f = 0$$

and in terms of currents, we get

$$I_1 - 10 - 5(I_4 - I_1) = 0;$$
$$2I_2 - 8(I_4 - I_2) = 0;$$
$$5I_3 - 10(I_4 - I_3) = 0; \text{ and}$$
$$5(I_4 - I_1) + 8(I_4 - I_2) + 10(I_4 - I_3) = 0$$

or

$$6I_1 - 5I_4 = 10;$$
$$10I_2 - 8I_4 = 0;$$
$$15I_3 - 10I_4 = 0; \quad \text{and}$$
$$-5I_1 - 8I_2 - 10I_3 + 23I_4 = 0$$

Problem No. 1.3 For the circuit shown in Fig. 1.7 write down (a) number of branches; (b) number of nodes; (c) number of links; (d) graph; (e) Cutset matrix; (f) incidence matrix.

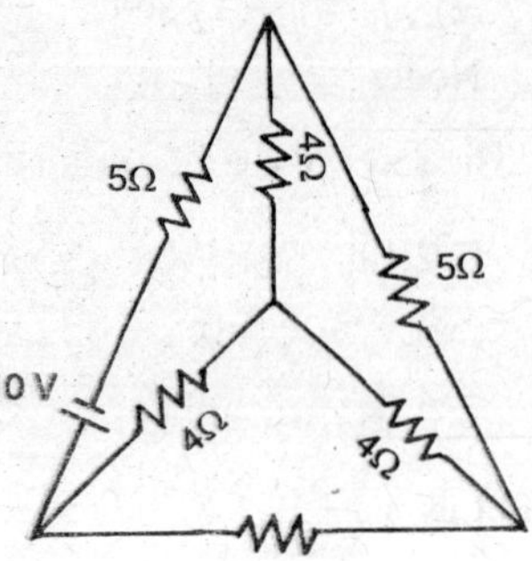

Fig. 1.7

Solution: The graph of the given electrical network is shown in Fig. 1.8.

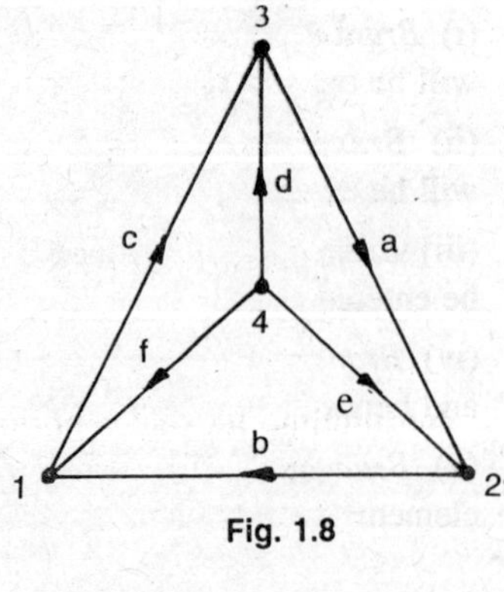

Fig. 1.8

(a) *No. of branches*: There are six branches namely a, b, c, d, e and f.

(b) *No. of nodes*: There are four nodes namely 1, 2, 3 and 4.

(e) *No. of links:*

$= b - (n-1)$

$= b - n + 1$

where b = No. of branches = 6

n = No. of nodes = 4

Therefore,

$$\text{No. of links (chords)} = 6-4+1 = 3.$$

(d) *Incidence Matrix*: Incidence matrix defines which branches are incident at which nodes and with what orientation. Thus incidence matrix gives out the incidence information in a tabulated or matrix form. For a graph having 'n' nodes and 'b' branches incidence matrix has an order of 'n×b', and has the following elements.

(a) Element 1, if the branch is oriented away from the node.

(b) Element −1, if the branch is oriented towards the node.

(c) Element zero (0), if the branch is not associated with the node.

Keeping above in view let us formulate the incidence matrix. Table 1.7 shows structure of an incidence matrix for the graph shown in Fig. 1.8.

Table 1.7

	Branches					
Nodes	a	b	c	d	e	f
1						
2						
3						
4						

Let us now fill the elements against node 1.

Node 1

(i) *Branch 'a'*: It has no association with node 1 and hence element zero (0) will be entered against them.

(ii) *Branch 'b'*: It's incidence is towards the node 1 and hence element −1 will be entered against them.

(iii) *Branch 'c'*: It incidence is away from the node and hence element 1 will be entered against them.

(iv) *Branch 'd' and 'e'*: There is no association of the branches with node 1 and hence element zero (0) will be entered against them.

(v) *Branch 'f'*: The incidence of the branch is towards the node and hence element 1 will be entered against them.

From the above deductions let us fill the incidence matrix of Fig. 1.7 against node 1 as shown in Table 1.8.

Table 1.8

	Branches					
Nodes	a	b	c	d	e	f
1	0	–1	1	0	0	–1
2						
3						
4						

By applying the same logic, let us complete the incidence matrix of Fig. 1.7. The complete incidence matrix is shown in Table 1.9.

Table 1.9

	Branches					
Nodes	a	b	c	d	e	f
1	0	–1	1	0	0	–1
2	–1	1	0	0	–1	0
3	1	0	–1	–1	0	0
4	0	0	0	–1	–1	–1

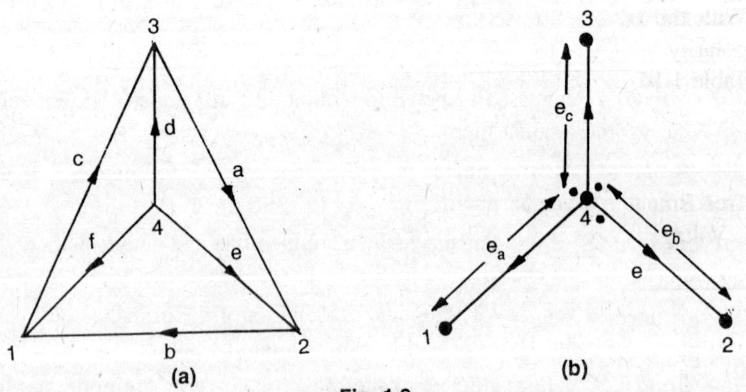

Fig. 1.9

(e) *Cut Set Matrix*: It gives the relation between branch voltages and tree branch voltages. A tree is a subgraph of a network which consists of all the nodes as in the graph, but has no closed paths.

Graph of the given network is shown in Fig. 1.8 and is redrawn in Fig. 1.9 (a) and the tree is shown in Fig. 1.9 (b).

Short circuit of all the branches except branch 'f' reduces the tree to as shown in Fig. 1.10 (a). The remaining set of branches is referred to as a cut-set.

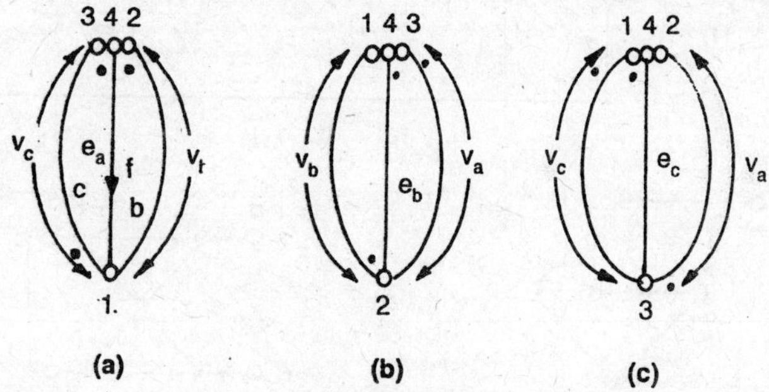

Fig. 1.10

Based on the above, entries are made in the first row of matrix (Table 1.10). The polarity has been marked with the help of a dot (.). Fig. 1.10 (a) indicates that branches 'b', 'c' and 'f' have relations with node 1 only. Branch 'b' joining nodes 1 & 2 have the same polarity as that e_a and branch 'c' joining nodes 1 & 3 has opposite polarity to e_a. Therefore, entry of +1 will be made in the first row of Table 1.10 against column 'b' and column 'f' and entry of −1 will be made against column 'c'. Since other branches have no relation with node 1, entry of 0 is made against columns 'a', 'd' and 'e'.

With the same logic draw figures 1.10 (b) and 1.10 (c) and indicate the polarity with the help of dots (.) and complete second and third rows of Table 1.10.

Table 1.10

Tree Branch Voltage	Branches					
	a	b	c	d	e	f
e_a	0	+1	−1	0	0	+1
e_b	+1	−1	0	0	+1	0
e_c	−1	0	+1	+1	0	0

Problem No. 1.4. In Fig. 1.11 (a), a resistive network is shown. In Fig. 1.11 (b) and 1.11 (c) are shown graphs with two of the four nodes identified. For these two graphs, assign resistors to the branches and identify the two remaining nodes such that the resulting networks are topologically identical to the one shown in Fig. 1.11 (a).

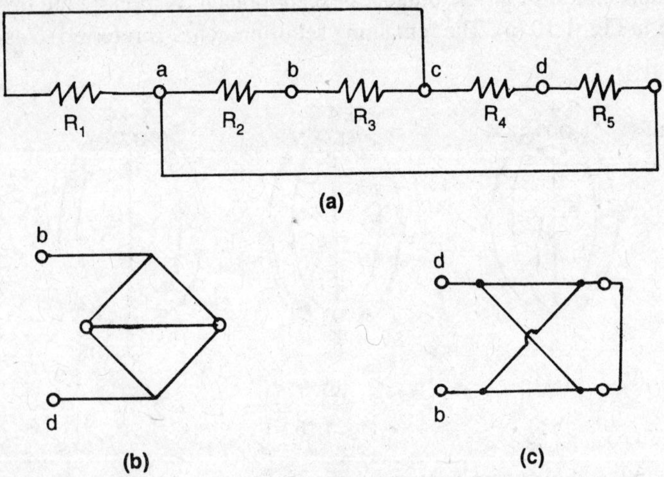

Fig. 1.11

Solution

Fig. 1.11 (b): Assign two nodes as 'a' and 'c' as shown in Fig. 1.12 (a) and assign the resistors as shown in Fig. 1.12 (b) which clearly indicates that resistive network of Fig. 1.11 (a) and Fig. 1.12 (b) are topologically identical.

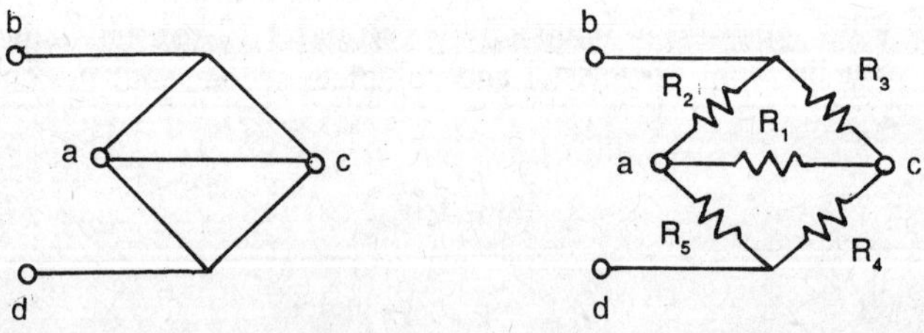

Fig. 1.12

Similarly assign two nodes to Fig. 1.11 (b) as shown in Fig. 1.13 (a) and assign resistors as shown in Fig. 1.13 (b) which clearly indicates that resistive networks of Fig. 1.11 (a) and Fig. 1.13 (b) are topologically identical.

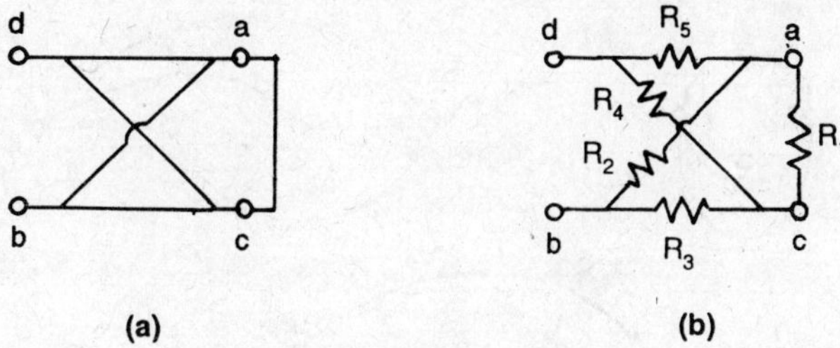

Fig. 1.13

Problem No. 1.5. Three graphs are shown in Fig. 1.14. Classify each of the graphs as planer or nonplaner.

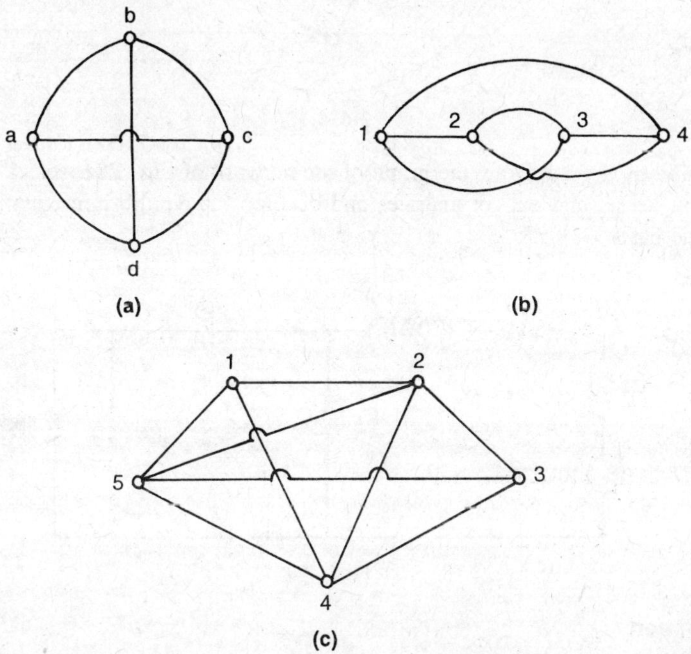

Fig. 1.14

Solution: Planer graphs are those which can be drawn on a sheet of paper without crossing lines where as non-planer graphs cannot be drawn on a sheet of paper without crossing lines.

Keeping above in view, let us try to draw the graphs without crossing lines. It is seen that graphs can be drawn without crossing lines as shown in Fig. 1.15.

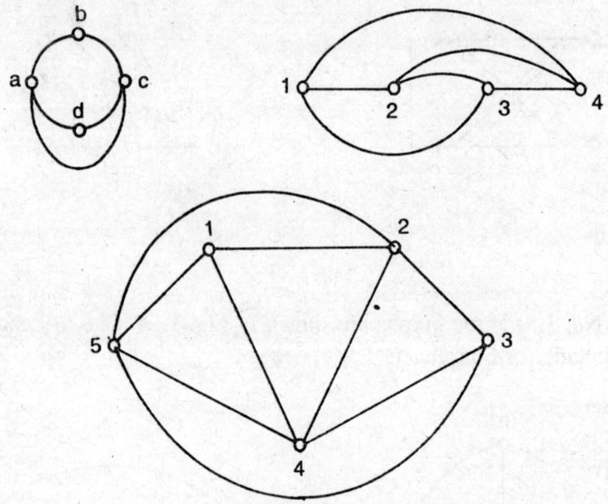

Fig. 1.15

Problem No. 1.5 Draw the graph of the network of Fig. 1.16, select a tree, form tie set and cut-set matrices and develop the equilibrium equation on node basis.

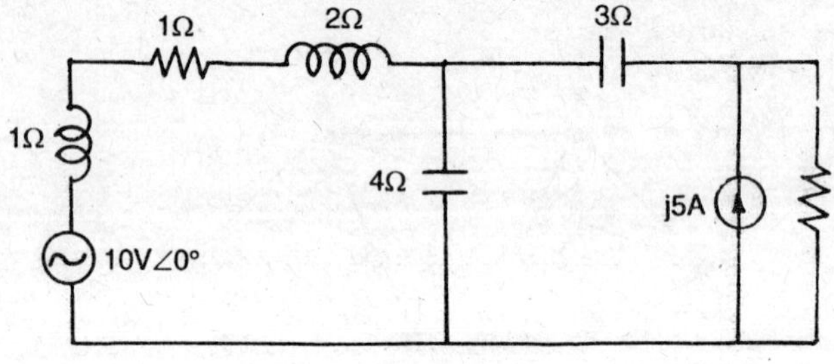

Fig. 1.16

Solution

Mark the nodes as shown in Fig. 1.17 and draw the graph and number the branches for easy identification as a, b, c, d and e as shown in Fig. 1.18. The number of nodes are three and hence number of tree branches

$$= n-1$$

$$= 3-1$$

$$= 2$$

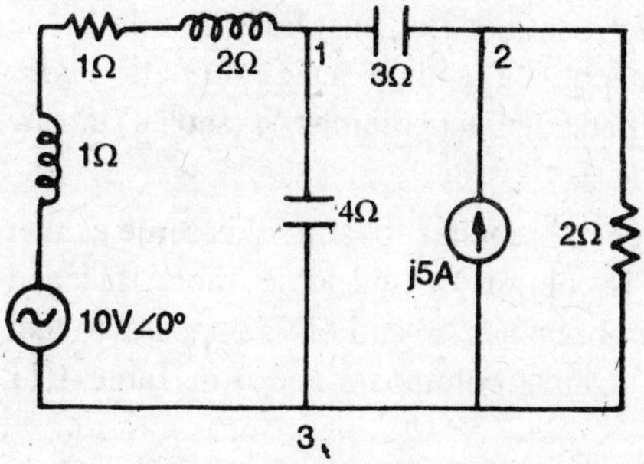

Fig. 1.17

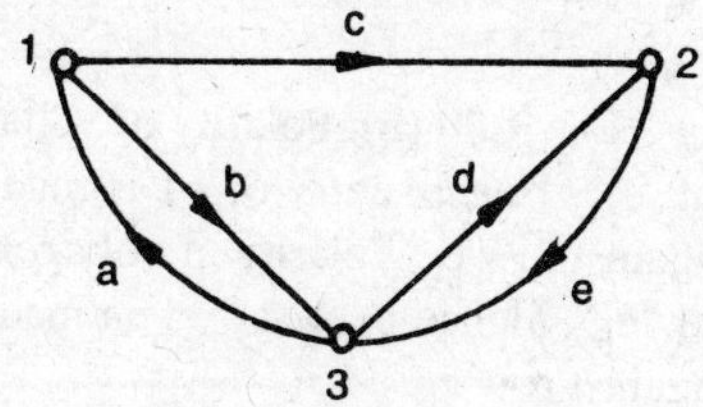

Fig. 1.18

The selected tree from the drawn graph is shown in Fig. 1.19. The polarity is marked with dots (.). First short circuiting branch 'c' of Fig. 1.19, we draw Fig. 1.20 (a) and mark the polarity by dots (.) and short circuiting branch "b', we draw Fig. 1.20 (b) and mark the polarity by dots (.). With the help of Fig. 1.20 (a) and Fig. 1.20 (b) we complete the first and second row of cutset matrix as shown in Table 1.11.

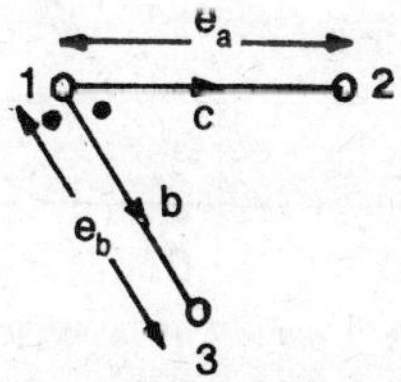

Fig. 1.19

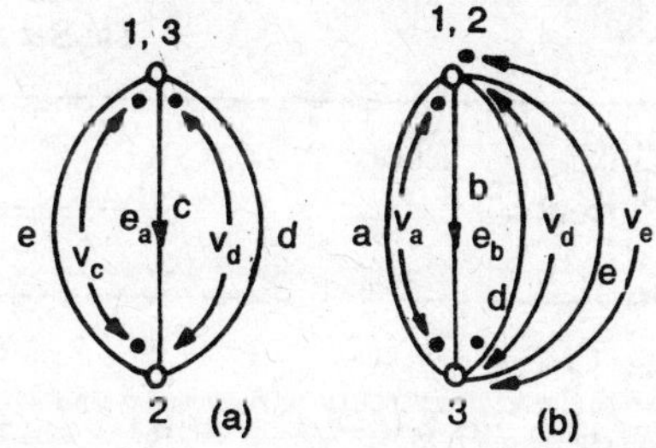

Fig. 1.20

Table 1.11
Cut-Set Schedule

Tree Branch Voltage	Branches a	b	c	d	e
e_a	0	0	+1	+1	−1
e_b	−1	+1	0	−1	+1

Fig. 1.20 (a). Polarity of voltage 'v_d' in branch 'd' is same as that of voltage 'e_a' in 'c', while voltage polarity of 'v_e' in branch 'e' is opposite to that of e_a in branch 'c'. Similarly, polarity of voltage in branch 'c' is same as that of 'e_a'. Hence entry of +1 is made against columns c and d in Table 1.11,

being of same polarity and entry of −1 is made against column e in Table 1.11, being of opposite polarity. Branches 'a' and 'b' have no relation with 'e_a'. Therefore, entry of zero (0) is made against columns 'a' and 'b' in row 'e_a' of Table 1.11.

Fig. 1.20 (b). Polarity of voltages in branches 'b' and 'e' is same as that of 'e_b'. Hence, entry of +1 is made is column 'b' and 'e' in Table 1.11 and against row e_b. Polarity of voltages in branches 'a' and 'd' is opposite to that of 'e_b'. Hence entry of −1 is made against columns a and d in Table 1.11 against row e_b.

Branch 'c' has no relation with 'e_b' and has entry of zero (0) is made against column c against row 'e_b'. This completes row e_b of Table 1.11.

Tie Set Matrix Let us draw the graph again duly marked with circulating currents and branch currents (Fig. 1.21).

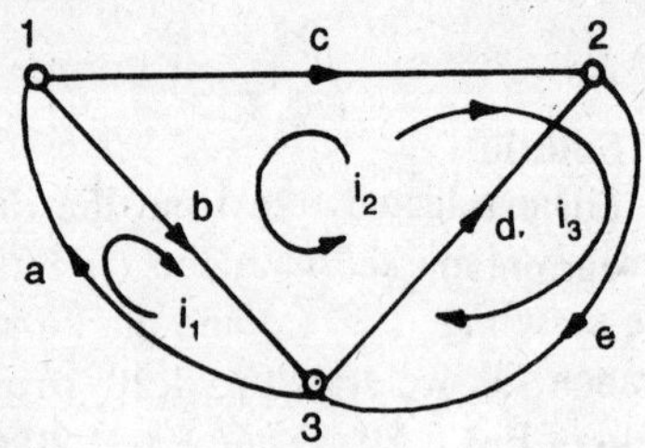

Fig. 1.21

Table 1.12
Tie-Set Schedule

Link Current	Loop	Branch a	b	c	d	e
i_1	a–b–a	+1	+1	0	0	0
i_2	c–b–d–c	0	+1	−1	+1	0
i_3	c–e–b–c	0	−1	0	+1	+1

Note: Refer Problem No. 1.1 in which complete explanation is given for formulation of tie-set matrix.

Equilibrium equations formed on node-basis based on Cut-Set Matrix

From row e_a

$$j_c + j_d - j_e = 0$$

From row e_b

$$-j_a + j_b - j_c + j_d = 0$$

where j_a, j_b, j_c, j_d and j_e are branch currents.

Problem No. 1.6. For the network shown in Fig. 1.22, write down the tie-set matrix and obtain the network equilibrium equations in matrix form using KVL and calculate the loop currents.

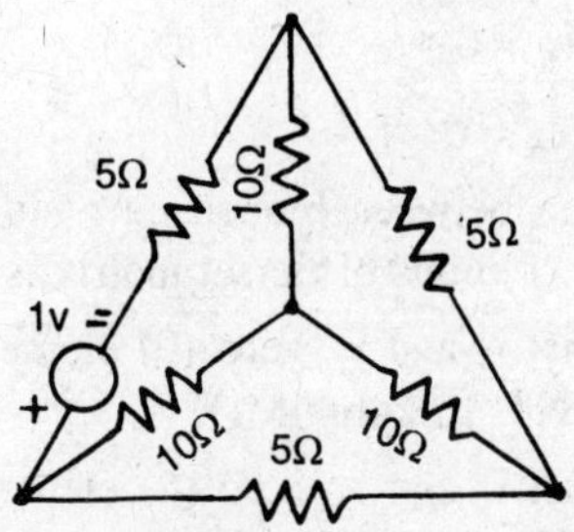

Fig. 1.22

Solution. For the network shown in Fig. 1.22, graph is drawn in Fig. 1.23(a). For easy identification branch numbers are marked 1 to 6 and circulating currents indicated in Fig. 1.23 (b).

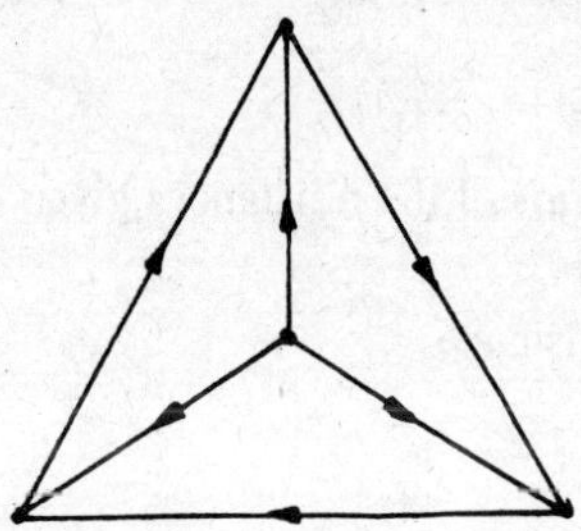

Fig. 1.23 (a)

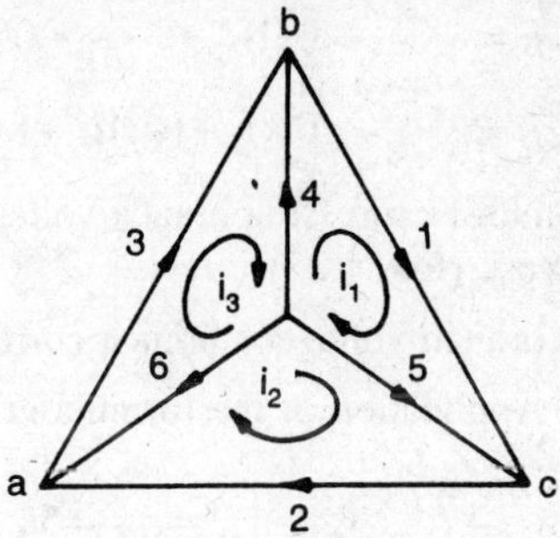

Fig. 1.23 (b)

Tie-set matrix is shown in Table 1.13

Table 1.13
Tie-Set Matrix

Link Current	Loop	Branches					
		1	2	3	4	5	6
		j_1	j_2	j_3	j_4	j_5	j_6
i_1	1–5–4–1	+1	0	0	+1	–1	0
i_2	2–6–5–2	0	+1	0	0	+1	–1
i_3	6–3–4–6	0	0	+1	–1	0	+1

From tie-set matrix KVL equations, are:

$$v_1 + v_4 - v_5 = 0 \qquad \text{(row 1)}$$

$$v_2 + v_5 - v_6 = 0 \qquad \text{(row 2)}$$

$$v_3 - v_4 + v_6 = 0 \qquad \text{(row 3)}$$

Also writing relation between branch currents and link currents. For this, reading against the columns of tie-set matrix is made which gives

(a) $j_1 = i_1$, as against i_2 and i_3, zero (0) elements are entered against Branch No. 1 (Column 1).

(b) $j_2 = i_2$, as against i_1 and i_3, zero (0) elements are entered against Branch No. 2 (Column 2).

(c) $j_3 = i_3$, as against i_1 and i_2, zero (0) elements are entered against Branch No. 3 (Column 3).

(d) $j_4 = 1 \times i_1 + 0 \times i_2 + (-1)i_3 = i_1 - i_3$

(e) $j_5 = (-1)i_1 + 1 \times i_2 + 0 \times i_3 = -i_1 + i_2 = i_2 - i_1$

(f) $j_6 = 0 \times i_1 + (-1)i_2 + 1 \times i_3 = -i_2 + i_3 = i_3 - i_2$

Now, expressing branch voltages in terms of the resistances given in the network (Fig. 1.22).

branch voltage = branch current × resistance.

Application of the formula gives

$$v_1 = r_1 j_1 = r_1 i_1 = 5i_1$$

$$v_2 = r_2 j_2 = r_2 i_2 = 5i_2$$

$$v_3 = r_3 j_3 - 1 = r_3 i_3 - 1 = 5i_3 - 1.$$

$$v_4 = r_4 j_4 = r_4 (i_1 - i_3) = 10(i_1 - i_3).$$

$$v_5 = r_5 j_5 = r_5 (i_2 - i_1) = 10(i_2 - i_1).$$

$$v_6 = r_6 j_6 = r_6 (i_3 - i_2) = 10(i_3 - i_2)$$

substituting the values of branch voltages so obtained in KVL equations derived earlier, we get

(a) $v_1 - v_5 + v_4 = 0,$ (found earlier)

or $5i_1 - 10(i_2 - i_1) + 10(i_1 - i_3) = 0$

or $25i_1 - 10i_2 - 10i_3 = 0$ ...(1)

(b) $v_5 + v_2 - v_6 = 0$ (found earlier)

or $10(i_2 - i_1) + 5i_2 - 10(i_3 - i_2) = 0$

or $\quad -10i_1 + 25i_2 - 10i_3 = 0$ ...(2)

(c) $\quad v_3 - v_4 + v_6 = 0$

or $\quad (5i_3 - 1) - 10(i_1 - i_3) + 10(i_3 - i_2) = 0$

or $\quad -10i_1 - 10i_2 + 25i_1 = 1$ ...(3)

In matrix form the equations (1), (2) and (3) can be written as

$$\begin{bmatrix} i_1 & i_2 & i_3 \end{bmatrix} \begin{bmatrix} 25 & -10 & -10 \\ -10 & 25 & -10 \\ -10 & -10 & 25 \end{bmatrix} = \begin{bmatrix} 0 \\ 0 \\ 1 \end{bmatrix}$$

For calculating the values of link (circulating), current or mesh currents i.e. values of i_1, i_2 and i_3, equations (1), (2) and (3) will have to be solved

Multiply equation (2) by a factor 2.5 and adding to equation (1), we get

$$52.5i_2 - 35i_3 = 0$$

or $\quad i_2 = \frac{35}{52.5} i_3 = 0.67 i_3$ ...(4)

Subtracting equation (3) from equation (2) gives

$$35i_2 - 35i_3 = -1 \quad \text{...(3)}$$

Putting value of i_2 from equation (4) in equation (5), we get

$$35 \times \frac{35}{52.5} i_3 - 35i_3 = -1$$

or $\quad 35i_3 - 52.5i_3 = \frac{-52.5}{35}$

or $\quad i_3 = \frac{-52.5}{35 \times 17.5} = \frac{3}{35} = 0.086$ amp

Putting the value $i_3 = 0.086$ in equation (4) we get

$$i_2 = 0.67 \times 0.086 = 0.057 \text{ amp.}$$

Putting the value of i_2 and i_3 in equation (1), gives

$$25i_1 - 10 \times 0.057 - 10 \times 0.086 = 0$$

or $\quad i_1 = 0.057$ amp.

Problem No. 1.7. Write the fundamental loop and fundamental cut-set matrices for the graph in Fig. 1.24 for the tree (1, 2, 3, 4)

Solution. Tree 1, 2, 3, 4 of the above graph is shown in Fig. 1.25.

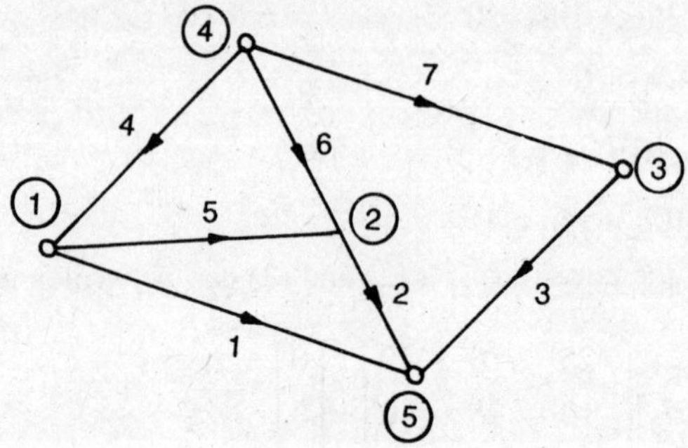

Fig. 1.24

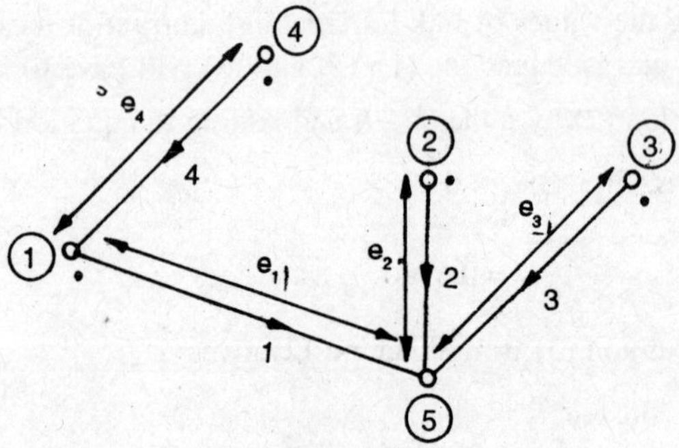

Fig. 1.25

Fundamental Loop matrix In the tree shown in Fig. 1.25, branches (1, 2, 3, 4) are the twigs and branches (5, 6, 7) are the links. The fundamental loops are formed by replacing one by one the links and then defining the closed path.

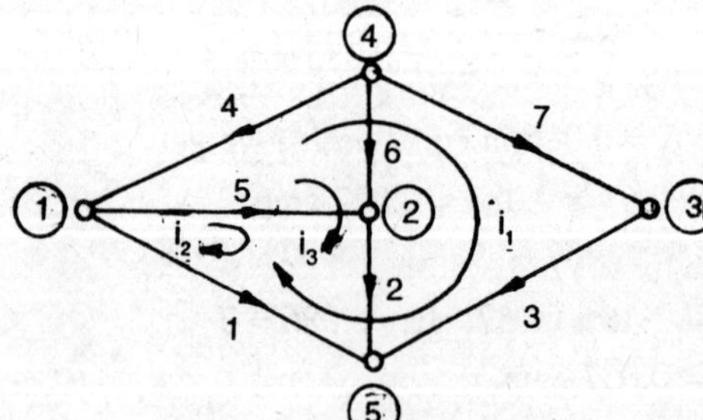

Fig. 1.26

(a) Replacement of link 7 gives a closed path defined by branches (1, 4, 7, 3) and closed by nodes (1, 4, 3, 5). Let the circulating current be i_1.

(b) Replacement of link 5 gives a closed path defined by branches (1, 5, 2) and closed by nodes (1, 2, 5). Let the circulating current be i_2.

(c) Replacement of link 6 gives a closed path defined by branches (1, 4, 6, 2) and closed by nodes (1, 4, 2, 5). Let the circulating current be i_3.

Deductions obtained above are marked on the graph shown in Fig. 1.26.

The fundamental loop matrix is shown in Table 1.14. The elements in the table have been based on the directions of the circulating current and branch currents.

Table 1.14
Fundamental Loop Matrix

Link Current	Branches						
	1	2	3	4	5	6	7
i_1	–1	0	+1	–1	0	0	+1
i_2	–1	+1	0	0	+1	0	0
i_3	–1	+1	0	–1	0	+1	0

Fundamental Cut–Set matrix: The fundamental cut-set matrix is one in which each row of the matrix represents a cut-set with respect to a given tree of the graph. The graph is redrawn in Fig. 1.27. The given tree (1, 2, 3, 4) is drawn with dark lines and the remaining branches with broken lines. The directed cut-sets are also shown.

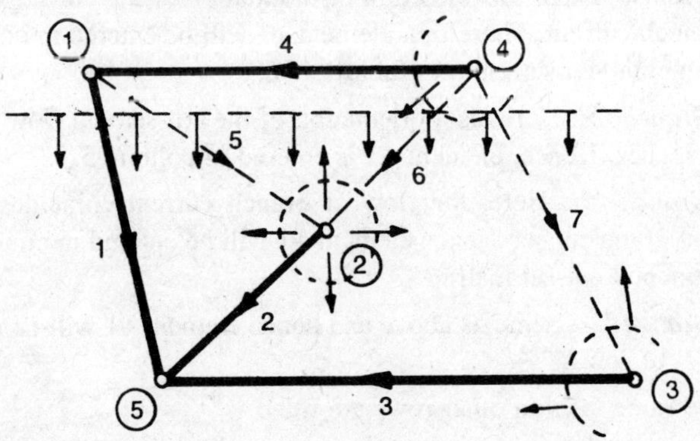

Fig. 1.27

The fundamental cut-set is shown in Table 1.15.

Table 1.15
Fundamental Cut-Set Matrix

f-Cut-Sets	Branches						
	1	2	3	4	5	6	7
1-5-6-7	+1	0	0	0	+1	+1	+1
2-5-6	0	+1	0	0	−1	−1	0
3-7	0	0	+1	0	0	0	−1
4-6-7	0	0	0	+1	0	+1	+1

The entries in the fundamental cut-set-matrix is made based on the direction of the cut-set and direction of flow of current in branches. Let us fill row No. 1 of f-cut-set matrix.

Node 1

The direction of the cut-set matrix is shown by arrows shown on the straight line (broken) drawn at node 1, intersecting branches 1, 5, 6 & 7. This cut-set semi circle, cuts branches 1, 5 and 4 and hence, entries of the elements will be made in the columns 1, 4 and 5 only i.e. either element +1 will be entered or −1 will be entered based on orientation of cut-set and flow of currents in the branches. Other columns in the first row will be entered by zero (0) element.

(a) *Branch '1'*: In this direction or orientation of the cut-set is same as that of branch current. Therefore, element +1 will be entered in column 1 of the fundamental cut set matrix (Table 1.15).

(b) *Branch '5'*: Here also, orientation of the cut-set and flow of branch current matches. Hence, element +1 is entered in column 5.

(c) *Branch '6'*: Here, the flow of branch current coincides with the orientation of the cut set. Hence element +1 will be entered in column '6' of the fundamental cut-set matrix.

(d) *Branch '7'*: Same as above and hence element +1 will be entered in column '7' of row '1'.

In the same fashion other rows are filled.

Problem No. 1.8. Select the fundamental loop and fundamental cut-sets corresponding to a tree of the network graph which is shown by solid lines in Fig. 1.28. Hence write the KCL and KVL equations for the network in matrix form.

Solution

Fundamental Loop From Fig. 1.28, it can be seen that

(a) Twigs are branches 3, 4, 5 and 6.

(b) Links are branches 1, 2 and 7.

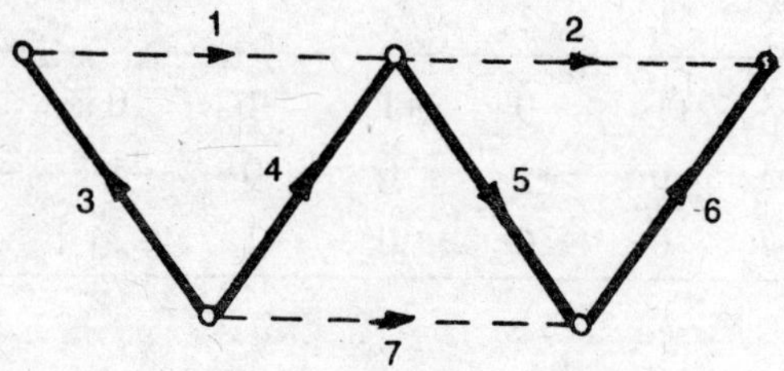

Fig. 1.28

If links are removed, tree defined by branches (3, 4, 5, 6) is formed and is shown by thick lines. The loops are obtained if links are replaced one by one.

(a) Replacement of link 1 defines the loop formed by branches (1, 3, 4). Let i_1 is the link current

(b) Replacement of link 2 defines the loop formed by branches (2, 5, 6). Let i_2 be the link current.

(c) Replacement of link 7 defines the loop formed by branches (4, 5, 7). Let i_3 be the circulating current.

The circulating currents, i_1, i_2 and i_3 are shown in Fig. 1.29 in the loops obtained above.

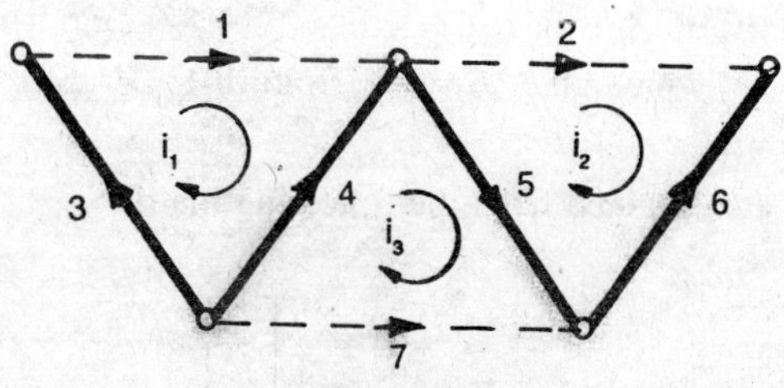

Fig. 1.29

The fundamental loop matrix is shown in Table 1.16. Method to fill the elements has been explained in previous problems.

Table 1.16
f-Loop Matrix

Link Current	f-loop	Branches j_1 1	j_2 2	j_3 3	j_4 4	j_5 5	j_6 6	j_7 7
i_1	1–4–3	+1	0	+1	–1	0	0	0
i_2	2–6–5	0	+1	0	0	–1	–1	0
i_3	4–5–7	0	0	0	–1	–1	0	+1

In matrix form

$$\begin{bmatrix} j_1 \\ j_2 \\ j_3 \\ j_4 \\ j_5 \\ j_6 \\ j_7 \end{bmatrix} = \begin{bmatrix} 1 & 0 & 1 & -1 & 0 & 0 & 0 \\ 0 & 1 & 0 & 0 & -1 & -1 & 0 \\ 0 & 0 & 0 & -1 & -1 & 0 & +1 \end{bmatrix}^T \begin{bmatrix} i_1 \\ i_2 \\ i_3 \end{bmatrix}$$

From the matrix

(a) $j_1 = i_1$

(b) $j_2 = i_2$

(c) $j_3 = i_1$ which gives $j_1 = j_3$ (from (a))

(d) $j_4 = -i_3 - i_1$

(e) $j_5 = -i_3 - i_2$

(f) $j_6 = -i_2$ which gives $j_2 = -j_6$ (from (b))

(g) $j_7 = -i_3$

KCL equations are obtained from the following matrix

$$\begin{bmatrix} 1 & 0 & 1 & -1 & 0 & 0 & 0 \\ 0 & 1 & 0 & 0 & -1 & -1 & 0 \\ 0 & 0 & 0 & -1 & -1 & 0 & +1 \end{bmatrix} \begin{bmatrix} j_1 \\ j_2 \\ j_3 \\ j_4 \\ j_5 \\ j_6 \\ j_7 \end{bmatrix} = 0$$

which gives

$$j_1 + j_3 - j_4 = 0 \quad ...(1)$$

$$j_2 - j_5 - j_6 = 0 \quad ...(2)$$

$$-j_4 - j_5 + j_7 = 0 \quad ...(3)$$

Writing the above derived equations in terms of circulating current i_1, i_2 and i_3, we get

$$i_1 + i_1 + i_3 + i_1 = 0$$

or $$3i_1 + i_3 = 0\,; \quad ...(4)$$

$$i_2 + i_3 + i_2 + i_2 = 0$$

or $$3i_2 + i_3 = 0 \quad ...(5)$$

$$+i_3 + i_1 + i_3 + i_2 + i_3 = 0$$

$$i_1 + i_2 + 3i_3 = 0 \quad ...(6)$$

Fundamental Cut-set matrix

Selection of Cut-sets: Consider the given tree defined by branches (3, 4, 5, 6) as shown in Fig. 1.28 by thick lines.

(a) First remove twig 3 from the tree. This disconnects the tree into two pieces. Therefore, the twig 3 and the link 1 which go from one part of this disconnected tree to the other will constitute a cut-set. Therefore, (3, 1) is a f-cut-set.

(b) Now remove twig 4 from the tree. This also disconnects the tree into two pieces. Therefore, the twig 4 and the links 1 and 7 which go from one part of the disconnected tree to the other will constitute a cut-set. Therefore, (4, 1, 7) is another f-cut-set.

(c) After this remove twig 5 from the tree. This also disconnects the tree into two pieces. Therefore, twig 5 and the links 2 and 7, which go from one part of the disconnected tree to the other form a cut-set. Hence, (5, 2, 7) is a f-cut-set.

(d)Similarly, removal of twig 6 from the tree disconnects the tree into two pieces. Therefore, twig 6 alongwith link 2 which go from one part of the disconnected tree to the other form the fourth cut-set. Hence, (6, 2) is also a f-cut-set.

Orientation of Cut-set: The directions of the cut-sets have been chosen to coincide with that of the branches of the tree defining the cut-set.

The graph is redrawn (Fig. 1.30) showing twigs defining the tree and the **links alongwith the** cut-sets and their orientation.

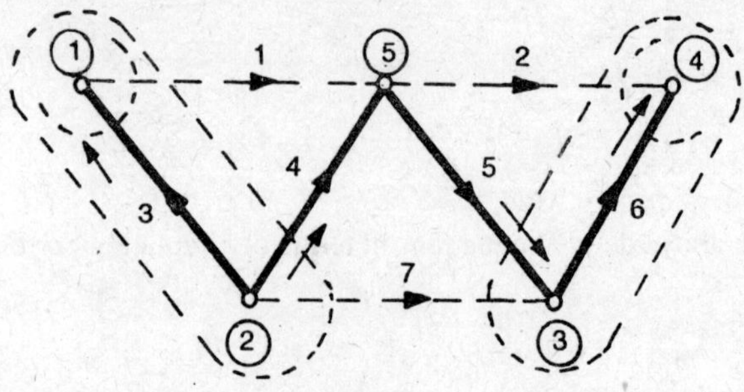

Fig. 1.30

Fundamental cut-set matrix related to the given tree is shown in Table 1.17.

Table 1.17
Fundamental Cut-set Matrix

f-Cut Set	Branches						
	1	2	3	4	5	6	7
(3, 1)	–1	0	+1	0	0	0	0
(4, 1, 7)	+1	0	0	+1	0	0	+1
(5, 2, 7)	0	+1	0	0	+1	0	+1
(6, 2)	0	+1	0	0	0	+1	0

In matrix form

$$\begin{bmatrix} v_1 \\ v_2 \\ v_3 \\ v_4 \\ v_5 \\ v_6 \\ v_7 \end{bmatrix} = \begin{bmatrix} -1 & 0 & +1 & 0 & 0 & 0 & 0 \\ 1 & 0 & 0 & +1 & 0 & 0 & +1 \\ 0 & +1 & 0 & 0 & +1 & 0 & +1 \\ 0 & +1 & 0 & 0 & 0 & +1 & 0 \end{bmatrix}^T \begin{bmatrix} e_1 \\ e_2 \\ e_3 \\ e_4 \end{bmatrix}$$

From the above matrix

(a) $v_1 = -e_1 + e_2$

(b) $v_2 = e_3 + e_y$

(c) $v_3 = e_1$

(d) $v_4 = e_2$

(e) $v_5 = e_3$

(f) $v_6 = e_4$

(g) $v_7 = e_2 + e_3$

KVL equations are obtained from the following matrix

$$\begin{bmatrix} -1 & 0 & +1 & 0 & 0 & 0 & 0 \\ +1 & 0 & 0 & +1 & 0 & 0 & +1 \\ 0 & +1 & 0 & 0 & +1 & 0 & +1 \\ 0 & +1 & 0 & 0 & 0 & +1 & 0 \end{bmatrix} \begin{bmatrix} v_1 \\ v_2 \\ v_3 \\ v_4 \\ v_5 \\ v_6 \\ v_7 \end{bmatrix} = 0$$

The equations are

(a) $-v_1 + v_3 = 0$

(b) $v_1 + v_4 + v_7 = 0$

(c) $v_2 + v_5 + v_7 = 0$

(d) $v_2 + v_6 = 0$

Problem 1.9 Draw the graph of the network shown in Fig. 1.31. Select a tree consisting of elements 1 and 2 and write the equilibrium equations taking branch voltages as variable.

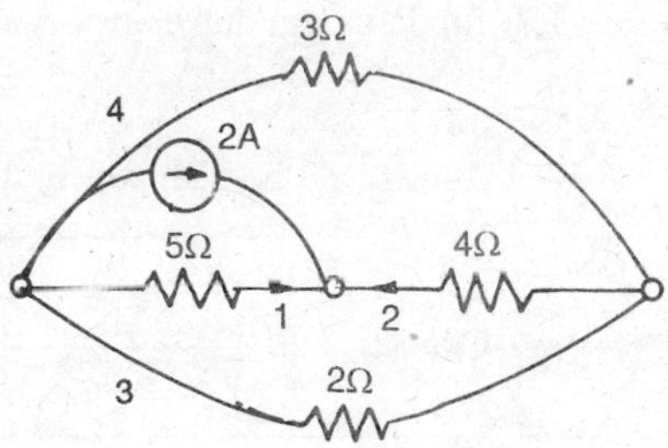

Fig. 1.31

Solution Redrawing the given electrical network after converting the current source into voltage source, we get the network shown in Fig. 1.32.

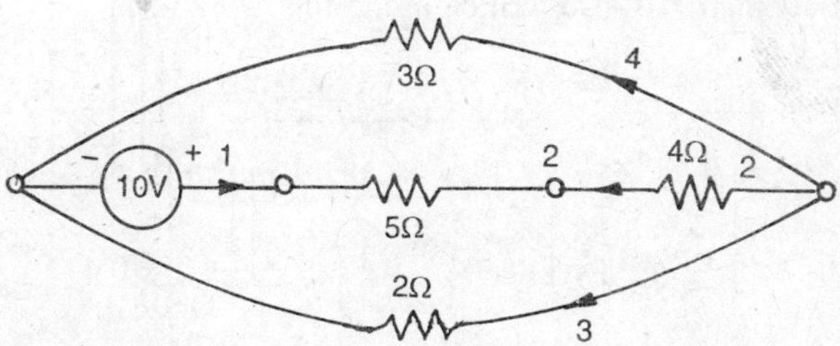

Fig. 1.32

Graph of the network is shown in Fig. 1.33 (a) and tree of the graph consisting of element 1 and 2 is shown in Fig. 1.33 (b).

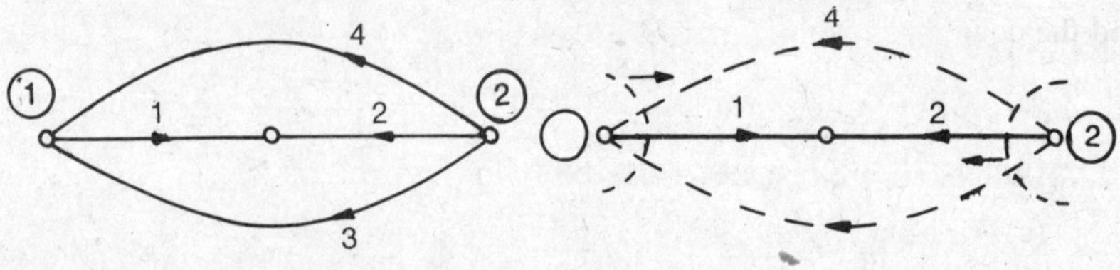

Fig. 1.33

The twigs of the tree are branches 1 and 2 and links are branches (3, 4). Removal of twig 1 disconnects the tree into two pieces. Therefore, the twig 1 and the two links 3 and 4 constitute cut-set. Hence, f-cut-set is (1, 3, 4). Similarly, removal of twig 2 also disconnects the tree into two pieces. Therefore, twig 2 and links 3 and 4 constitute another cut-set. Hence (2, 3, 4) is another f-cut-set. Orientation of the f-cut sets is marked in Fig. 1.33 (b) with the help of semi-circles drawn with broken lines and coincides with the orientation of the branches defining the f-cut-sets respectively.

The fundamental cut-matrix is shown in Table 1.18.

Table 1.18
Fundamental Cut-Set Matrix

f-Cut-Set		Branches 1	2	3	4
e_1	(1, 3, 4)	1	0	−1	−1
e_2	(2, 3, 4)	0	1	1	1

The f-Cut-set matrix is also written in the following form

$$\underline{\text{f-Cut-Sets}} \quad \underline{1 \quad 2 \quad 3 \quad 4}$$

$$\begin{bmatrix} 1, 3, 4 \\ 2, 3, 4 \end{bmatrix} \begin{bmatrix} 1 & 0 & -1 & -1 \\ 0 & 1 & 1 & 1 \end{bmatrix} \begin{bmatrix} v_1 \\ v_2 \\ v_3 \\ v_4 \end{bmatrix} = 0$$

KVL equations are

(a) $v_1 - v_3 - v_4 = 0$, and

(b) $v_2 + v_3 + v_4 = 0$

and the equations relating branch voltage with e_1 and e_2 are

(a) $e_1 = v_1$,

(b) $e_2 = v_2$,

(c) $v_3 = e_2 - e_1$, and

(d) $v_4 = e_2 - e_1$.

Problem No. 1.10 The incidence matrix of a network graph is given in Fig. 1.34. Draw its oriented graph.

Branches

$$A = \begin{array}{c} \begin{matrix} 1 & 2 & 3 & 4 & 5 & 6 & 7 & 8 \end{matrix} \\ \begin{bmatrix} 1 & 0 & 0 & 0 & 1 & 0 & 0 & 1 \\ 0 & 1 & 0 & 0 & -1 & 1 & 0 & 0 \\ 0 & 0 & 1 & 0 & 0 & -1 & 1 & -1 \\ 0 & 0 & 0 & 1 & 0 & 0 & -1 & 0 \end{bmatrix} \end{array}$$

Fig. 1.34

Solution Incidence matrix gives an indication of the way in which branches are oriented towards a node. The sign convention for filling the elements has been explained in Problem No. 1.1. Refer Problem No. 1.16 also to draw the oriented graph.

Row 1 It indicates branches 1, 5 and 8 are oriented or directed away from the node 1 and is shown in Fig. 1.35 (a).

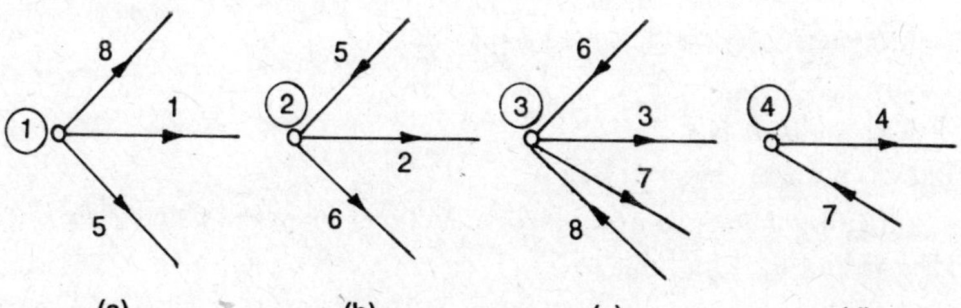

Fig. 1.35

From the incidence matrix, one deduces that there are four nodes. However, analysis of Fig. 1.35 indicates presence of another node whose details have not been entered in the incidence matrix.

Fig. 1.35 (a) It is seen that branch No. 1 is oriented away from node 1 but is not directed towards the nodes 2, 3 and 4 because Figs. 1.35 (b), 1.35 (c) and 1.35 (d) does not show any relation of nodes 2, 3 and 4 with branch 1.

Fig. 1.35 (b) Branch No. 2 is oriented away from the node 2 but does not incident on any of the nodes (1, 3 and 4).

Fig. 1.35 (c) Branch No. 3 is directed away from the node 3 but does not incident on any of the nodes (1, 2 and 4).

Fig. 1.35 (d) Branch No. 4 is directed away from the node 4 but does not incident on any of the nodes (1, 2 and 3).

It means there is another node, say node No. 5. Branches 1, 2, 3 and 4 which are directed away from nodes 1, 2, 3 and 4, are directed towards node 5.

Fig. 1.36 (a) shows five nodes and orientation of branches 1, 2, 3 and 4.

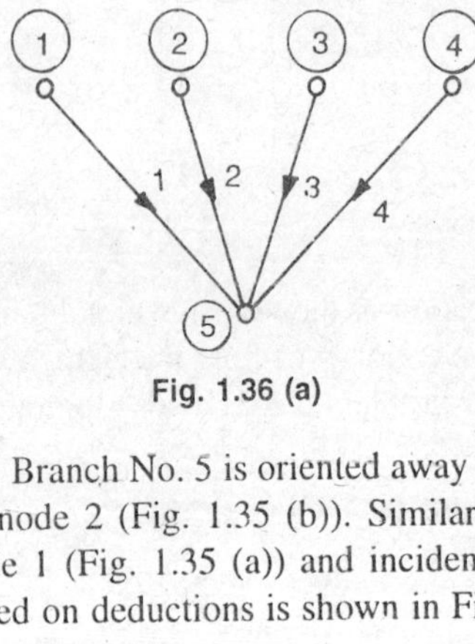

Fig. 1.36 (a)

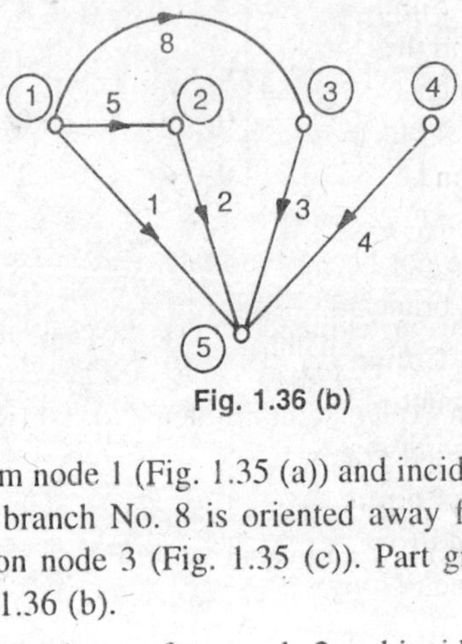

Fig. 1.36 (b)

Branch No. 5 is oriented away from node 1 (Fig. 1.35 (a)) and incidents on node 2 (Fig. 1.35 (b)). Similarly, branch No. 8 is oriented away from node 1 (Fig. 1.35 (a)) and incidents on node 3 (Fig. 1.35 (c)). Part graph based on deductions is shown in Fig. 1.36 (b).

Branch No. 6 (Fig. 1.35 (b)) is directed away from node 2 and incidents on node 3 (Fig. 1.35 (c)).

Also, branch No. 7 is directed away from node 3 (Fig. 1.35 (c)) and incidents on node 4 Fig. ((1.35 (d)).

Based on the above mentioned deductions, branches 6 and 7 are drawn as shown in Fig. 1.36 (c).

Complete directed graph is drawn in Fig. 1.36 (d). This is obtained by combining graphs of Figs. 1.36 (c) and 1.36 (b).

Similar deductions can be drawn from incidence matrix directly.

Branch 1 (Column 1) Entries (1, 0, 0, 0) are existing against node Nos. 1 to 4 respectively which indicates branch No. 1 is oriented away from node 1 but does not incident on nodes No. 2, 3 and 4. This means, branch No. 1

incidents on another node say node 5, whose details have not been given incidence matrix.

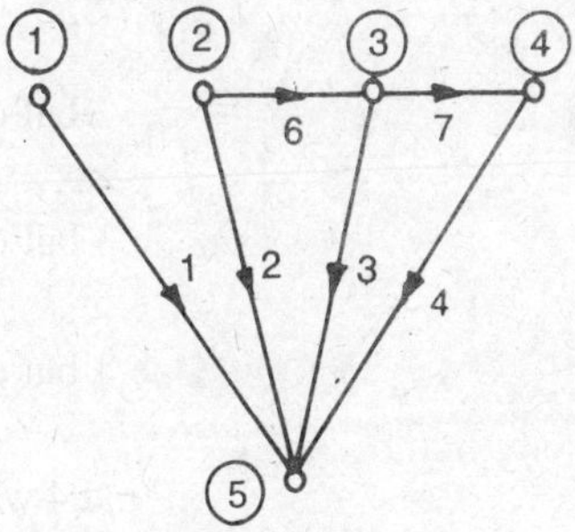

Fig. 1.36 (c)

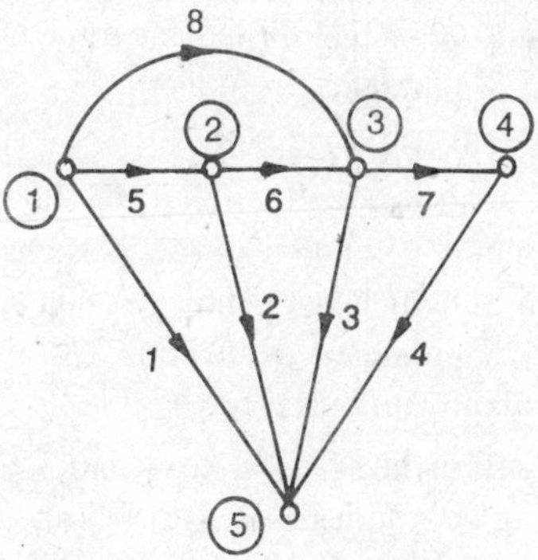

Fig. 1.36 (d)

Similarly, analysis of column Nos. 2, 3 and 4 related to branch Nos. 2, 3 and 4 respectively indicate that branch No. 2, eminates from node No. 2, branch No. 3 eminates from node 3 and branch No. 4 eminates from node No. 4 but do not incident on the given nodes in the incidence matrix. Hence, it can be safely deduced that branches Nos. 1, 2, 3 and 4 eminate from nodes Nos. 1, 2, 3 and 4 respectively and incident on node No. 5 whose details have not been given in incidence matrix. Part graph showing nodes 1 to 5 and branches Nos. 1, 2, 3 and 4 is shown in Fig. 1.36 (a).

Column 5 (Branch No. 5) Branch No. 5 eminates from node 1 as element +1 is entered against row No. 1 and branch No. 5 and incidents on node No. 2 as element –1 is exists in row No. 2 Column No. 5.

Column 6 (Branch No. 6) Existence of elements +1 and –1 in row No. 2 and row No. 3 respectively against column No. 6 indicate orientation of branch No. 6 away from node No. 2 and its incidence on node No. 3.

Branches No. 1 to 4 and 5 and 6 is shown in Fig. 1.37 (a).

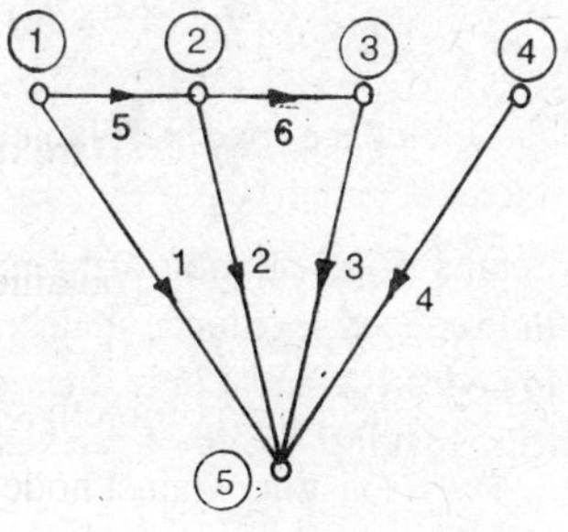

(a)

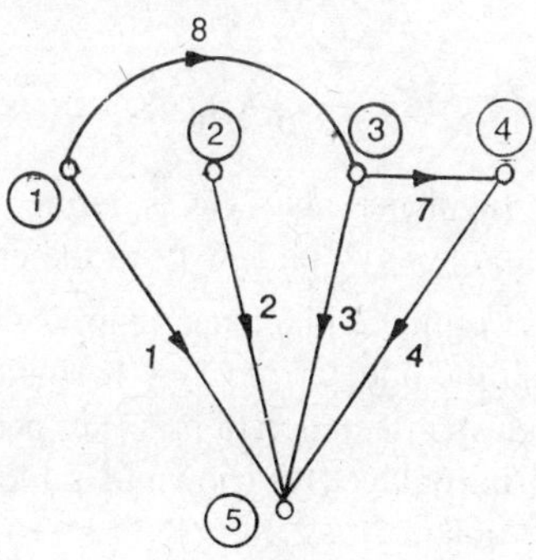

(b)

Fig. 1.37

Column 7 (Branch 7) Existence of elements +1 and −1 in rows Nos. 3 and 4 against branch No. 7 indicates that branch No. 7 eminates from node No. 3 and incidents on node No. 4.

Column 8 (Branch 8) Existence of elements +1 and −1 in row Nos. 1 and 3 against branch No. 8 indicates that branch No. 8 eminates from node No. 1 and incidents at node No. 3.

Branches Nos. 1 to 4 and 7 and 8 alongwith nodes Nos. 1 to 5 is shown in Fig. 1.37 (b).

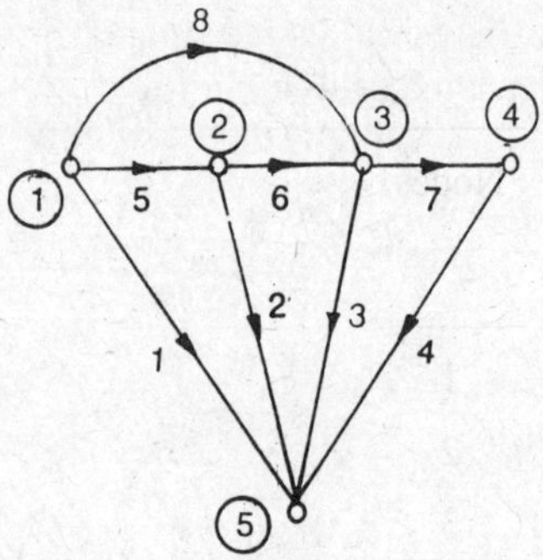

(c)

Fig. 1.37

Complete graph which is combination of graphs shown in Figs. 1.37 (a) and 1.37 (b) is shown in Fig. 1.37 (c).

Note: When incidence matrix is defined by alphabet A, it means it is reduced incidents matrix. In such a matrix one row is omitted.

Problem No. 1.11 Write the incidence matrix A and the fundamental cut-set matrix Q_f for the graph shown in Fig. 1.38 and for the tree (1, 3, 5).

Solution Tree of the graph is shown in Fig. 1.39 in thick lines and the links are shown with broken lines. Tree is defined by branch (1, 3, 5).

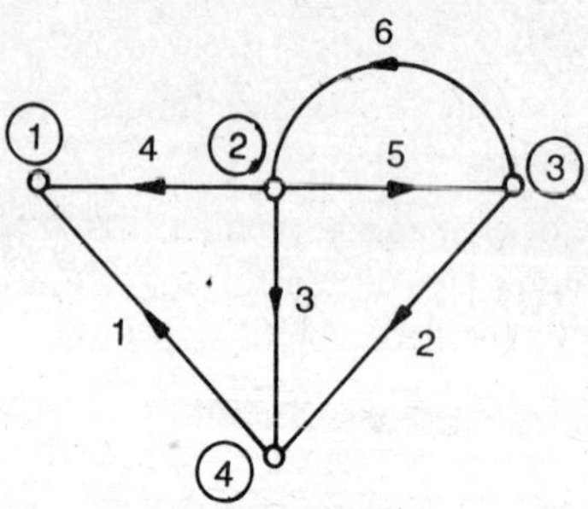

Fig. 1.38

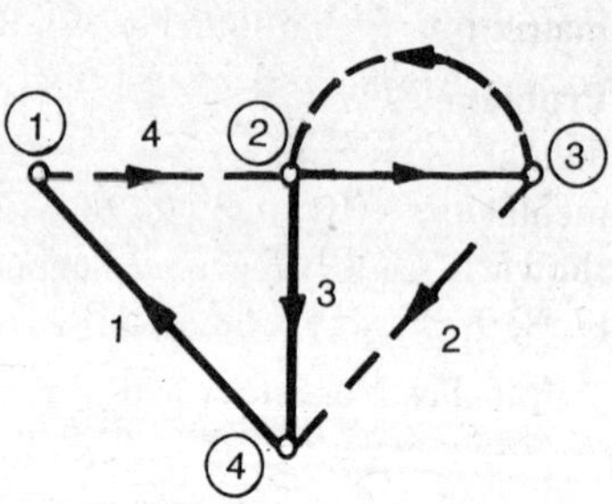

Fig. 1.39

Incidence matrix A is defined as reduced incidence matrix obtained by removing any one row from the complete incidence matrix A_a.

Complete incidence matrix will have details of orientation of branches of all the nodes from 1 to 4. However, in the reduced incidence matrix, we will give orientation of branches pertaining to nodes 1, 2 and 3 only. Incidence matrix (reduced) is shown in Table 1.19, formed with the help of rules given in Problem No. 1.1.

Table 1.19
Reduced Incidence Matrix

Nodes	Branch No.					
	1	2	3	4	5	6
1	−1	0	0	−1	0	0
2	0	0	+1	+1	+1	−1
3	0	+1	0	0	−1	+1

Fundamental Cut-Set Matrix (Q_f)

Twigs of the tree are branches (1, 3, 5) and links are branches (2, 4, 6). Fundamental cut-set (f-cut-sets) selected are

(a) f-cut-set 1 = (1, 4), Orientation as per branch No. 1.
(b) f-cut-set 2 = (3, 2, 4), Orientation as per branch No. 3.
(c) f-cut-set 3 = (5, 2, 6), Orientation as per branch No. 5.

Selection of f-cut-sets is explained in Problem No. 1.8.

Fig. 1.40 shows f-cut-sets and their orientation based on which f-cut-set matrix is formed (Table 1.20).

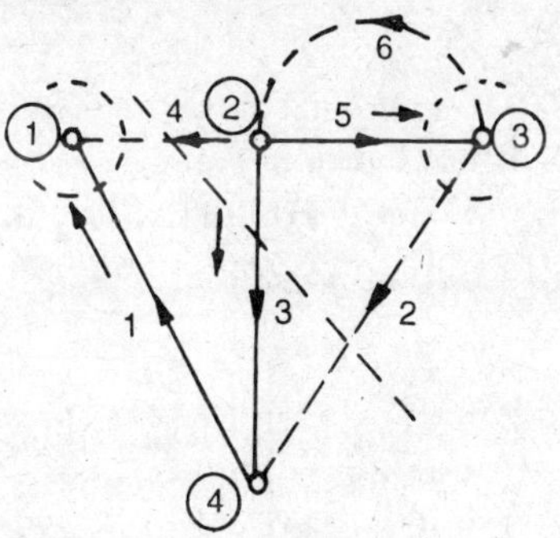

Fig. 1.40

Problem No. 1.12 Write the fundamental tie-set matrix and the fundamental cut-set matrix for the graph shown in Fig. 1.41 choosing the tree (7, 8, 9, 10).

Table 1.20
Fundamental Cut-Set Matrix

f-Cut-Set	Branch					
	1	2	3	4	5	6
1-4	+1	0	0	+1	0	0
3-2-4	0	+1	+1	+1	0	0
5-2-6	0	−1	0	0	+1	−1

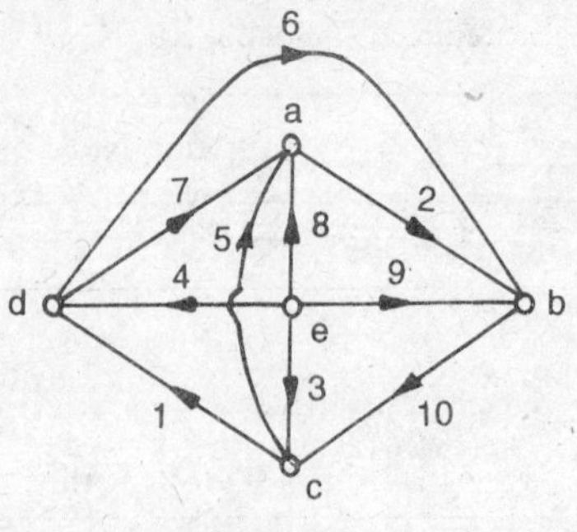

Fig. 1.41

Solution

Fundamental Tie-Set Matrix In the given graph

b = No. of branches = 10

n = No. of nodes = 5

No. of f-circuits = b – (n–1)

= 10 – (5–1)

= 6

Fundamental tie-set or circuits are found by first drawing the tree (7, 8, 9, 10) as shown in Fig. 1.42 (a).

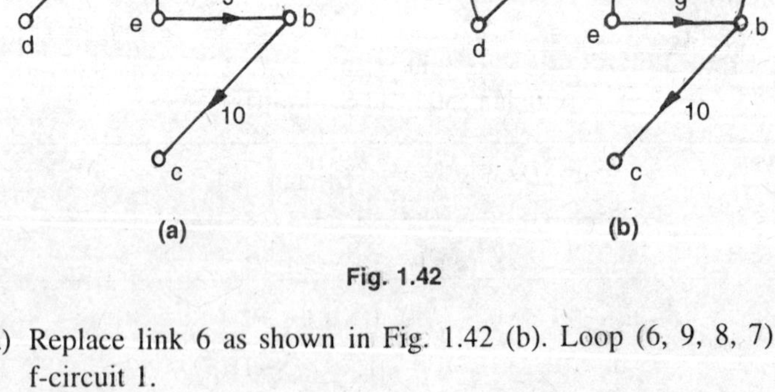

Fig. 1.42

(a) Replace link 6 as shown in Fig. 1.42 (b). Loop (6, 9, 8, 7) gives f-circuit 1.

(b) Replace link 2 as shown in Fig. 1.42 (c). Loop (2, 9, 8) gives f-circuit 2.

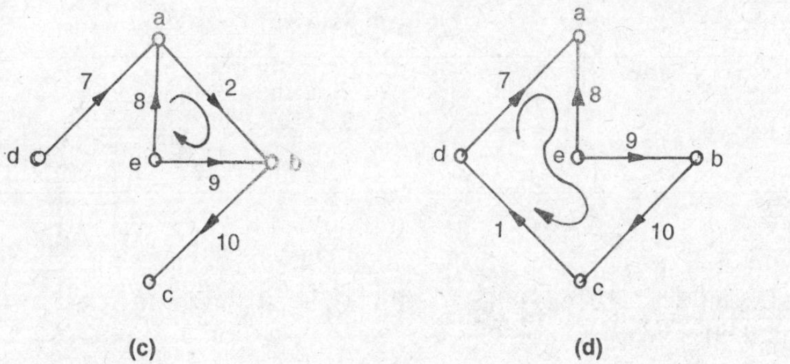

Fig. 1.42

(c) Replace link 1 as shown in Fig. 1.42 (d). Loop (1, 7, 8, 9, 10) gives f-circuit 3.

(d) Replace link 5 as shown in Fig. 1.42 (e). Loop (5, 8, 9, 10) gives f-circuit 4.

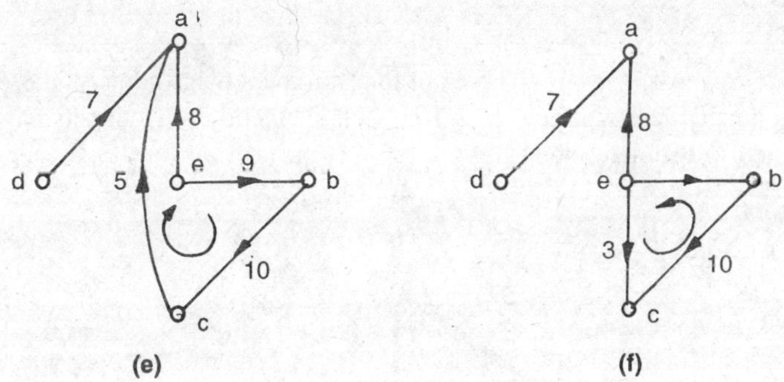

Fig. 1.42

(e) Replace link 3 as shown in Fig. 1.42 (f). Loop (3, 9, 10) gives f-circuit 5.

(f) Replace link 4 as shown in Fig. 1.42 (g). Loop (4, 7, 8) gives f-circuit 6.

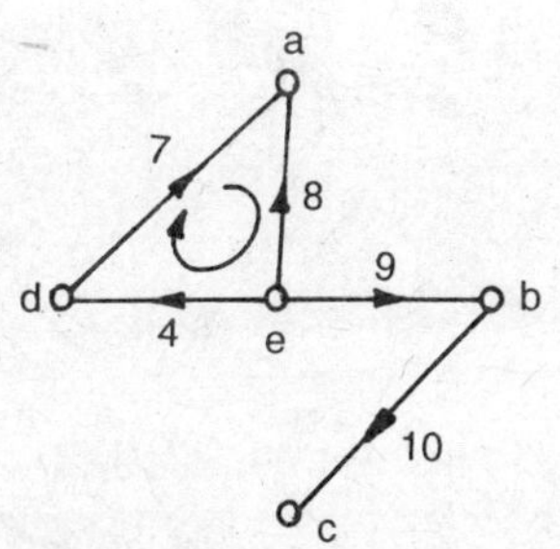

(g)

The f-circuit or tie-matrix is shown in Table 1.20.

Table 1.20
F-Circuit Matrix

f-Circuit	Branches 1	2	3	4	5	6	7	8	9	10
f-Circuit 1 (6, 9, 8, 7)	0	0	0	0	0	+1	−1	+1	−1	0
f-Circuit 2 (2, 9, 8)	0	+1	0	0	0	0	0	+1	−1	0
f-Circuit 3 (1, 7, 8, 9, 10)	+1	0	0	0	0	0	+1	−1	+1	+1
f-Circuit 4 (5, 8, 9, 10)	0	0	0	0	+1	0	0	−1	+1	+1
f-Circuit 5 (3, 9, 10)	0	0	+1	0	0	0	0	0	−1	−1
f-Circuit 6 (4, 7, 8)	0	0	0	+1	0	0	+1	−1	0	0

Fundamental Cut-Sets There are four numbers twigs namely 7, 8, 9 and 10 and hence there will be 4 Nos. f-cut-sets. Selection of f-cut-sets has been explained in Problem No. 1.8.

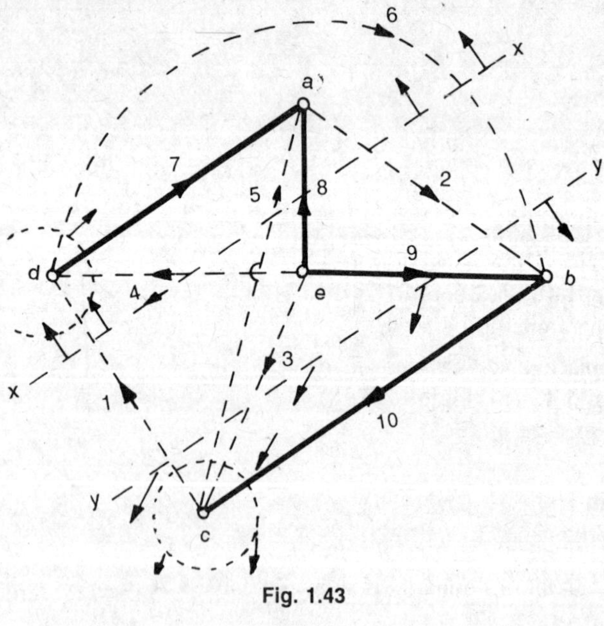

Fig. 1.43

The f-cut-sets selected are

(a) *f-Cut-Set 1*: (7, 1, 4, 6) and its orientation is as per branch No. 7.

(b) *f-Cut-Set 2*: (8, 1, 2, 4, 5, 6) and to orientation is as per branch No. 8.

(c) *f-Cut-Set 3*: (9, 1, 2, 3, 5, 6) and its orientation is as per branch No. 9.

(d) *f-Cut-Set 4*: (10, 1, 3, 5) and its orientation is as per branch No. 10.

Fig. 1.43 shows complete graph in which tree (7, 8, 9, 10) is marked with bold lines and links (1, 2, 3, 4, 5, 6) are marked with broken lines.

Orientation of f-cut-set 1 is marked at node d with a semi-circle drawn with broken line and an arrow.

Orientation of f-cut-set 4 is marked at node c with a circle drawn with broken line and an arrow.

Broken line X-X gives orientation of f-cut-set 2 and broken lines Y-Y gives orientation of f-Cut-Set 3.

With the help of above given orientation f-cut-set matrix is formed as shown in Table 1.21.

Table 1.21
Fundamental Cut-Set Matrix

f-cut-set	Branches									
	1	2	3	4	5	6	7	8	9	10
f-Cut Set 1 (7, 1, 4, 6)	−1	0	0	−1	0	+1	+1	0	0	0
f-Cut-Set 2 (8, 1, 2, 4, 5, 6)	+1	−1	0	+1	+1	+1	0	+1	0	0
f-Cut-Set 3 (9, 1, 2, 3, 5, 6)	−1	+1	+1	0	−1	+1	0	0	+1	0
f-Cut-Set 4 (10, 1, 3, 5)	−1	0	+1	0	−1	0	0	0	0	+1

Problem No. 1.13 Draw the oriented graph for the circuit shown in Fig. 1.44. Write network equilibrium equations.

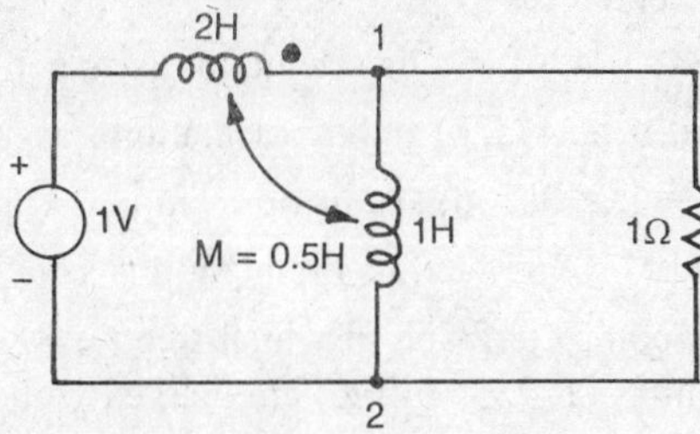

Fig. 1.44

Solution Graph of the given network is shown in Fig. 1.45.

Tree (2) is shown in bold line and links 1 and 3 with broken lines in Fig. 1.46.

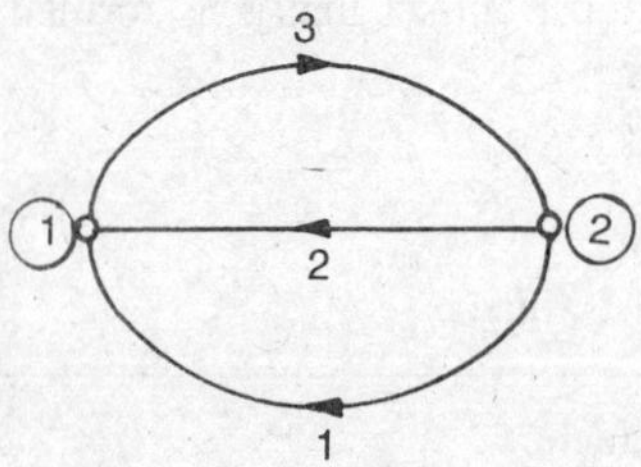

Fig. 1.45

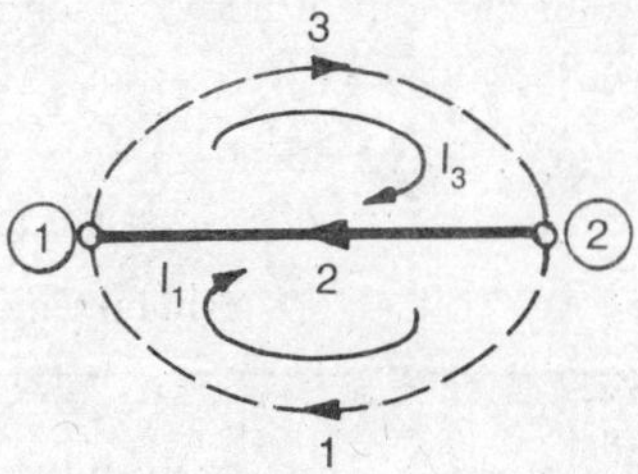

Fig. 1.46

Tie-Set matrix is shown in Table 1.22.

Table 1.22
Tie-Set Matrix

	Nodes	Branch		
		1	2	3
B =	1	+1	−1	0
	2	0	+1	+1

or

$$B = \begin{bmatrix} +1 & -1 & 0 \\ 0 & +1 & 1 \end{bmatrix}$$

The network equilibrium equations on loop basis is

$$[Z][I_L]=[B]\{[V_s]-[Z_b][I_s]\} \quad ...(1)$$

where

$$[Z]=[B][Z_b][B^T]$$

Now

$$[B]=\begin{bmatrix} +1 & -1 & 0 \\ 0 & +1 & 1 \end{bmatrix}$$

$$[Z_b] = \text{Branch Impedance Matrix} = \begin{bmatrix} 1 & 0 & 0 \\ 0 & s & 0.5s \\ 0 & 0.5s & 2s \end{bmatrix}$$

$$[B^T]=\begin{bmatrix} 1 & 0 \\ -1 & 1 \\ 0 & 1 \end{bmatrix}$$

$$[I_L]=\begin{bmatrix} I_1 \\ I_3 \end{bmatrix}$$

Therefore,

$$[Z]=[B][Z_b][B^T]=\begin{bmatrix} 1+s & -1.5s \\ -1.5s & 4s \end{bmatrix}$$

Also

$[I_s]$ = [0] where $[I_s]$ is branch input current source vector. Substituting $[I_s]$ = [0] in equation (1), we get

$$[Z][I_L]=[B][V_s]$$

or

$$\begin{bmatrix} 1+s & -1.5s \\ -1.5s & 4s \end{bmatrix}\begin{bmatrix} I_1 \\ I_3 \end{bmatrix}=\begin{bmatrix} 1 & -1 & 0 \\ 0 & 1 & 1 \end{bmatrix}\begin{bmatrix} 0 \\ 0 \\ 1 \end{bmatrix}$$

or

$$\begin{bmatrix} 1+s & -1.5s \\ -1.5s & 4s \end{bmatrix}\begin{bmatrix} I_1 \\ I_3 \end{bmatrix}=\begin{bmatrix} 0 \\ 1 \end{bmatrix}$$

Therefore, the network equilibrium equations are

$$(1+s)I_1 - 1.5s\,I_3 = 0$$

$$-1.5s\,I_1 + 4s\,I_3 = 1$$

Problem No. 1.14 The fundamental cut-set matrix is given as

Twigs				Links		
1	2	3	4	5	6	7
1	0	0	0	–1	0	0
0	1	0	0	1	0	1
0	0	1	0	0	1	1
0	0	0	1	0	1	0

Draw the oriented graph of the network.

Solution The fundamental-cut-sets as seen from the matrix are

(a) f-Cut-Set 1 : [1, 5]

(b) f-Cut-Set 2 : [2, 5, 7]

(c) f-Cut-Set 3 : [3, 6, 7]

(d) f-Cut-Set 4: [4, 6]

Also, the twigs constitute the tree. Since there are four Nos. twigs consisting of branches 1, 2, 3 and 4, there has to be five nodes. The five nodes constituting the tree is shown in Fig. 1.47.

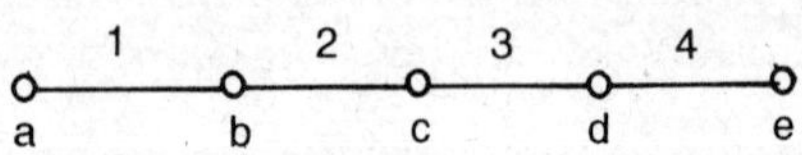

Fig. 1.47

(a) *f-Cut-Set 1 [1, 5]* Two branches emanate from this and are branch Nos. 1 and 5. Orientation of the f-Cut-Set 1 coincides with branch No. 1. Orientation of branch No. 5 is away from the f-Cut-Set 1 orientation or opposite to that of branch No. 1.

(b) *f-Cut-Set 4 [4, 6]* Orientation of the f-Cut-Set 4 coincides with that of branch No. 4 and orientation of branch No. 6 is also same.

Let us develop the graph with the help of above deductions (Fig. 1.48).

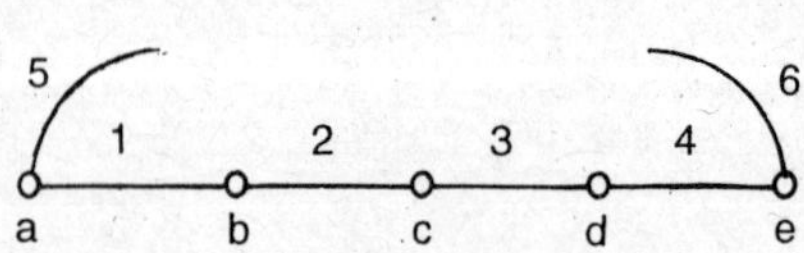

Fig. 1.48

Let us assume the orientations of f-Cut-Set-1 and f-Cut-Set-4 as shown in Fig. 1.49 (a). Then the directions of branches No. 5 and 6 will be as shown in Fig. 1.49 (b).

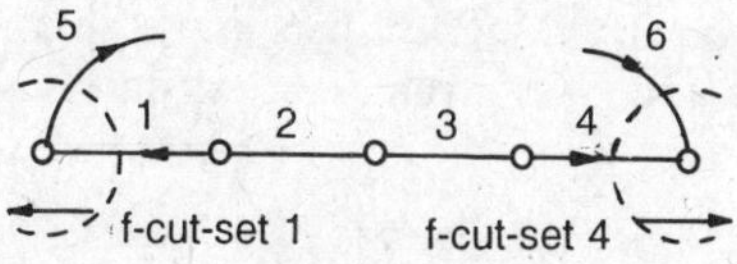

Fig. 1.49 (a)

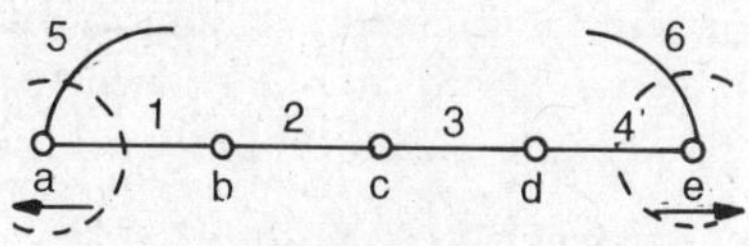

Fig. 1.49 (b)

(c) *f-Cut-Set 2 [2, 5, 7]*. The f-Cut-Set 2 cuts the branches No. 2, 5 and 7 and orientation of branches is same.

(d) *f-Cut-Set 3 [3, 6, 7]*. The f-Cut-Set 4 cuts the branches No. 3, 6 and 7 and orientation of branches is same.

Note: Inference of orientation is obtained from the positiveness of the elements entered in row Nos. 2 and 3. Fig. 1.50 shows further developed graph.

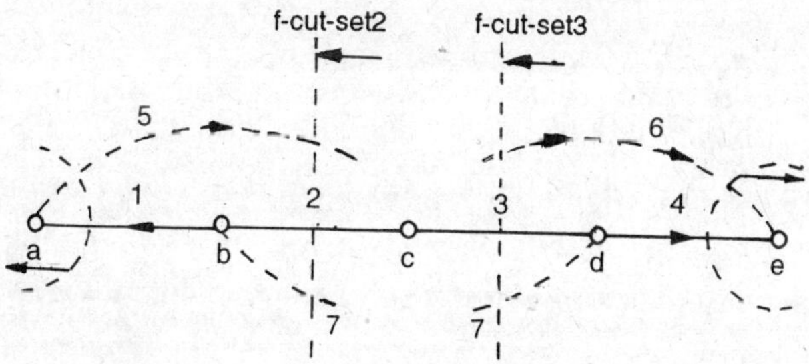

Fig. 1.50

Orientation of f-Cut-Set-2 and branch No. 5 is same. Therefore, as seen in para 'c' above, direction or orientation of branches No. 2 and 7 will also be same as that of branch No. 5.

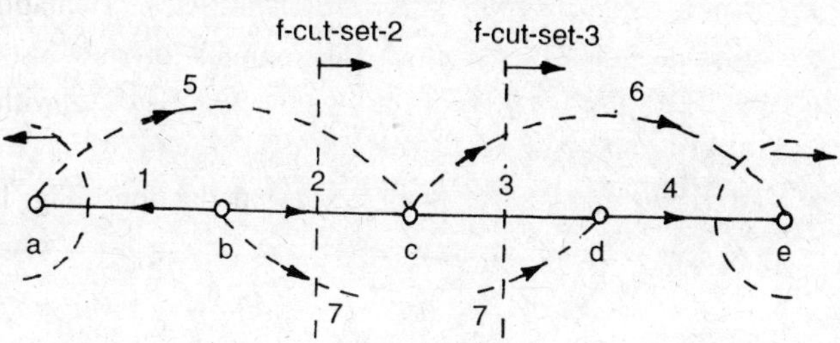

Fig. 1.51

Similarly, orientation of f-Cut-Set 3 is same as that of branch No. 6. Therefore, as seen in para 'd' above, orientation of branch No. 3 and 7 will be same as that of branch No. 6.

Also, since graph is a closed path, branches No. 5 and 6 must terminate, or start from a node without affecting other f-Cut-Sets. This is possible if branches No. 5 and 6 are connected with node 'c'. From above deductions, the graph is developed further (Fig. 1.51).

As stated earlier, since graph is a closed path, branch No. 7 must join with other. This will not disturb the f-Cut-Sets.

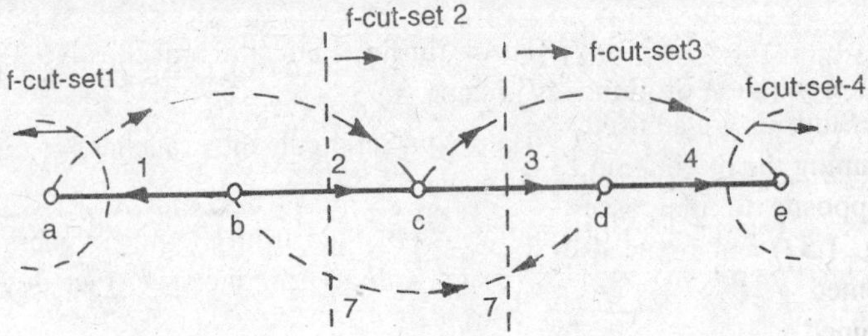

(a)

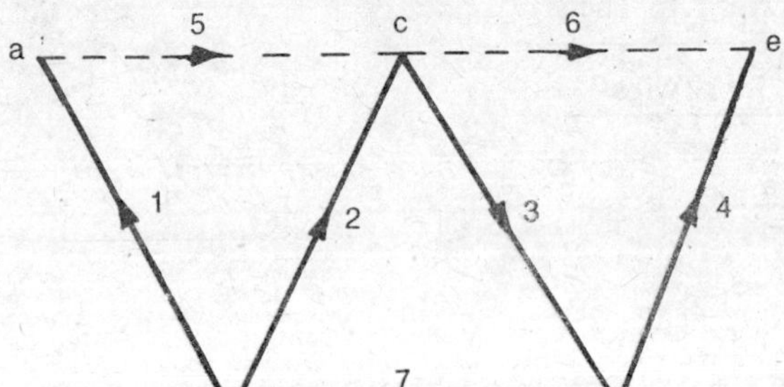

(b)

Fig. 1.52

Fig. 1.52(a) and Fig. 1.52(b) shows the complete oriented graph.

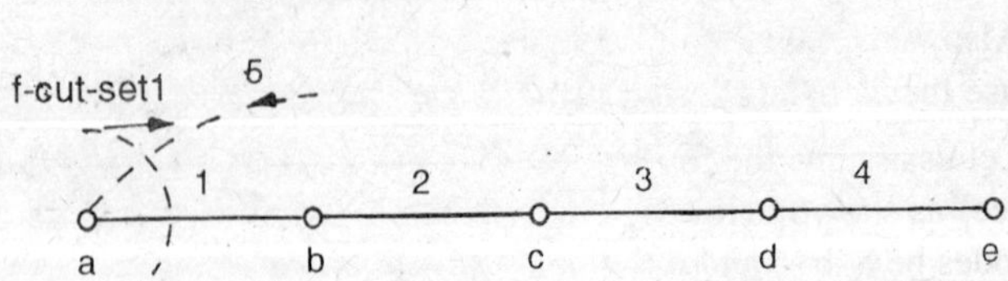

Fig. 1.53

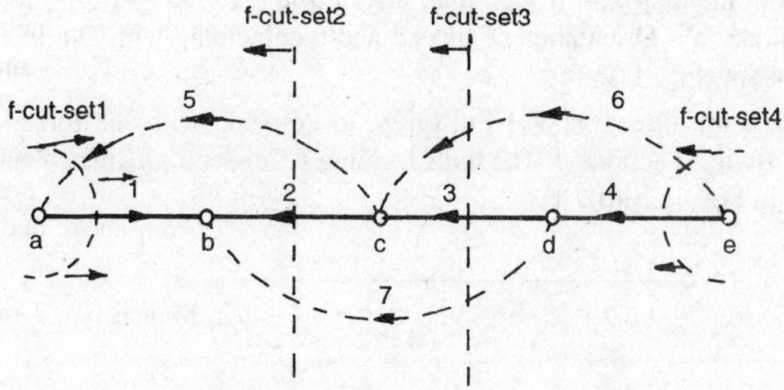

Fig. 1.54

Students are advised to start with assuming the orientation of f-Cut-Set 1 opposite to that assumed earlier (Fig. 1.53) and proceed to draw the oriented graph. The oriented graph obtained will be as per Fig. 1.54(a) and Fig. 1.54(b).

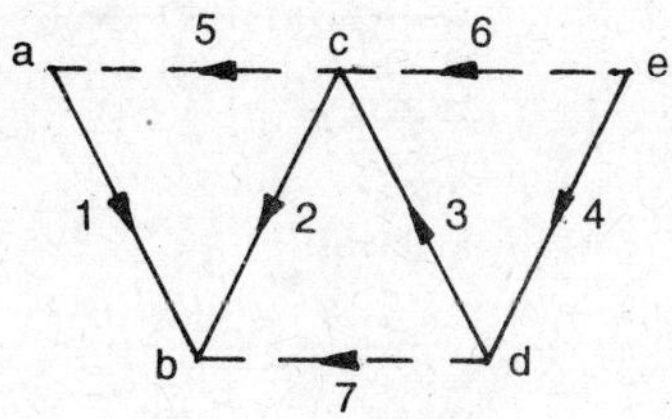

Fig. 1.54 (b)

Problem No. 1.15 The fundamental cut-set matrix is given as

Twigs			Links		
1	3	5	2	4	6
1	0	0	1	0	1
0	1	0	0	1	1
0	0	1	1	1	1

Draw the oriented graph of the network.

Solution Various Cut-Sets as seen from the given matrix are

(a) f-Cut-Set 1: (1, 2, 6)
(b) f-Cut-Set 2: (3, 4, 6)
(d) f-Cut-Set 3: (5, 2, 4, 6)

Also, the orientation of the branches and links is same as that of Cut-Set because the elements entered in the given matrix are all positive.

Let us assume the four nodes (because there are three twigs constituting the tree) as shown below and draw the branches (twigs) (Fig. 1.55 (a)). Let the nodes be a, b, c and d. Let a-b be twig 1, Mark the direction of the Cut-Set and twig 1 to coincide with each other (Fig. 1.55 (b)).

From the matrix, it is seen that, links 2 and 6 and twig 1 form the Cut-Set at node 'a'. Orientation of links 2 and 6 coinciding with that of twig 1 and Cut-set (Fig. 1.55 (a)).

Mark the direction of f-Cut-set 3, to coincide with the direction of branch (twig 3) at node d. The links forming f-Cut-set 2 are links 4 and 6 as shown in Fig. 1.55 (d).

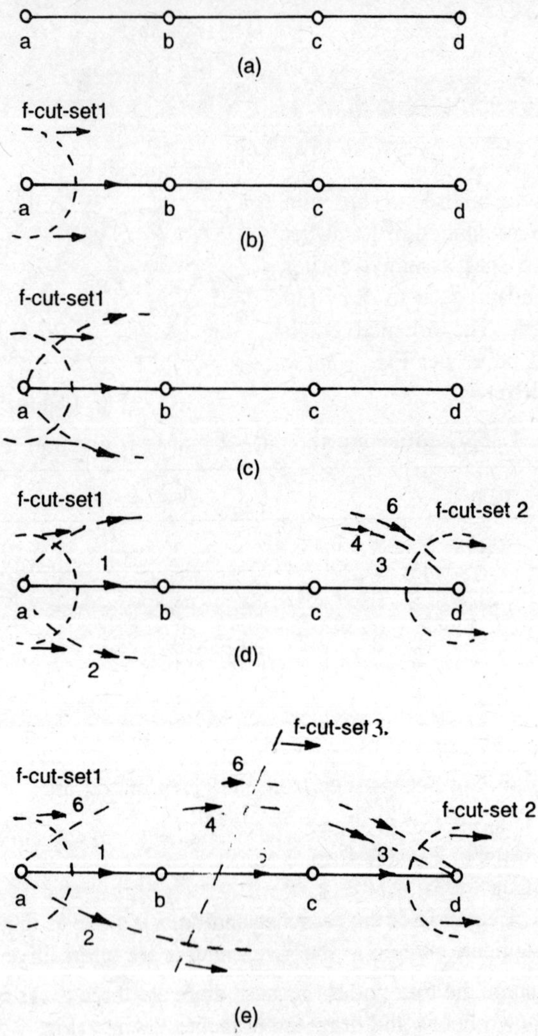

Fig. 1.55

Mark the direction of f-Cut-set 2 to coincide with direction of branch (twig) 5 between nodes b and c. The links (2, 4, 6) forming f-Cut-set 3 alongwith twig 5 are shown in Fig. 1.55 (e).

Joining branch (link 6) and joining with nodes a and d, joining branch (link 2) with nodes a and c and joining branch (link 4) with nodes b and d will not disturb the f-Cut-Sets defined by the given matrix and will complete the oriented graph as shown in Fig. 1.56.

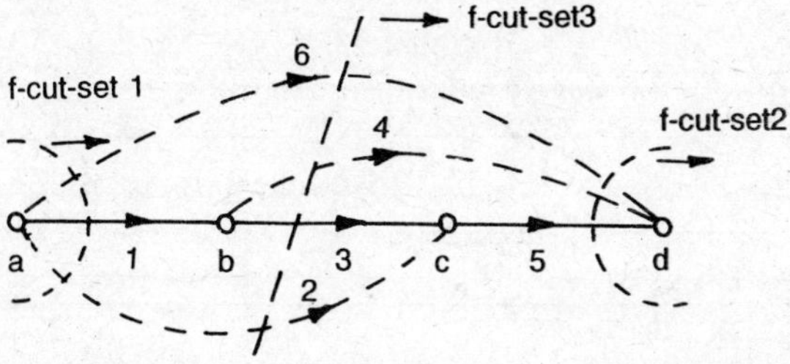

Fig. 1.56

The oriented graph can be redrawn (Fig. 1.57) by pulling node c down directly under node b and straightening branches 2 and 4. Node d is drawn in line with nodes a and b.

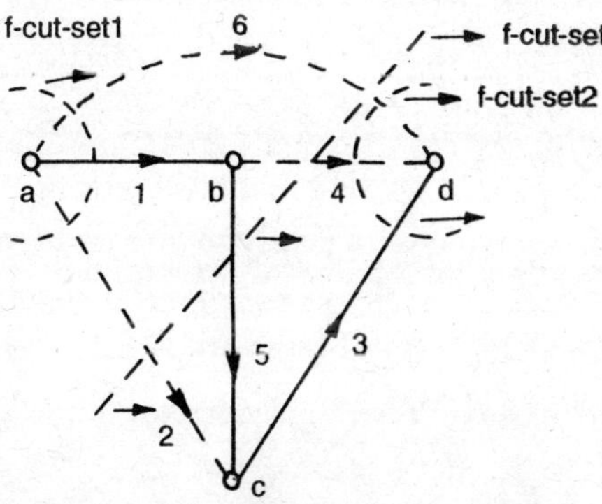

Fig. 1.57

Problem No. 1.16 The incidence matrix is given below as

$$\text{Branches} \Rightarrow \quad \begin{matrix} 1 & 2 & 3 & 4 & 5 & 6 & 7 & 8 \end{matrix}$$

$$A = \begin{bmatrix} 1 & 0 & 0 & 0 & 1 & 0 & 0 & 1 \\ 0 & 1 & 0 & 0 & -1 & 1 & 0 & 0 \\ 0 & 0 & 1 & 0 & 0 & -1 & 1 & -1 \\ 0 & 0 & 0 & 1 & 0 & 0 & -1 & 0 \end{bmatrix}$$

Draw the oriented graph.

Solution Given matrix A is a reduced incidence matrix as the sum of entries in columns 1, 2, 3 and 4 is not zero. In order to draw the oriented graph one method has been explained in Problem No. 1.10. Let us adopt a different method here.

Let us add another row in the given incidence matrix and enter the elements in such a way that sum of the elements entered in the columns is zero. This will give us the following matrix (Table 1.23).

Table 1.23
Complete Incidence Matrix

Node	Branches							
	1	2	3	4	5	6	7	8
a	1	0	0	0	1	0	0	1
b	0	1	0	0	–1	1	0	0
c	0	0	1	0	0	–1	1	–1
d	0	0	0	1	0	0	–1	0
e	–1	–1	–1	–1	0	0	0	0

Also

$$b = 8$$

$$n = 5$$

Therefore

$$\text{No. of links } = b-(n-1)$$

$$= 8-(5-1)$$

$$= 4.$$

Hence, there are four links and b – 4 i.e. 8 – 4 = 4 twigs. Let branch Nos. 1, 2, 3, 4 be the twigs and branch Nos. 5, 6, 7, 8 be the links.

Let the assumed nodes be a, b, c, d and e as shown in Table 1.23. Now we will proceed to draw the oriented graph as given below:

(a) Draw the nodes as shown in Fig. 1.58 (a)

(b) *Column 1* It indicates that there is branch 1 between node a and node e. It originates from node a and terminates at node e.

(c) *Column 2* It indicates that branch 2 originates from node b and terminates at node e.

(d) *Column 3* It indicates that branch 3 originates from node c and terminates at node f.

(e) *Column 4* It indicates that branch 4 originates from node d and terminates at node e.

Let us now draw the branches as per above deductions (Fig. 1.58 (b).

(f) *Column 5* Branch 5 originates from node a and terminates at node b.

(g) *Column 6* Branch 6 originates from node b and terminates at node c.

(h) *Column 7* Branch 7 originates from node c and terminates at node d.

(i) *Column 8* Branch 8 originates from node a and terminates at node c.

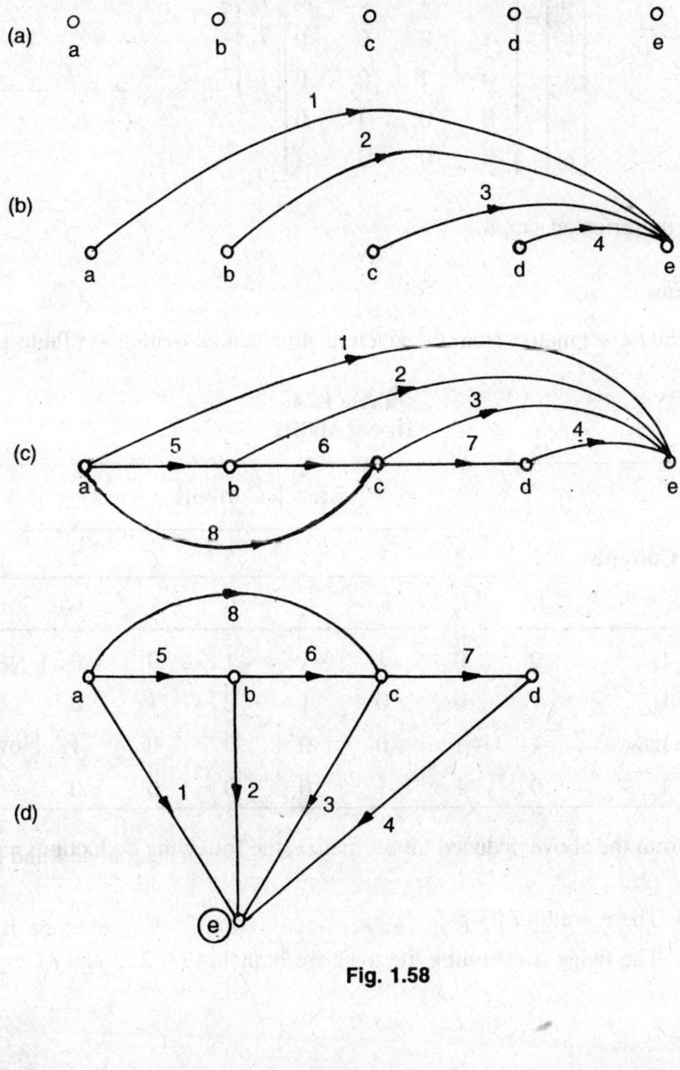

Fig. 1.58

Let us now complete the oriented graph with the help of above concluded results (Fig. 1.58 (c)).

Oriented graph can also be drawn as shown in Fig. 1.58 (d).

Problem No. 1.17 Branch current and loop current relations are expressed in matrix form as

$$\begin{bmatrix} i_1 \\ i_2 \\ i_3 \\ i_4 \\ i_5 \\ i_6 \\ i_7 \\ i_8 \end{bmatrix} = \begin{bmatrix} 0 & 1 & -1 & 0 \\ 0 & 0 & -1 & 1 \\ 1 & 0 & 0 & -1 \\ -1 & 1 & 0 & 0 \\ 1 & 0 & 0 & 0 \\ 0 & 1 & 0 & 0 \\ 0 & 0 & 1 & 0 \\ 0 & 0 & 0 & 1 \end{bmatrix} \begin{bmatrix} I_5 \\ I_6 \\ I_7 \\ I_8 \end{bmatrix}$$

Draw the oriented graph.

Solution

The tie-set matrix from the given relation can be written as (Table 1.24).

Table 1.24
Tie-Set Matrix

	Branch Currents							
Link Currents	1	2	3	4	5	6	7	8
	i_1	i_2	i_3	i_4	i_5	i_6	i_7	i_8
I_5	0	0	1	–1	1	0	0	0
I_6	1	0	0	1	0	1	0	0
I_7	–1	–1	0	0	0	0	1	0
I_8	0	1	–1	0	0	0	0	1

From the above deduced tie-set matrix, the following deductions can be made

(a) There are four loops

(b) The twigs constituting the trees are branches (1, 2, 3 and 4).

(c) The links are branches (5, 6, 7 and 8)

(d) Loop Currents (I_5, I_6, I_7 and I_8) coincide with the orientation of the corresponding links i.e., links 5, 6, 7 and 8 respectively.

(e) **Number of Nodes**

$b-(n-1) =$ No. of Loops

or $8-(n-1)=4$

or $n-1=4$

or $n=5$ i.e., there are five nodes.

(f) No. of branches in a tree

$= n-1$

$= 5-1$

$= 4.$

Consider the rows of the tie matrix given in Table 1.24. The rows give us the following relations

(a) $v_3 - v_4 + v_5 = 0$, ...(1)

(b) $v_1 - v_4 + v_6 = 0$, ...(2)

(c) $-v_1 - v_2 + v_7 = 0$, and ...(3)

(d) $v_2 - v_3 + v_8 = 0$. ...(4)

Addition of the above relations give

$$v_5 + v_6 + v_7 + v_8 = 0. \quad ...(5)$$

where $v_1, v_2, v_3, v_4, v_5, v_6, v_7$ and v_8 are the voltages in the branches 1 to 8 respectively.

Relation given in equation 1 forms a mesh consisting of branches 3, 4 and 5 as shown in Fig. 1.59(a). Current circuiting in the mesh is I_5 and coincides with direction marked on branch 5. Branch current in branch 5 is i_5.

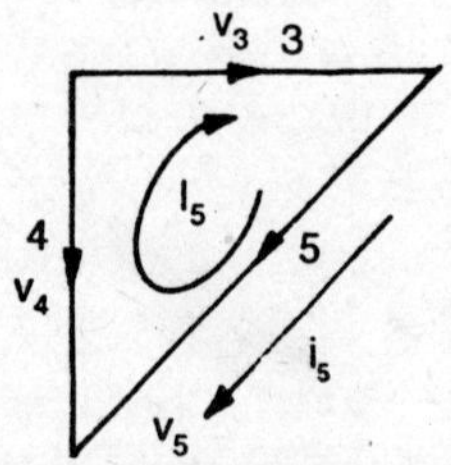

Fig. 1.59 (a)

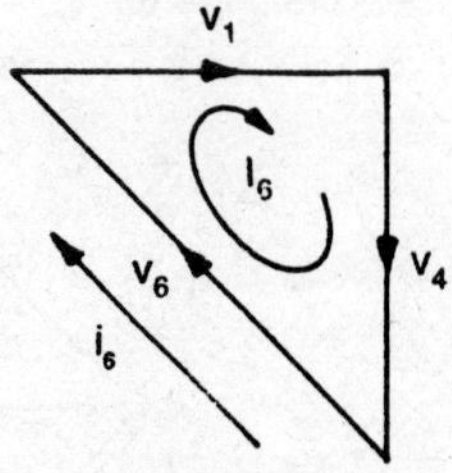

Fig. 1.59 (b)

Similarly, relation given in equation 2 is diagrammatically represented in Fig. 1.59 (b). Note that

(a) directions of branch 4 in Figs. 1.59 (a) and 1.59 (b) is same, and

(b) orientation of branch current i_5 and mesh current I_5 is same.

Equation (3) is diagrammatically represented in Fig. 1.59 (c). Note that

(a) directions of branch 3 in Figs. 1.59 (c) and 1.59 (b) is same, and

(b) Orientation of branch current i_7 and mesh current I_7 is same.

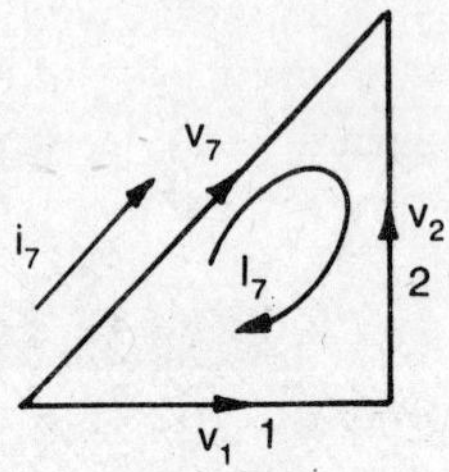

Fig. 1.59 (c)

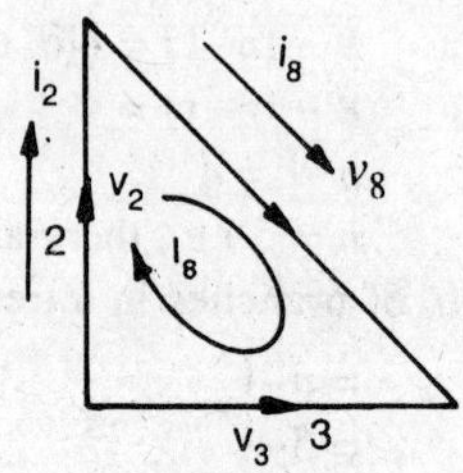

Fig. 1.59 (d)

Fig. 1.59 (d) shows diagrammatic representation of relation given in equation (4). Note that

(a) directions of branch 3 is same in Figs. 1.59 (d) and 1.59 (b), and

(b) orientation of branch current i_8 and mesh current I_8 is same.

Combining Figs. 1.59 (a) and 1.59 (b) we get a graph as shown in Fig. 1.60 (a) and combining Figs. 1.59 (c) and 1.59 (d), we get a graph as shown in Fig. 1.60 (b).

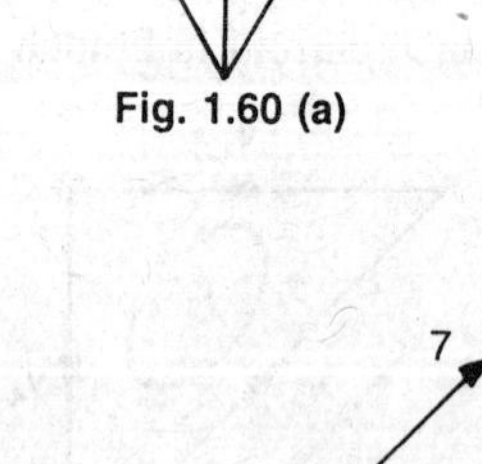

Fig. 1.60 (a)

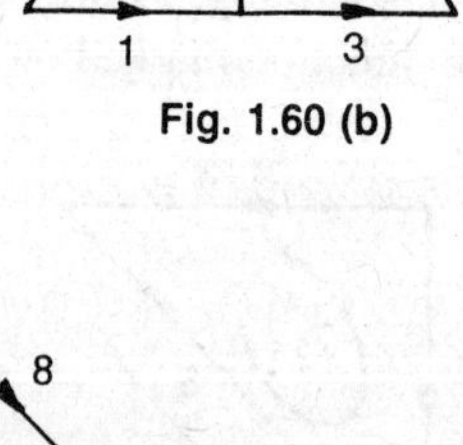

Fig. 1.60 (b)

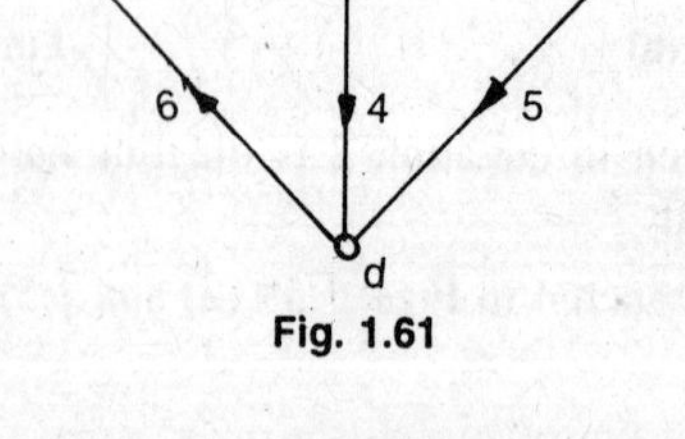

Fig. 1.61

By combining Figs. 1.60 (a) and 1.60 (b), we get the complete oriented graph as shown in Fig. 1.61.

Problem No. 1.18 Branch current and loop current relations are expressed in matrix form as

$$\begin{bmatrix} i_1 \\ i_2 \\ i_3 \\ i_4 \\ i_5 \\ i_6 \\ i_7 \\ i_8 \end{bmatrix} = \begin{bmatrix} 1 & 0 & 0 & -1 \\ 0 & 1 & 0 & -1 \\ 0 & 1 & 1 & 0 \\ 0 & 1 & 1 & 0 \\ 1 & -1 & 0 & 0 \\ 0 & 0 & -1 & 0 \\ -1 & 0 & 0 & 0 \\ 0 & 0 & 0 & 1 \end{bmatrix} \begin{bmatrix} I_1 \\ I_2 \\ I_3 \\ I_4 \end{bmatrix}$$

Draw the oriented graph.

Solution The above relation can be in a tie-matrix form as shown in Table 1.25.

Table 1.25
Tie-Matrix

	Branch Currents							
Mesh Current	1	2	3	4	5	6	7	8
	i_1	i_2	i_3	i_4	i_5	i_6	i_7	i_8
I_1	1	0	0	0	1	0	−1	0
I_2	0	1	1	1	−1	0	0	0
I_3	0	0	1	1	0	−1	0	0
I_4	−1	−1	0	0	0	0	0	1

No. of loops = 4 (as these four Nos. mesh current I_1 I_2 I_3 and I_4)
b = 8.
Therefore

$$b - (n-1) = \text{No. of loops}$$

or $\quad 8 - (n-1) = 4$

or $\quad n = 5$ i.e., number of nodes are five.

Number of branches in a tree = $n-1 = 5-1 = 4$.

From the rows of the tie-matrix, following relations are obtained

(a) $v_1 + v_5 - v_7 = 0$ ...(1)

(b) $v_2 + v_3 + v_4 - v_5 = 0$ ...(2)

(c) $v_3 + v_4 - v_6 = 0$ ...(3)

(d) $-v_1 - v_2 + v_8 = 0$ ...(4)

The diagrammatic representation of the relations is shown in Fig. 1.61.

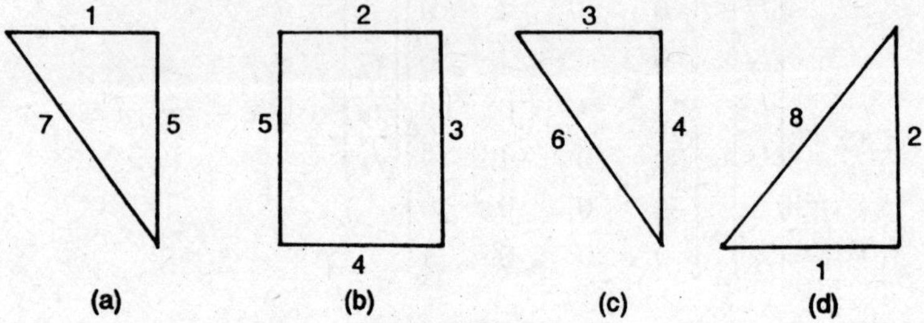

Fig. 1.61

Combining Figs. 1.61 (a) and 1.61 (b), we get Fig. 1.62

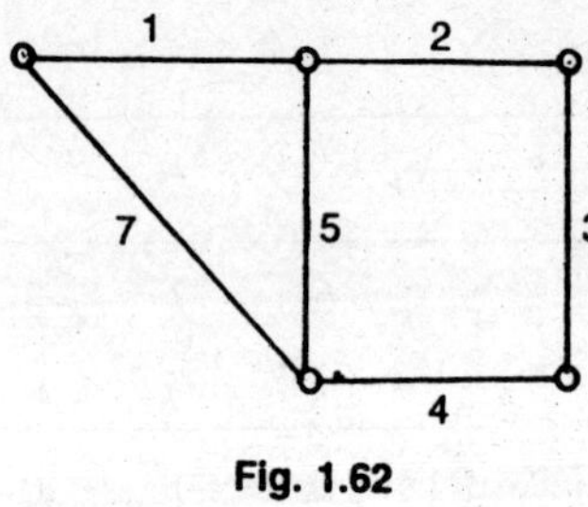

Fig. 1.62

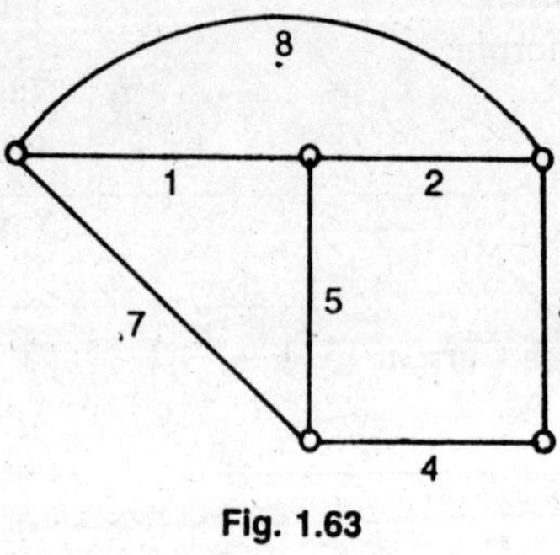

Fig. 1.63

Combining Fig. 1.62 and Fig. 1.61 (d), we get Fig. 1.63

Combining Fig. 163 with Fig. 1.61 (c), we get Fig. 1.64

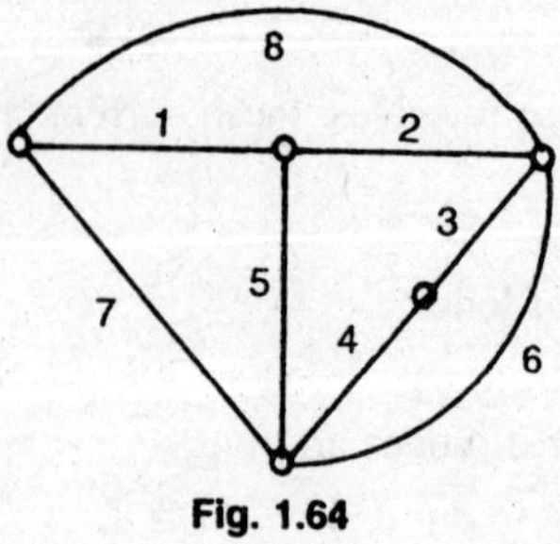

Fig. 1.64

Solution After short circuiting the voltage source and open-circuiting the current source, the graph is drawn and is shown in Fig. 1.68.

Since there are four number nodes, the tree will have three number branches. Let us select tree consisting of branches (4, 5, 6). The links are (1, 2, 3).

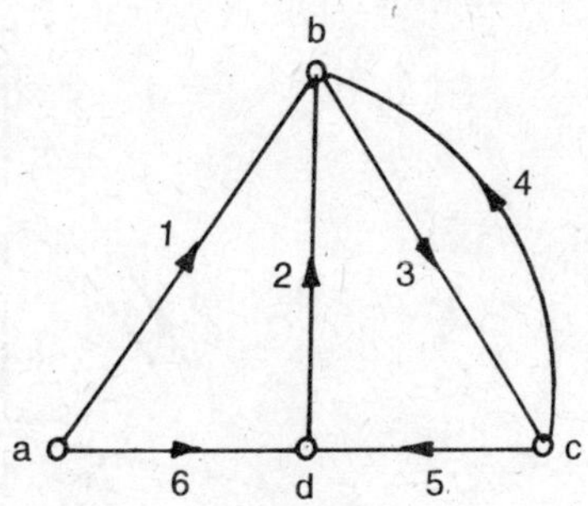

Fig. 1.68

Tree of the network is drawn in Fig. 1.69

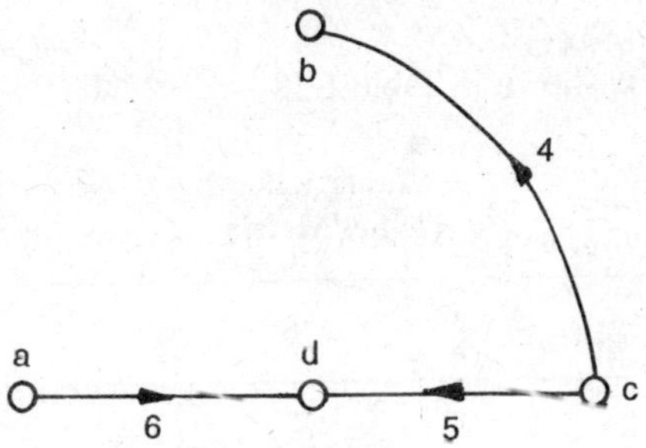

Fig. 1.69

Replacement of link 1 between nodes a and b, as shown in Fig. 1.70 (a) gives the first loop current I_1. Replacement of link 2 between nodes b and d as shown in Fig. 1.70 (b) gives the second loop current I_2 and replacement of link 3 as shown in Fig. 1.70 (c) gives the third loop current I_3.

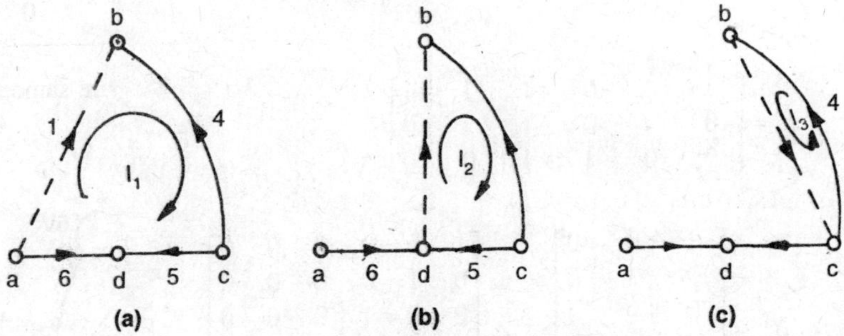

Fig. 1.70

The complete oriented graph with loop currents is shown in Fig. 1.71.

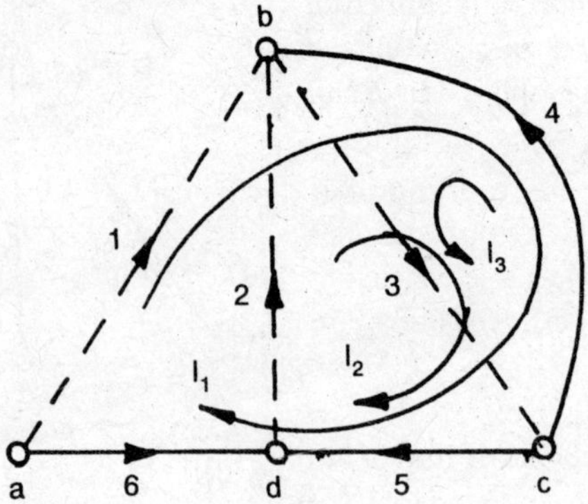

Fig. 1.71

Tie-set matrix is shown in Table 1.28

Table 1.28
Tie-Set Matrix

Link Current		Branches					
		1	2	3	4	5	6
		i_1	i_2	i_3	i_4	i_5	i_6
I_1		1	0	0	−1	1	−1
I_2	B =	0	1	0	−1	1	0
I_3		0	0	1	1	0	0

Now

$$B = \begin{bmatrix} 1 & 0 & 0 & -1 & 1 & -1 \\ 0 & 1 & 0 & -1 & 1 & 0 \\ 0 & 0 & 1 & 1 & 0 & 0 \end{bmatrix}$$

$$B^T = \begin{bmatrix} 1 & 0 & 0 \\ 0 & 1 & 0 \\ 0 & 0 & 1 \\ -1 & -1 & 1 \\ 1 & 1 & 0 \\ -1 & 0 & 0 \end{bmatrix}; \quad Z_b \begin{bmatrix} 1 & 0 & 0 & 0 & 0 & 0 \\ 0 & 1 & 0 & 0 & 0 & 0 \\ 0 & 0 & 1 & 0 & 0 & 0 \\ 0 & 0 & 0 & 0 & 0 & 0 \\ 0 & 0 & 0 & 0 & 1 & 0 \\ 0 & 0 & 0 & 0 & 0 & 1 \end{bmatrix}$$

$$[B][Z_b][B^T] = \begin{bmatrix} 1 & 0 & 0 & -1 & 1 & -1 \\ 0 & 1 & 0 & -1 & 1 & 0 \\ 0 & 0 & 1 & 1 & 0 & 0 \end{bmatrix} \text{X}$$

$$\begin{bmatrix} 1 & 0 & 0 & 0 & 0 & 0 \\ 0 & 1 & 0 & 0 & 0 & 0 \\ 0 & 0 & 1 & 0 & 0 & 0 \\ 0 & 0 & 0 & 0 & 0 & 0 \\ 0 & 0 & 0 & 0 & 1 & 0 \\ 0 & 0 & 0 & 0 & 0 & 1 \end{bmatrix} \begin{bmatrix} 1 & 0 & 0 \\ 0 & 1 & 0 \\ 0 & 0 & 1 \\ -1 & -1 & 1 \\ 1 & 1 & 0 \\ -1 & 0 & 0 \end{bmatrix}$$

$$= \begin{bmatrix} 1 & 0 & 0 & 0 & 1 & -1 \\ 0 & 1 & 0 & 0 & 1 & 0 \\ 0 & 0 & 1 & 0 & 0 & 0 \end{bmatrix} \begin{bmatrix} 1 & 0 & 0 \\ 0 & 1 & 0 \\ 0 & 0 & 1 \\ -1 & -1 & 1 \\ 1 & 1 & 0 \\ -1 & 0 & 0 \end{bmatrix}$$

$$= \begin{bmatrix} 3 & 1 & 0 \\ 1 & 2 & 0 \\ 0 & 0 & 1 \end{bmatrix}$$

Also

$$V_s = \begin{bmatrix} 0 \\ 0 \\ 0 \\ 5 \\ 0 \\ 0 \end{bmatrix}; \text{ and } I_s = \begin{bmatrix} -2 \\ 0 \\ 0 \\ 0 \\ 0 \\ 0 \end{bmatrix}$$

Therefore,

$$[V_s] - [Z_b][I_s] = \begin{bmatrix} 0 \\ 0 \\ 0 \\ 5 \\ 0 \\ 0 \end{bmatrix} - \begin{bmatrix} 1 & 0 & 0 & 0 & 0 & 0 \\ 0 & 1 & 0 & 0 & 0 & 0 \\ 0 & 0 & 1 & 0 & 0 & 0 \\ 0 & 0 & 0 & 0 & 0 & 0 \\ 0 & 0 & 0 & 0 & 1 & 0 \\ 0 & 0 & 0 & 0 & 0 & 0 \end{bmatrix} \begin{bmatrix} -2 \\ 0 \\ 0 \\ 0 \\ 0 \\ 0 \end{bmatrix} = \begin{bmatrix} 2 \\ 0 \\ 0 \\ 5 \\ 0 \\ 0 \end{bmatrix}$$

The equilibrium equation on loop basis is

$$[B][Z_b][B^T][I_L]=[B]\{[V_s]-[Z_b]\}$$

substituting the values obtained above, we get

$$\begin{bmatrix}3 & 1 & 0\\ 1 & 2 & 0\\ 0 & 0 & 1\end{bmatrix}\begin{bmatrix}I_1\\ I_2\\ I_3\end{bmatrix}=\begin{bmatrix}1 & 0 & 0 & -1 & 1 & -1\\ 0 & 1 & 0 & -1 & 1 & 0\\ 0 & 0 & 1 & 1 & 0 & 0\end{bmatrix}\begin{bmatrix}2\\ 0\\ 0\\ 5\\ 0\\ 0\end{bmatrix}$$

or
$$\begin{bmatrix}3 & 1 & 0\\ 1 & 2 & 0\\ 0 & 0 & 1\end{bmatrix}\begin{bmatrix}I_1\\ I_2\\ I_3\end{bmatrix}=\begin{bmatrix}-3\\ -5\\ 5\end{bmatrix}$$

The above relation results in the following set of equations

$$3I_1+I_2=-3 \quad ...(1)$$

$$I_1+2I_2=-5 \quad ...(2)$$

$$I_3=5 \quad ...(3)$$

Solving equation (1) and (2), we get

$$I_1=-\frac{1}{5}\text{ amp.}$$

$$I_2=-\frac{12}{5}\text{ amp.}$$

and $I_3=5$ from equation (3).

Branch Currents

Branch currents are found out from

$$[I_b]=[B^T][I_L]+[I_s]$$

or
$$\begin{bmatrix}i_1\\ i_2\\ i_3\\ i_4\\ i_5\\ i_6\end{bmatrix}=\begin{bmatrix}1 & 0 & 0\\ 0 & 1 & 0\\ 0 & 0 & 1\\ -1 & -1 & 1\\ 1 & 1 & 0\\ -1 & 0 & 0\end{bmatrix}\begin{bmatrix}-1/5\\ -12/5\\ 5\end{bmatrix}+\begin{bmatrix}-2\\ 0\\ 0\\ 0\\ 0\\ 0\end{bmatrix}$$

or

$$\begin{bmatrix} i_1 \\ i_2 \\ i_3 \\ i_4 \\ i_5 \\ i_6 \end{bmatrix} = \begin{bmatrix} -1/5 \\ 112/5 \\ 5 \\ 38/5 \\ -13/5 \\ 1/5 \end{bmatrix} + \begin{bmatrix} -2 \\ 0 \\ 0 \\ 0 \\ 0 \\ 0 \end{bmatrix} = \begin{bmatrix} -11/5 \\ -12/5 \\ 5 \\ 38/5 \\ -13/5 \\ 1/5 \end{bmatrix}$$

Therefore,

$$i_1 = \frac{-11}{5} \text{ amp}; \qquad i_4 = \frac{38}{5} \text{ amp}$$

$$i_2 = \frac{-12}{5} \text{ amp}; \qquad i_5 = \frac{-13}{5} \text{ amp}$$

$$i_3 = 5 \text{ amp}; \qquad i_6 = \frac{1}{5} \text{ amp}$$

Branch Voltages

Branch voltages is obtained from the following equation.

$$[v_b] = [Z_b][I_b] - [v_s]$$

or

$$\begin{bmatrix} v_1 \\ v_2 \\ v_3 \\ v_4 \\ v_5 \\ v_6 \end{bmatrix} = \begin{bmatrix} 1 & 0 & 0 & 0 & 0 & 0 \\ 0 & 1 & 0 & 0 & 0 & 0 \\ 0 & 0 & 1 & 0 & 0 & 0 \\ 0 & 0 & 0 & 0 & 0 & 0 \\ 0 & 0 & 0 & 0 & 1 & 0 \\ 0 & 0 & 0 & 0 & 0 & 1 \end{bmatrix} \begin{bmatrix} -11/5 \\ -12/5 \\ 5 \\ 38/5 \\ -13/5 \\ 1/5 \end{bmatrix} - \begin{bmatrix} 0 \\ 0 \\ 0 \\ 5 \\ 0 \\ 0 \end{bmatrix}$$

or

$$\begin{bmatrix} v_1 \\ v_2 \\ v_3 \\ v_4 \\ v_5 \\ v_6 \end{bmatrix} = \begin{bmatrix} -11/5 \\ -12/5 \\ 5 \\ 0 \\ -13/5 \\ 1/5 \end{bmatrix} - \begin{bmatrix} 0 \\ 0 \\ 0 \\ 5 \\ 0 \\ 0 \end{bmatrix} = \begin{bmatrix} -11/5 \\ -12/5 \\ 5 \\ -5 \\ -13/5 \\ 1/5 \end{bmatrix}$$

Therefore

$$v_1 = \frac{-11}{5} \text{ volts}; \qquad v_4 = -5 \text{ volts}$$

$$v_2 = \frac{-12}{5} \text{ volts}; \qquad v_5 = \frac{-13}{5} \text{ volts}$$

$$v_3 = 5 \text{ volts}; \qquad v_6 = \frac{1}{5} \text{ volts}$$

Sum of voltages in any Loop is Zero

This is obtained from the following relation

$$[B][V_b] = 0$$

[B][V$_b$]

$$= \begin{bmatrix} 1 & 0 & 0 & -1 & 1 & -1 \\ 0 & 1 & 0 & -1 & 1 & 0 \\ 0 & 0 & 1 & 1 & 0 & 0 \end{bmatrix} \begin{bmatrix} -11/5 \\ -12/5 \\ 5 \\ -5 \\ -13/5 \\ 1/5 \end{bmatrix}$$

$$= \begin{bmatrix} \dfrac{-11}{5} + 5 - \dfrac{13}{5} - \dfrac{1}{5} \\ \dfrac{-12}{5} + 5 - \dfrac{13}{5} \\ 5-5 \end{bmatrix} = \begin{bmatrix} 0 \\ 0 \\ 0 \end{bmatrix}$$

Substituting the value of $[B]\,[V_b]$ in equation (4) satisfies the equation (4). Hence, the sum of voltages in any loop is zero.

Problem No. 1.21 Write the incidence matrix of the graph in Fig. 1.72 and then express branch voltages in terms of node voltages. Write the loop matrix B and express branch currents in terms of loop currents.

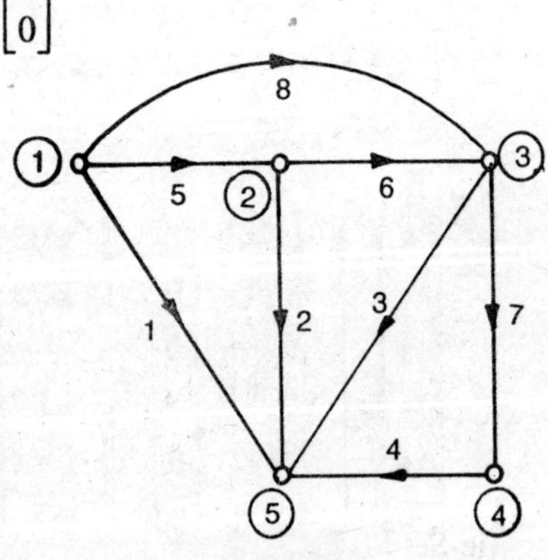

Fig. 1.72

Solution Since number of nodes are five, the branches of the tree will be

$= 5-1$
$= 4.$

The tree consisting of branches (5, 6, 2, 4) is selected. The links are (1, 3, 7, 8).

Incidence Matrix Incidence matrix is shown in Table 1.29.

Table 1.29
Incidence Matrix

	Branches							
Nodes	1	2	3	4	5	6	7	8
1	1	0	0	0	1	0	0	1
2	0	1	0	0	−1	1	0	0
3	0	0	1	0	0	−1	1	−1
4	0	0	0	1	0	0	−1	0
5	−1	−1	−1	−1	0	0	0	0

Branch Currents and Loop Currents

Tree branches in thick (bold) lines and links in broken lines is shown in Fig. 1.73.

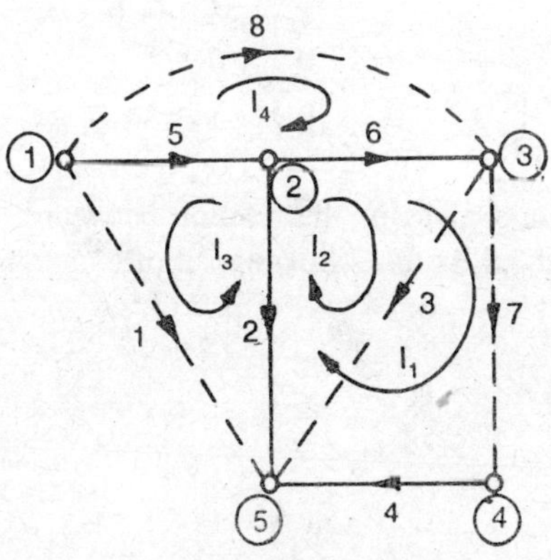

Fig. 1.73

Tie-Set Matrix is shown in Table 1.30.

Table 1.30
Tie-Set Matrix

Link Currents		Branch Currents							
		1	2	3	4	5	6	7	8
		i_1	i_2	i_3	i_4	i_5	i_6	i_7	i_8
I_1		0	−1	0	1	0	1	1	0
I_2		0	−1	1	0	0	1	0	0
I_3	B =	1	−1	0	0	−1	0	0	0
I_4		0	0	0	0	−1	−1	0	1

The relation which expresses branch currents in terms of loop currents is obtained from

$$[I_b] = [B^T][I_L]$$

$$\begin{bmatrix} i_1 \\ i_2 \\ i_3 \\ i_4 \\ i_5 \\ i_6 \\ i_7 \\ i_8 \end{bmatrix} = \begin{bmatrix} 0 & 0 & 1 & 0 \\ -1 & -1 & -1 & 0 \\ 0 & 1 & 0 & 0 \\ 1 & 0 & 0 & 0 \\ 0 & 0 & -1 & -1 \\ 1 & 1 & 0 & -1 \\ 1 & 0 & 0 & 0 \\ 0 & 0 & 0 & 1 \end{bmatrix} \begin{bmatrix} I_1 \\ I_2 \\ I_3 \\ I_4 \end{bmatrix}$$

The above relation gives us the relation between branch currents and loops and is represented by the following set of equations

(a) $i_1 = I_3$

(b) $i_2 = -I_1 - I_2 - I_3$

(c) $i_3 = I_2$

(d) $i_4 = I_1$

(e) $i_5 = -I_3 - I_4$

(f) $i_6 = I_1 + I_2 - I_4$

(g) $i_7 = I_1$

(h) $i_8 = I_4$

Branch Voltage and Node Voltages

Let us first write down the cut-set matrix

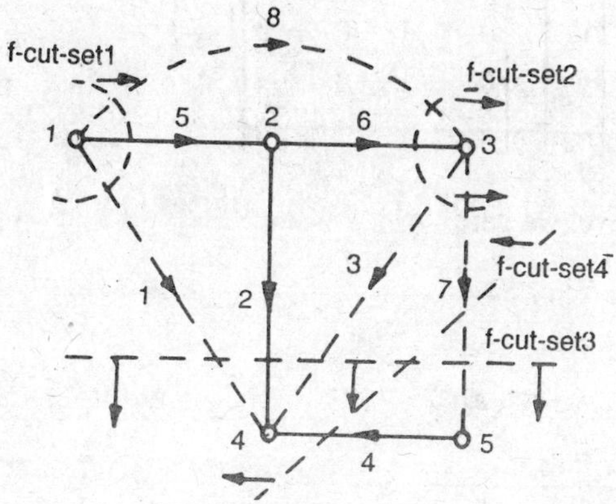

Fig. 1.74

f-Cut-Sets with their orientations are shown in Fig. 1.74. The Cut-Set Matrix is shown in Table 1.31.

Table 1.31
Fundamental Cut-Set Matrix

f-Cut-Sets		Branches							
		1	2	3	4	5	6	7	8
1 (1, 5, 8)		1	0	0	0	1	0	0	1
3 (3, 6, 7, 8)	Q =	0	0	–1	0	0	1	–1	1
4 (1, 2, 3, 7)		1	1	1	0	0	0	0	1
5 (4, 7)		0	0	0	1	0	0	–1	0

In matrix form

$$[V_b] = [Q^T][V_n]$$

$$\begin{bmatrix} v_1 \\ v_2 \\ v_3 \\ v_4 \\ v_5 \\ v_6 \\ v_7 \\ v_8 \end{bmatrix} = \begin{bmatrix} 1 & 0 & 1 & 0 \\ 0 & 0 & 1 & 0 \\ 0 & -1 & 1 & 0 \\ 0 & 0 & 0 & 1 \\ 1 & 0 & 0 & 0 \\ 0 & 1 & 0 & 0 \\ 0 & -1 & 0 & -1 \\ 1 & 1 & 1 & 0 \end{bmatrix} \begin{bmatrix} e_1 \\ e_3 \\ e_4 \\ e_5 \end{bmatrix}$$

From above, we get

(a) $v_1 = e_1 - e_4$

(b) $v_2 = e_4$

(c) $v_3 = -e_3 + e_4$

(d) $v_4 = e_5$

(e) $v_5 = e_1$

(f) $v_6 = e_3$

(g) $v_7 = -e_3 - e_5$

(h) $v_8 = e_1 + e_3 + e_4$

Alternate Tree Let twigs forming another tree and links be branches (1, 2, 3, 4) and (5, 6, 7, 8) respectively (Fig. 1.75).

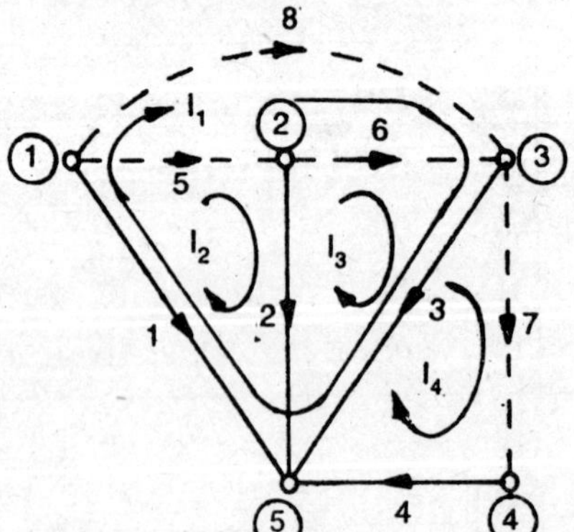

Fig. 1.75

This gives the following tie-set matrix (Table 1.32)

Table 1.32
Tie-Set Matrix

Loop Current		Branch Currents							
		i_1	i_2	i_3	i_4	i_5	i_6	i_7	i_8
I_1 (8, 3, 1)		−1	0	+1	0	0	0	0	+1
I_2 (5, 2, 1)	B =	−1	+1	0	0	+1	0	0	0
I_3 (6, 3, 2)		0	−1	+1	0	0	+1	0	0
I_4 (7, 4, 3)		0	0	−1	+1	0	0	+1	0

The following branch current and loop current relations are obtained from the above tie-set matrix shown in Table 1.32.

(a) $i_1 = -I_1 - I_2$

(b) $i_2 = -I_2 - I_3$

(c) $i_3 = -I_1 + I_3 - I_4$

(d) $i_4 = I_4$

(e) $i_5 = I_2$

(f) $i_6 = I_3$

(g) $i_7 = I_4$

(h) $i_8 = I_1$

f-Cut-Set matrix is obtained from the f-Cut-Sets shown in Fig. 1.76.

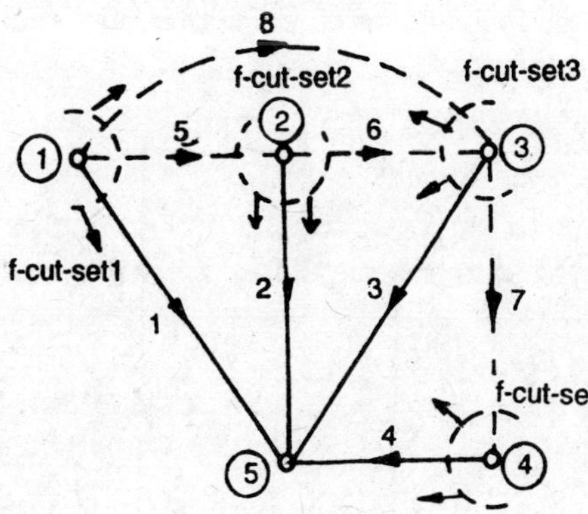

Fig. 1.76

Table 1.33
Fundamental Cut-Set Matrix

f-Cut-Sets	Branches							
	1	2	3	4	5	6	7	8
	v_1	v_2	v_3	v_4	v_5	v_6	v_7	v_8
1 (1, 5, 8)	+1	0	0	0	+1	0	0	+1
2 (2, 5, 6)	0	+1	0	0	–1	+1	0	0
3 (3, 6, 7, 8)	0	0	+1	0	0	–1	+1	–1
4 (4, 7)	0	0	0	+1	0	0	–1	0

The following relations between branch voltages and node voltages is obtained from the f-Cut-Set matrix shown in Table 1.33.

(a) $v_1 = e_1$

(b) $v_2 = e_2$

(c) $v_3 = e_3$

(d) $v_4 = e_4$

(e) $v_5 = e_1 - e_2$

(f) $v_6 = e_2 - e_3$

(g) $v_7 = e_3 - e_4$

(h) $v_8 = e_1 - e_3$

Problem 1.22 Draw the graph of the network shown in Fig. 1.77. Select a tree and construct the f-circuit matrix. Using f-circuit matrix, obtain the loop impedance parameter matrix.

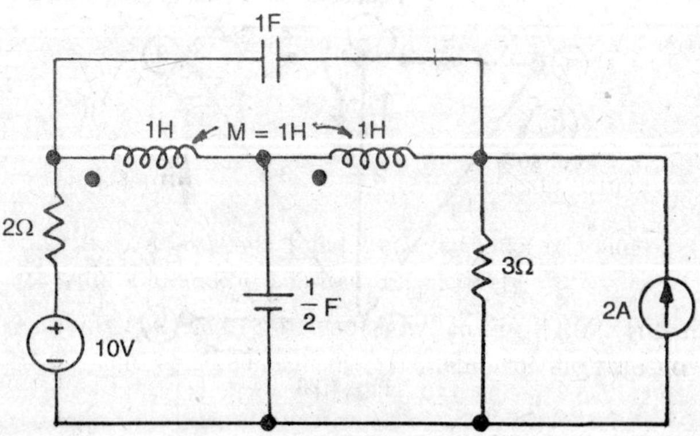

Fig. 1.77

Solution Short-circuiting the voltage source and open-circuiting the current source, graph of the network shown in Fig. 1.78 is drawn.

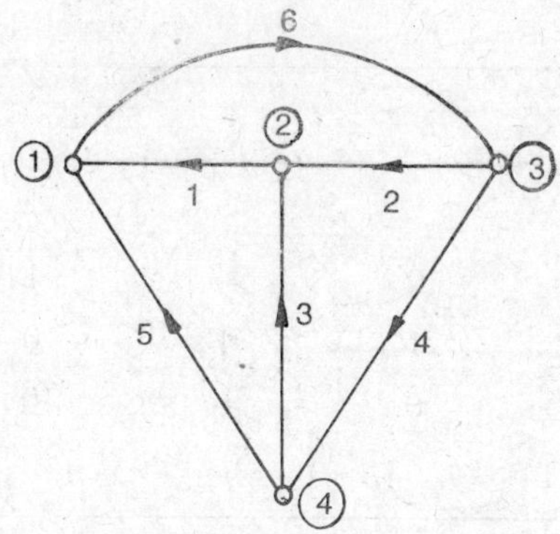

Fig. 1.78

There are four nodes and hence branches in the tree will be three. Let the tree selected consists of twigs (1, 2, 3) and therefore, the links are 4, 5 and 6. Replacement of links (4, 5, 6), one by one gives the f-circuits as shown in Fig. 1.79.

The three f-circuits are

(a) f-Circuit 4 = (4, 3, 2), I_1 aligned with link 4.

(b) f-Circuit 5 = (5, 3, 1), I_2 aligned with link 5.

(c) f-Circuit 6 = (6, 2, 1), I_3 aligned with link 6.

The f-Circuit matrix or tie-set matrix as obtained from Fig. 1.79 is shown in Table 1.34.

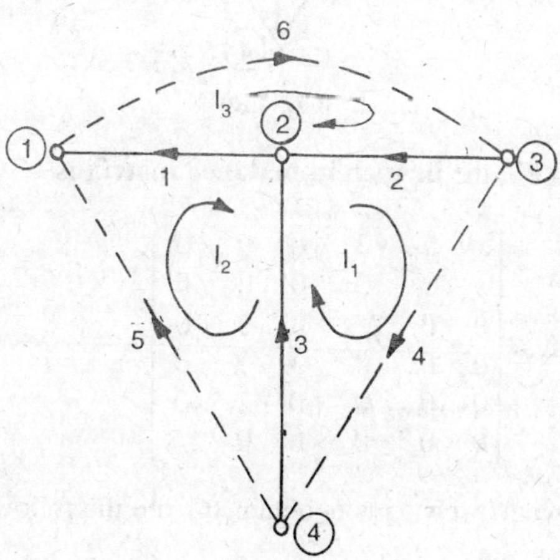

Fig. 1.79

Table 1.34
F-Circuit Matrix

f-Circuits		Branches					
		1	2	3	4	5	6
f-Circuit 4 (4, 3, 2)		0	–1	1	1	0	0
f-Circuit 5 (5, 3, 1)	B =	–1	0	–1	0	1	0
f-Circuit 6 (6, 2, 1)		1	1	0	0	0	1

Now, let us obtain branch impedance matrix. Refer Fig. 1.80 where the graph is drawn with network elements values.

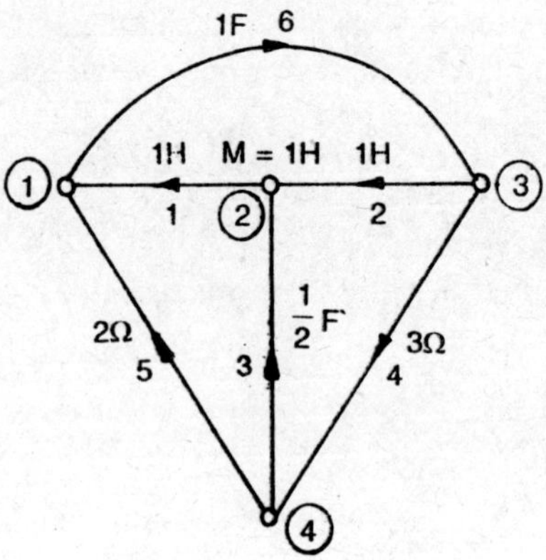

Fig. 1.80

From Fig. 1.80, the **branch impedance matrix** is

$$Z_b = \begin{bmatrix} s & s & 0 & 0 & 0 & 0 \\ s & s & 0 & 0 & 0 & 0 \\ 0 & 0 & 2/s & 0 & 0 & 0 \\ 0 & 0 & 0 & 3 & 0 & 0 \\ 0 & 0 & 0 & 0 & 2 & 0 \\ 0 & 0 & 0 & 0 & 0 & 1/s \end{bmatrix}$$

Loop Impedance Matrix This is obtained from the following relation

$$Z = BZ_b\, B^T$$

where

Z = loop impedance matrix.

Matrix B is given in Table 1.34. Therefore,

$$Z = \begin{bmatrix} 0 & -1 & 1 & 1 & 0 & 0 \\ -1 & 0 & -1 & 0 & 1 & 0 \\ 1 & 1 & 0 & 0 & 0 & 1 \end{bmatrix} \begin{bmatrix} s & s & 0 & 0 & 0 & 0 \\ s & s & 0 & 0 & 0 & 0 \\ 0 & 0 & 2/s & 0 & 0 & 0 \\ 0 & 0 & 0 & 3 & 0 & 0 \\ 0 & 0 & 0 & 0 & 2 & 0 \\ 0 & 0 & 0 & 0 & 0 & 1/s \end{bmatrix}$$

$$\times \begin{bmatrix} 0 & -1 & 1 \\ -1 & 0 & 1 \\ 1 & -1 & 0 \\ 1 & 0 & 0 \\ 0 & 1 & 0 \\ 0 & 0 & 1 \end{bmatrix}$$

or $$Z = \begin{bmatrix} 0 & -1 & 1 & 1 & 0 & 0 \\ -1 & 0 & -1 & 0 & 1 & 0 \\ 1 & 1 & 0 & 0 & 0 & 1 \end{bmatrix} \begin{bmatrix} -s & -s & 2s \\ -s & -s & 2s \\ 2/s & -2/s & 0 \\ 3 & 0 & 0 \\ 0 & 2 & 0 \\ 0 & 0 & 1/s \end{bmatrix}$$

or $$Z = \begin{bmatrix} \dfrac{s^2+3s+2}{s} & \dfrac{s^2-2}{s} & -2s \\ \dfrac{s^2-2}{s} & \dfrac{s^2+2s+2}{s} & -2s \\ -2s & -2s & \dfrac{4s^2+1}{s} \end{bmatrix}$$

Problem No. 1.23 The network in Fig. 1.81 is the voltage equivalent of a grounded grid amplifier. Draw the graph and find a tree. Determine the loop matrix B, branch impedance matrix Z_b and source voltage matrix v_g.

Solution Graph of the network is shown in Fig. 1.82.

$n = 4$

$b = 5$

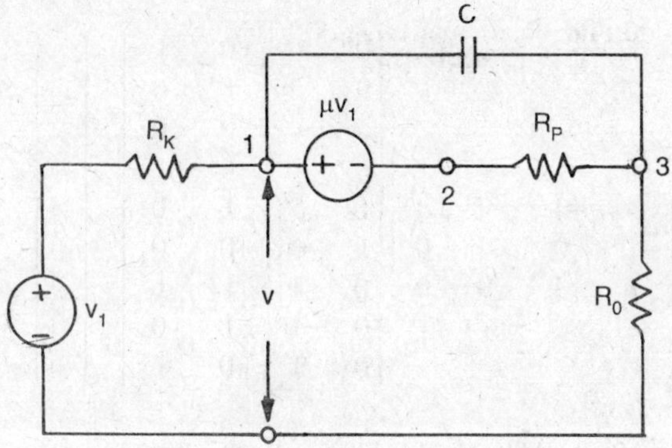

Fig. 1.81

Number of branches in the tree $= n - 1$
$= 4 - 1$
$= 3$

Let the tree consists of branches (1, 2, 3) and links be (4, 5) as shown in Fig. 1.83.

Let the links (4, 5) be removed. Loop currents are found out as given below

(a) First replace link 4. This gives the first loop circuit consisting of branches (4, 3, 2). Let I_1 be the loop current flowing as shown in Fig. 1.82.

The direction of I_1 is marked so as to follow the orientation of link 4.

(b) Now, replace only link 5. This gives the second loop circuit consisting of branch (5, 2, 1). Let I_2 be the loop current flowing as shown in Fig. 1.82. The direction of I_2 is marked so as to coincide with the orientation of link 5.

The loop matrix is obtained from Fig. 1.82 and is shown in Table 1.35.

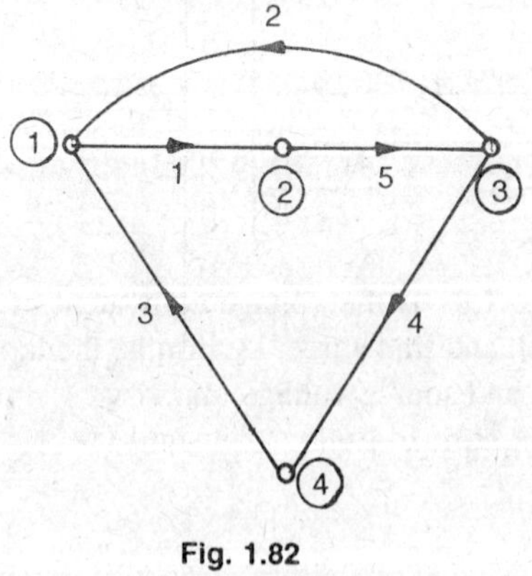

Fig. 1.82

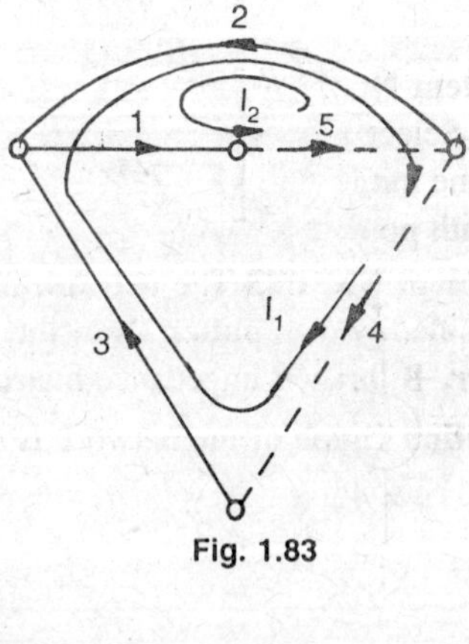

Fig. 1.83

Table 1.35
Loop Matrix

Loop Current		Branches				
		1	2	3	4	5
I_1	B =	0	-1	+1	+1	0
I_2		+1	+1	0	0	+1

Voltage Matrix (V_g)

Source voltage matrix V_g is as obtained from the network is

$$V_g = \begin{bmatrix} \mu v_1 \\ 0 \\ v_1 \\ 0 \\ 0 \end{bmatrix} = [\mu v_1 \quad 0 \quad v_1 \quad 0 \quad 0]^T$$

Note that entries in the column are based on input voltages related to the branches.

Branch Impedance Matrix

$$Z_b = \begin{bmatrix} 0 & 0 & \mu R_K & 0 & 0 \\ 0 & 1/sc & 0 & 0 & 0 \\ 0 & 0 & R_K & 0 & 0 \\ 0 & 0 & 0 & R_0 & 0 \\ 0 & 0 & 0 & 0 & R_P \end{bmatrix}$$

Problem No. 1.24 Draw the oriented graph of the network shown in Fig. 1.83. Select a tree, formulate the tie-set matrix and write loop equations. Assume initial charge on the capacitors C_1 and C_2 equivalent to v_1 (0) and v_2 (0) with potential marked.

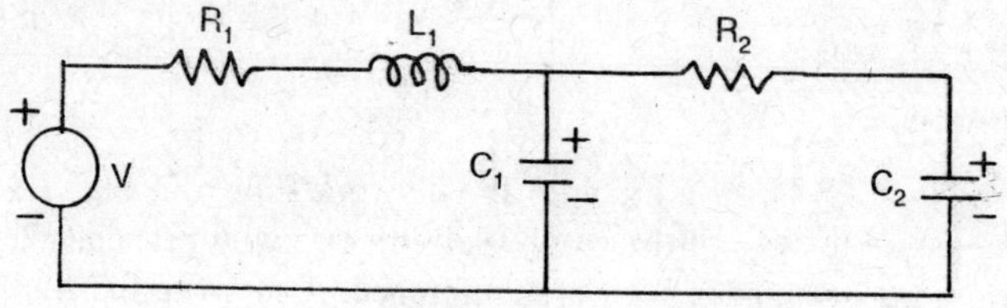

Fig. 1.83

Solution

Oriented graph of the network is shown in Fig. 1.84. Since, there are two nodes, there will be only one branch in the tree.

Let the branch selected be branch No. 1. Therefore, the links will be branches 2 and 3. Fig. 1.85 shows the tree and the links.

Replacement of links 2 and 3, one by one, gives two loop circuits in which currents I_1 and I_2 flow respectively as shown in Fig. 1.86.

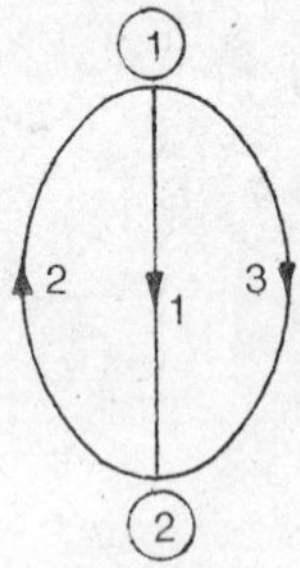

Fig. 1.84

Fig. 1.85

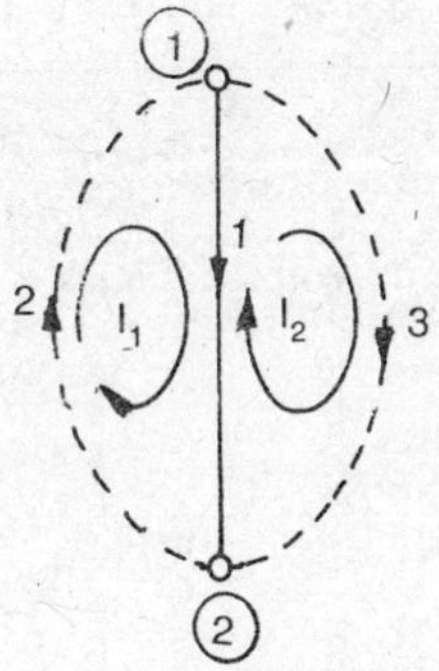

Fig. 1.86

From Fig. 1.86, tie-set matrix is shown in Table 1.36.

Table 1.36
Tie-Set Matrix

Loop Current	Branches		
	1	2	3
I_1	+1	+1	0
I_2	−1	0	+1

From the tie-set matrix following equations are obtained

(a) $i_1 = I_1 - I_2$

(b) $i_2 = I_1$

(c) $i_3 = I_2$

Problem No. 1.25 Draw a graph of the network shown in Fig. 1.87. Draw any three possible trees of the graph. Select a tree with branches 2, 3 and 6 and form a cut-set matrix with 0 as reference. Also write the fundamental circuit matrix.

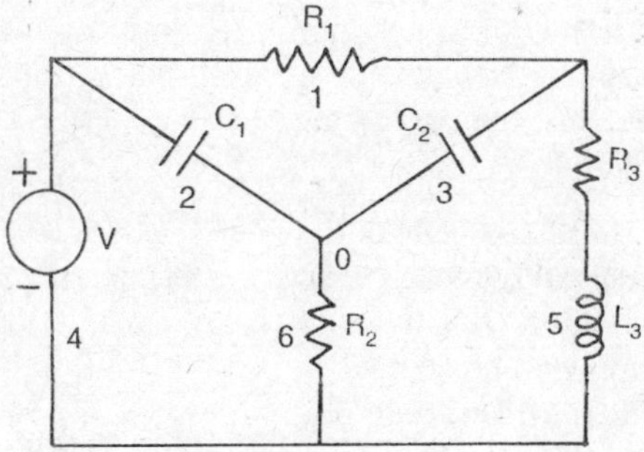

Fig. 1.87

Solutions

After short-circuiting the voltage source v in branch No. 4 and removing the circuit elements in other branches, the oriented graph is shown in Fig. 1.88.

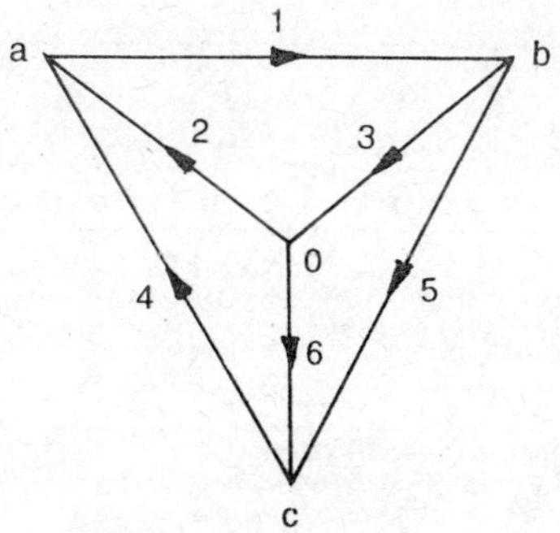

Fig. 1.88

Three Possible Trees Three possible trees are shown in Fig. 1.89. Since there are four nodes, a tree will have three branches touching all the nodes.

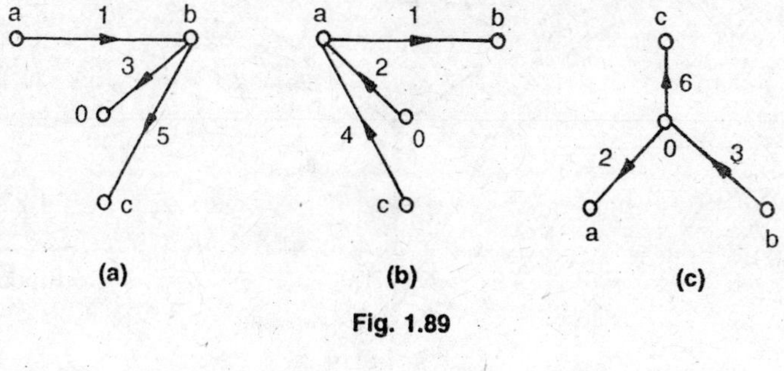

Fig. 1.89

Fundamental Circuit Matrix Tree with branches 2, 3 and 6 and links (1, 4, 5) is drawn in Fig. 1.90. Replacements of links one by one, give us the following fundamental-circuits.

(a) *f-Circuit 1*: (1, 2, 3); Direction of I_1 in orientation with branch No. 1.
(b) *f-Circuit 4*: (4, 2, 6); Direction of I_2 in orientation with branch No. 4.
(c) *f-Circuit 5*: (5, 3, 6); Direction of I_3 coincides with branch No. 5.

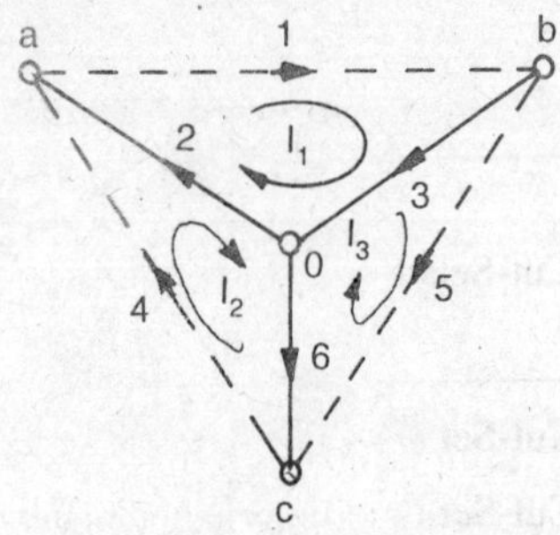

Fig. 1.90

Fundamental-circuit matrix is shown in Table 1.37.

Table 1.37
Fundamental Circuit Matrix

f-Circuits	Branches					
	1	2	3	4	5	6
f-Circuit 1	1	1	1	0	0	0
f-Circuit 4	0	−1	0	1	0	1
f-Circuit 5	0	0	−1	0	1	−1

Fundamental Cut-Sets Since there are four nodes, f-cut-sets are (n−1) i.e. 3. The three f-cut-sets are

(a) *f-Cut-Set-2*: (2, 1, 4)
(b) *f-Cut-Set-3*: (3, 1, 5)
(c) *f-Cut-Set-6*: (6, 4, 5)

Fundamental-Cut-Sets with their orientation are shown in Fig. 1.91. Fundamental Cut-Set matrix is shown in Table 1.38.

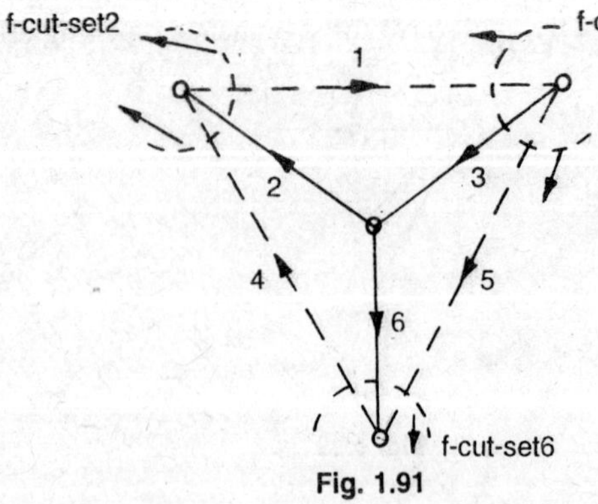

Fig. 1.91

Table 1.38
Fundamental Cut-Set Matrix

f-Cut-Sets	Branches					
	1	2	3	4	5	6
f-Cut-Set (2)	–1	1	0	1	0	0
f-Cut-Set (3)	–1	0	1	0	1	0
f-Cut-Set (6)	0	0	0	–1	1	1

Problem No. 1.26 For the network shown in Fig. 1.92

(a) Draw network graph

(b) Draw network tree with branches R_1, R_3, R_4, R_6 and R_7.

(c) Find the number of loop current variable and number of node voltage variables necessary to solve the network by the respective methods.

(d) Write tie-set matrix corresponding to above tree.

(e) Obtain loop-impedance matrix using tie-set matrix and branch impedance matrix.

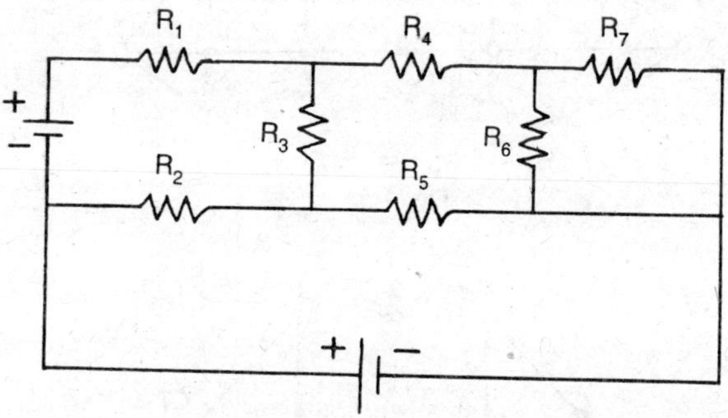

Fig. 1.92

Solution

Network Graph: Network graph is shown in Fig. 1.93

Tree: Tree with links is shown in Fig. 1.94.

Loop Current Variables

Replacement of links 9, 2, 5 and 8, one by one gives the loop current variables as shown in Fig. 1.95. Direction of loop currents coincides with respective links. Fundamental circuit-matrix is shown in Table 1.39.

Table 1.39
Fundamental Circuit Matrix

f-Circuit	Branches								
	1	2	3	4	5	6	7	8	9
f-Circuit 9 (9, 1, 4 7)	1	0	0	1	0	0	1	0	1
f-Circuit 2 (2, 1 3)	1	1	–1	0	0	0	0	0	0
f-Circuit 5 (5, 3, 4 6)	0	0	1	1	1	–1	0	0	0
f-Circuit 8	0	0	0	0	0	1	1	1	0

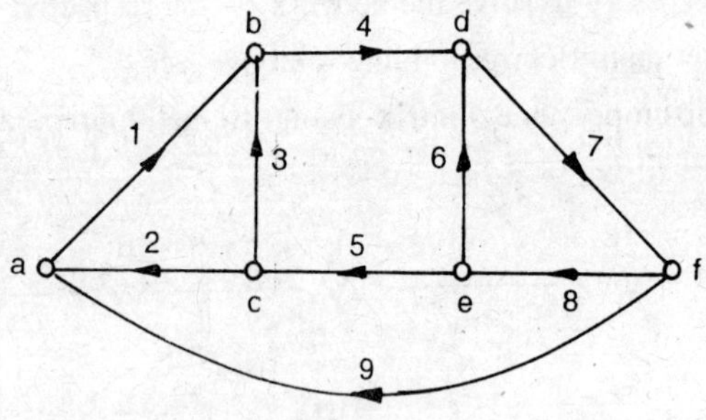

Fig. 1.93

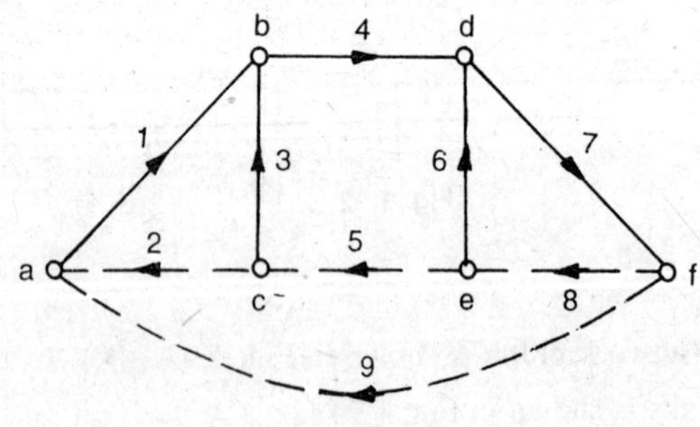

Fig. 1.94

Number of loop current variables are four as there are four links and have been shown in Fig. 1.95 marked as I_1, I_2, I_3 and I_4. Based on this the tie-set matrix or fundamental circuit matrix is shown in Table 1.39. The relation between loop currents and branch currents can be expressed as

$$[I_b] = [B^T][I_L]$$

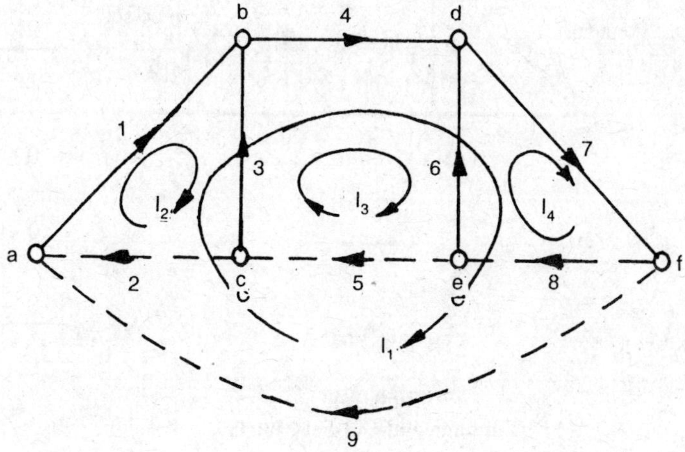

Fig. 1.95

$$\begin{bmatrix} i_1 \\ i_2 \\ i_3 \\ i_4 \\ i_5 \\ i_6 \\ i_7 \\ i_8 \\ i_9 \end{bmatrix} = \begin{bmatrix} 1 & 1 & 0 & 0 \\ 0 & 1 & 0 & 0 \\ 0 & -1 & 1 & 0 \\ 1 & 0 & 1 & 0 \\ 0 & 0 & 1 & 0 \\ 0 & 0 & -1 & 1 \\ 1 & 0 & 0 & 1 \\ 0 & 0 & 0 & 1 \\ 1 & 0 & 0 & 0 \end{bmatrix} \begin{bmatrix} I_1 \\ I_2 \\ I_3 \\ I_4 \end{bmatrix}$$

Number of Node-Voltage Variables A n node-circuit requires $n-1$ unknown voltages and same number of equations to solve the network. Also, number of f-cut-sets will also be $n-1$. In the network under consideration, number of nodes are six and hence five number voltage variables and same number of equations are required to solve the circuit. Number of f-cut-sets are also five.

Five number f-cut-sets along with their orientation are shown in Fig. 1.96. Fundamental cut-set matrix is shown in Table 1.40.

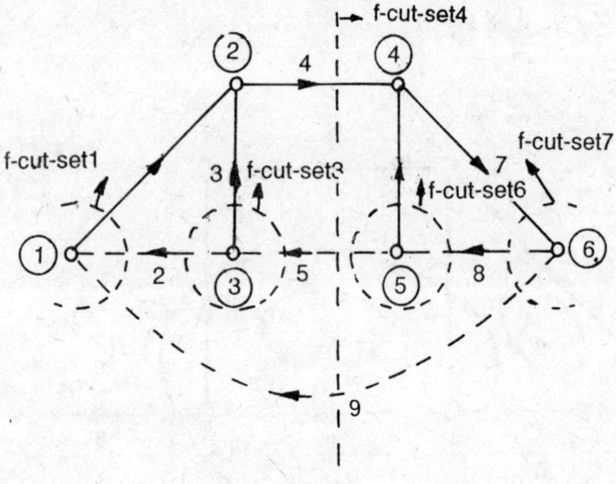

Fig. 1.96

Table 1.40
Fundamental Cut-Set Matrix

f-Cut Sets		Branches								
		1	2	3	4	5	6	7	8	9
f-Cut Set 1 (1, 2, 9)		1	−1	0	0	0	0	0	0	−1
f-Cut Set 3 (3, 2, 5)		0	1	1	0	−1	0	0	0	0
f-Cut Set 4 (4, 5, 9)	Q =	0	0	0	1	−1	0	0	0	−1
f-Cut-Set 6 (6, 5, 8)		0	0	0	0	1	1	0	−1	0
f-Cut-Set 7 (7, 8, 9)		0	0	0	0	0	0	1	−1	−1

The relation is expressed as

$$\left[V_b\right]=\left[Q^T\right]\left[V_n\right]$$

$$\begin{bmatrix} v_1 \\ v_2 \\ v_3 \\ v_4 \\ v_5 \\ v_6 \\ v_7 \\ v_8 \\ v_9 \end{bmatrix} = \begin{bmatrix} 1 & 0 & 0 & 0 & 0 \\ -1 & 1 & 0 & 0 & 0 \\ 0 & 1 & 0 & 0 & 0 \\ 0 & 0 & 1 & 0 & 0 \\ 0 & -1 & -1 & 1 & 0 \\ 0 & 0 & 0 & 1 & 0 \\ 0 & 0 & 0 & 0 & 1 \\ 0 & 0 & 0 & -1 & -1 \\ -1 & 0 & -1 & 0 & -1 \end{bmatrix} \begin{bmatrix} e_1 \\ e_3 \\ e_4 \\ e_6 \\ e_8 \end{bmatrix}$$

which gives the relation

(a) $v_1 = e_1$

(b) $v_2 = e_3 - e_1$

(c) $v_3 = e_3$

(d) $v_4 = e_4$

(e) $v_5 = e_6 - e_4 - e_3$

(f) $v_6 = e_6$

(g) $v_7 = e_8$

(h) $v_8 = -e_6 - e_8$

(j) $v_9 = -e_1 - e_4 - e_8$

Branch Impedance Matrix

Branch impedance matrix is given below

$$Z_b = \begin{bmatrix} R_1 & 0 & 0 & 0 & 0 & 0 & 0 & 0 & 0 \\ 0 & R_2 & 0 & 0 & 0 & 0 & 0 & 0 & 0 \\ 0 & 0 & R_3 & 0 & 0 & 0 & 0 & 0 & 0 \\ 0 & 0 & 0 & R_4 & 0 & 0 & 0 & 0 & 0 \\ 0 & 0 & 0 & 0 & R_5 & 0 & 0 & 0 & 0 \\ 0 & 0 & 0 & 0 & 0 & R_6 & 0 & 0 & 0 \\ 0 & 0 & 0 & 0 & 0 & 0 & R_7 & 0 & 0 \\ 0 & 0 & 0 & 0 & 0 & 0 & 0 & 0 & 0 \\ 0 & 0 & 0 & 0 & 0 & 0 & 0 & 0 & 0 \end{bmatrix}$$

Loop Impedance Matrix (Z)

$$Z = B\,Z_b\,B^T$$

where B = f-Circuit or tie-set matrix

Z_b = Branch impedance matrix

Therefore,

$$Z = \begin{bmatrix} 1 & 0 & 0 & 1 & 0 & 0 & 1 & 0 & 1 \\ 1 & 1 & -1 & 0 & 0 & 0 & 0 & 0 & 0 \\ 0 & 0 & 1 & 1 & 1 & -1 & 0 & 0 & 0 \\ 0 & 0 & 0 & 0 & 0 & 1 & 1 & 1 & 0 \end{bmatrix} \times$$

$$\begin{bmatrix} R_1 & 0 & 0 & 0 & 0 & 0 & 0 & 0 & 0 \\ 0 & R_2 & 0 & 0 & 0 & 0 & 0 & 0 & 0 \\ 0 & 0 & R_3 & 0 & 0 & 0 & 0 & 0 & 0 \\ 0 & 0 & 0 & R_4 & 0 & 0 & 0 & 0 & 0 \\ 0 & 0 & 0 & 0 & R_5 & 0 & 0 & 0 & 0 \\ 0 & 0 & 0 & 0 & 0 & R_6 & 0 & 0 & 0 \\ 0 & 0 & 0 & 0 & 0 & 0 & R_7 & 0 & 0 \\ 0 & 0 & 0 & 0 & 0 & 0 & 0 & 0 & 0 \\ 0 & 0 & 0 & 0 & 0 & 0 & 0 & 0 & 0 \end{bmatrix} \begin{bmatrix} 1 & 1 & 0 & 0 \\ 0 & 1 & 0 & 0 \\ 0 & -1 & 1 & 0 \\ 1 & 0 & 1 & 0 \\ 0 & 0 & 1 & 0 \\ 0 & 0 & -1 & 1 \\ 1 & 0 & 0 & 1 \\ 0 & 0 & 0 & 1 \\ 1 & 0 & 0 & 0 \end{bmatrix}$$

$$Z = \begin{bmatrix} 1 & 0 & 0 & 1 & 0 & 0 & 1 & 0 & 1 \\ 1 & 1 & -1 & 0 & 0 & 0 & 0 & 0 & 0 \\ 0 & 0 & 1 & 1 & 1 & -1 & 0 & 0 & 0 \\ 0 & 0 & 0 & 0 & 0 & 1 & 1 & 1 & 0 \end{bmatrix} \begin{bmatrix} R_1 & R_1 & 0 & 0 \\ 0 & R_2 & 0 & 0 \\ 0 & -R_3 & R_3 & 0 \\ R_4 & 0 & R_4 & 0 \\ 0 & 0 & R_5 & 0 \\ 0 & 0 & -R_6 & R_6 \\ R_7 & 0 & 0 & R_7 \\ 0 & 0 & 0 & 0 \\ 0 & 0 & 0 & 0 \end{bmatrix}$$

$$Z = \begin{bmatrix} R_1 + R_4 + R_7 & R_1 & R_4 & R_7 \\ R_1 & R_1 + R_2 + R_3 & -R_3 & 0 \\ R_4 & -R_3 & R_3 + R_4 + R_5 + R_6 & -R_6 \\ R_7 & 0 & -R_6 & R_6 + R_7 \end{bmatrix}$$

Problem 1.27 Formulate the cut-set matrix of the network shown in Fig. 1.97 and hence obtain the node equations.

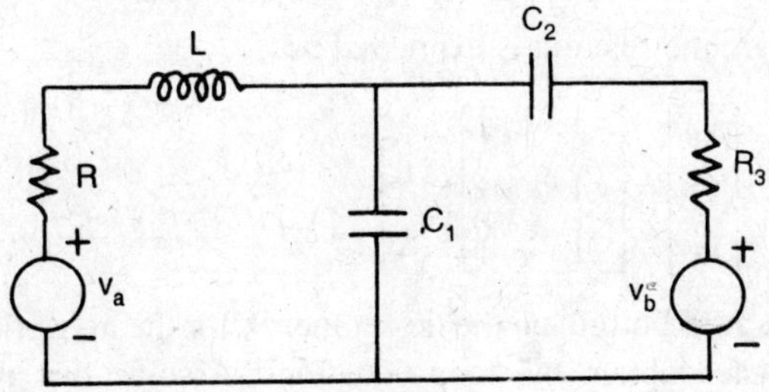

Fig. 1.97

Solution Short circuiting the voltage sources, the graph is shown in Fig. 1.98. Since there are two nodes, tree will have only one branch (2) and is shown in Fig. 1.99. Cut-Set matrix is shown in Table 1.41.

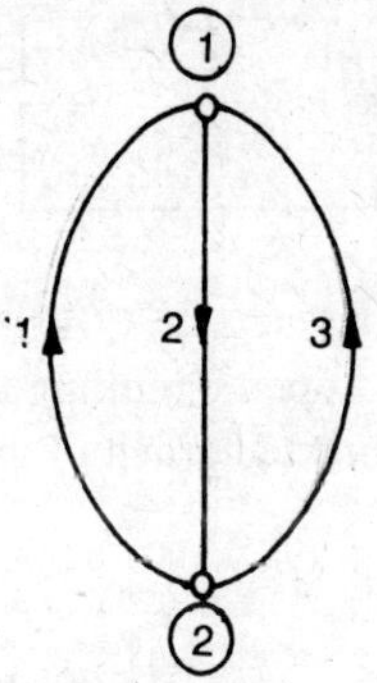

Fig. 1.98

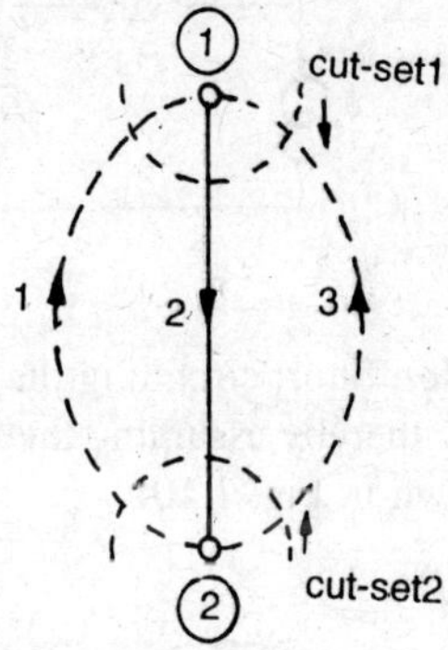

Fig. 1.99

Table 1.41
Cut-Set Matrix

Cut-Sets	Branches		
	1	2	3
Cut-Set 1	–1	1	–1
Cut-Set 2	1	–1	1

The node equations are

(a) $v_1 = e_2 - e_1$,

(b) $v_2 = e_1 - e_2$, and

(c) $v_3 = e_2 - e_1$.

In matrix form, the relation is expressed as

$$\begin{bmatrix} v_1 \\ v_2 \\ v_3 \end{bmatrix} = \begin{bmatrix} -1 & 1 \\ 1 & -1 \\ -1 & 1 \end{bmatrix} \begin{bmatrix} e_1 \\ e_2 \end{bmatrix}$$

Problem No. 1.28 Formulate the tie-set matrix for the network shown in fig. 1.100 and hence obtain the loop equations. Assume that the network is relaxed initially.

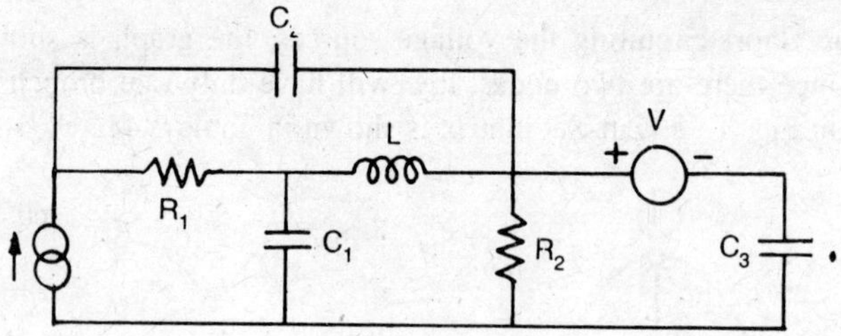

Fig. 1.100

Solution Short circuiting the voltage source and open-circuiting the current source thereby assuming that it is not present, directed graph of the network is shown in Fig. 1.101.

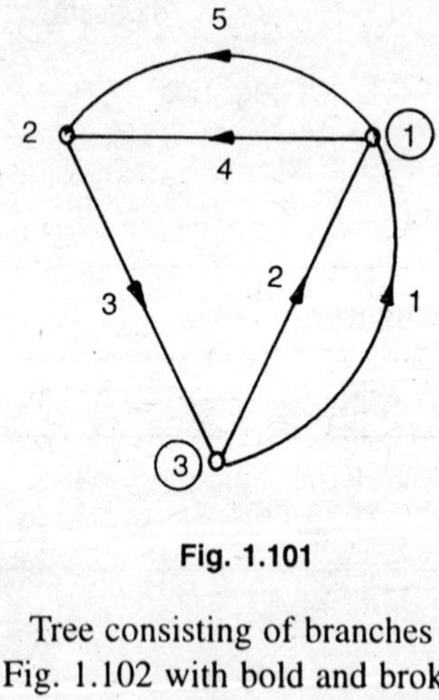

Fig. 1.101

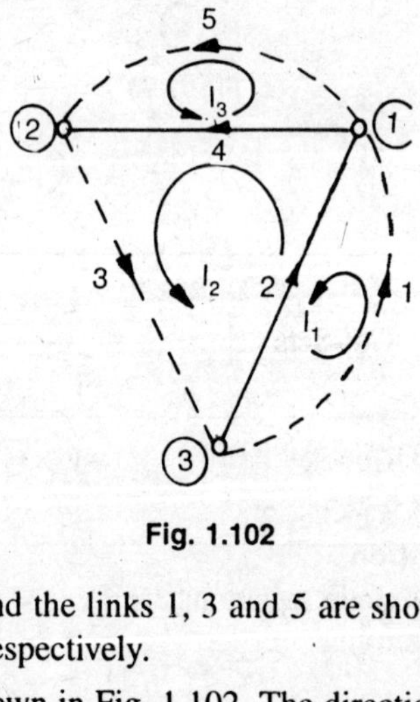

Fig. 1.102

Tree consisting of branches 4 and 2 and the links 1, 3 and 5 are shown in Fig. 1.102 with bold and broken lines respectively.

Mesh currents I_1, I_2 and I_3 are also shown in Fig. 1.102. The directions of the mesh currents coincide with the orientation of the respective links.

Tie-set matrix deduced from Fig. 1.102 is shown in Table 1.42.

Table 1.42
Tie-Set Matrix

Loop Currents	Branches				
	1	2	3	4	5
I_1 (1, 2)	1	−1	0	0	0
I_2 (3, 2, 4) B =	0	1	1	1	0
I_3 (5, 4)	0	0	0	−1	1

Loop equations in the matrix form is expressed as

$$[I_b] = [B^T][I_L]$$

$$\begin{bmatrix} i_1 \\ i_2 \\ i_3 \\ i_4 \\ i_5 \end{bmatrix} = \begin{bmatrix} 1 & 0 & 0 \\ -1 & 1 & 0 \\ 0 & 1 & 0 \\ 0 & 1 & -1 \\ 0 & 0 & 1 \end{bmatrix} \begin{bmatrix} I_1 \\ I_2 \\ I_3 \end{bmatrix}$$

Set of equations relating branch currents and mesh currents are

(a) $i_1 = I_1$,

(b) $i_2 = I_2 - I_1$,

(c) $i_3 = I_2$,

(d) $i_4 = I_2 - I_3$, and

(f) $i_5 = I_3$.

Problem No. 1.29 Draw the graph of the resistive network shown in Fig. 1.103. Formulate equilibrium equations on loop basis selecting branches 4, 5, 7, 8 and 9 as tree branches. Resistances given are in ohms.

Solution Tree of the network consisting of branches 4, 5, 7, 8 and 9 and the links consisting of branches 1, 2, 3 and 6 are shown in Fig. 1.104 as bold and broken lines respectively.

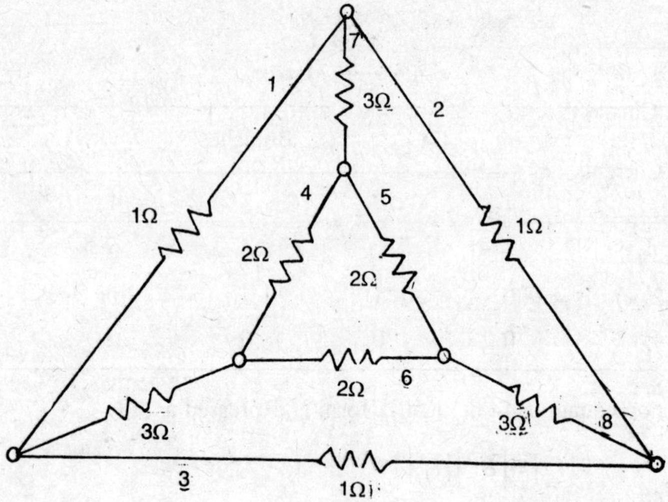

Fig. 1.103

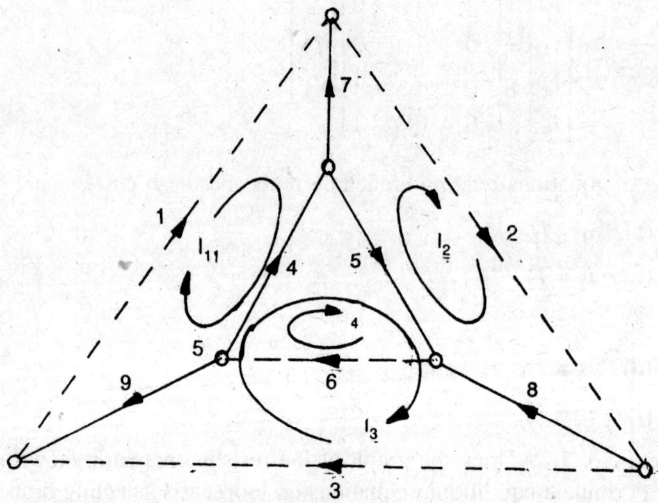

Fig. 1.104

Let the loop currents be defined by I_1, I_2, I_3 and I_4 whose directions coincide with that of orientation of links 1, 2, 3 and 6 respectively as shown in Fig. 1.104. Tie-Set schedule deduced from Fig. 1.104 is shown in Table 1.43.

Table 1.43
Tie-Set Matrix

Link Current		Branches								
		1	2	3	4	5	6	7	8	9
		i_1	i_2	i_3	i_4	i_5	i_6	i_7	i_8	i_9
I_1 (1, 7, 4, 9)		1	0	0	-1	0	0	-1	0	1
I_2 (2, 8, 5, 7)	B =	0	1	0	0	-1	0	1	-1	0
I_3 (3, 9, 4, 5, 8)		0	0	1	1	1	0	0	1	-1
I_4 (6, 4, 5)		0	0	0	1	1	1	0	0	0

KVL equations for the loops are expressed as

$$[B][V_b] = [0],$$

or

$$\begin{bmatrix} 1 & 0 & 0 & -1 & 0 & 0 & -1 & 0 & 1 \\ 0 & 1 & 0 & 0 & -1 & 0 & 1 & -1 & 0 \\ 0 & 0 & 1 & 1 & 1 & 0 & 0 & 1 & -1 \\ 0 & 0 & 0 & 1 & 1 & 1 & 0 & 0 & 0 \end{bmatrix} \begin{bmatrix} v_1 \\ v_2 \\ v_3 \\ v_4 \\ v_5 \\ v_6 \\ v_7 \\ v_8 \\ v_9 \end{bmatrix} = \begin{bmatrix} 0 \\ 0 \\ 0 \\ 0 \end{bmatrix}$$

Relation $[I_b] = [B^T][I_L]$ gives relation between branch currents and link currents.

$$\begin{bmatrix} i_1 \\ i_2 \\ i_3 \\ i_4 \\ i_5 \\ i_6 \\ i_7 \\ i_8 \\ i_9 \end{bmatrix} = \begin{bmatrix} 1 & 0 & 0 & 0 \\ 0 & 1 & 0 & 0 \\ 0 & 0 & 1 & 0 \\ -1 & 0 & 1 & 1 \\ 0 & -1 & 1 & 1 \\ 0 & 0 & 0 & 1 \\ -1 & 1 & 0 & 0 \\ 0 & -1 & 1 & 0 \\ 1 & 0 & -1 & 0 \end{bmatrix} \begin{bmatrix} I_1 \\ I_2 \\ I_3 \\ I_4 \end{bmatrix}$$

Also,

$$[V_b] = [Z_b][I_b] - [V_s]$$

where

V_b = branch voltage vector of order (9 × 1),
Z_b = impedance matrix of order (9 × 9),
I_b = branch current vector of order (9 × 1), and
V_s = branch input voltage source vector of order (9 × 1).

Here,

$[V_s] = 0$. Therefore,

$$[V_b] = [Z_b][I_b]$$

$$\begin{bmatrix} v_1 \\ v_2 \\ v_3 \\ v_4 \\ v_5 \\ v_6 \\ v_7 \\ v_8 \\ v_9 \end{bmatrix} = \begin{bmatrix} 1 & 0 & 0 & 0 & 0 & 0 & 0 & 0 & 0 \\ 0 & 1 & 0 & 0 & 0 & 0 & 0 & 0 & 0 \\ 0 & 0 & 1 & 0 & 0 & 0 & 0 & 0 & 0 \\ 0 & 0 & 0 & 2 & 0 & 0 & 0 & 0 & 0 \\ 0 & 0 & 0 & 0 & 2 & 0 & 0 & 0 & 0 \\ 0 & 0 & 0 & 0 & 0 & 2 & 0 & 0 & 0 \\ 0 & 0 & 0 & 0 & 0 & 0 & 3 & 0 & 0 \\ 0 & 0 & 0 & 0 & 0 & 0 & 0 & 3 & 0 \\ 0 & 0 & 0 & 0 & 0 & 0 & 0 & 0 & 3 \end{bmatrix} \begin{bmatrix} i_1 \\ i_2 \\ i_3 \\ i_4 \\ i_5 \\ i_6 \\ i_7 \\ i_8 \\ i_9 \end{bmatrix}$$

Loop Impedance matrix is given by

$$[Z] = [B][Z_b][B^T]$$

or $$Z = \begin{bmatrix} 1 & 0 & 0 & -1 & 0 & 0 & -1 & 0 & 1 \\ 0 & 1 & 0 & 0 & -1 & 0 & 1 & -1 & 0 \\ 0 & 0 & 1 & 1 & 1 & 0 & 0 & 1 & -1 \\ 0 & 0 & 0 & 1 & 1 & 1 & 0 & 0 & 0 \end{bmatrix} \times$$

$$\begin{bmatrix} 1 & 0 & 0 & 0 & 0 & 0 & 0 & 0 & 0 \\ 0 & 1 & 0 & 0 & 0 & 0 & 0 & 0 & 0 \\ 0 & 0 & 1 & 0 & 0 & 0 & 0 & 0 & 0 \\ 0 & 0 & 0 & 2 & 0 & 0 & 0 & 0 & 0 \\ 0 & 0 & 0 & 0 & 2 & 0 & 0 & 0 & 0 \\ 0 & 0 & 0 & 0 & 0 & 2 & 0 & 0 & 0 \\ 0 & 0 & 0 & 0 & 0 & 0 & 3 & 0 & 0 \\ 0 & 0 & 0 & 0 & 0 & 0 & 0 & 3 & 0 \\ 0 & 0 & 0 & 0 & 0 & 0 & 0 & 0 & 3 \end{bmatrix} \begin{bmatrix} 1 & 0 & 0 & 0 \\ 0 & 1 & 0 & 0 \\ 0 & 0 & 1 & 0 \\ -1 & 0 & 1 & 1 \\ 0 & -1 & 1 & 1 \\ 0 & 0 & 0 & 1 \\ -1 & 1 & 0 & 0 \\ 0 & -1 & 1 & 0 \\ 1 & 0 & -1 & 0 \end{bmatrix}$$

$$Z = \begin{bmatrix} 1 & 0 & 0 & -1 & 0 & 0 & -1 & 0 & 1 \\ 0 & 1 & 0 & 0 & -1 & 0 & 1 & -1 & 0 \\ 0 & 0 & 1 & 1 & 1 & 0 & 0 & 1 & -1 \\ 0 & 0 & 0 & 1 & 1 & 1 & 0 & 0 & 0 \end{bmatrix} \begin{bmatrix} 1 & 0 & 0 & 0 \\ 0 & 1 & 0 & 0 \\ 0 & 0 & 1 & 0 \\ -2 & 0 & 2 & 2 \\ 0 & -2 & 2 & 2 \\ 0 & 0 & 0 & 2 \\ -3 & 3 & 0 & 0 \\ 0 & -3 & 3 & 0 \\ 3 & 0 & -3 & 0 \end{bmatrix}$$

or $$Z = \begin{bmatrix} 9 & -3 & -5 & -2 \\ -3 & 9 & -5 & -2 \\ -5 & -5 & 11 & 4 \\ -2 & -2 & 4 & 6 \end{bmatrix}$$

Kirchoff's voltage law equations are now written as

$$[Z][I_L] = \begin{bmatrix} 9 & -3 & -5 & -2 \\ -3 & 9 & -5 & -2 \\ -5 & -5 & 11 & 4 \\ -2 & -2 & 4 & 6 \end{bmatrix} \begin{bmatrix} I_1 \\ I_2 \\ I_3 \\ I_4 \end{bmatrix} = [0]$$

which gives

$$9I_1 - 3I_2 - 5I_3 - 2I_4 = 0$$

$$-3I_1 + 9I_2 - 5I_3 - 2I_4 = 0$$

$$-5I_1 - 5I_2 + 11I_3 + 4I_4 = 0$$

$$-2I_1 - 2I_2 + 4I_3 + 6I_4 = 0$$

Alternatively

By making use of co-efficients in the rows of the tie-set matrix, Kirchoff's voltage law equations for the loops are

$$\left. \begin{aligned} v_1 - v_4 - v_7 + v_9 &= 0 \\ v_2 - v_5 + v_7 - v_8 &= 0 \\ v_3 + v_4 + v_5 + v_8 - v_9 &= 0 \\ v_4 + v_3 + v_6 &= 0 \end{aligned} \right\} \quad ...(1)$$

The columns of the tie-set matrix express the branch currents in terms of loop currents

$$\left. \begin{aligned} j_1 &= I_1 \\ j_2 &= I_2 \\ j_3 &= I_3 \\ j_4 &= -I_1 + I_3 + I_4 \\ j_5 &= -I_2 + I_3 + I_4 \\ j_6 &= I_4 \\ j_7 &= -I_1 + I_2 \\ j_8 &= -I_2 + I_3 \\ j_9 &= I_1 - I_3 \end{aligned} \right\} \quad ...(2)$$

Also

$$\left.\begin{aligned} v_1 &= r_1 j_1 \\ v_2 &= r_2 j_2 \\ v_3 &= r_3 j_3 \\ v_4 &= r_4 j_4 \\ v_5 &= r_5 j_5 \\ v_6 &= r_6 j_6 \\ v_7 &= r_7 j_7 \\ v_8 &= r_8 j_8 \\ v_9 &= r_9 j_9 \end{aligned}\right\} \qquad ...(3)$$

It is given that

$$\left.\begin{aligned} r_1 &= r_2 = r_3 = 1 \text{ ohm} \\ r_4 &= r_5 = r_6 = 2 \text{ ohm} \\ r_7 &= r_8 = r_9 = 3 \text{ ohm.} \end{aligned}\right\} \qquad ...(4)$$

substituting the values of branch currents j_1 to j_9 from equation (2) and resistance from equation (4) in equation (3), we get

$$\left.\begin{aligned} v_1 &= j_1 ; & v_4 &= 2j_4 ; & v_7 &= 3j_7 \\ v_2 &= j_2 ; & v_5 &= 2j_5 ; & v_8 &= 3j_8 \\ v_3 &= j_3 ; & v_6 &= 2j_6 ; & v_9 &= 3j_9 \end{aligned}\right\} \qquad ...(4)$$

substituting the values of j_1 to j_9 from equation (2) into equation (4), we get

$$\left.\begin{aligned} v_1 &= I_1 ; & v_4 &= 2(-I_1 + I_3 + I_4) ; & v_7 &= 3(-I_1 + I_2) \\ v_2 &= I_2 ; & v_5 &= 2(-I_2 + I_3 + I_4) ; & v_8 &= 3(-I_2 + I_3) \\ v_3 &= I_3 ; & v_6 &= 2I_4 \qquad ; & v_9 &= 3(I_1 - I_3) \end{aligned}\right\} \qquad ...(5)$$

substituting the values of branch voltages v_1 to v_9 from equation (5) in equation (1), we get

$$I_1 - 2(-I_1 + I_3 + I_4) - 3(-I_1 + I_2) + 3(I_1 - I_3) = 0$$

$$I_2 - 2(-I_2 + I_3 + I_4) + 3(-I_1 + I_2) - 3(-I_2 + I_3) = 0$$

$$I_3 + 2(-I_1 + I_3 + I_4) + 2(-I_2 + I_3 + I_4) - 3(I_1 - I_3) = 0$$

$$2(-I_1 + I_3 + I_4) + 2(-I_2 + I_3 + I_4) + 2(I_4) = 0$$

Rearranging, it gives

$$9I_1 - 3I_2 - 5I_3 - 2I_4 = 0$$
$$-3I_1 + 9I_2 - 5I_3 - 2I_4 = 0$$
$$-5I_1 - 5I_2 + 11I_3 + 4I_4 = 0$$
$$-2I_1 - 2I_2 + 4I_3 + 6I_4 = 0$$

which is same as found out earlier.

Problem No. 1.30 Given that the tie-set schedule shown was obtained from set of mesh currents represent the geometrical network numbering all branches and including all reference arrows.

Loop No.	Branch No.								
	1	2	3	4	5	6	7	8	9
1	1	1	0	0	0	0	1	–1	0
2	0	0	1	1	0	0	0	1	–1
3	0	0	0	0	1	1	–1	0	1

Solution There are three loops and nine branches as seen from the tie-set schedule.

We know that

b – (n–1) = number of loops

which gives

9 – (n–1) = 3

or n = 7 i.e. seven nodes

and the tree will have n–1 i.e. 7 – 1 = 6 branches and hence three links.

The rows of the tie set schedule give the following relations:

(1) *First Row*: First row gives $v_1 + v_2 + v_7 - v_8 = 0$. This can be represented by a closed mesh as shown in Fig. 1.105 (a).

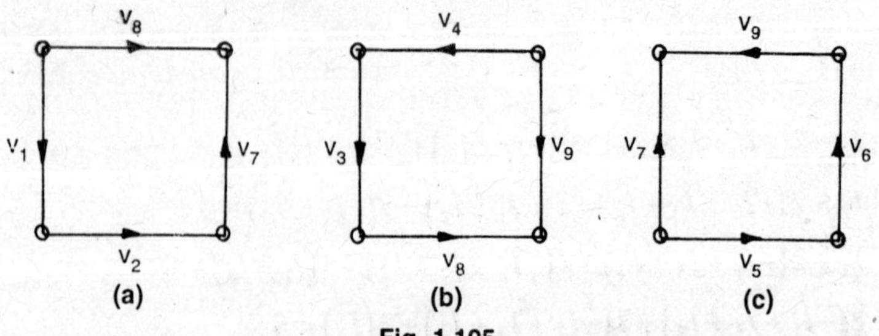

Fig. 1.105

(2) *Second Row*: Second row gives $v_3 + v_4 + v_8 - v_9 = 0$. This can be represented by a closed mesh as shown in Fig. 1.106 (b).

(3) *Third Row:* Third row gives $v_5 + v_6 - v_7 + v_9 = 0$. This can be represented by a closed mesh as shown in Fig. 1.106 (c).

Combining Figs. 1.105 (a) and (c), we get Fig. 1.106.

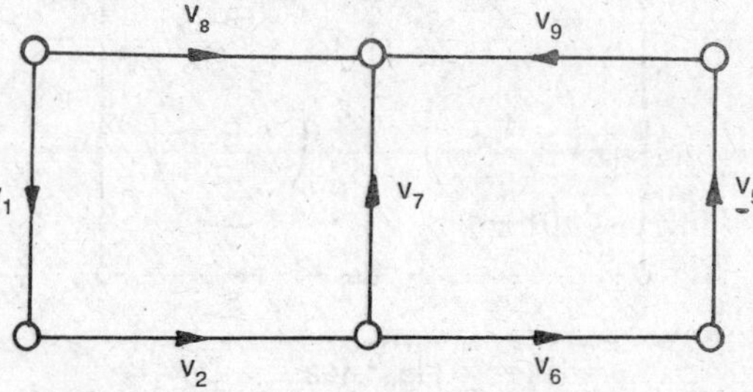

Fig. 1.106

Combining Fig. 1.106 and Fig. 1.105 (b), we get complete directed graph as shown in Fig. 1.107. Nodes 1 to 9 are also marked.

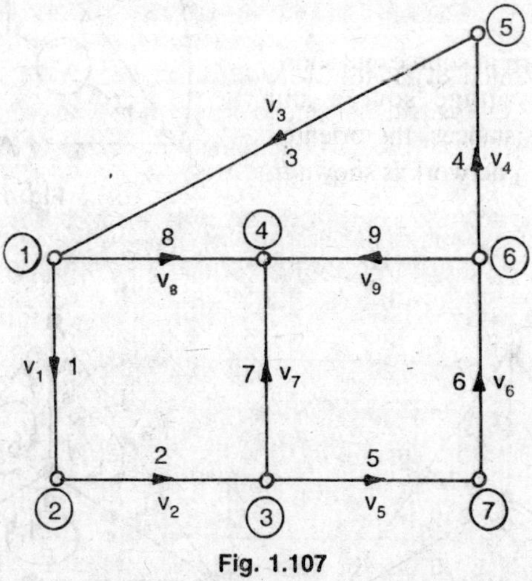

Fig. 1.107

Tree consisting of branches 1, 8, 7, 9, 6 and 4 marked with bold lines and links 2, 3 and 5 marked with broken lines along with mesh currents I_1, I_2 and I_3 are shown in Fig. 1.108.

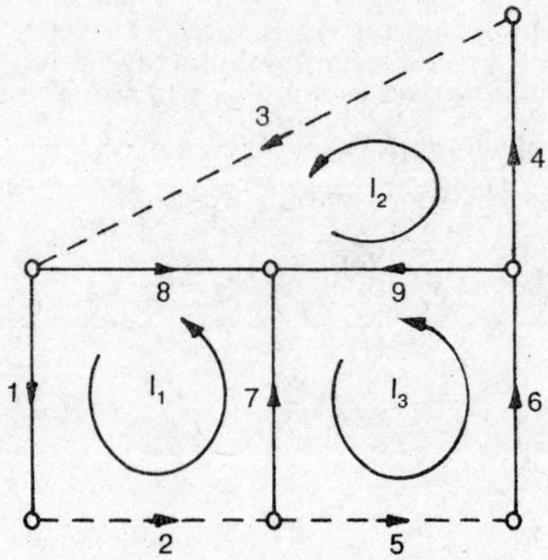

Fig. 1.108

Problem No. 1.31 Draw a tree for the circuit shown in Fig. 1.109. For this tree, tie-set and cut-set schedule.

(VTU Mar 2000, Electric Circuits)

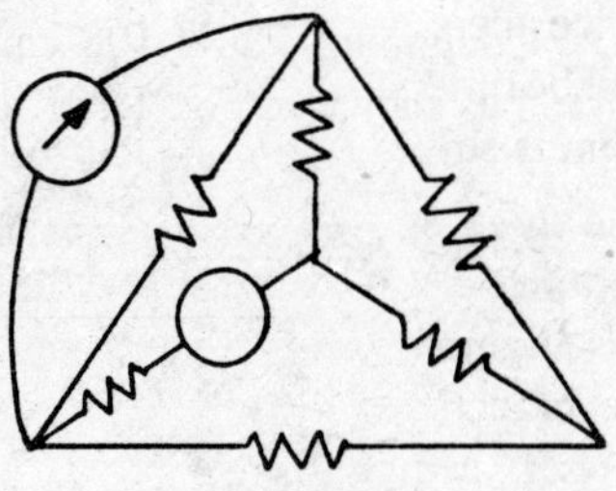

Fig. 1.109

Solution

Removing the current source and short-circuiting the voltage source and removing the resistances, the oriented graph of the given network is shown in Fig. 1.110.

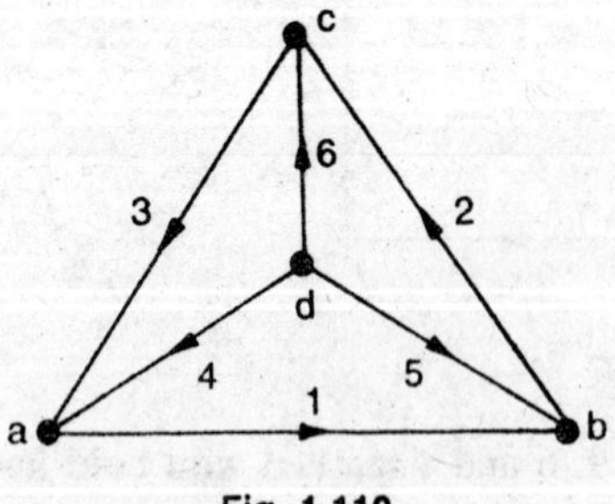

Fig. 1.110

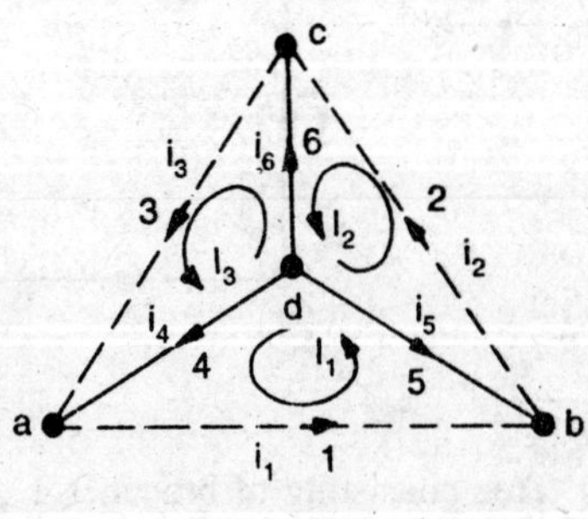

Fig. 1.111

Since these are four nodes, number of branches in a tree will be one less i.e., three. Let the tree selected has branches 4, 5 and 6.

Tie-Set Matrix/Schedule

Mark link currents I_1, I_2 and I_3 and branch currents i_1, i_2, i_3, i_4, i_5 and i_6 as shown in Fig. 1.111. Refer Problem No. 1.1 in which how to make tie-set schedule has been explained. This is shown in Table 1.44.

Table 1.44: Tie-Set Schedule

Link Current	Loop	Branch					
		1	2	3	4	5	6
I_1	1-4-5	1	0	0	1	–1	0
I_2	2-5-6	0	1	0	0	1	–1
I_3	3-4-6	0	0	1	–1	0	1

Cut-Set Schedule

Since there are four nodes, cut sets will be one less i.e. three. Similar problem is solved in Problem No. 1.24.

The three cut-sets selected at points a, b and c and directions of the cut-sets align with that of the direction of the branch of the tree defining the cut-set. Method for selecting cut-sets has been explained in Problem No. 1.8.

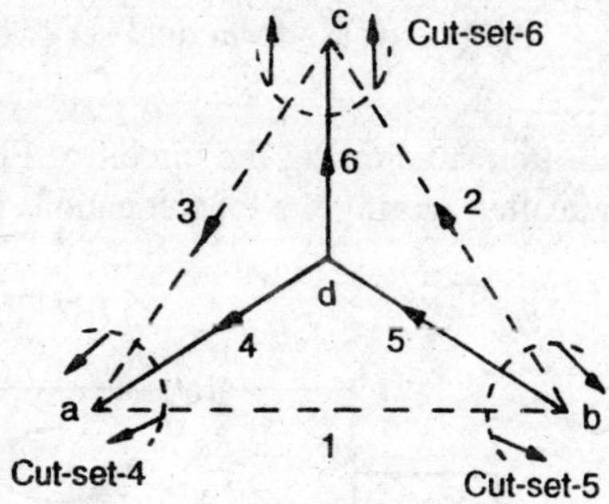

Fig. 1.112

Cut-sets have been shown in Fig.1.112 and cut-set schedule is shown in Table 1.45.

Table 1.45: Cut-Set Schedule

Cut-Sets	Branches					
	1	2	3	4	5	6
Cut-Set-4 (4, 3, 1)	–1	0	1	1	0	0
Cut-Set-5 (5, 2, 1)	1	–1	0	0	1	0
Cut-Set-6 (6, 2, 3)	0	1	–1	0	0	1

Problem 1.32 For the network shown in Fig. 1.113

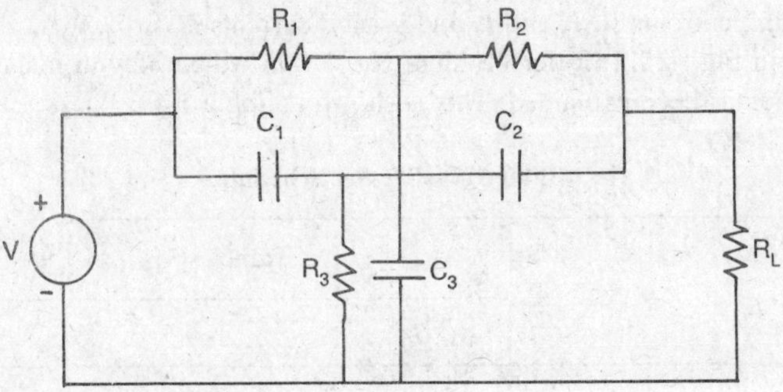

Fig. 1.113

(a) Write loop equations

(b) Draw dual of the network

(c) Write loop equations for the dual network and show the original and dual networks form dual set of equations

(VTU March 2000, Electric Circuits)

Solution. Redrawing the circuit of Fig. 1.113 as shown in Fig. 1.114 and thereafter, writing the loop equations based on Kirchoff's Laws, we get

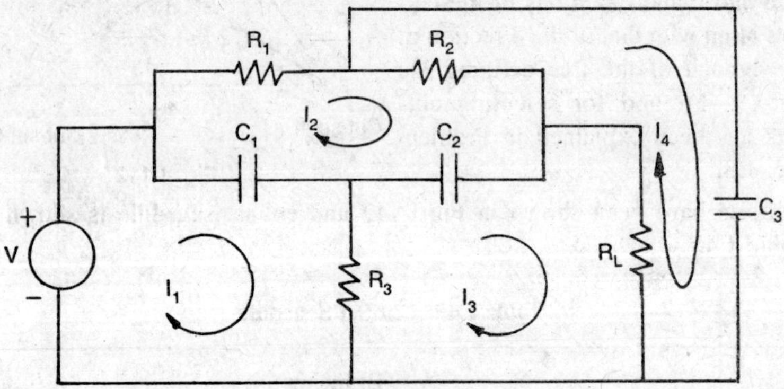

Fig. 1.114

Mesh 1

$$R_3 I_1 + \frac{I_1}{sC_1} - \frac{I_2}{sC_1} - I_3 R_3 = V$$

or $$\left(R_3 + \frac{1}{sC_1}\right) I_1 - \frac{I_2}{sC_1} - I_3 R_3 = V \qquad ...(1)$$

Mesh 2

$$R_1 I_2 + R_2 I_2 + \frac{I_2}{sC_2} + \frac{I_2}{sC_1} - \frac{I_1}{sC_1} - \frac{I_3}{sC_2} - I_4 R_2 = 0$$

or $$-\frac{I_1}{sC_1} + \left(R_1 + R_2 + \frac{1}{sC_1} + \frac{1}{sC_2}\right) I_2 - \frac{I_3}{sC_2} - I_4 R_2 = 0 \qquad \text{...(2)}$$

Mesh '3'

$$\frac{I_3}{sC_2} + R_1 I_3 + I_3 R_3 - I_1 R_3 - \frac{I_2}{sC_2} - I_4 R_L = 0$$

or $$-I_1 R_3 - \frac{I_2}{sC_2} + \left(R_L + R_3 + \frac{1}{sC_2}\right) I_2 - I_4 R_L = 0 \qquad \text{...(3)}$$

Mesh '4'

$$I_4 R_2 + \frac{I_4}{sC_3} + I_4 R_L - I_3 R_L - I_2 R_2 = 0$$

or $$-I_2 R_2 - I_3 R_L + \left(R_2 + R_L + \frac{1}{sC_3}\right) I_4 = 0 \qquad \text{...(4)}$$

(b) Dual of the network

Let us redraw the network with points 1 to 4 and datum selected as shown in Fig. 1.115.

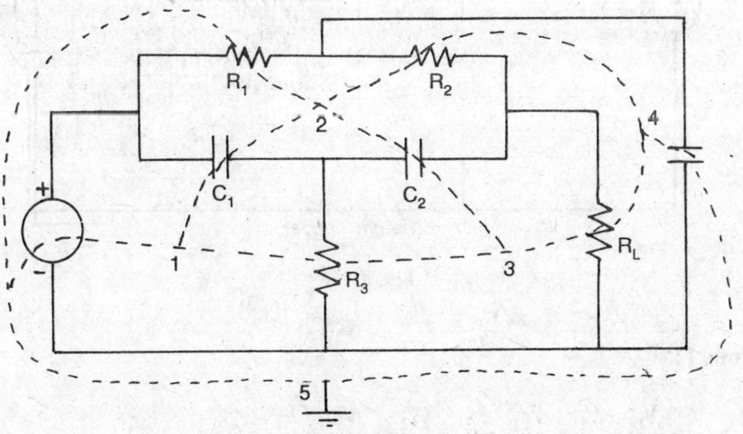

Fig. 1.115

Dual of the network is shown in Fig. 1.116 and is drawn by

(a) Inversing the resistances;

(b) by replacing inductances in lieu of capacitors and

(c) by replacing voltage source with current source

Node equations are:

Node '1'

$$\frac{V_1}{R_3}+\frac{V_1}{sL_1}-\frac{V_2}{sL_1}-\frac{V_3}{R_3}=I$$

$$\text{or} \quad \left(\frac{1}{R_3}+\frac{1}{sL_1}\right)V_1-\frac{V_2}{sL_1}-\frac{V_3}{R_3}=I \qquad \text{...(5)}$$

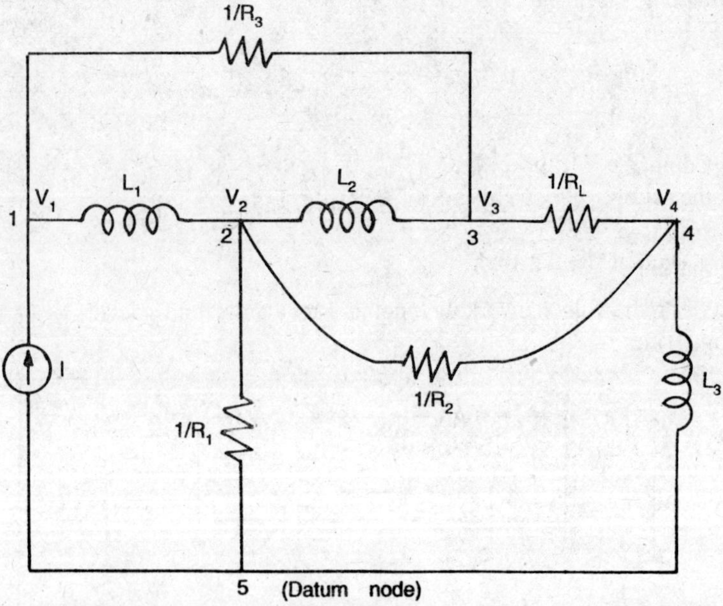

Fig. 1.116

Node '2'

$$-\frac{V_1}{sL_1}+\frac{V_2}{R_1}+\frac{V_2}{R_2}+\frac{V_2}{sL_1}+\frac{V_2}{sL_2}-\frac{V_3}{sL_3}-\frac{V_4}{R_2}=0$$

or $$-\frac{V_1}{sL_1}+\left(\frac{1}{R_1}+\frac{1}{R_2}+\frac{1}{sL_1}+\frac{1}{sL_2}\right)V_2-\frac{V_3}{sL_3}-\frac{V_4}{R_2}=0 \qquad ...(6)$$

Node '3'

$$-\frac{V_1}{R_3}-\frac{V_2}{sL_2}-\frac{V_4}{R_L}+\frac{V_3}{sL_3}+\frac{V_3}{R_3}+\frac{V_3}{R_L}=0$$

or $$-\frac{V_1}{R_3}-\frac{V_2}{sL_2}+\left(\frac{1}{R_L}+\frac{1}{R_3}+\frac{1}{sL_3}\right)V_3-\frac{V_4}{R_L}=0 \qquad ...(7)$$

Node '4'

$$-\frac{V_2}{R_2}-\frac{V_3}{R_L}+\frac{V_4}{R_2}+\frac{V_4}{R_L}+\frac{V_4}{sL_3}=0$$

or $$-\frac{V_2}{R_2}-\frac{V_3}{R_L}+\left(\frac{1}{R_2}+\frac{1}{R_L}+\frac{1}{sL_3}\right)=0 \qquad ...(8)$$

Comparison of equations 1 to 4 with equations 5 to 8 respectively, show that the set of equations form dual sets.

Problem No. 1.33 Draw the dual of the network shown in Fig. 1.117. Show that mesh equations of the given network and node equations of the dual network are dual equations.

(VTU Aug 2000, Electric Circuits)

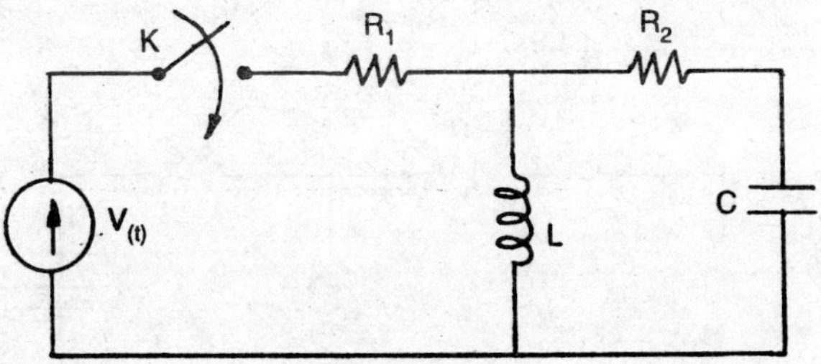

Fig. 1.117

Solution

Mesh Equations

The given network is redrawn with mesh currents I_1 and I_2 marked as shown in Fig. 1.118.

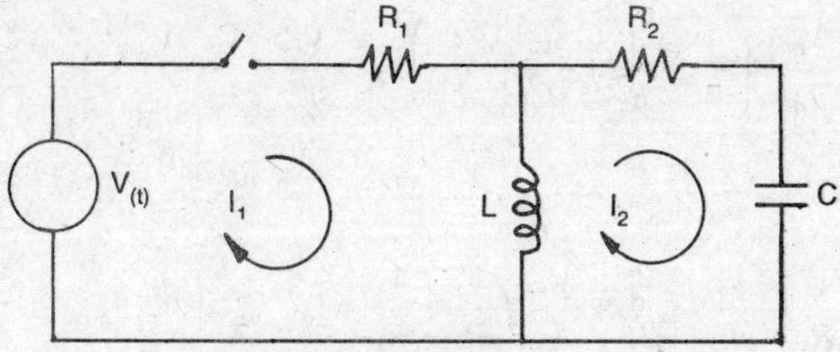

Fig. 1.118

Mesh '1'

$$R_1 I_1(s) + sL\left(I_1(s) - I_2(s)\right) = V(s)$$

or $$\left(R_1 + sL\right) I_1(s) - sL\, I_2(s) = V(s) \qquad ...(1)$$

Mesh '3'

$$R_2 I_2(s) + \frac{I_2(s)}{sC} + sL\, I_2(s) - sL\, I_1(s) = 0$$

or $$-sL\, I_1(s) + \left(R_2 + sL + \frac{1}{sC}\right) I_2(s) = 0 \qquad ...(2)$$

Dual Network

Select nodes '1' and '2' and datum node '3' as shown in Fig. 1.119. Draw the dual of the original network (Fig. 1.120).

(a) by replacing current source with voltage source and inductor with capacitor and vice versa and

(b) by inversing the resistors.

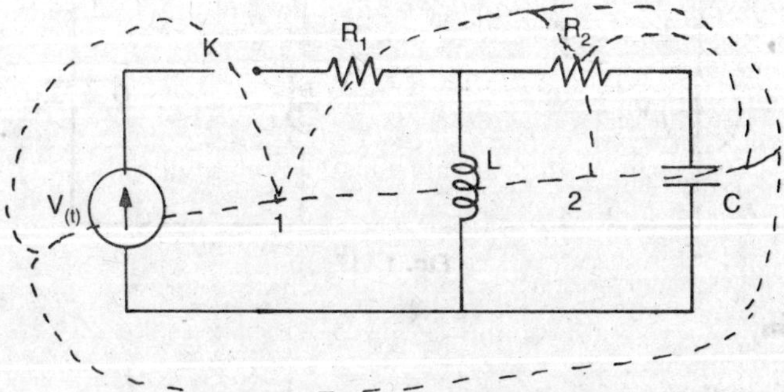

Fig. 1.119

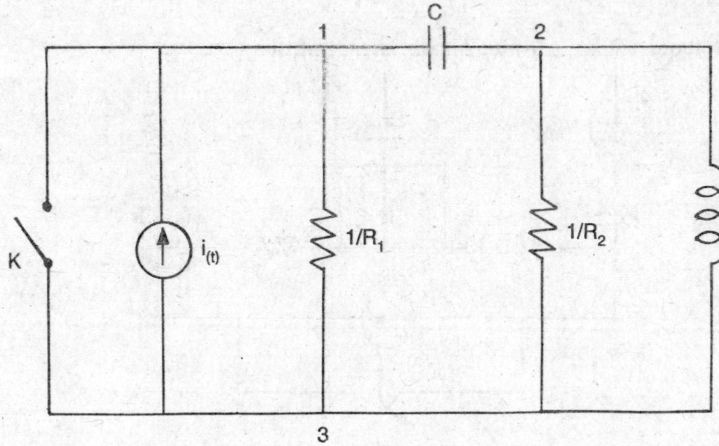

Fig. 1.120

Node Equations

Node '1'

$$\frac{V_1}{R_1} + sC\,V_1(s) - s\,C\,V_2(s) = I(s)$$

or $$\left(\frac{1}{R_1} + sC\right)V_1(s) - sC\,V_2(s) = I(s) \qquad ...(3)$$

Node '2'

$$\frac{V_2(s)}{R_2} + \frac{V_2(s)}{sL} + sC\,V_2(s) - sC\,V_1(s) = 0$$

or $$-sC\,V_1(s) + \left(\frac{1}{R_2} + \frac{1}{sL} + sC\right)V_2(s) = 0 \qquad ...(4)$$

Comparing equations 1 and 2 with equations 3 and 4 respectively, we see that equations are similar and hence form dual sets.

Problem 1.34 For the network shown in Fig 1.121, drawn the dual network and the equilibrium equations of the dual network.

(VTU Mar 2001, Electrical Circuits)

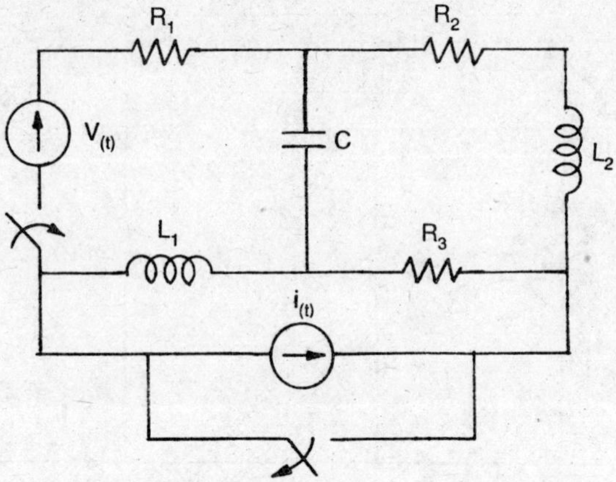

Fig. 1.121

Solution. Select nodes 1, 2, 3 and 4 as shown in Fig. 1.122 and draw the network by

(a) replacing current source with voltage source and vice versa,

(b) inversing the resistances, and

(c) replacing inductor with capacitor and vice-versa.

Dual network so obtained is shown in Fig. 1.123.

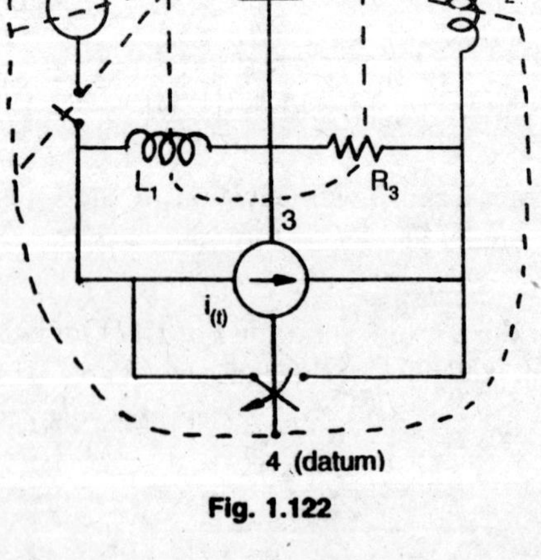

Fig. 1.122

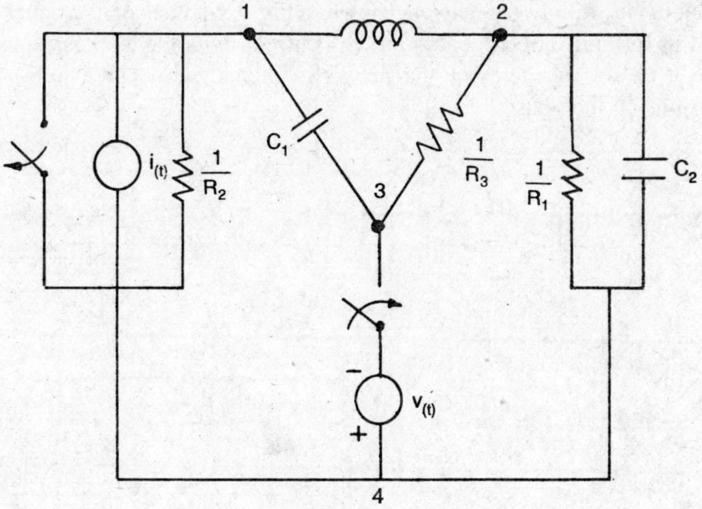

Fig. 1.123

Node equations are

Node 'a'

$$\frac{V_1(s)}{R_1}+\frac{V_1(s)}{sL}-\frac{V_2(s)}{sL}+sC_1\,V_1(s)-sC_1\,V_3(s)=I(s)$$

or $$\left(\frac{1}{R_1}+\frac{1}{sL}+sC_1\right)V_1(s)-\frac{V_2(s)}{sL}-sC_1\,V_3(s)=I(s) \quad ...(1)$$

Node 'b'

$$\frac{V_2(s)}{R_2}+sC_2\,V_2(s)+\frac{V_2(s)-V_3(s)}{R_3}+\frac{(V_2(s)-V_1(s))}{sL}=0$$

or $$\left(\frac{1}{R_2}+\frac{1}{R_3}+\frac{1}{sL}+sC_2\right)V_2(s)-\frac{V_1(s)}{sL}-\frac{V_3(s)}{R_3}=0 \quad ...(2)$$

Node 'c'

$$V(s)=V_3(s) \quad ...(3)$$

Problem No. 1.35 In the network shown in Fig. 1.24, consider branches 1, 3 and 4 as forming a tree. Identify the link currents with the loop currents and write a tie-set schedule for the network. Thereafter, obtain equili-brium equations on loop basis.

(VTU, Electric Circuits)

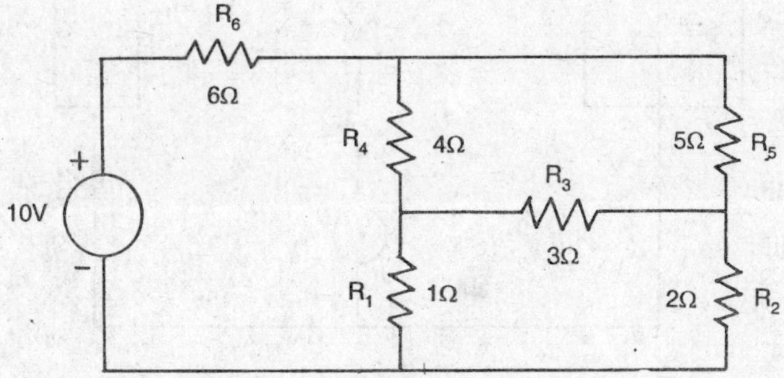

Fig. 1.124

Solution. Considering that branches of the network 1 to 6 contain resistance R_1 to R_6 respectively, tree is drawn with branches 1, 3 and 4 as given in problem. While drawing the tree resistances are reinored and voltage source is short-circuited. Tree consisting of branches 1, 3 and 4 are shown in thick lines and the remaining branches constituting links are shown with broken lines in the Fig. 1.124. Oriented graph shows direction of loop currents I_1, I_2 and I_3 and branch currents i_1 to i_6.

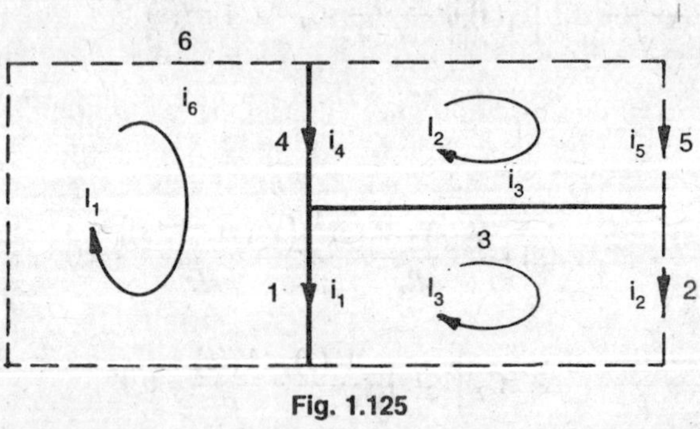

Fig. 1.125

Tie-Set Schedule

Tie Set schedule is shown in Table 1.46. Method adopted to fill the tie-set schedule has been explained in earlier problems.

Table 1.46: Tie-Set Schedule

Link Current	Loop		Branch 1	2	3	4	5	6
I_1	1, 4, 6		1	0	0	1	0	1
I_2	3, 4, 5	B =	0	0	–1	–1	1	0
I_3	1, 2, 3		–1	1	1	0	0	0

Equilibrium equation on Loop basis

Equilibrium equation on loop basis is given by

$$[B][Z_b][B^T][I_L]=[B]\{[V_s]-[Z_b][I_s]\} \quad ...(1)$$

$$[B]=\begin{bmatrix}1 & 0 & 0 & 1 & 0 & 1\\ 0 & 0 & -1 & -1 & 1 & 0\\ -1 & 1 & 1 & 0 & 0 & 0\end{bmatrix}$$

$$[B^T]=\begin{bmatrix}1 & 0 & -1\\ 0 & 0 & 1\\ 0 & -1 & 1\\ 1 & -1 & 0\\ 0 & 1 & 0\\ 1 & 0 & 0\end{bmatrix}; \qquad I_L=\begin{bmatrix}I_1\\ I_2\\ I_3\end{bmatrix} \quad ...(2)$$

$$[Z_b]=\begin{bmatrix}R_1 & 0 & 0 & 0 & 0 & 0\\ 0 & R_2 & 0 & 0 & 0 & 0\\ 0 & 0 & R_3 & 0 & 0 & 0\\ 0 & 0 & 0 & R_4 & 0 & 0\\ 0 & 0 & 0 & 0 & R_5 & 0\\ 0 & 0 & 0 & 0 & 0 & R_6\end{bmatrix}=\begin{bmatrix}1 & 0 & 0 & 0 & 0 & 0\\ 0 & 2 & 0 & 0 & 0 & 0\\ 0 & 0 & 3 & 0 & 0 & 0\\ 0 & 0 & 0 & 4 & 0 & 0\\ 0 & 0 & 0 & 0 & 5 & 0\\ 0 & 0 & 0 & 0 & 0 & 6\end{bmatrix}$$

[B] [Z_b] [B^T]

$$\begin{bmatrix}1 & 0 & 0 & 1 & 0 & 1\\ 0 & 0 & -1 & -1 & 1 & 0\\ -1 & 1 & 1 & 0 & 0 & 0\end{bmatrix}\begin{bmatrix}1 & 0 & 0 & 0 & 0 & 0\\ 0 & 2 & 0 & 0 & 0 & 0\\ 0 & 0 & 3 & 0 & 0 & 0\\ 0 & 0 & 0 & 4 & 0 & 0\\ 0 & 0 & 0 & 0 & 5 & 0\\ 0 & 0 & 0 & 0 & 0 & 6\end{bmatrix}\begin{bmatrix}1 & 0 & -1\\ 0 & 0 & 1\\ 0 & -1 & 1\\ 1 & -1 & 0\\ 0 & 1 & 0\\ 1 & 0 & 0\end{bmatrix}$$

$$= \begin{bmatrix} 1 & 0 & 0 & 1 & 0 & 1 \\ 0 & 0 & -1 & -1 & 1 & 0 \\ -1 & 1 & 1 & 0 & 0 & 0 \end{bmatrix} \begin{bmatrix} 1 & 0 & -1 \\ 0 & 0 & 2 \\ 0 & -3 & 3 \\ 4 & -4 & 0 \\ 0 & 5 & 0 \\ 6 & 0 & 0 \end{bmatrix} = \begin{bmatrix} 11 & -4 & -1 \\ -4 & 12 & -3 \\ -1 & -3 & 6 \end{bmatrix} \quad ...(3)$$

{[V$_s$ - [Z$_b$] [I$_s$]}

$$V_s = \begin{bmatrix} 0 \\ 0 \\ 0 \\ 0 \\ 0 \\ 10 \end{bmatrix}; \qquad I_s = \begin{bmatrix} 0 \\ 0 \\ 0 \\ 0 \\ 0 \\ 0 \end{bmatrix}$$

Therefore,

$$[V_s]-[Z_b][I_s] = \begin{bmatrix} 0 \\ 0 \\ 0 \\ 0 \\ 0 \\ 10 \end{bmatrix} \begin{bmatrix} 1 & 0 & 0 & 0 & 0 & 0 \\ 0 & 2 & 0 & 0 & 0 & 0 \\ 0 & 0 & 3 & 0 & 0 & 0 \\ 0 & 0 & 0 & 4 & 0 & 0 \\ 0 & 0 & 0 & 0 & 5 & 0 \\ 0 & 0 & 0 & 0 & 0 & 6 \end{bmatrix} \begin{bmatrix} 0 \\ 0 \\ 0 \\ 0 \\ 0 \\ 0 \end{bmatrix} = \begin{bmatrix} 0 \\ 0 \\ 0 \\ 0 \\ 0 \\ 10 \end{bmatrix}$$

and $$[B]\{[V_s]-[Z_b][I_s]\} = \begin{bmatrix} 1 & 0 & 0 & 1 & 0 & 1 \\ 0 & 0 & -1 & -1 & 1 & 0 \\ -1 & 1 & 1 & 0 & 0 & 0 \end{bmatrix} \begin{bmatrix} 0 \\ 0 \\ 0 \\ 0 \\ 0 \\ 10 \end{bmatrix} = \begin{bmatrix} 10 \\ 0 \\ 0 \end{bmatrix} \quad ...(4)$$

Putting the values from equations (2), (3) and (4) in equation (1), we get the equilibrium equaitons on loop basis as given below:

$$\begin{bmatrix} 11 & -4 & -1 \\ -4 & 12 & -3 \\ -1 & -3 & 6 \end{bmatrix} \begin{bmatrix} I_1 \\ I_2 \\ I_3 \end{bmatrix} = \begin{bmatrix} 10 \\ 0 \\ 0 \end{bmatrix}$$

Problem No. 1.36 Verify reciprocity theorem by computing voltage V_{ab} in Fig. 1.126.

(VTU, Electric Circuits)

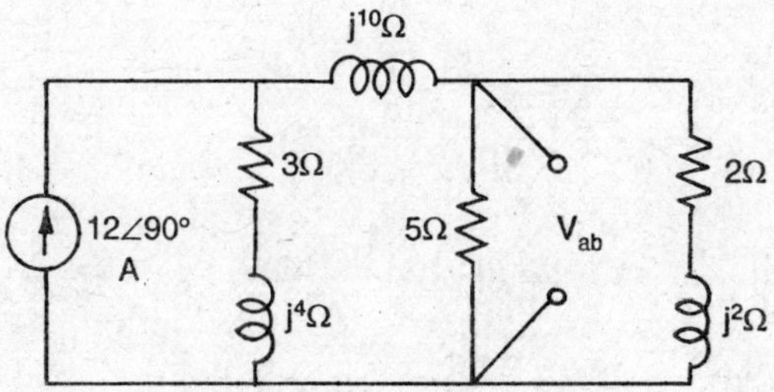

Fig. 1.126

Solution. Select nodes as shown in Fig. 1.127

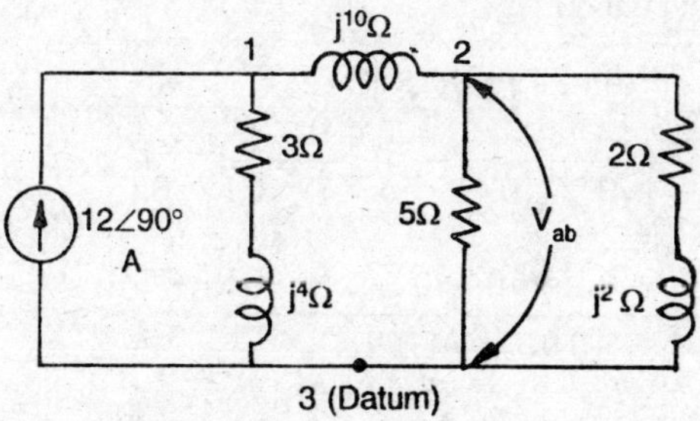

Fig. 1.127

Fig. 1.127 Let the node voltages be V_1 and V_2. Therefore

$$V_2 = V_{ab}$$

Node '1'

$$\left(\frac{1}{3+j4}+\frac{1}{j10}\right)V_1 - \frac{V_2}{j10} = 12\angle 90^\circ$$

or $$\left(\frac{3-j4}{3+j4}+\frac{j}{-10}\right)V_1 + j0.1V_2 = 12\angle 90^\circ$$

or $$\left(\frac{3-j4}{25}-j0.1\right)V_1 + j0.1V_2 = 12\angle 90^\circ$$

or $$(0.12-j0.16-j0.1)V_1 + j0.1V_2 = 12\angle 90^\circ$$

or $$(0.12-j0.26)V_1 + j0.1V_2 = 12\angle 90^\circ \qquad ...(1)$$

Node '2'

$$\left(\frac{1}{2+j2}+\frac{1}{5}+\frac{1}{j10}\right)V_2-\frac{V_1}{j10}=0$$

or
$$\left(\frac{2-j2}{8}+0.2-j0.1\right)V_2-\frac{V_1}{j10}=0$$

or
$$(0.25-j0.25+0.2-j0.1)\,V_2+j0.1\,V_1=0$$

or
$$(0.45-j0.35)\,V_2+j0.1\,V_1=0$$

or
$$j0.1\,V+(0.45-j0.35)\,V_2=0 \qquad ...(2)$$

From equations (1) and (2),

$$V_2=\frac{\begin{vmatrix}0.12-j0.16 & 12\angle 90^\circ\\ j0.1 & 0\end{vmatrix}}{(0.12-j0.26)(0.45-j0.35)-j0.1\times j0.1}$$

or
$$V_2=\frac{(0.12-j0.16)\times 0-j0.1\times 12\angle 90^\circ}{-0.027-j0.159}$$

or
$$V_2=\frac{-(j0.1)(j12)}{-0.027-j0.159}=\frac{1.2}{\sqrt{0.027^2+0.159}\left(\angle\tan^{-1}\dfrac{0.159}{0.027}+180^\circ\right)}$$

$$=\frac{1.2}{0.161\angle 260^\circ{\cdot}36^\circ}$$

$$=7.453\angle-260^\circ{\cdot}36^\circ$$

Shifting the current source as shown in Fig. 1.128 and application of nodal analysis gives

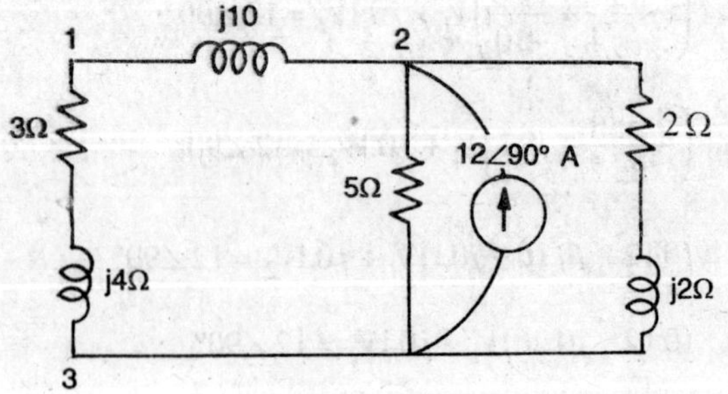

Fig. 1.128

Node '1'

$$\left(\frac{1}{3+j4}+\frac{1}{j10}\right)V_1-\frac{V_2}{j10}=0$$

or $$(0.12-j0.26)\,V_1+j0.1V_2=0 \quad ...(3)$$

Node '2'

$$\left(\frac{1}{2+j2}+\frac{1}{5}+\frac{1}{j10}\right)V_2-\frac{V_1}{j10}=12\angle 90°$$

or $$j0.1V_1+(0.45-j0.35)\,V_2=12\angle 90° \quad ...(4)$$

Solving equations (3) and (4) with the help of Cramer's rule, we get

$$V_1=\frac{\begin{vmatrix}0 & j0.1\\ 12\angle 90° & 0.45-j0.35\end{vmatrix}}{(0.12-j0.26)(0.45-j0.35)(j0.1\times j0.1)}$$

$$=\frac{1.2}{0.157\angle 260.36°}$$

$$=7.453\angle -260°{\cdot}36°$$

Problem No. 1.37 Write the tie-set schedule for the network shown in Fig. 1.129 and using the tie-set schedule, obtain the equilibrium equations on loop current basis. **(VTU, Network Analysis Mar 2000)**

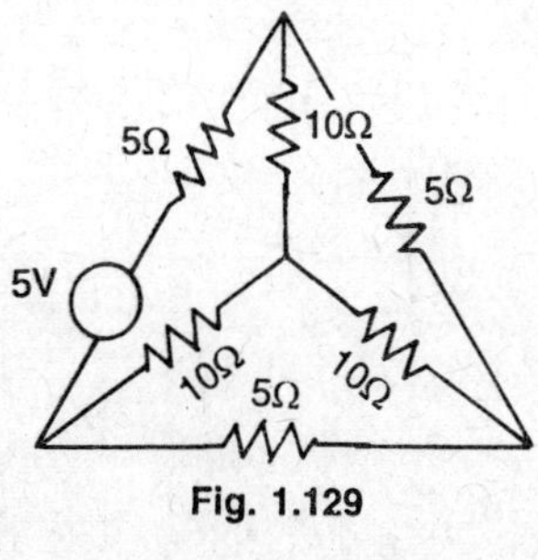

Fig. 1.129

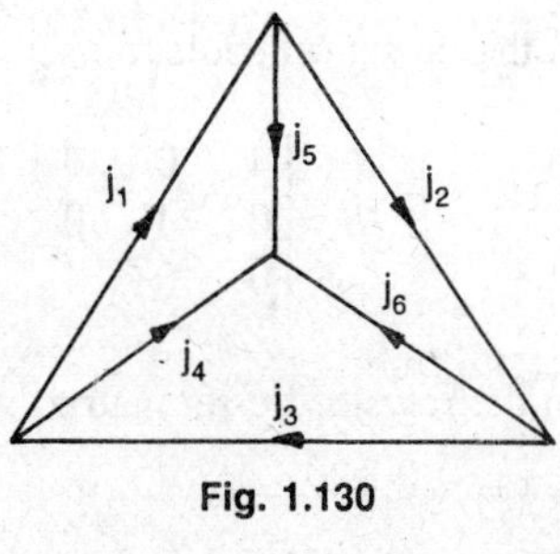

Fig. 1.130

Solution. Refer Problem No. 1.6 where similar problem has been solved. Oriented graph is shown in Fig. 1.30 and tree with branch currents and loop currents is shown in Fig. 1.31.

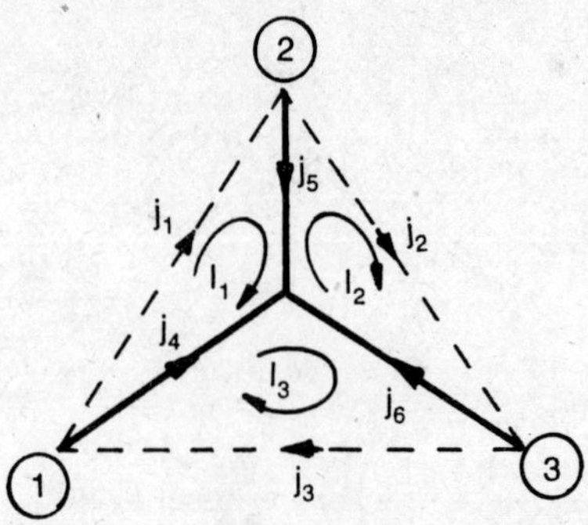

Fig. 1.131

Tie-Set Schedule

It is shown in Table 1.47.

Table 1.47
Tie-Set Schedule

Loop Current	Loop branches	Branches					
		j_1	j_2	j_3	j_4	j_5	j_6
I_1	1–5–4–1	1	0	0	–1	1	0
I_2	2–6–5–2 B =	0	1	0	0	–1	1
I_3	3–4–6–3	0	0	1	1	0	–1

From the tie-set schedule, tie-set matrix is

$$B = \begin{bmatrix} 1 & 0 & 0 & -1 & 1 & 0 \\ 0 & 1 & 0 & 0 & -1 & 1 \\ 0 & 0 & 1 & 1 & 0 & -1 \end{bmatrix}$$

From the rows of tie-set matrix, we get

$$V_1 - V_4 + V_5 = 0 \quad ...(1)$$

$$V_2 - V_5 + V_6 = 0 \quad ...(2)$$

$$V_3 + V_4 - V_6 = 0 \quad ...(3)$$

Columns of the tie-set matrix give us

$$\left.\begin{aligned} j_1 &= I_1 \\ j_2 &= I_2 \\ j_3 &= I_3 \\ j_4 &= -I_1 + I_3 \\ j_5 &= I_1 - I_2 \\ j_6 &= I_2 - I_3 \end{aligned}\right\} \quad ...(4)$$

Relating branch currents and branch voltages, we get

$$\left.\begin{aligned} V_1 &= r_1 j_1 - 5 = 5 \times j_1 - 5 = 5j_1 - 5 \\ V_2 &= r_2 j_2 \quad = 5 \times j_2 \quad = 5j_2 \\ V_3 &= r_3 j_3 \quad = 5 \times j_3 \quad = 5j_3 \\ V_4 &= r_4 j_4 \quad = 10 \times j_4 \quad = 10j_4 \\ V_5 &= r_5 j_5 \quad = 10 \times j_5 \quad = 10j_5 \\ V_6 &= r_6 j_6 \quad = 10 \times j_6 \quad = 10j_6 \end{aligned}\right\} \quad ...(5)$$

Subtituting the values from equations (5) in equation (1) to (3), we get

$$\left.\begin{aligned} 5j_1 - 5 - 10j_4 + 10j_5 &= 0 \\ 5j_2 - 10j_5 + 10j_6 &= 0 \\ 5j_3 + 10j_4 - 10j_6 &= 0 \end{aligned}\right\} \quad ...(6)$$

and in terms of branch currents, by substituting values from equation (4) in (6), we get

$$5I_1 - 5 - 10I_1 - 10I_3 + 10I_1 - 10I_2 = 0$$

or $$25I_1 - 10I_2 - 10I_3 = 5, \quad ...(7)$$

$$5I_2 - 10I_1 + 10I_2 + 10I_2 - 10I_3 = 0$$

or $$-10I_1 - 25I_2 - 10I_3 = 0 \quad ...(8)$$

and $$5I_3 - 10I_1 + 10I_3 - 10I_2 + 10I_3 = 0$$

or $$-10I_1 - 10I_2 + 25I_3 = 0 \quad ...(9)$$

Problem No. 1.38 Draw the oriented graph of the network shown in Fig. 1.132 and write the incidence matrix.

(VTU, Network Analysis Aug. 2000)

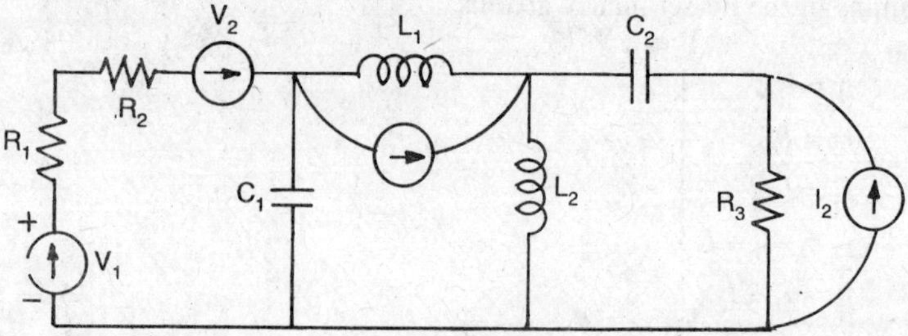

Fig. 1.132

Solution. Short-circuiting the voltage sources, open-circuiting the current sources and removing the circuit elements, the given network reduces to as shown in Fig. 1.133.

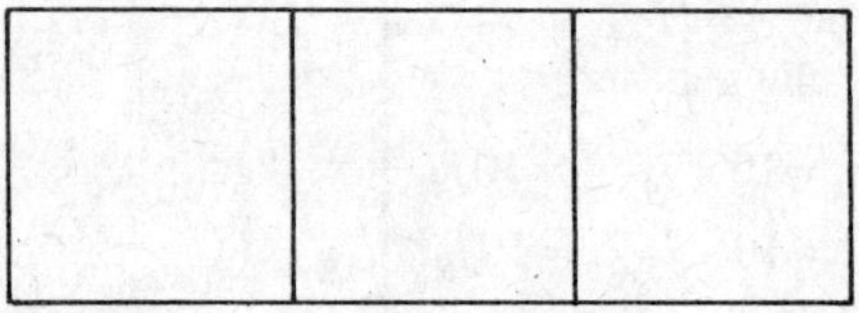

Fig. 1.133

Circuit can be redrawn as shown in Fig. 1.134

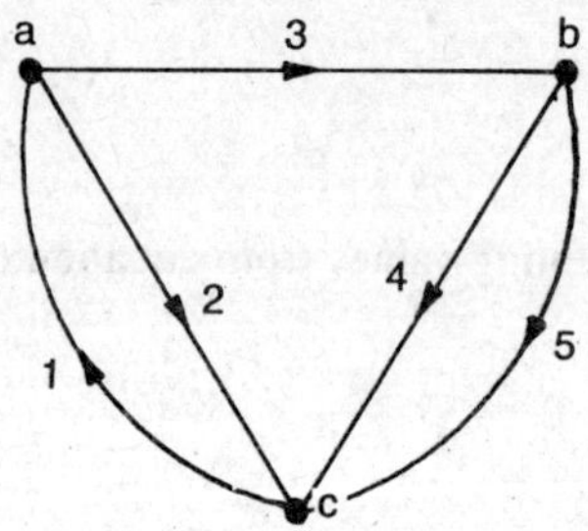

Fig. 1.134

Nodes a, b, c are selected and branches 1 to 5 marked. Incidence matrix is shown in Table 1.48. Refer problem No. 1.3 in which method to tabulate incidence matrix is explained.

Table 1.48: Incidence Matrix

Nodes	Branches				
	1	2	3	4	5
a	–1	1	1	0	0
b	0	0	–1	1	1
c	1	–1	0	–1	–1

Problem No. 1.39 For the network shown in Fig. 1.135, draw the oriented graph, select a tree and write the tie-set schedule.

(VTU, Network Analysis Mar 2001)

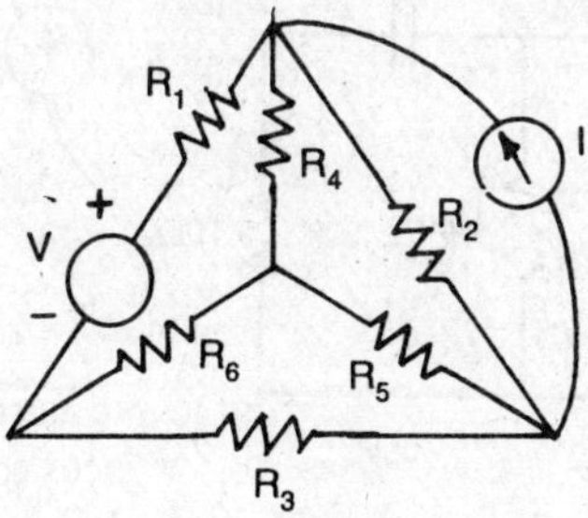

Fig. 1.135

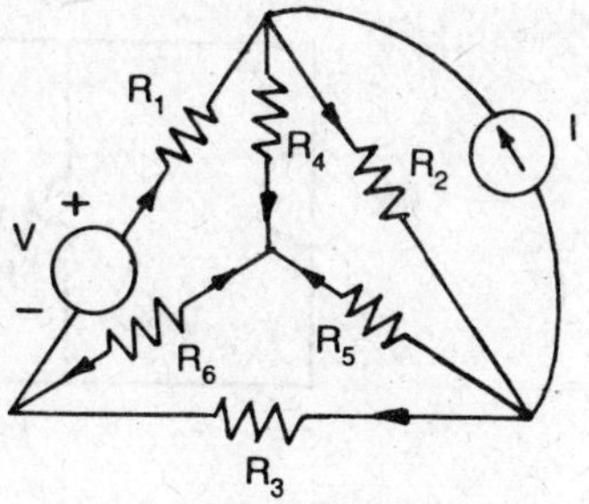

Fig. 1.136

Solution. Let us mark the current distribution as shown in Fig. 1.136. Short-circuiting the voltage source and open-circuiting the current source and removing the circuit elements, we get the oriented graph as shown in Fig. 1.137.

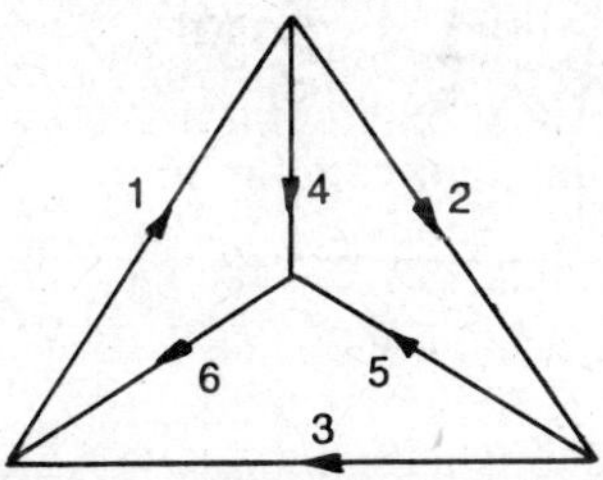

Fig. 1.137

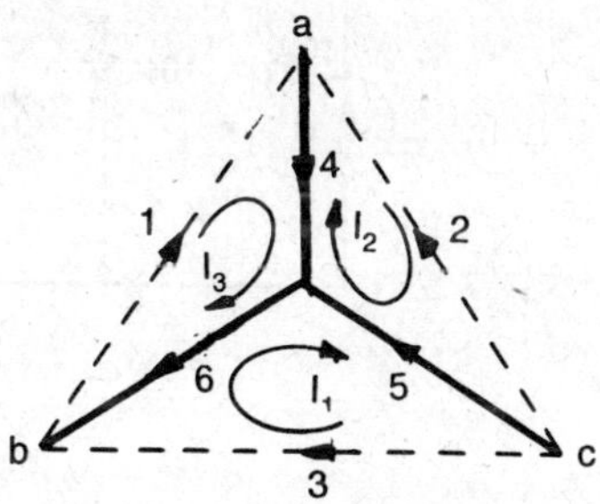

Fig. 1.138

Since there are four nodes, three branches will consititute the tree as shown in Fig. 1.138

Tie-Set Schedule It is tabulated in Table 1.49

Table 1.49: Tie-Set Schedule

Loop Current	Loop	Branch					
		1	2	3	4	5	6
I_1	3-6-5-3	0	0	1	0	–1	–1
I_2	2-5-4-2	0	1	0	–1	1	0
I_3	1-4-6-1	1	0	0	1	0	1

Problem No. 1.40 Construct the dual of the network shown in Fig. 1.139

(VTU, Network Analysis, Mar 2001)

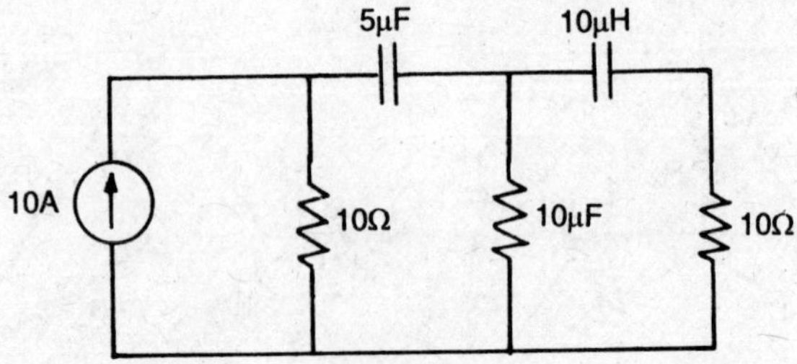

Fig. 1.139

Solution. Select points/nodes a, b, c and d as shown in Fig. 1.140. Draw the dual of network

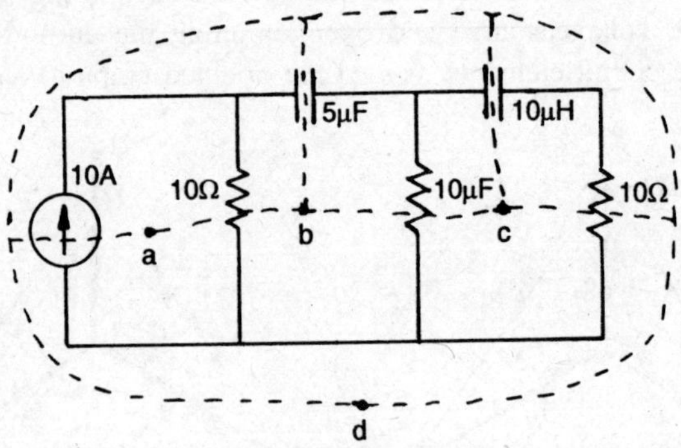

Fig. 1.140

(a) by replacing capacitors with inductors with similar values and vice-versa,

(b) by replacing resistances with their inverse, and

(c) by replacing current source with voltage source.

It should be noted that as we move from node to node along the broken lines, circuit elements encountered are taken in account while constructing dual network as given below:

(a) between nodes a and b there will be resistance of value $\frac{1}{10}\Omega$,

(b) between nodes b and c there will be a capacitor of value 10μf.

(c) between nodes c and d, there will be a resistance of value $\frac{1}{10}\Omega$ and inductor of value 10μH,

(d) between nodes b and d, there will be an inductor of value 5μH and

(e) between nodes a and d, there will be voltage source of 10V.

Dual network so obtained is shown in Fig. 1.141

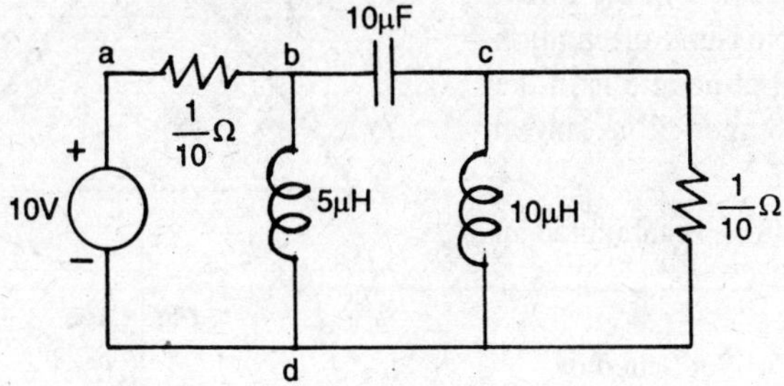

Fig. 1.141

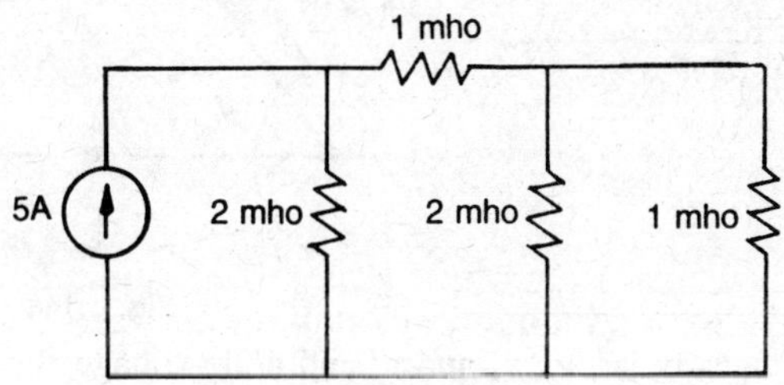

Fig. 1.142

Problem No. 1.41 For the network shown in Fig. 1.142 write the fundamental cut-set matrix and hence find KCL equations in the matrix form.

(VTU, Network Analysis, March 2001)

Solution Open circuiting the currrent source, and selecting the nodes a, b and c as shown in Fig. 1.143, oriented graph is drawn as shown in Fig. 1.144 after removing the circuit elements.

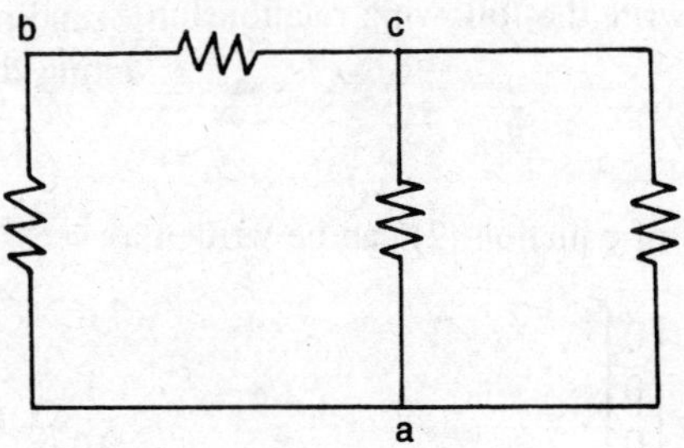

Fig. 1.143

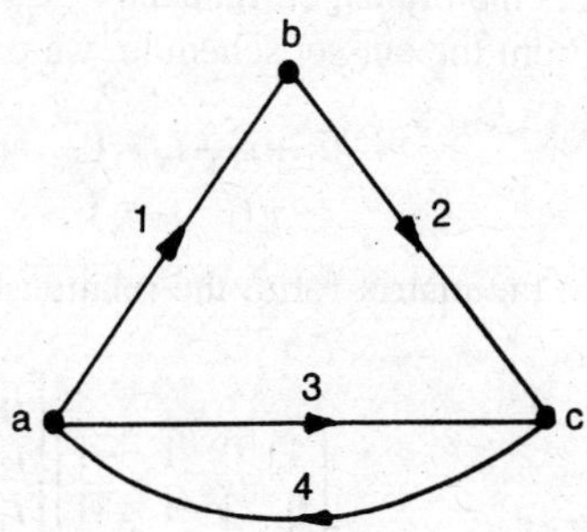

Fig. 1.144

Since there are three nodes, two branches will constitute the tree. Let branches 1 and 2 form a tree and branches 3 and 4 be links as shown in Fig. 1.145.

The two cut-sets are drawn at nodes a and c. Orientation of the cut-set at node a is in the direction of branch 1 and orientation of the cut-set at node c is in the direction of branch '2' as shown in Fig. 1.146.

Cut-set schedule is tabulated in Table 1.50.

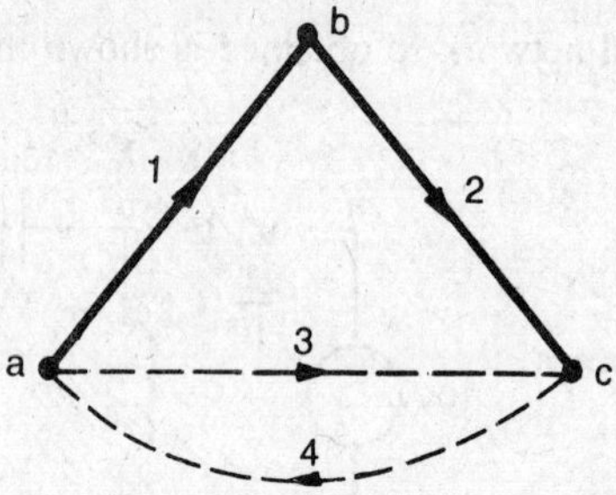

Fig. 1.145

Table 1.50: Cut-Set Schedule

Cut-Sets	Branch 1	2	3	4
f–cut–set1 (1, 3, 4)	1	0	1	−1
f–cut–set 2 (2, 3, 4)	0	1	1	−1

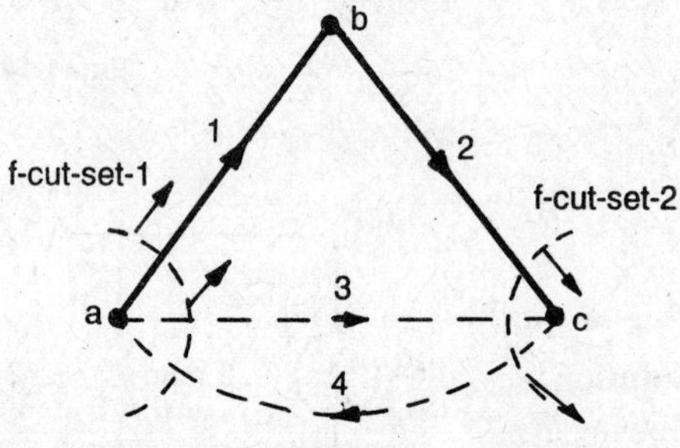

Fig. 1.146

Let branch voltages be v_1, v_2, v_3 and v_4 and node voltages be V_a and V_b at nodes a and b respectively. From the cut-set schedule, we write the following voltage relationships

$$\left.\begin{aligned} v_1 &= V_a \\ v_2 &= V_b \\ v_3 &= V_a + V_b \\ v_4 &= -V_a - V_b \end{aligned}\right\} \qquad ...(1)$$

Let the branch currents be i_1, i_2, i_3 and i_4 in the branches 1 to 4 respectively. From the cut-set schedule, we can write the following relationship

$$\left.\begin{aligned} i_1 + i_3 - i_4 &= 0, \quad \text{and} \\ i_2 + i_3 - i_4 &= 0 \end{aligned}\right\} \qquad ...(2)$$

In the matrix form, the relationship of equation (2) can be written as

$$\begin{bmatrix} 1 & 0 & 1 & -1 \\ 0 & 1 & 1 & -1 \end{bmatrix} \begin{bmatrix} i_1 \\ i_2 \\ i_3 \\ i_4 \end{bmatrix} = \begin{bmatrix} 0 \\ 0 \\ 0 \\ 0 \end{bmatrix}$$

Problem No. 1.42 For the network shown in Fig. 1.147, draw the graph, tree and write a tie-set schedule and equilibrium equation on current basis and obtain the branch voltages and branch currents.

(VTU, Network Analysis, Aug. 2001)

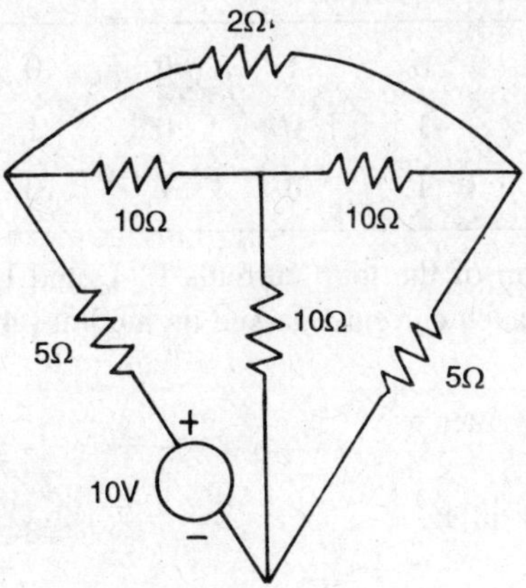

Fig. 1.147

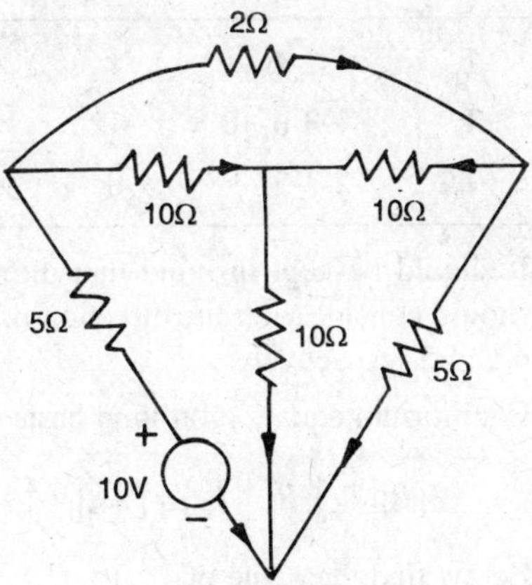

Fig. 1.148

Solution. Fig. 1.148 shows the network in which direction of the currents flowing in the branches is marked tentatively. Oriented graph is shown in Fig. 1.149.

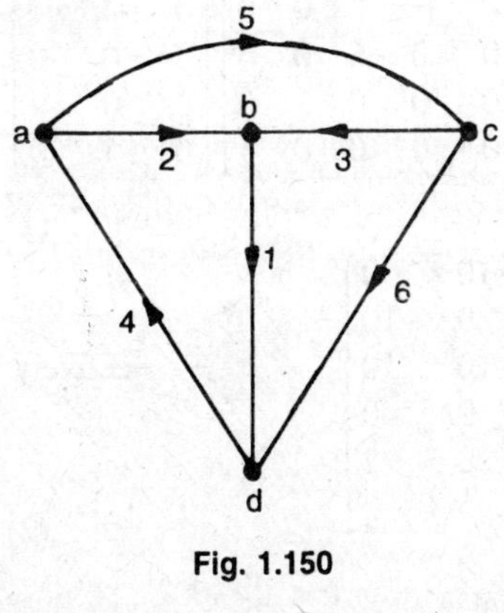

Fig. 1.150

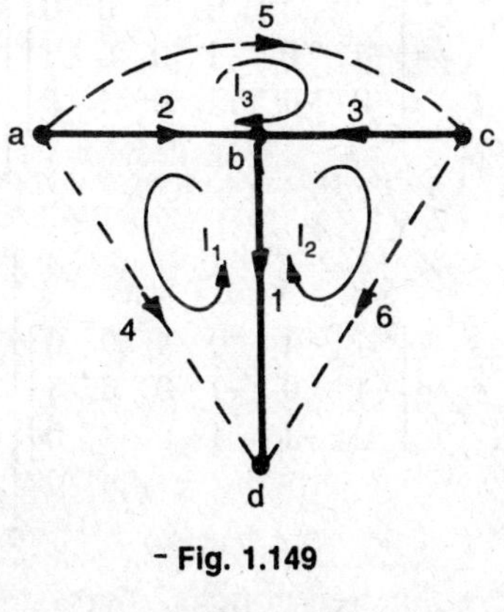

- Fig. 1.149

Since there are four nodes, three branches will constitute the tree. Let the branches 1, 2 and 3 constitute the tree and branches 4, 5 and 6 be the links as shown in Fig. 1.50 Tie-set schedule is tabulated in Table 1.51.

Table 1.51: Tie-Set Schedule

Loop Current	Loop		Branches 1	2	3	4	5	6
I_1	1-2-4		1	1	0	1	0	0
I_2	1-3-6	B =	−1	0	−1	0	0	1
I_3	2-3-5		0	−1	1	0	1	0

It should be kept in mind that direction of the loop currents I_1, I_2 and I_3 should concide with the direction of branch currents marked on the links 4, 6 and 5 respectively.

Equilibrium equation on loop basis is written as

$$[B][Z_b][B^T][I_L]=[B]\{[V_s]-[Z_b][I_s]\} \qquad \text{...(1)}$$

Let us find the value of

$[B][Z_b][B^T]$

$$= \begin{bmatrix} 1 & 1 & 0 & 1 & 0 & 0 \\ -1 & 0 & -1 & 0 & 0 & 1 \\ 0 & -1 & 1 & 0 & 1 & 0 \end{bmatrix} \begin{bmatrix} 10 & 0 & 0 & 0 & 0 & 0 \\ 0 & 10 & 0 & 0 & 0 & 0 \\ 0 & 0 & 10 & 0 & 0 & 0 \\ 0 & 0 & 0 & 5 & 0 & 0 \\ 0 & 0 & 0 & 0 & 2 & 0 \\ 0 & 0 & 0 & 0 & 0 & 5 \end{bmatrix} \begin{bmatrix} 1 & -1 & 0 \\ 1 & 0 & -1 \\ 0 & -1 & 1 \\ 1 & 0 & 0 \\ 0 & 0 & 1 \\ 0 & 1 & 0 \end{bmatrix}$$

$$= \begin{bmatrix} 1 & 1 & 0 & 1 & 0 & 0 \\ -1 & 0 & -1 & 0 & 0 & 1 \\ 0 & -1 & 1 & 0 & 1 & 0 \end{bmatrix} \begin{bmatrix} 10 & -10 & 0 \\ 10 & 0 & -10 \\ 0 & -10 & 10 \\ 5 & 0 & 0 \\ 0 & 0 & 2 \\ 0 & 5 & 0 \end{bmatrix}$$

$$= \begin{bmatrix} 25 & -10 & -10 \\ -10 & 25 & -10 \\ -10 & -10 & 22 \end{bmatrix} \qquad \text{...(2)}$$

$$[I_L] = \begin{bmatrix} I_1 \\ I_2 \\ I_3 \end{bmatrix} \qquad ...(3)$$

$$[I_s] = \begin{bmatrix} 0 \\ 0 \\ 0 \\ 0 \\ 0 \\ 0 \end{bmatrix}, \text{ Therefore, } [Z_b][I_s] = [Z_b]\begin{bmatrix} 0 \\ 0 \\ 0 \\ 0 \\ 0 \\ 0 \end{bmatrix} = 0 \qquad ...(4)$$

Therefore

$$[B]\{[V_s] - [Z_b][I_s]\}$$

$$= [B][V_s]$$

$$[V_s] = \begin{bmatrix} 0 \\ 0 \\ 0 \\ 10 \\ 0 \\ 0 \end{bmatrix}, \text{ Therefore,}$$

$$[B][V_s]$$

$$= \begin{bmatrix} -1 & -1 & 0 & 1 & 0 & 0 \\ -1 & 0 & -1 & 0 & 0 & 1 \\ 0 & -1 & 1 & 0 & 1 & 0 \end{bmatrix} \begin{bmatrix} 0 \\ 0 \\ 0 \\ 10 \\ 0 \\ 0 \end{bmatrix} - \begin{bmatrix} 10 \\ 0 \\ 0 \end{bmatrix} \qquad ...(5)$$

Substituting the values so obtained from equations (2) to (5) in equation (1), we get

$$= \begin{bmatrix} 25 & -10 & -10 \\ -10 & 25 & -10 \\ -10 & -10 & 22 \end{bmatrix} \begin{bmatrix} I_1 \\ I_2 \\ I_3 \end{bmatrix} = \begin{bmatrix} 10 \\ 0 \\ 0 \end{bmatrix}$$

which gives

$$25I_1 - 10I_2 - 10I_3 = 10$$

or $5I_1 - 2I_2 - 2I_3 = 2$ ...(6)

$-10I_1 + 25I_2 - 10I_3 = 0$

or $2I_1 - 5I_2 + 2I_3 = 0$ ...(7)

and $-10I_1 - 10I_2 + 22I_3 = 0$

or $5I_1 + 5I_2 - 11I_3 = 0$ ...(8)

Solving equations (6), (7) and (8) gives

$I_1 = 0.989A,$

$I_2 = 0.7035A,$ and

$I_3 = 0.769A.$

Branch Currents These are obtained from

$$[I_b] = [B^T][I_L] + [I_s]$$

or
$$\begin{bmatrix} i_1 \\ i_2 \\ i_3 \\ i_4 \\ i_5 \\ i_6 \end{bmatrix} = \begin{bmatrix} 1 & -1 & 0 \\ 1 & 0 & -1 \\ 0 & -1 & 1 \\ 1 & 0 & 0 \\ 0 & 0 & 1 \\ 0 & 1 & 0 \end{bmatrix} \begin{bmatrix} 0.989 \\ 0.7035 \\ 0.769 \end{bmatrix} + \begin{bmatrix} 0 \\ 0 \\ 0 \\ 0 \\ 0 \\ 0 \end{bmatrix}$$

which gives

$i_1 = 0.989 - 0.7035 = 0.2855A$

$i_2 = 0.989 - 0.769 = 0.22A$

$i_3 = -0.7035 + 0.769 = 0.0655A$

$i_4 = 0.989A$

$i_5 = 0.769A$

$i_6 = 0.7035A$

Branch Voltages

$$[V_b] = [Z][I_b] - [V_s]$$

$$\begin{bmatrix} v_1 \\ v_2 \\ v_3 \\ v_4 \\ v_5 \\ v_6 \end{bmatrix} = \begin{bmatrix} 10 & 0 & 0 & 0 & 0 & 0 \\ 0 & 10 & 0 & 0 & 0 & 0 \\ 0 & 0 & 10 & 0 & 0 & 0 \\ 0 & 0 & 0 & 5 & 0 & 0 \\ 0 & 0 & 0 & 0 & 2 & 0 \\ 0 & 0 & 0 & 0 & 0 & 5 \end{bmatrix} \begin{bmatrix} 0.2855 \\ 0.22 \\ 0.0655 \\ 0.989 \\ 0.769 \\ 0.7035 \end{bmatrix} - \begin{bmatrix} 0 \\ 0 \\ 0 \\ 10 \\ 0 \\ 0 \end{bmatrix} = \begin{bmatrix} 2.855 \\ 2.2 \\ 0.655 \\ -5.055 \\ 1.538 \\ 3.5175 \end{bmatrix}$$

Check

$$[B][V_b] = 0$$

i.e.,

$$\begin{bmatrix} -1 & 1 & 0 & 1 & 0 & 0 \\ -1 & 0 & -1 & 0 & 0 & 1 \\ 0 & -1 & 1 & 0 & 1 & 0 \end{bmatrix} \begin{bmatrix} 2.855 \\ 2.2 \\ 0.655 \\ -5.055 \\ 1.538 \\ 3.5175 \end{bmatrix} \text{ should be zero}$$

Multiplying we get

$$= \begin{bmatrix} 2.855 + 2.2 - 5.055 \\ -2.855 - 0.655 + 3.5175 \\ -2.2 + 0.655 + 1.538 \end{bmatrix} = \begin{bmatrix} 0 \\ 0 \\ 0 \end{bmatrix}$$

and hence the results obtained are verified.

Problem No. 1.43 In the circuit shown in Fig. 1.151, construct a tree in which V_1 and V_2 are tree branch voltages. Write a cut-set schedule and there from obtain the equilibrium node equations.

(VTU, Electric Circuits, Mar 2001)

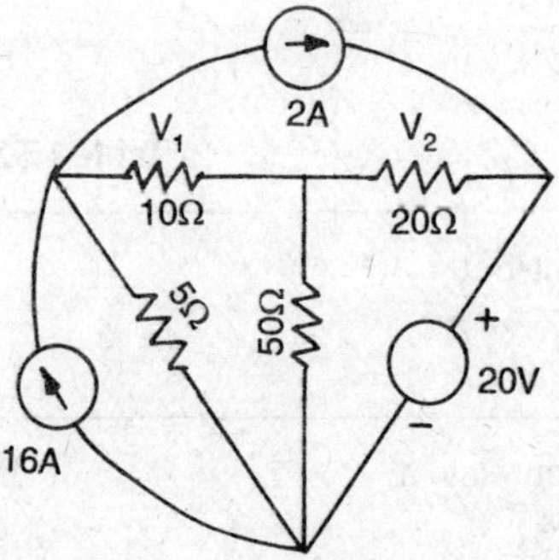

Fig. 1.151

Solution. Open-circuiting the current sources and short-circuiting the voltage source, the given network seduces to a network shown in Fig. 1.152. Removing the circuit elements and selecting branches 1 and 2 to form the tree, the oriented graph of the network in shown in Fig. 1.152.

The two cut-sets at nodes 'a' and 'c' with orientation as per branches 1 and 2 respectively is shown in Fig. 1.154.

Cut-set schedule is shown in Table 1.52.

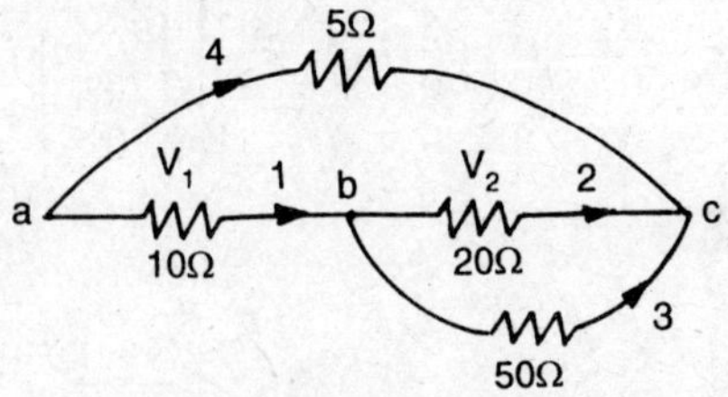

Fig. 1.152

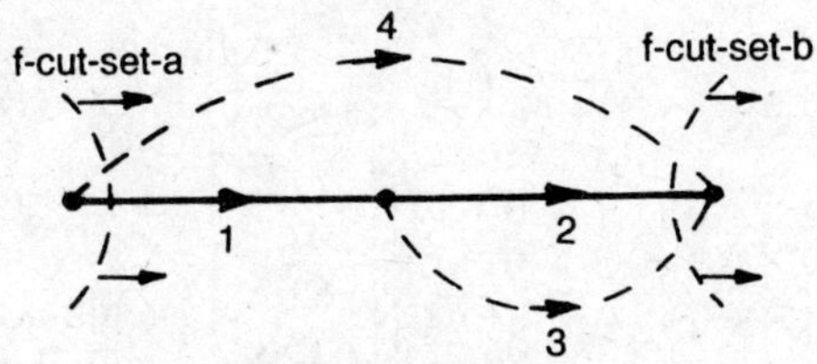

Fig. 1.153

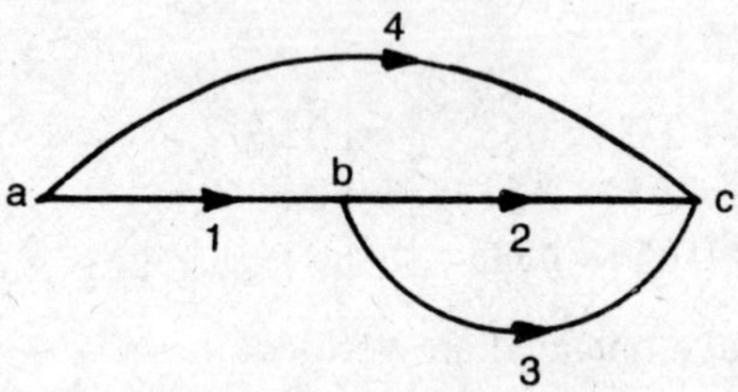

Fig. 1.154

Table 1.52: Cut-Set Schedule

Cut-Set		Branches			
		1	2	3	4
f-cut-set 'a' 1-4	Q	1	0	0	1
f cut-set 'c' 2-3-4		0	1	1	1

Equilibrium equation in matrix form on node-basis may be written as

$$[Q][Y_b][Q^T][V_t] = [Q]\{[I_s] - [Y_b][V_s]\} \qquad ...(1)$$

$$Q = f - \text{cut set matrix} = \begin{bmatrix} 1 & 0 & 0 & 1 \\ 0 & 1 & 1 & 1 \end{bmatrix}$$

$$Q^T = \begin{bmatrix} 1 & 0 \\ 0 & 1 \\ 0 & 1 \\ 1 & 1 \end{bmatrix}; Y_b = \text{branch admittance matrix} = \begin{bmatrix} 0.1 & 0 & 0 & 0 \\ 0 & 0.05 & 0 & 0 \\ 0 & 0 & 0.02 & 0 \\ 0 & 0 & 0 & 0.2 \end{bmatrix}$$

$$[V_t] = \text{Independent tree-branch voltage column matrix} = \begin{bmatrix} V_1 \\ V_2 \end{bmatrix}$$

Therefore,

$$[Q][Y_b][Q^T][V_t]$$

$$= \begin{bmatrix} 1 & 0 & 0 & 1 \\ 0 & 1 & 1 & 1 \end{bmatrix} \begin{bmatrix} 0.1 & 0 & 0 & 0 \\ 0 & 0.05 & 0 & 0 \\ 0 & 0 & 0.02 & 0 \\ 0 & 0 & 0 & 0.2 \end{bmatrix} \begin{bmatrix} 1 & 0 \\ 0 & 1 \\ 0 & 1 \\ 1 & 1 \end{bmatrix} \begin{bmatrix} V_1 \\ V_2 \end{bmatrix}$$

$$= \begin{bmatrix} 1 & 0 & 0 & 1 \\ 0 & 1 & 1 & 1 \end{bmatrix} \begin{bmatrix} 0.1 & 0 \\ 0 & 0.05 \\ 0 & 0.02 \\ 0.2 & 0.2 \end{bmatrix} \begin{bmatrix} V_1 \\ V_2 \end{bmatrix} = \begin{bmatrix} 0.3 & 0.2 \\ 0.2 & 0.27 \end{bmatrix} \begin{bmatrix} V_1 \\ V_2 \end{bmatrix} \qquad ...(2)$$

$$[Q]\{[I_s] - [Y_b][V_s]\}$$

$[I_s]$ = branch input current source column matrix

$$= \begin{bmatrix} -2 \\ -2 \\ 0 \\ 16 \end{bmatrix} \qquad ...(3)$$

$[V_s]$ = branch input voltage source column matrix

$$= \begin{bmatrix} 0 \\ -20 \\ 0 \\ 0 \end{bmatrix} \qquad ...(4)$$

Therefore,

$$[Q]\{[I_s]-[Y_b][V_s]\} = [Q][I_s]-[Q][Y_b][V_s] \qquad \text{...(5)}$$

$$[Q][I_s] = \begin{bmatrix} 1 & 0 & 0 & 1 \\ 0 & 1 & 1 & 1 \end{bmatrix} \begin{bmatrix} -2 \\ -2 \\ 0 \\ 16 \end{bmatrix} = \begin{bmatrix} 14 \\ 14 \end{bmatrix} \qquad \text{...(6)}$$

and

$[Q][Y_b][V_s]$

$$= \begin{bmatrix} 1 & 0 & 0 & 1 \\ 0 & 1 & 1 & 1 \end{bmatrix} \begin{bmatrix} 0.1 & 0 & 0 & 0 \\ 0 & 0.05 & 0 & 0 \\ 0 & 0 & 0.02 & 0 \\ 0 & 0 & 0 & 0.2 \end{bmatrix} \begin{bmatrix} 0 \\ -20 \\ 0 \\ 0 \end{bmatrix}$$

$$= \begin{bmatrix} 1 & 0 & 0 & 1 \\ 0 & 1 & 1 & 1 \end{bmatrix} \begin{bmatrix} 0 \\ -1 \\ 0 \\ 0 \end{bmatrix} = \begin{bmatrix} 0 \\ -1 \end{bmatrix} \qquad \text{...(7)}$$

Substituting the values of equations (6) and (7) in equation (5), we get

$$[Q][I_s]-[Q][Y_b][V_s] = \begin{bmatrix} 14 \\ 14 \end{bmatrix} - \begin{bmatrix} 0 \\ -1 \end{bmatrix} = \begin{bmatrix} 14 \\ 15 \end{bmatrix} \qquad \text{...(8)}$$

Substituting the values obtained in equations (2) and (8) in equation (1), we get

$$\begin{bmatrix} 0.3 & 0.2 \\ 0.2 & 0.27 \end{bmatrix} \begin{bmatrix} V_1 \\ V_2 \end{bmatrix} = \begin{bmatrix} 14 \\ 15 \end{bmatrix}$$

i.e., $0.3V_1 + 0.2V_1 = 14;$ and $0.2V_1 + 0.27V_2 = 15$

2

Laplace Transform & Applications

Problem No. 2.1 Define Laplace transform. What is the condition for a function to be Laplace transformable.

Solution Let us consider a function $f(t)$ which is function of time. Then, the Laplace transform of the function $f(t)$ is denoted by $F(s)$ equals time integration of $f_{(t)}e^{-st}$ as t varies from zero to infinity and value of s being equal to $\sigma + j\omega$. Laplace transform is denoted by letter 'L'.

Mathematically,

$$L\, f_{(t)} = F(s) = \int_0^\infty f_{(t)} e^{-st}\, dt$$

Letter 'L' means "Laplace transform of".

Condition for Laplace transformation For a function $f_{(t)}$ to be Laplace transformable, it is necessary that

$$\int_0^\infty \left| f_{(t)} \right| e^{-\sigma_1 t}\, dt \;<\; \infty \text{, for a real positive } \sigma_1.$$

Problem No. 2.2 State and prove initital and final value theorems.

Solution

Initial Value Theorem If $f_{(t)}$ is a function of time and its first derivative is

Laplace transformable, then the initial value of $f_{(t)}$ i.e., $f_{(0+)}$, is given by

$$f_{(0+)} = \underset{t \to 0+}{Lt} f_{(t)} = \underset{S \to \infty}{Lt} [s\,F(s)]$$

Proof Let $f_{(t)}$ is function of time and its first derivative is Laplace transformable

first derivative of $f_{(t)} = \frac{d}{dt} f_{(t)}$.

then its Laplace transform

$$L\left[\frac{d}{dt} f_{(t)}\right] = \int_0^\infty \left[\frac{d}{dt} f_{(t)}\right] e^{-st}\,dt = s\,F(s) - f_{(0+)}$$

or

$$\int_0^\infty \left[\frac{d}{dt} f_{(t)}\right] e^{-st}\,dt = s\,F(s) - f_{(0+)}\,.$$

Let $s \to \infty$, which gives

$$0 = \underset{s \to \infty}{Lt} \left[s\,F(s) - f_{(0+)}\right]$$

or

$$f_{(0+)} = \underset{s \to \infty}{Lt} [s\,F(s)]$$

Thus, it can be seen that initial value of a function $f(t)$ can be obtained by

(a) finding out its Laplace transform $F(s)$

(b) multiplying $F(s)$ by 's', and

(c) making 's' tend to reach infinity

Initial value theorem helps us in determining the initial value of $f_{(t)}$ and also its derivative.

Final Value Theorem

It states that if a function $f_{(t)}$ and its derivative are Laplace transformable, then the final value of function $f_{(t)}$ i.e $f_{(\infty)}$ is given by

$$\underset{t \to \infty}{Lt} f_{(t)} = \underset{s \to 0}{Lt}\, s\,F(s)$$

Proof Let us consider a function $f_{(t)}$ whose Laplace transform is $F(s)$. Its first derivative is $\frac{d}{dt} f_{(t)}$. Then,

$$L\left[\frac{d}{dt} f_{(t)}\right] = \int_0^\infty \left[\frac{d}{dt} f_{(t)}\right] e^{-st} dt$$

$$= s\,F(s) - f_{(0+)}.$$

or

$$\int_0^\infty \left[\frac{d}{dt} f_{(t)}\right] e^{-st} = s\,F(s) - f_{(0+)}.$$

Let $s \to 0$, then the above expression becomes

$$\underset{S\to 0}{Lt} \int_0^\infty \left[\frac{d}{dt} f_{(t)}\right] e^{-st} dt = \underset{S\to 0}{Lt}\left[s\,F(s) - f_{(0+)}\right]$$

or

$$\int_0^\infty \frac{d}{dt} f_{(t)}\, dt = \underset{S\to 0}{Lt}\left[s\,F(s) - f_{(0+)}\right]$$

or

$$\underset{t\to\infty}{Lt}\left[\,f_{(t)} - f_{(0+)}\right] = \underset{S\to 0}{Lt}\left[s\,F(s) - f_{(0+)}\right]$$

or

$$\underset{t\to\infty}{Lt}\ f_{(t)} = \underset{s\to 0}{Lt}\ s\,F(s).$$

Problem No. 2.3 Find Laplace transform of the following

(a) damped sine function

(b) unit impulse function

Solution (a) A damped sine function is of the following form

$$f_{(t)} = e^{-at} \sin \omega t$$

$$L\,f_{(t)} = \int_0^\infty e^{-at} \sin \omega t\, e^{-st}\, dt$$

$$= \int_0^\infty e^{-at} \sin\omega t\, e^{-(a+s)t} dt$$

$$= \int_0^\infty \frac{e^{+j\omega t} - e^{-j\omega t}}{2j} \times e^{-(a+s)t} dt$$

$$= \frac{1}{2j}\int_0^\infty \left(e^{j\omega t} - e^{-j\omega t}\right) e^{-(a+s)t} dt$$

$$= \frac{1}{2j}\int_0^\infty \left[e^{(-a-s+j\omega)t} - e^{(-a-s-j\omega)t}\right] dt$$

$$= \frac{1}{2j}\left(\left[\frac{e^{(-a-s+j\omega)t}}{-a-s+j\omega}\right]_0^\infty - \left[\frac{e^{(-a-s-j\omega)t}}{-a-s-j\omega}\right]_0^\infty\right)$$

$$= \frac{1}{2j}\left(\left[0 - \frac{1}{-a-s+j\omega}\right] - \left[0 - \frac{1}{-a-s-j\omega}\right]\right)$$

$$= \frac{1}{2j}\left(\frac{1}{(s+a)-j\omega} - \frac{1}{(s+a)+j\omega}\right)$$

$$= \frac{1}{2j}\left[\frac{(s+a+j\omega)-(s+a-j\omega)}{[(s+a)-j\omega][(s+a)+j\omega]}\right]$$

$$= \frac{2j\omega}{2j[(s+a)-j\omega][(s+a)+j\omega]}$$

$$= \frac{\omega}{(s+a)^2 - (j\omega)^2}$$

$$= \frac{\omega}{(s+a)^2 + \omega^2}$$

(b) Unit Impulse Function

It is given by

$$\delta(t) = \frac{d}{dt} u_{(t)}$$

where $u_{(t)}$ is unit step function.

$$L\left[\frac{d}{dt} f_{(t)}\right] = sF(s) - f_{(0+)}$$

$$F(s) = \frac{d}{dt} u_{(t)} = \frac{1}{s}; \text{ and}$$

$$f_{(0+)} = 0 \text{ for an impulse function.}$$

Therefore,

$$L[\delta(t)] = s\,F(s)$$

$$= s \times \frac{1}{s}$$

$$= 1.$$

Problem No. 2.4 Obtain Laplace transform of (a) unit Ramp function (b) gate function shown in Fig. 2.1.

Solution (a) A ramp function is represented by

$$f_{(t)} = t$$

Therefore, $F(s) = \int_0^\infty t\, e^{-st}\, dt$

$$= \left.\frac{-te^{-st}}{s}\right|_0^\infty - \int_0^\infty \frac{e^{-st}}{s} dt$$

(Integration by part)

$$= \frac{1}{s^2}$$

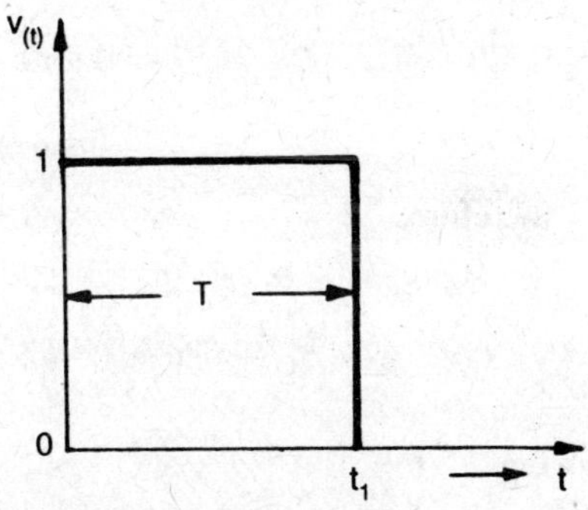

Fig. 2.1

(b) The given gatefunction of Fig 2.1 is represented mathematically as

$$f_{(t)} = 1\left[u_{(t)} - u_{(t-t_1)}\right]$$

$$L f_{(t)} = L\left[u_{(t)} - u_{(t-t_1)}\right]$$

$$= \frac{1}{s} - \frac{e^{-t_1 s}}{s}$$

$$= \frac{1}{s}\left(1 - e^{-t_1 s}\right)$$

Note: Use of shifting theorem has been made to find Laplace transform of $u_{(t-t_1)}$.

Problem No. 2.5 State and prove shifting theorem.

Solution It states that if a function $f_{(t)}$ is Laplace transformable i.e., $L f_{(t)} = F(s)$ then the Laplace transform of $f_{(t-a)}\, u_{(t-a)}$ is $e^{-as} F(s)$

Proof

Mathematically, the theorem can be defined as

$$L\left[f_{(t-a)}\, u_{(t-a)}\right] = \int_0^\infty \left[f_{(t-a)}\, u_{(t-a)}\right] e^{-st}\, dt$$

Now

$$f_{(t-a)} u_{(t-a)} = 0 \text{ for } t < a$$

$$= f_{(t-a)} \text{ for } t > a$$

therefore,

$$L\left[f_{(t-a)} u_{(t-a)}\right] = \int_0^\infty f_{(t-a)}\, e^{-st}\, dt$$

Let

$$t - a = \tau \quad \text{or} \quad \tau + a = t$$

also

$$dt = d\tau$$

Therefore,

$$L\left[f_{(t-a)}\, u_{(t-a)}\right] = \int_0^\infty f_{(\tau)}\, e^{-s(\tau+a)} d\tau$$

$$= e^{-as}\int_0^\infty f_{(\tau)}\, e^{-s\tau}\, d\tau$$

$$= e^{-as} F(s).$$

Problem No. 2.6 Find the Laplace transform of the function $f(t)$ shown in Fig. 2.2.

Solution The saw-tooth waveform can be defined as

$$f_{(t)} = \frac{A}{T} t \quad \text{for} \quad 0 < t < T;$$

$$= 0 \ \text{for} \ t > T$$

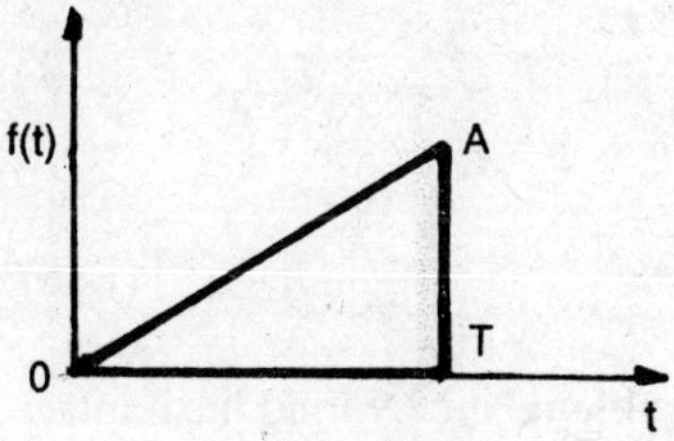

Fig. 2.2

By definition

$$L\, f_{(t)} = \int_0^\infty f_{(t)}\, e^{-st}\, dt$$

$$= \int_0^T f_{(t)}\, e^{-st}\, dt + \int_T^\infty f_{(t)}\, e^{-st}\, dt$$

$$= \int_T^\infty \frac{A}{T} t\, e^{-st}\, dt + 0$$

$$= \int_0^T \frac{A}{T} t\, e^{-st}\, dt$$

$$= \frac{A}{T}\int_0^T t\, e^{-st}\, dt$$

$$= \frac{A}{T}\left[t\int_0^T e^{-st}\, dt - \int_0^T \frac{e^{-st}}{-s}\, dt \right]$$

$$= \frac{A}{T}\left[\left. \frac{t\, e^{-st}}{-s} \right|_0^T - \left. \frac{e^{-st}}{s^2} \right|_0^T \right]$$

$$= \frac{A}{T}\left[\left(\frac{Te^{-sT}}{-s} - 0\right) - \left(\frac{e^{-sT}}{s^2} - \frac{1}{s^2}\right)\right]$$

$$= \frac{A}{T}\left[-\frac{Te^{-sT}}{s} - \frac{e^{-sT}}{s^2} + \frac{1}{s^2}\right]$$

$$= \frac{A}{T}\left[\frac{1}{s^2} - \frac{e^{-sT}}{s^2} - \frac{Te^{-sT}}{s}\right]$$

$$= \frac{A}{Ts^2}\left[1 - e^{-sT} - sTe^{-sT}\right]$$

$$= \frac{A}{Ts^2}\left[1 - e^{-sT}(1 + sT)\right]$$

Problem No. 2.7 Find the Laplace transform of periodic waveform shown in Fig. 2.3.

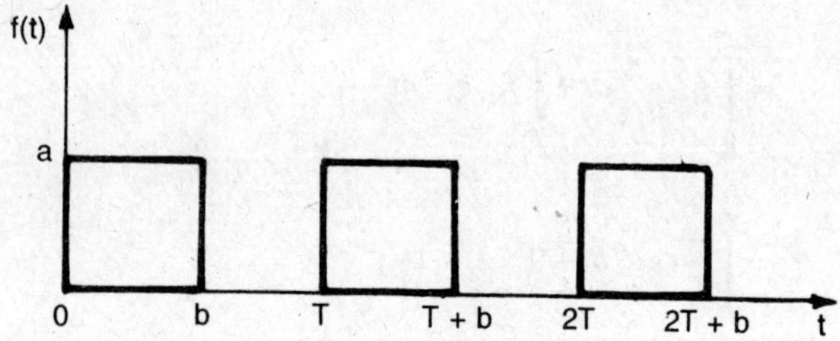

Fig. 2.3

Solution

For the time period T, the definition of the periodic waveform shown in Fig. 2.3 is

$$f(t) = a, \qquad 0 < t \le b$$

$$= 0, \qquad b < t \le T$$

Laplace of a periodic function which satisfies the conditions

$f(t) = f(t + T)$ for all $t > 0$ where T is the period of the function, is given by

$$F(s) = \frac{1}{1 - e^{-sT}} \int_0^T f(t) e^{-st} dt$$

For the given waveform,

$$L f_{(t)} = F(s) = \frac{1}{1-e^{-sT}}\left[\int_0^b a e^{-st} dt + \int_b^T 0 \times e^{-st} dt\right]$$

$$= \frac{1}{1-e^{-sT}}\left[\int_0^b a e^{-st} dt + 0\right]$$

$$= \frac{a}{1-e^{-sT}} \int_0^b e^{-st} dt$$

$$= \frac{a}{1-e^{-sT}}\left[\frac{e^{-st}}{-s}\right]_0^b$$

$$= \frac{a}{1-e^{-sT}}\left[\frac{e^{-sb}}{-s} - \frac{e^{-s\times 0}}{-s}\right]$$

$$= \frac{a}{1-e^{-sT}}\left[\frac{e^{-sb}}{-s} + \frac{1}{s}\right]$$

$$= \frac{a}{s\left(1-e^{-sT}\right)}\left[\frac{1-e^{-sb}}{1}\right]$$

$$= \frac{a\left(1-e^{-sb}\right)}{s\left(1-e^{-sT}\right)}.$$

Problem No. 2.8 Find Laplace transform of the triangular waveform shown in Fig. 2.4.

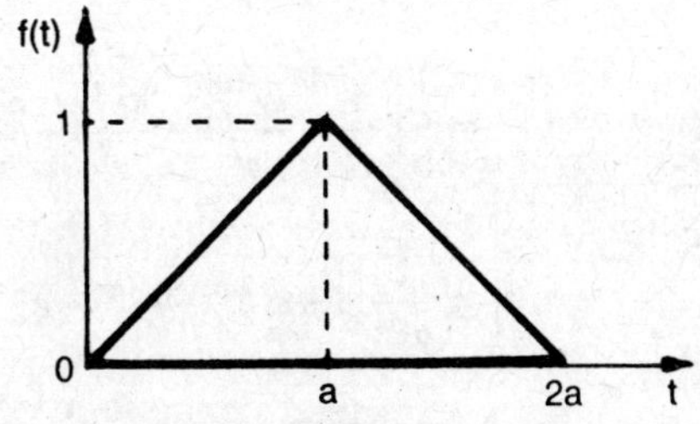

Fig. 2.4

Solution The waveform is defined as

$$f_{(t)} = 0 \text{ for } t < 0$$

$$= \frac{1}{a} t \quad \textit{for} \quad 0 \leq t \leq a$$

$$= \frac{-1}{a}(t-2a) \quad \text{for} \quad a \le t \le 2a$$

Therefore,

$$F(s) = L f_{(t)} = \int_0^\infty f_{(t)} e^{-st} dt$$

$$= \int_0^a \frac{1}{a} t e^{-st} dt + \int_a^{2a} -\frac{1}{a}(t-2a) e^{-st} dt$$

$$= \frac{1}{a}\left[\int_0^a t e^{-st} dt - \int_a^{2a} (t-2a) e^{-st} dt\right]$$

$$= \frac{1}{a}\left[\left(\frac{t e^{-st}}{-s}\Bigg|_0^a - \int_0^a \frac{e^{-st}}{-s} dt\right) - \left(\frac{(t-2a)e^{-st}}{-s}\Bigg|_0^{2a} - \int_0^{2a} \frac{e^{-st}}{-s} dt\right)\right]$$

$$= \frac{1}{a}\left[\left(\frac{a e^{-as}}{-s} - 0 - \left[\frac{e^{-st}}{s^2}\right]_0^a\right) - \left(0 - \frac{(-a)e^{-as}}{-s} - \left[\frac{e^{-st}}{s^2}\right]_0^{2a}\right)\right]$$

$$= \frac{1}{a}\left[\frac{a e^{-as}}{-s} - \left(\frac{e^{-as}}{s^2} - \frac{1}{s^2}\right) - \left(\frac{a e^{-as}}{s} - \left(\frac{e^{-2as}}{s^2} - \frac{e^{-as}}{s^2}\right)\right)\right]$$

$$= \frac{1}{a}\left[\frac{-a e^{-as}}{s} - \frac{e^{-as}}{s^2} + \frac{1}{s^2} + \frac{a e^{-as}}{s} + \frac{e^{-2as}}{s^2} - \frac{e^{-as}}{s^2}\right]$$

$$= \frac{1}{a}\left[\frac{1}{s^2} + \frac{e^{-2as}}{s^2} - \frac{e^{-as}}{s^2} - \frac{e^{-as}}{s^2}\right]$$

$$= \frac{1}{as^2}\left[1 + e^{-2as} - e^{-as} - e^{-as}\right]$$

$$= \frac{1 + e^{-2as} - 2e^{-as}}{as^2} \qquad ...(1)$$

$$= \frac{\left(1-e^{-as}\right)^2}{as^2} \qquad \ldots(2)$$

Alternatively

This is obtained by constructing the triangular wave with the help of three ramp functions as shown in Fig. 2.5.

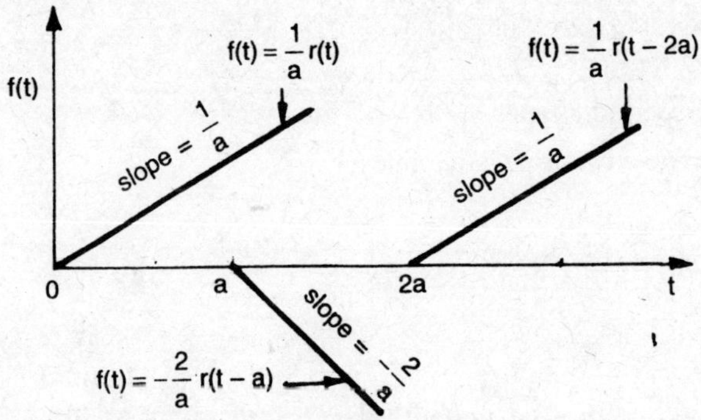

Fig. 2.5

The given triangular wave is obtained by adding the three ramp functions shown in Fig. 2.5. Therefore

$$f_{(t)} = \frac{1}{a} r(t) - \frac{2}{a} r(t-a) + \frac{1}{a} r(t-2a)$$

Laplace transform is thus

$$F(s) = \frac{1}{a} \times \frac{1}{s^2} - \frac{2}{a} \times \frac{e^{-as}}{s^2} + \frac{1}{a} \times \frac{e^{-2as}}{s^2}$$

$$= \frac{1}{as^2}\left(1 - 2e^{-as} + e^{-2as}\right)$$

$$= \frac{\left(1-e^{-as}\right)^2}{as^2}.$$

Note: Refer table 1 and 2 to obtain Laplace transform directly.

Problem No. 2.9 Prove that Laplace transform of any function that repeats itself is given by

$$F(s) = \frac{1}{1-e^{-sT}} F_1(s)$$

Solution

Consider a function $f_{(t)}$ which is function of time t.

Let

$$f_{(t+nT)} = f_{(t)}$$

where

T = periodic time

n = positive or negative integer

$$F(s) = \int_0^{\infty} f_{(t)}\, e^{-st}\, dt$$

$$= \int_0^{T} f_{(t)}\, e^{-st}\, dt + \int_T^{2T} f_{(t)}\, e^{-st}\, dt + \int_{2T}^{3T} f_{(t)}\, e^{-st}\, dt + \ldots\ldots + \int_{nT}^{(n+1)T} f_{(t)}\, e^{-st}\, dt$$

$$= \int_0^{T} f_{(t)}\, e^{-st}\, dt + \int_0^{T} f_{(t)}\, e^{-st} e^{-sT}\, dt + \ldots\ldots + \int_0^{T} f_{(t)}\, e^{-st} e^{-nsT}\, dt$$

$$= \left(1 + e^{-sT} + e^{-2sT} + \ldots\ldots\ldots + e^{-nsT}\right) \cdot \int_0^{T} f_{(t)}\, e^{-st}\, dt$$

$$= \frac{1}{1-e^{-sT}} \int_0^{T} f_{(t)}\, e^{-st}\, dt$$

$$= \frac{1}{1-e^{-sT}} F_1(s).$$

where

$$F_1(s) = \int_0^{T} f_{(t)}\, e^{-st}\, dt.$$

Problem No. 2.10 Find out the Laplce trnsform of the square wave shown in Fig. 2.6.

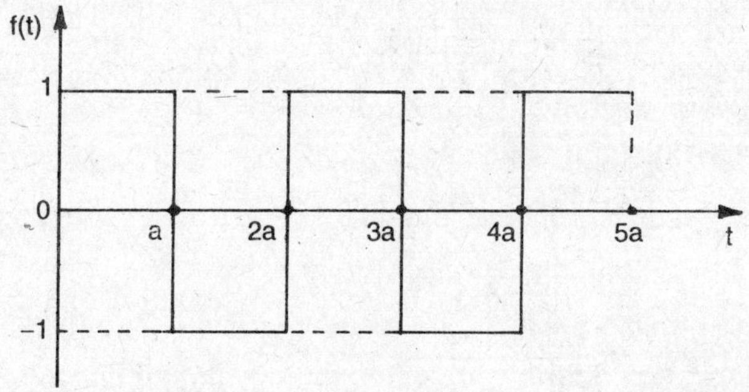

Fig. 2.6

Solution The given square wave has been constructed with the help of step functions. Amplitude is unity and periodic time 2a. The square waveform may be represented by the following infinite series of step functions.

$$f_{(t)} = u_{(t)} - 2u_{(t-a)} + 2u_{(t-2a)} - 2u_{(t-3a)} + \ldots\ldots$$

From Table 2, Laplace transforms can be written directly

$$F(s) = \frac{1}{s} - \frac{2}{s}e^{-as} + \frac{2}{s}e^{-2as} - \frac{2}{s}e^{-3as} + \ldots\ldots$$

$$= \frac{1}{s}\left[1 - 2e^{-as}\left(1 - e^{-as} + e^{-2as} - \ldots\ldots + \ldots\ldots\right)\right]$$

$$= \frac{1}{s}\left[1 - 2e^{-as}\left(\frac{1}{1+e^{-as}}\right)\right]$$

$$= \frac{1}{s}\left[1 - \frac{2e^{-as}}{1+e^{-as}}\right]$$

$$= \frac{1}{s}\left[\frac{1-e^{-as}}{1+e^{-as}}\right]$$

$$= \frac{1}{s}\tanh\frac{as}{2}.$$

Alternatively

Laplace transform of a function that repeats itself is given by,

$$F(s) = \frac{1}{\left(1+e^{-2as}\right)} F_1(s) \qquad ...(1)$$

$$F_1(s) = \int_0^a 1 \times e^{-st}\,dt + \int_a^{2a} -1 \times e^{-st}\,dt$$

$$= \left.\frac{e^{-st}}{-s}\right|_0^a + \left.\frac{-e^{-st}}{-s}\right|_a^{2a}$$

$$= \frac{-e^{-as}}{s} + \frac{1}{s} + \frac{e^{-2as}}{s} - \frac{e^{-as}}{s}$$

$$= \frac{1}{s}\left[1+e^{-2as} - 2e^{-as}\right]$$

$$= \frac{1}{s}\left(1-e^{-as}\right)^2$$

Substituting in equation (1), we get

$$F(s) = \frac{1}{s} \times \frac{\left(1-e^{-as}\right)^2}{\left(1+e^{-2as}\right)}$$

$$= \frac{\left(1-e^{-as}\right)^2}{s\left(1-e^{-as}\right)\left(1+e^{-as}\right)}$$

$$= \frac{\left(1-e^{-as}\right)}{s\left(1+e^{-as}\right)}$$

$$= \frac{1}{s}\tanh\cdot\frac{as}{2}$$

Problem No. 2.11 Find the Laplace transform of the waveform shown in Fig. 2.7.

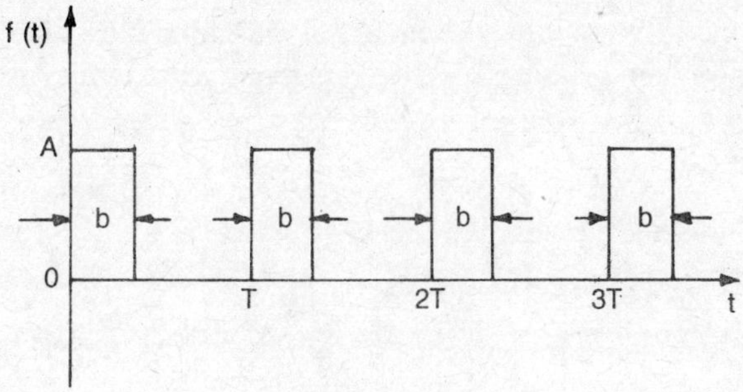

Fig. 2.7

Solution

Periodic time = T

Amplitude = A

$$F(s) = \frac{1}{1-e^{-sT}} F_1(s) \qquad ...(1)$$

$$F_1(s) = \int_0^T f(t) e^{-sT} dt$$

The first pulse is defined by the equation

$$f_{(t)} = A\left(u_{(t)} - u_{(t-b)}\right)$$

Therefore,

$$F_1(s) = L f_{(t)} = LA\left[u_{(t)} - u_{(t-b)}\right]$$

With the help of Table 2, Laplace can be written directly

Therefore,

$$F_1(s) = L f_{(t)} = A\left[\frac{1}{s} - \frac{1}{s} \times e^{-bs}\right]$$

$$= \frac{A}{s}\left(1 - e^{-bs}\right)$$

Substituting in equation (1), we get

$$F(s) = \frac{A}{s}\left(\frac{1-e^{-bs}}{1-e^{-sT}}\right)$$

Problem No. 2.12 Write the equation for the wave form shown in Fig. 2.8 in terms of linear combinations of step functions. Also write its Laplace transform.

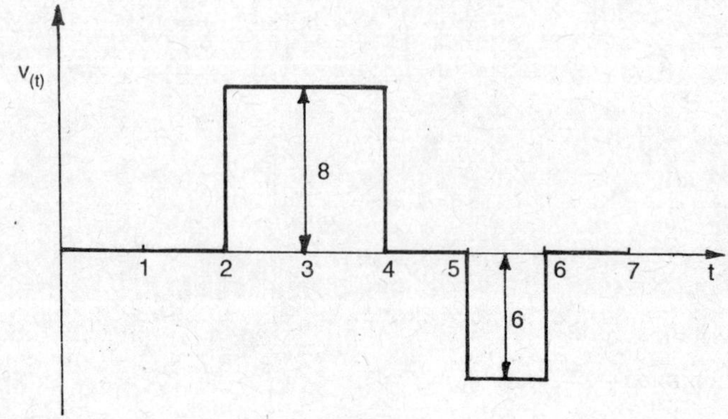

Fig. 2.8

Solution

(a) **Between $t = 2$ and $t = 4$** The amplitude is 8 and hence in terms of step functions it can be represented as shown in Fig. 2.9 which

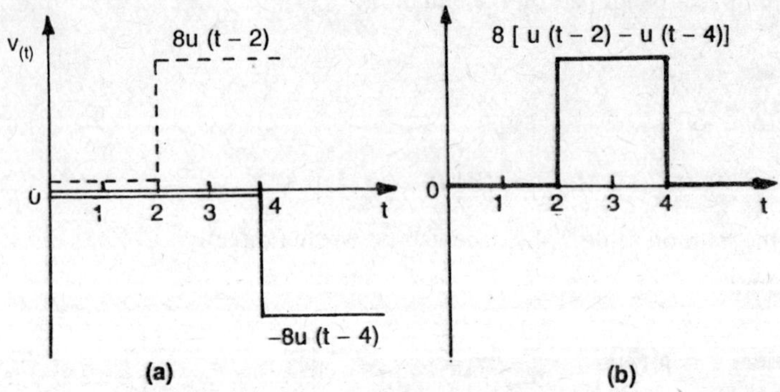

Fig. 2.9

brings out how a pulse of amplitude 8 has been built by taking difference between two step functions. Step function $8u\ (t–2)$ is combined with $-8u\ (t-u)$, the inverted step function delayed by another 2 seconds resulting in the waveform as shown in Fig. 2.9(b) having equation

$$v_{(t)} = 8u_{(t-2)} - 8u_{(t-4)} \cdot$$

(b) Between $t = 5$ and $t = 6$ Pictorial representation in shown in Fig. 2.10 The resultant waveform

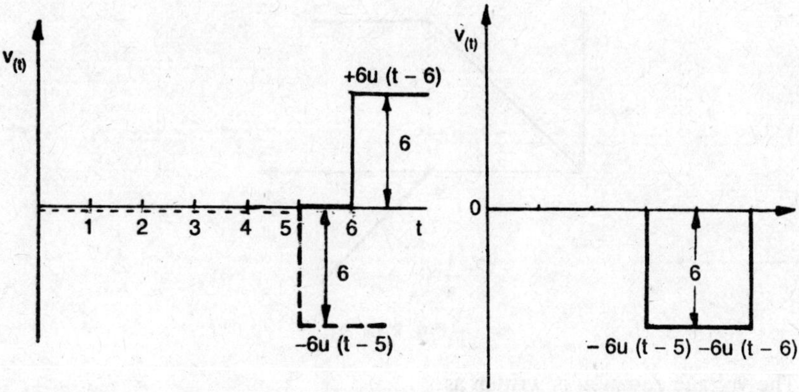

(a)

(b)

Fig. 2.10

as seen in Figure 2.10(b) is as a result of step function – $6u\ (t–5)$ and delayed inverted step function $6u\ (t–6)$

Waveform Final waveform is thus represented by

$$v_{(t)} = 8u_{(t-2)} - 8u_{(t-4)} - 6u_{(t-5)} + 6u_{(t-6)}$$

Laplace Transforms: With the help of Table 2, Laplace transform can be directly written as

$$V(s) = \frac{8e^{-2s}}{s} - \frac{8}{s}e^{-4s} - \frac{6}{s}e^{-5s} - \frac{6e^{-6s}}{s}$$

$$V(s) = \frac{8}{s}\left(e^{-2s} - e^{-4s}\right) - \frac{6}{s}\left(e^{-5s} - e^{-6s}\right)$$

Problem No. 2.13 Write the equation of the waveform shownin Fig. 2.11 in terms of linear combination of ramp function and then

Solution

The given waveform has been constructed with the help of two ramps as shown in Fig. 2.12.

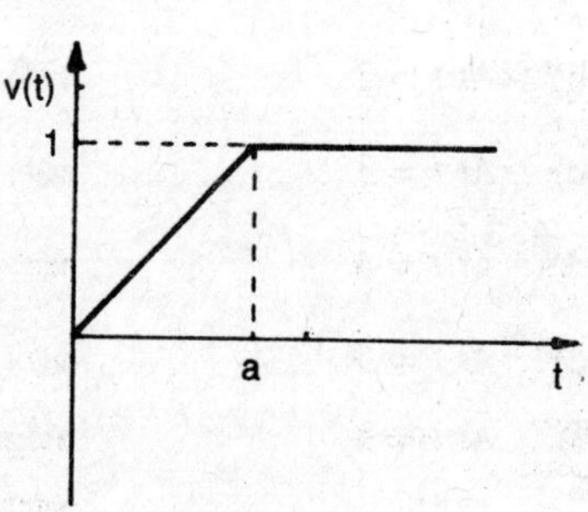

Fig. 2.11

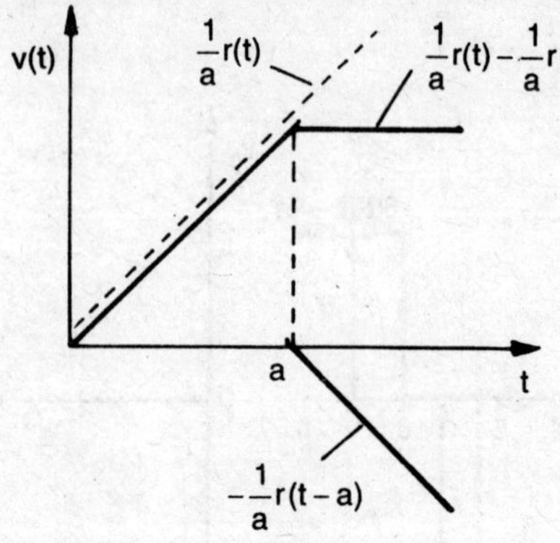

Fig. 2.12

The voltage equation is written as

$$v_{(t)} = \frac{1}{a}r(t) - \frac{1}{a}r(t-a)$$

Laplace can be written with the help of Table 2 and is

$$V(s) = \frac{1}{as^2} - \frac{1}{as^2}e^{-as}$$

or $$V(s) = \frac{1}{as^2}\left(1 - e^{-as}\right)$$

Problem No. 2.14 Write the equation of voltage wave form (Fig. 2.13) in terms of step function and then write its Laplace transform.

Solution The given voltage wave form has been constructed with the help of step function

(a) *At t = 1* $\quad v(t) = 1 \times u(t-1)$

(b) *At t = 2* $\quad v(t) = 1 \times u(t-2)$

(c) *At t = 3* $\quad v(t) = 1 \times u(t-3)$

(d) *At t = 4* $\quad v(t) = 1 \times u(t-4)$

(e) *At t = 5* $\quad v(t) = 1 \times u(t-5)$

(f) *At t = 6* $\quad$ magnitude of voltage of –5. Therefore,

$$v(t) = -5 \times u(t-6)$$

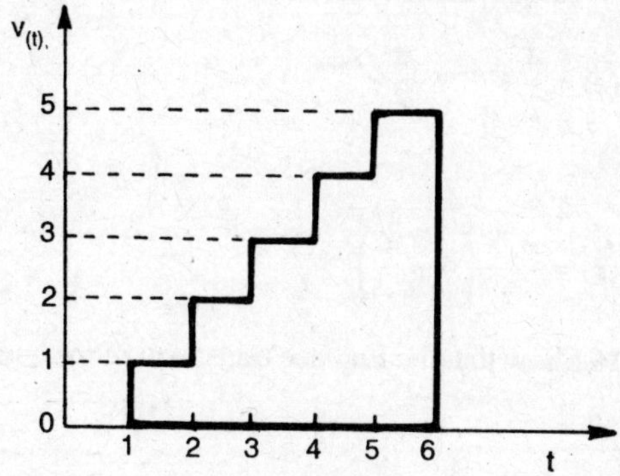

Fig. 2.13

Therefore, the given voltage is represented by combination of above step-voltages and is

$$v_{(t)} = u(t-1) + u(t-2) + u(t-3) + u(t-4) + u(t-5) - 5u(t-6)$$

Writing Laplce transform directly with the help of Table 2, we get

$$V(s) = \frac{e^{-s}}{s} + \frac{e^{-2s}}{s} + \frac{e^{-3s}}{s} + \frac{e^{-4s}}{s} + \frac{e^{-5s}}{s} - \frac{5e^{-6s}}{s}$$

or $$V(s) = \frac{1}{s}\left(e^{-s} + e^{-2s} + e^{-3s} + e^{-4s} + e^{-5s} - e^{-6s}\right)$$

Problem No. 2.15 Write the voltage equation of the raised ramp function shown in Fig. 2.14. Also, find the Laplace transform.

Solution

The given voltage waveform has been constructed with

(a) step function of magnitude 4 at $t = a$, which gives voltage equation.

$$v_{(t)} = 4u(t-a), \text{ and}$$

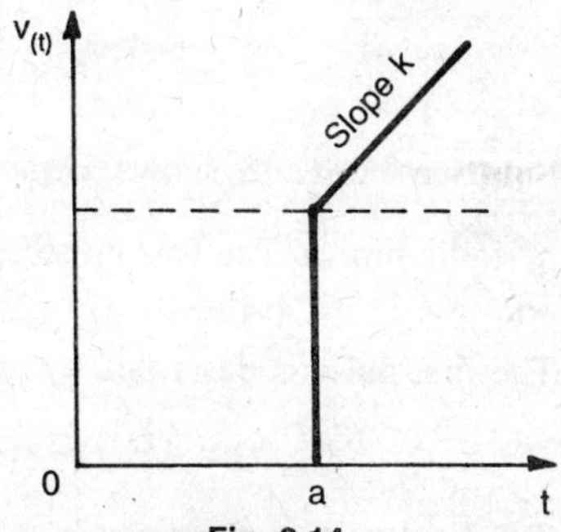

Fig. 2.14

(b) ramp function of slope K at $t = a$, which gives voltage equation

$$v_{(t)} = kr(t-a)$$

The resultant voltage comprises of combination of above two functions and is represented by

$$v_{(t)} = 4u(t-a) + Kr(t-a)$$

Laplace transform can be written directly from Table 2 and is

$$V(s) = \frac{4}{s}e^{-as} + \frac{K}{s^2}e^{-as}$$

or

$$V(s) = \frac{e^{-as}}{s}\left(4 + \frac{K}{s}\right)$$

Problem No. 2.16 Show that the Laplace transform of the square wave given in Fig. 2.15 is

$$V(s) = \frac{A}{s\left[1 + e^{-as/2}\right]}$$

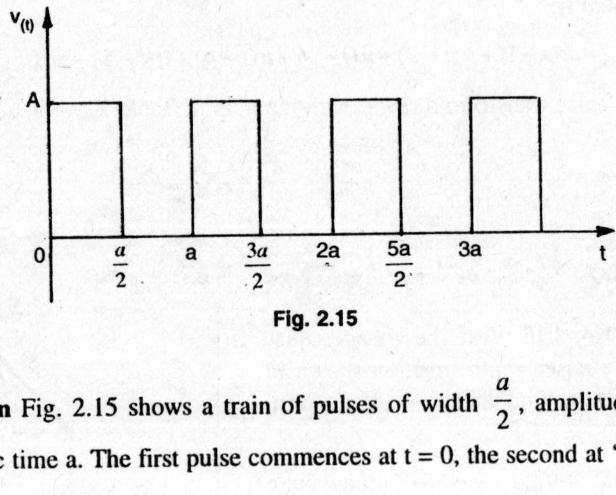

Fig. 2.15

Solution Fig. 2.15 shows a train of pulses of width $\frac{a}{2}$, amplitude A and periodic time a. The first pulse commences at t = 0, the second at 'a' and so on.

The first pulse is described by the equation

$$v_1(t) = A\left[u(t) - u(t - a/2)\right]$$

The Laplace transform can be written directly from Table 2 and is

$$V_1(s) = A\left(\frac{1}{s} - \frac{e^{-as/2}}{s}\right)$$

or

$$V_1(s) = \frac{A}{s}\left(1 - e^{-as/2}\right)$$

Laplace transform of train of pulses which repeates itself is given by

$$V_1(s) = \frac{1}{1-e^{-as}} V_1(s)$$

Substituting the value of $V_1(s)$, we get

$$V(s) = \frac{A}{s}\left(\frac{1-e^{-as/2}}{1-e^{-as}}\right)$$

or

$$V(s) = \frac{A}{s}\left(\frac{1}{1-e^{-as/2}}\right)$$

$$= \frac{A}{s(1+e^{-as/2})}$$

Alternatively

$$v_{(t)} = Au(t) - Au\left(t-\frac{a}{2}\right) + Au(t-a) - Au\left(t-\frac{3a}{2}\right)$$

$$+ Au(t-2a) - Au\left(t-\frac{5a}{2}\right) + Au(t-3a)$$

Therefore

$$V(s) = \frac{A}{s} - \frac{A}{s}e^{-as/2} + \frac{A}{s}e^{-as} - \frac{A}{s}e^{-3as/2} + \frac{A}{s}e^{-2as} - \frac{A}{s}e^{-5as/2} + \frac{A}{s}e^{-3as}$$

$$= \frac{A}{s}\left[1-e^{-as/2} + e^{-as} - e^{-3as/2} + e^{-2as} - e^{-5as/2} + e^{-3as}\right]$$

$$= \frac{A}{s}\left[1-e^{-as/2}\left(1-e^{-as/2} + e^{\ as} - e^{-3as/2} + e^{-2as} - e^{-5as/2}\right)\right]$$

$$= \frac{A}{s}\left[1-\frac{e^{-as/2}}{1+e^{-as/2}}\right]$$

$$= \frac{A}{s}\left[\frac{1+e^{-as/2} - e^{-as/2}}{1+e^{-as/2}}\right]$$

$$= \frac{A}{s\left(1+e^{-as/2}\right)}$$

Problem No. 2.17 Show that the transform of the square wave is

$$F(s) = \frac{1}{s\left(1+e^{-as/2}\right)} \quad \text{(Fig 2.16)}$$

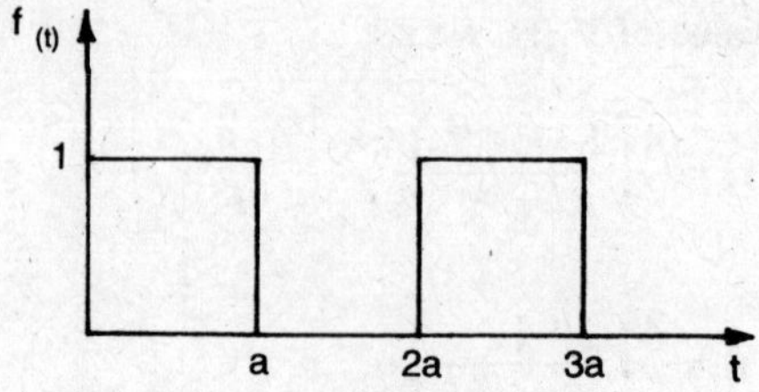

Fig. 2.16

Solution This problem is similar to Problem No. 2.16.

$$f_1(t) = 1\left(u(t) - u(t-a)\right) \quad \text{...(1)}$$

$$F_1(s) = \frac{1}{s} - \frac{e^{-as}}{s} \quad \text{...(2)}$$

$$= \frac{1}{s}\left(1-e^{-as}\right)$$

Now,

$$V(s) = \frac{1}{1-e^{-2as}} F_1(s)$$

$$= \frac{1}{1-e^{-2as}}\left[\frac{1-e^{-as}}{s}\right]$$

$$= \frac{1}{s}\left[\frac{1-e^{-as}}{1-e^{-2as}}\right]$$

$$= \frac{1}{s}\left[\frac{1-e^{-as}}{\left(1+e^{-as}\right)\left(1-e^{-as}\right)}\right]$$

$$= \frac{1}{s\left(1+e^{-as}\right)}$$

Problem No. 2.18 The waveform shown in Fig. 2.17 is a terminated truncated ramp. For this $v_{(t)}$, determine the transform V(s)

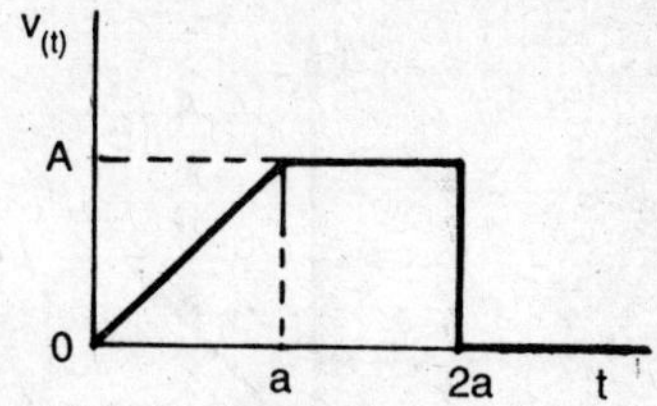

Fig. 2.17

Solution

The given waveform has been constructed with the help of two ramp functions and one step function as shown in Fig. 2.18. Therefore, the given voltage waveform can be represented as combination of two ramp functions and one step function as given below:

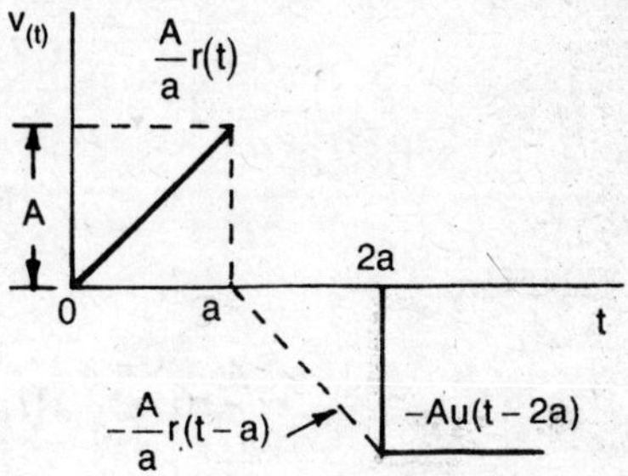

Fig. 2.18

$$v_{(t)} = \frac{A}{a} r(t) - \frac{A}{a} r(t-a) - Au(t-2a)$$

Laplace transform can be written directly with the help of Table 2 as given below:

$$V(s) = \frac{A}{a} \times \frac{1}{s^2} - \frac{A}{a} \times \frac{1}{s^2} \times e^{-as} - \frac{A}{s} e^{-2as}$$

$$= \frac{A}{as^2}\left(1 - e^{-as}\left(1 + ase^{-as}\right)\right)$$

Problem No. 2.19 Determine the Laplace transform of waveforms shown in Fig. 2.19 and compare their results

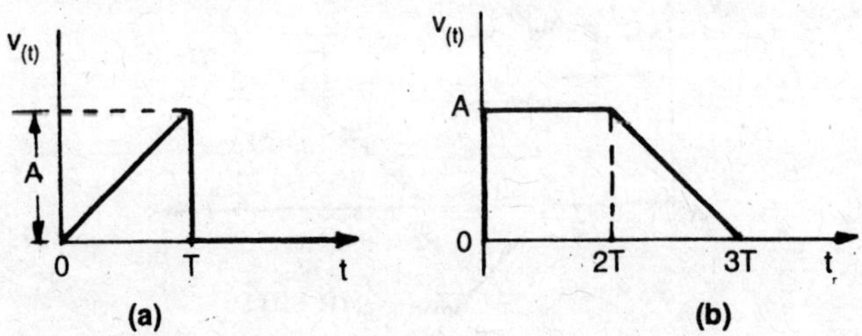

Fig. 2.19

Solution

Fig 2.19 (a) Voltage waveform has been constructed with combination of two ramp and one step functions as shown in Fig 2.20.

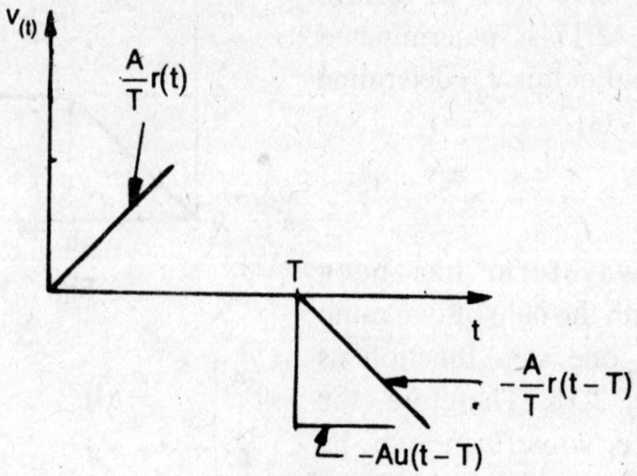

Fig.2.20

Therefore

$$v_{(t)} = \frac{A}{T}r(t) - \frac{A}{T}r(t-T) - Au(t-T)$$

The Laplace transform can be written directly with the help of Table 2, and is

$$V(s) = \frac{A}{Ts^2} - \frac{A}{Ts^2}e^{-sT} - \frac{A}{S}e^{-sT}$$

or

$$V(s) = \frac{A}{Ts^2}\left(1 - e^{-sT}(1+sT)\right)$$

Fig 2.19 (b) The voltage waveform has been constructed

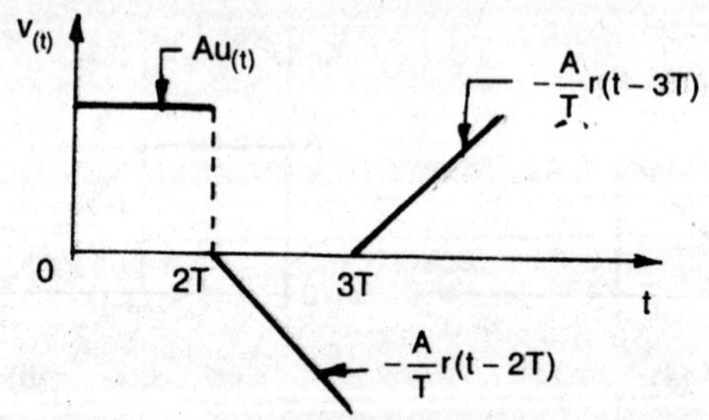

Fig. 2.21

with a step function and two ramp functions as shown in Fig. 2.21 Therefore,

$$V(s) = Au(t) - \frac{A}{T}r(t-2T) + \frac{A}{T}r(t-3T)$$

Laplace transform written directly with the help of Table 2 is:

$$V(s) = \frac{A}{s} - \frac{A}{Ts^2} e^{-2sT} + \frac{A}{Ts^2} e^{-3sT}$$

or

$$V(s) = \frac{A}{Ts^2} \left\{ sT - e^{-2sT} \left(1 - e^{-sT} \right) \right\}$$

Problem No. 2.20 Write the equation for the waveforms of the Fig. 2.22 as a linear combination of step functions.

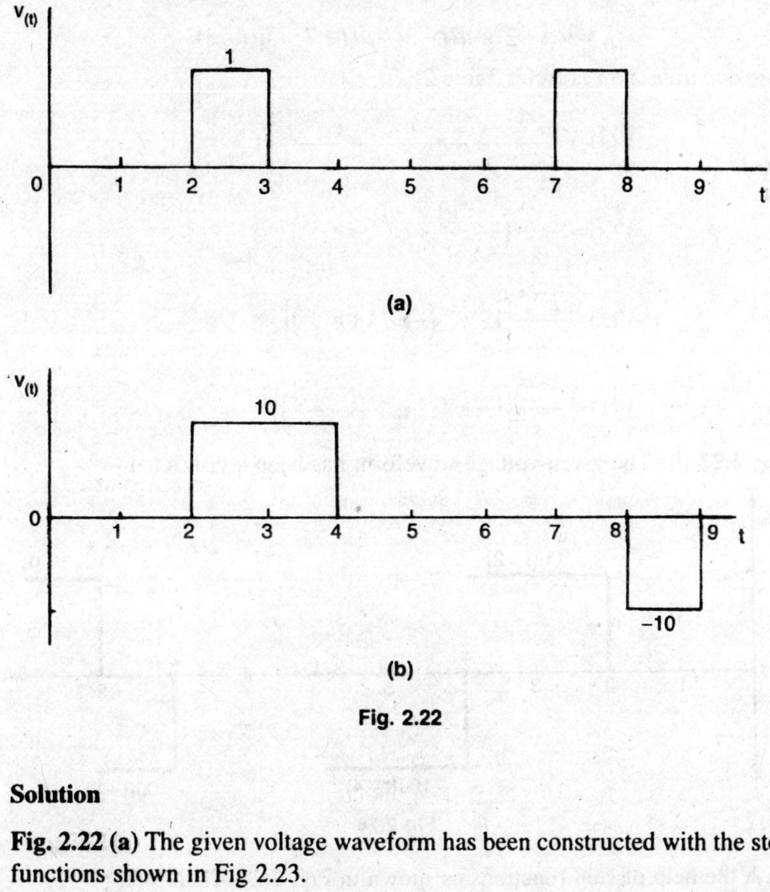

Fig. 2.22

Solution

Fig. 2.22 (a) The given voltage waveform has been constructed with the step functions shown in Fig 2.23.

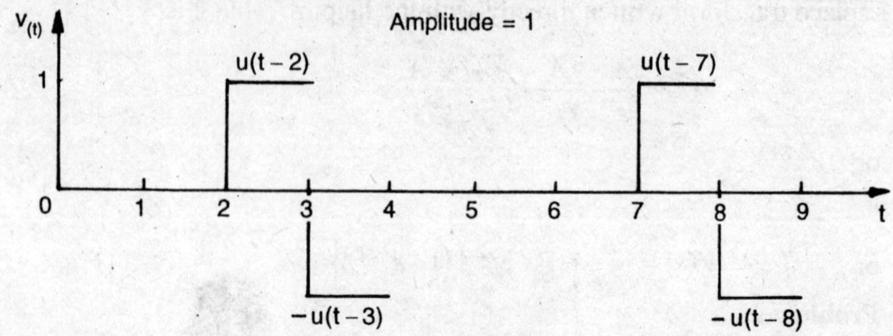

Fig. 2.23

Therefore,

$$v_{(t)} = u(t-2) - u(t-3) + u(t-7) - u(t-8)$$

Laplace transform is (refer Table 2)

$$V(s) = \frac{e^{-2s}}{s} - \frac{e^{-3s}}{s} + \frac{e^{-7s}}{s} - \frac{e^{-8s}}{s}$$

or $$V(s) = \frac{e^{-2s}}{s}\left(1 - e^{-s} + e^{-5s} - e^{-6s}\right)$$

or $$V(s) = \frac{e^{-2s}}{s}\left[1 - e^{-s}\left(-e^{-4s} + e^{-5s}\right)\right]$$

or $$V(s) = \frac{e^{-2s}}{s}\left[1 - e^{-s}\left\{-e^{-4s}\left(1 - e^{-s}\right)\right\}\right]$$

Fig 2.22 (b) The given voltage waveform has been constructed

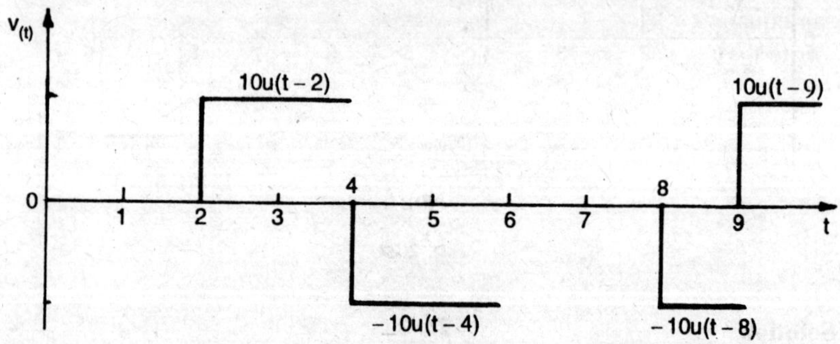

Fig 2.24

with the help of step functions as shown in Fig. 2.24. Therefore,

$$v_{(t)} = 10u(t-2) - 10u(t-4) - 10u(t-8) + 10u(t-9)$$

Laplace transform written directly by making use of Table 2 is

$$V(s) = \frac{10}{s}e^{-2s} - \frac{10}{s}e^{-4s} - \frac{10}{s}e^{-8s} + \frac{10}{s}e^{-9s}$$

or $$V(s) = \frac{10}{s}e^{-2s}\left(1 - e^{2s} - e^{-6s} + e^{-7s}\right)$$

or $$V(s) = \frac{10}{s}e^{-2s}\left[1 - e^{2s}\left\{1 + e^{-4s}\left(1 - e^{-s}\right)\right\}\right]$$

Problem No. 2.21 The waveform shown in Fig. 2.25 is non recurring. Write an equation for $v_{(t)}$ in terms of steps and ramps and related functions as needed.

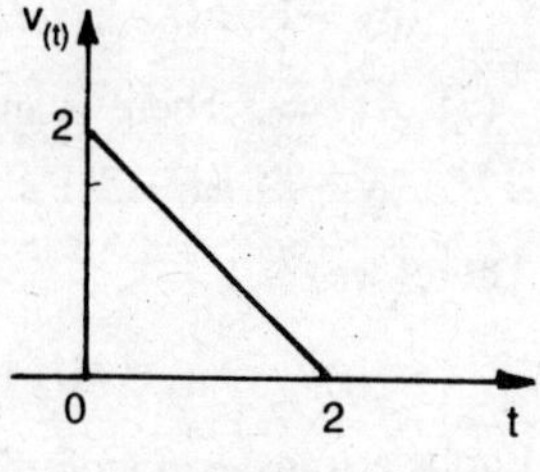

Fig. 2.25

Solution

Fig 2.26 shows as to to how the given voltage waveform has been constructed

Definition of $v_{(t)}$ as obtained from Fig. 2.26 is

$$v_{(t)} = 2u(t) - r(t) + r(t-2)$$

Laplace transform of $v_{(t)}$ is (refer Table 2)

$$V(s) = \frac{2}{s} - \frac{1}{s^2} + \frac{e^{-2s}}{s^2}$$

or $$V(s) = \frac{2}{s^2}\left(s - 1 + e^{-2s}\right)$$

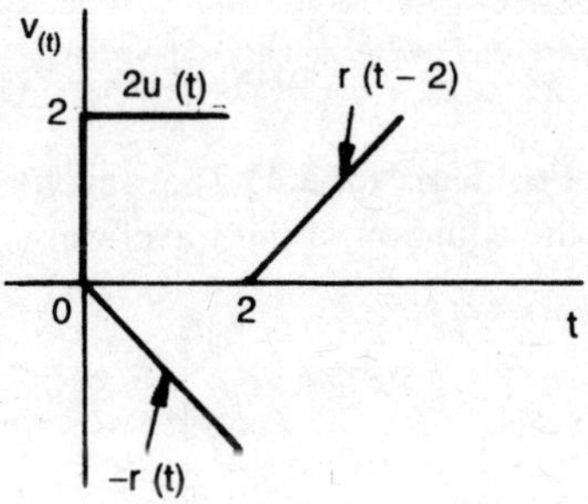

Fig 2.26

Problem No. 2.22 The waveform shown in Fig 2.27 occurs only once. Write an equation for $v_{(t)}$ in terms of steps and related functions as needed and write the Laplace transform.

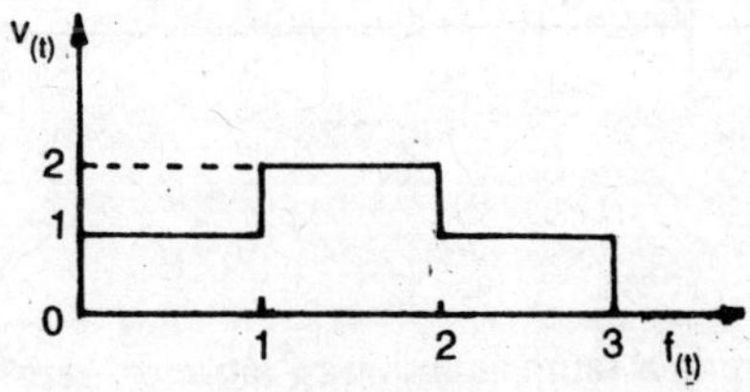

Fig 2.27

Solution

(a) **At $t = 0$** There is a step function magnitude of 1 represented by $1 \times u(t)$ i.e., $u(t)$.

(b) **At t = 1** There is a step function of magnitude 1 represented by $1 \times u(t-1)$ i.e. $u(t-1)$.

(c) **At t = 2** There is an inverted step function of magnitude 1 represented by $-1 \times u(t-2)$ i.e. $-u(t-2)$.

(d) **At t = 3** There is an inverted step function of magnitude 1 represented by $-1 \times u(t-3)$ i.e. $-u(t-3)$

Therefore,

$$v_{(t)} = u(t) + u(t-1) - u(t-2) - u(t-3)$$

Laplace transform (refer Table 2) is

$$V(s) = \frac{1}{s} + \frac{e^{-s}}{s} - \frac{e^{-2s}}{s} - \frac{e^{-3s}}{s}$$

or $$V(s) = \frac{1}{s}\left[1 + e^{-s}(1 - e^{-s}(1 + e^{-s}))\right]$$

Problem No. 2.23 The waveform shown in Fig 2.28 is nonrecurring. Write the equation of the waveform $v_{(t)}$ and determine $V(s)$.

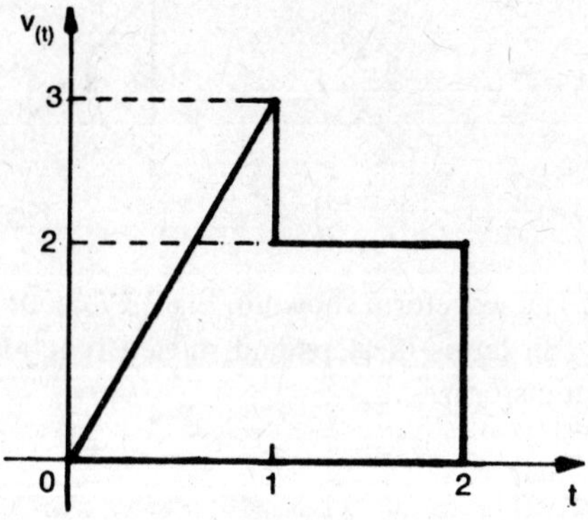

Fig. 2.28

Solution

(a) **At t = 0** There is a ramp function of slope $\frac{3}{1}$ represented by $3\,r(t)$.

(b) **At t = 1**

(i) There is a ramp function of slope $\left(-\frac{3}{1}\right)$ to bring the waveform parallel to x axis represented by $-3r(t-1)$, and

(ii) there is an inverted step function of magnitude (3 – 2) = 1 represented by $-1 \times u(t-1)$ i.e. $-u(t-1)$.

(c) **At t = 2** There is an inverted step function of magnitude (2 – 0) = 2 represented by $-2u(t-2)$.

The equation of waveform $v_{(t)}$ is thus:

$$v_{(t)} = 3r(t) - 3r(t-1) - u(t-1) - 2u(t-2)$$

Laplace transform (refer Table 2) is given by

$$V(s) = \frac{3}{s^2} - \frac{3e^{-s}}{s^2} - \frac{e^{-s}}{s} - \frac{2e^{-2s}}{s}$$

or $$V(s) = \frac{3}{s^2}\left[1 - e^{-s}\left(1 + \frac{s}{3} + \frac{2se^{-3}}{3}\right)\right]$$

Problem No. 2.24 The Fig. 2.29 shows a waveform made up of straight line segments. For this waveform, write an equation for $v_{(t)}$ in terms of steps, ramps, and related waveforms as needed.

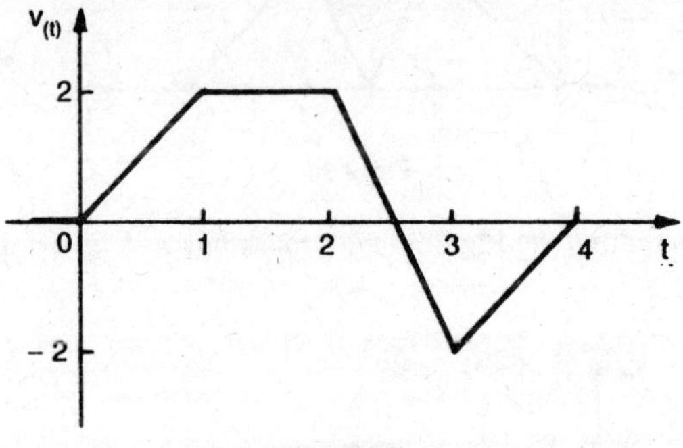

Fig. 2.29

Solution

(a) **At t = 0** There is a ramp function of slope 2 represented by 2r (t).

(b) **At t = 1** There is a ramp function of slope –2 which brings the waveform parallel to x axis and is represented by $-2r(t-1)$

(c) **At t = 2** There is a ramp function of slope –4 and is represented by $-4r(t-2)$

(d) **At t = 3** There is a ramp function represented by $4r$ (t – 3) which brings the waveform parallel to x axis and another ramp function represented by $2r$ (t – 3) which turns the waveform towards t = 4.

(e) **At $t = 4$** There is ramp function represented by $-2r(t-4)$ which brings the waveform along the x axis.

The given waveform is thus represented by

$$v_{(t)} = 2r(t) - 2r(t-1) - 4r(t-2) + 4r(t-3) + 2r(t-3) - 2r(t-4)$$

Laplace transform (refer Table 2) is given by

$$V(s) = \frac{2}{s^2} - \frac{2e^{-s}}{s^2} - \frac{4e^{-2s}}{s^2} + \frac{4e^{-3s}}{s^2} + \frac{2e^{-3s}}{s^2} - \frac{2e^{-4s}}{s^2}$$

or $$V(s) = \frac{2}{s^2}\left[1 - e^{-s} - 2e^{-2s} + 2e^{-3s} + e^{-3s} - e^{-4s}\right]$$

Problem No 2.24 Fig 2.30 shows a waveform made up of straight line segments. For this waveform, write an equation for $v_{(t)}$ in terms of steps, ramps and related waveforms as needed and determine $V(s)$.

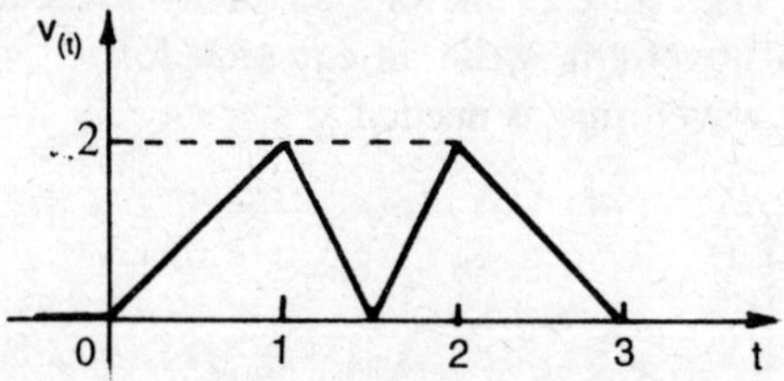

Fig 2.30

Solution Let us redraw the Fig 2.30 with markings *a, b, c, d* and *e* as shown in Fig. 2.31.

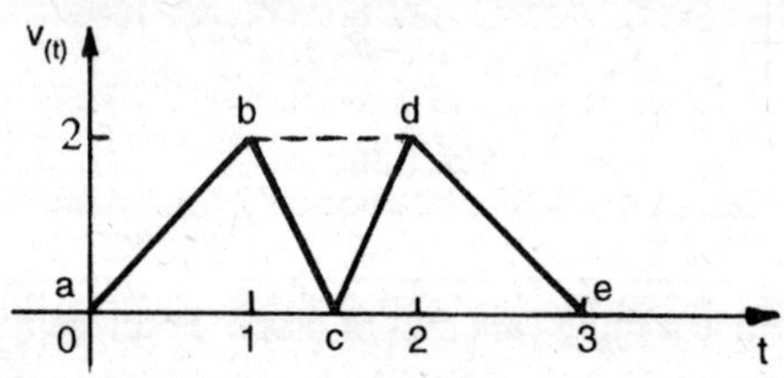

Fig. 2.31

(a) **Point '*a*', $t = 0$** There is a ramp function with slope $\frac{2}{1}$, directing the voltage wave form 'ab' and is represented by $2r(t)$.

(b) **Point '*b*', $t = 1$** At this point, the voltage waveform has changed direction towards x axis. This can be achieved by two ramp functions.

(i) At this point add a voltage waveform with negative slope which equals the slope of previous waveform to bring the waveform parallel to x axis along with the line bd represented by dotted line and represented by $-2r\ (t-1)$.

(ii) In order to change the direction of the waveform towards x axis add another waveform with negative slope along line bc, of $-\dfrac{2}{0.5}$ i.e., -4 represented by $-4r\ (t-1)$.

Note: With experience one can combine the two ramp function as done in Problem No. 2.8 and write directly the combined representation as $-6r\ (t-1)$ which is obtained by adding the two ramp functions obtained in para b(i) and b(ii) which is

$$= (-2r(t-1)) + (-4r(t-1))$$

$$= r(t-1)[-2-4]$$

$$= -6r(t-1).$$

(c) **Point 'c', $t = 1.5$** Two ramp functions, one represented by $\dfrac{+2}{0.5}r(t-1.5)$ brings the waveform along x-axis and another of similar value changes the direction away from x-axis along line cd. Both can be combined to represent it by a single equation as given below

$$= \frac{2}{0.5}r(t-1.5) + \frac{2}{0.5}r(t-1.5)$$

$$= r(t-1.5)\left[\frac{2}{0.5} + \frac{2}{0.5}\right]$$

$$= r(t-1.5)\left[\frac{2+2}{0.5}\right]$$

$$= r(t-1.5)\left(\frac{4\times 10}{5}\right)$$

$$= 8r(t-1.5)$$

(d) **Point 'd', $t = 2$** The direction de is obtained by the ramp function $-\dfrac{2}{0.5}r(t-2)$ which brings the waveform parallel to x-axis and another ramp function $-\frac{2}{1}r(t-2)$ which brings the waveform along line de. The combined equation is obtained by adding both as given below

$$= -\frac{2}{0.5}r(t-2) + \frac{2}{1}r(t-2)$$

$$= -4r(t-2) - 2r(t-2)$$

$$= -6r(t-2)$$

(e) **Point 'e', $t = 3$** At this point, the waveform has to bc brought along x-axis, which is achieved by a ramp function $2r(t-3)$.

The equation of the waveform is thus given by

$$v_{(t)} = 2r(t) - 6r(t-1) + 8r(t-1.5) - 6r(t-2) + 2r(t-3)$$

Laplace transform (refer Table 2) is

$$V(s) = \frac{2}{s^2} - \frac{6e^{-s}}{s^2} + \frac{8e^{-1.5s}}{s^2} - \frac{6e^{-2s}}{s^2} + \frac{2e^{-3s}}{s^2}$$

or $$V(s) = \frac{2}{s^2}\left[1 - e^{-s}(3 - 4e^{-0.5s} + 3e^{-s} - e^{-2s}\right]$$

Problem No. 2.25 The waveform shown in Fig. 2.31 is a sweep voltage used to deflect the beam in a cathode ray oscilloscope. Show that transform of this function is

$$F(s) = \frac{1}{as^2} - \frac{e^{-as}}{s(1-e^{-as})}$$

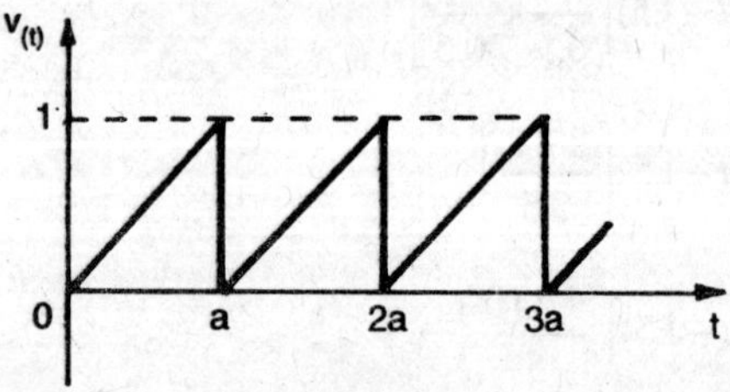

Fig. 2.31

Solution

Fig. 2.31 shows a voltage wave form which repeats itself. Laplace transform of such type of waveform is given by

$$F(s) = \frac{1}{1-e^{-Ts}}F_1(s)$$

Here time period = $T = a$

amplitude = 1

Therefore,

$$V(s) = \frac{1}{1-e^{-as}} V_1(s) \qquad ...(1)$$

$$V_1(t) = \frac{1}{a} r(t) - \frac{1}{a} r(t-a) - u(t-a)$$

Here $\frac{1}{a} r(t)$ is the equation of the waveform which represents a ramp function of slope $\frac{1}{a}$ at $t = 0$. The second term $-\frac{1}{a} r(t-a)$, brings the voltage waveform at $t = a$, parallel to x axis. The third term is inverted unit step function at $t = a$ which drops the voltage waveform vertically down.

Laplace transform (refer Table 2) is

$$V_1(s) = \frac{1}{as^2} - \frac{e^{-as}}{as^2} - \frac{e^{-as}}{s}$$

or $$V_1(s) = \frac{1}{as^2}\left[1 - e^{-as}\right] - \frac{e^{-as}}{s}$$

Substituting in equation (1) value of $V_1(s)$, we get

$$V(s) - \frac{1}{\left(1-e^{-as}\right)}\left[\frac{1}{as^2}\left(1-e^{-as}\right) - \frac{e^{-as}}{s}\right]$$

or $$V(s) = \frac{1}{as^2} - \frac{e^{-as}}{s(1-e^{-as})}.$$

Problem No. 2.26 Find the transform of the voltage waveform shown in Fig. 2.32.

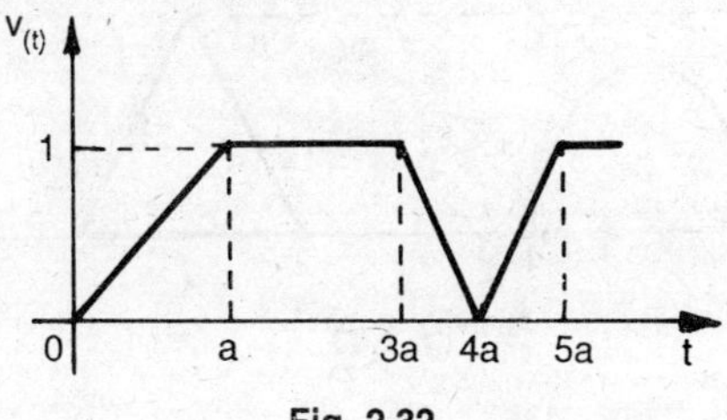

Fig. 2.32

Solution The given voltage waveform has

Time period = 4a;

Amplitude = 1;

and it repeats itself. Therefore, Laplace transform is

$$V(s) = \frac{1}{1-e^{-4as}} V_1(s) \qquad ...(1)$$

Voltage equation $v_{1(t)}$ is given by

$$v_{1(t)} = \frac{1}{a}r(t) - \frac{1}{a}r(t-a) - \frac{1}{a}r(t-3a) + \frac{1}{a}r(t-4a)$$

or
$$V_1(s) = \frac{1}{as^2} - \frac{e^{-as}}{as^2} - \frac{e^{-3as}}{as^2} + \frac{e^{-4as}}{as^2}$$

or
$$V_1(s) = \frac{1}{as^2}\left(1+e^{-4as}\right) - \frac{e^{-as}}{as^2}\left(1+e^{-2as}\right)$$

Substituting in equation (1), value of $V_1(s)$, we get

$$V(s) = \frac{(1+e^{-4as})}{as^2(1-e^{-4as})} - \frac{e^{-as}(1+e^{-2as})}{as^2(1-e^{-4as})}$$

or
$$V(s) = \frac{1}{as^2}\left[\frac{1+e^{-4as}}{1-e^{-4as}} - \frac{e^{-as}}{1-e^{-2as}}\right]$$

Problem No. 2.27 The waveform shown in Fig. 2.33 is sinosidal in the interval from $t = 0$ to $t = 1$ with peak value of K_1, and is an isoceles triangle in the interval from $t = 2$ to $t = 3$ with peak value of K_2. For all other t, v is zero. Sketch the waveform and write an equation $v_{(t)}$ using step, ramp and sine functions and determine $V(s)$.

Solution

Waveform is drawn in Fig 2.33

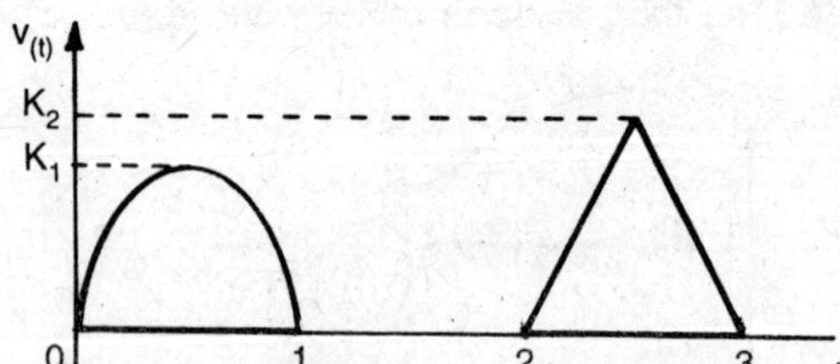

Fig. 2.33

The given half cycle of sine wave of periodic time $T = 2$ seconds and of amplitude K_l, beginning at $t = 0$ seconds is formed by linear combination of following two sine waves

(a) $K_1 \sin \frac{2\pi}{2}(t-0)u(t-0)$; and

(b) $K_1 \sin \frac{2\pi}{2}(t-1)u(t-1)$

Therefore,

$$v_{1(t)} = K_1 \sin \frac{2\pi}{2} t\, u(t) + K_1 \sin \frac{2\pi}{2}(t-1)u(t-1)$$

or

$$v_{1(t)} = K_1 \sin \pi t\, u(t) + K_1 \sin \pi t\,(t-1)u(t-1)$$

Therefore

$$V_1(s) = K_1 \frac{\pi}{s^2 + \pi^2}\left[1 + e^{-s}\right]$$

The given isosceles triangle is a linear combination of following three ramp functions

(a) $2K_2 r(t-2)$;

(b) $-4K_2 r(t-2.5)$;

(c) $2K_2 r(t-3)$

Therefore $v_{2(t)}$ is

$$v_{2(t)} = 2K_2 r(t-2) - 4K_2 r(t-2.5) + 2K_2 r(t-3)$$

Therefore,

$$V_2(s) = \frac{2K_2 e^{-2s}}{s^2} - \frac{4K_2 e^{-2.5s}}{s^2} + \frac{2K_2 e^{-3s}}{s^2}$$

$$= \frac{2K_2 e^{-2s}}{s^2}\left[1 - 2e^{-0.5s} + e^{-s}\right]$$

Therefore,

$$V(s) = V_1(s) + V_2(s)$$

$$= \frac{K_1 \pi}{s^2 + \pi^2}\left[1 + e^{-s}\right] + \frac{2K_2 e^{-2s}}{s^2}\left[1 - 2e^{-0.5s} + e^{-s}\right]$$

Problem No. 2.28 The waveform shown in Fig. 2.34 is that of full wave rectified voltage. The equation for the waveform is sin$t$ from 0 to π, –sin$t$ from π to 2π, etc. Show that the transform of their function is

$$V(s) = \frac{1}{s^2+1}\coth\frac{\pi s}{2}$$

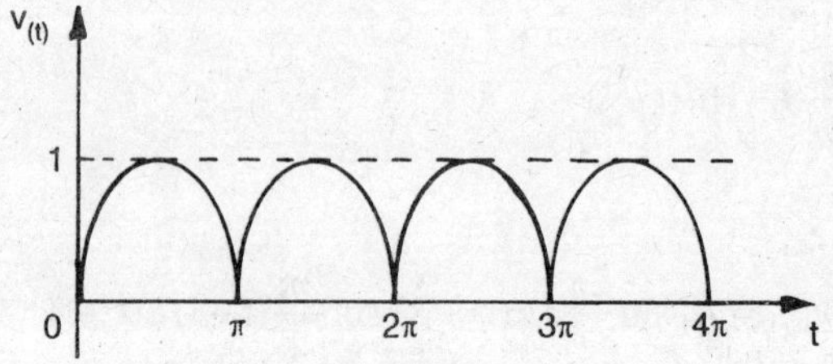

Fig 2.34

Solution A wave which repeats itself has a transform function as given below

$$F(s) = \frac{1}{1-e^{-Ts}} F_1(s)$$

Here T = π (time-period)

Amplitude = 1

$$v_{1(t)} = 1\times\sin\frac{2\pi}{2\pi} t\cdot u(t) + 1\times\sin\frac{2\pi}{2\pi}(t-\pi)u(t-\pi)$$

Note: Refer Problem No. 2.27 where equation for sine wave has been found out.

or $\quad v_{1(t)} = \sin t\cdot u(t) + \sin(t-\pi)\cdot u(t-\pi)$

Therefore,

$$V_1(s) = \frac{1}{1+s^2}\left[1+e^{-\pi s}\right]$$

Therefore,

$$F(s) = V_1(s) = \frac{1}{1+e^{-\pi s}}\left[\frac{1}{1+s^2}\left(1+e^{-\pi s}\right)\right]$$

or $$V(s) = \frac{1}{1+s^2}\left(\frac{1+e^{-\pi s}}{1-e^{-\pi s}}\right)$$

or $$V(s) = \frac{1}{1+s^2}\coth\frac{\pi s}{2}$$

Problem No.2.29 Determine the Laplace transform of the periodic rectified half-sine wave shown in Fig 2.35.

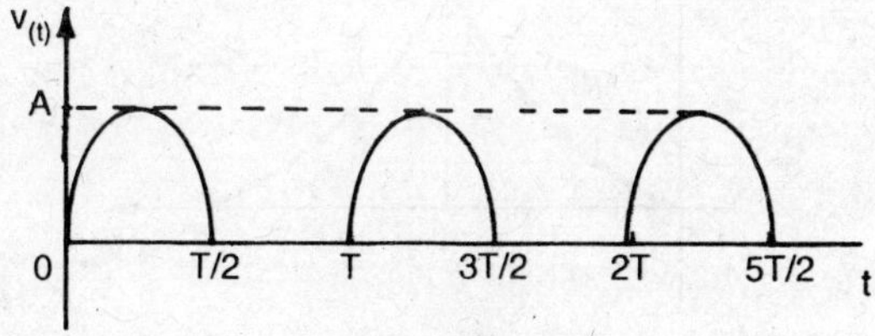

Fig. 2.35

Solution

Laplace transform of the periodic rectified half-sine wave is

$$V(s) = \frac{1}{1+e^{-Ts}} V_1(s)$$

$$V_1(t) = A\sin\frac{2\pi}{T}t \cdot u(t) + A\sin\frac{2\pi}{T}\left(t-\frac{T}{2}\right)\cdot u\left(t-\frac{T}{2}\right)$$

or
$$V_1(s) = \frac{A}{1} \times \frac{2\pi/T}{s^2+\left(\frac{2\pi}{T}\right)^2}\left(1+e^{-Ts/2}\right)$$

substituting the value of $V_1(s)$ in $V(s)$, we get

$$V(s) = \frac{A2\pi/T}{s^2+\left(\frac{2\pi}{T}\right)^2}\left[\frac{1+e^{-Ts/2}}{1-e^{-Ts}}\right]$$

or
$$V(s) = \frac{\frac{2\pi A}{T}}{s^2+\left(\frac{2\pi}{T}\right)^2}\left[\frac{1+e^{-Ts/2}}{\left(1+e^{-Ts/2}\right)\left(1-e^{-Ts/2}\right)}\right]$$

$$V(s) = \frac{\frac{2\pi A}{T}}{s^2+\left(\frac{2\pi}{T}\right)^2}\left[\frac{1}{\left(1-e^{-Ts/2}\right)}\right]$$

Problem No. 2.30 The parabolic pulse shown in Fig 2.36 occurs only once. Write an equation for $v_{(t)}$ using functions multiplied by the appropriate delayed step functions. Also determine Laplace transform

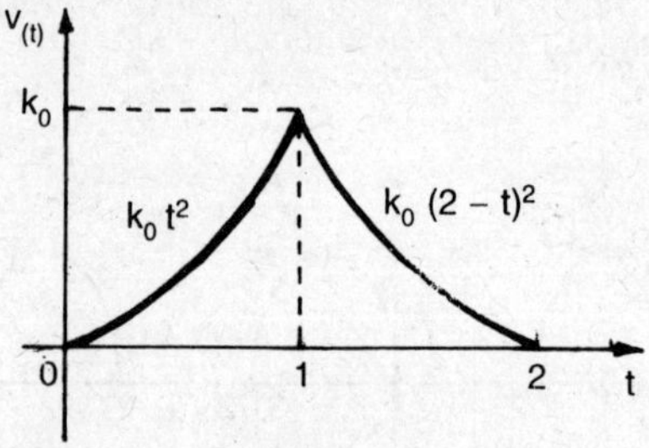

Fig. 2.36

Solution The given waveform has be constructed with the help of three functions as shown in Fig. 2.37

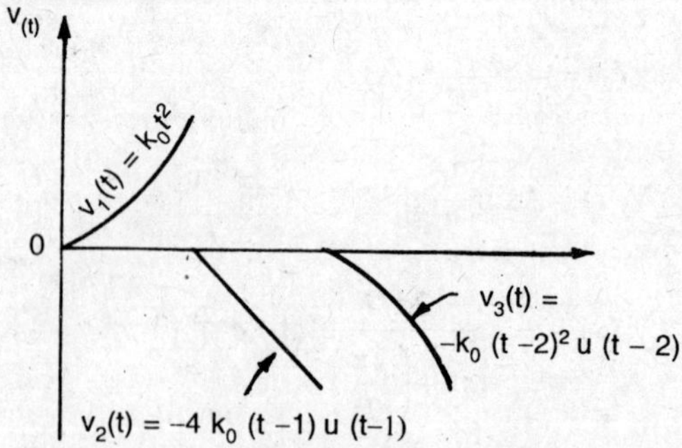

Fig. 2.37

Therefore

$$v_{(t)} = v_{1(t)} + v_{2(t)} + v_{3(t)}$$

or

$$v_{(t)} = K_0 t^2 - 4K_0(t-1)u(t-1) - K_0\,(t-2)^2 u(t-2)$$

Laplace can be written directly and is

$$V(s) = \frac{K_0}{s^3} - \frac{4K_0 e^{-s}}{s^2} - \frac{K_0 e^{-2s}}{s^3}$$

$$= \frac{K_0}{s^3}\left[1 - 4se^{-s} - e^{-2s}\right]$$

Problem No. 2.31 In the network shown in Fig. 2.38, switch is moved from position *'a'* to *'b'* at $t = 0$ At steady state position existing in position *'a'* before $t = 0$, solve for the current $i_{(t)}$ through the circuit using Laplace transform.

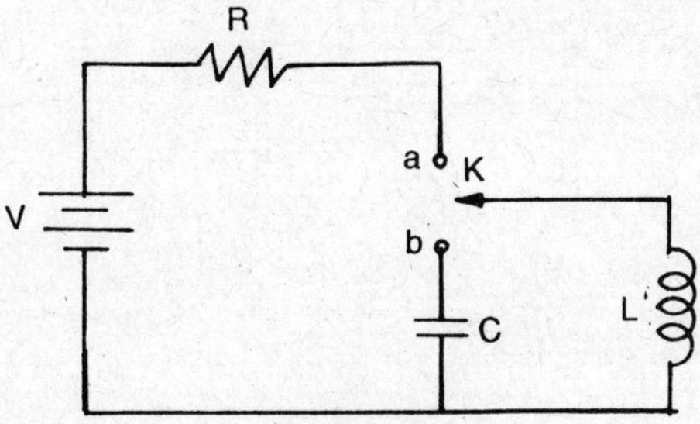

Fig. 2.38

Solution

At position *'a'*

When the key K is at position *'a'* the elements in the circuit are R & L and voltage V. At $t = 0$, the current in the circuit is

$$i_{(0)} = \frac{V}{R}$$

At position *'b'* KVL can be written as

$$L\frac{di}{dt} + \frac{1}{C}\int idt = 0$$

Taking Laplace transform, we get

$$L\left[sI(s) - i_{(0)}\right] + \frac{1}{sC}I(s) = 0$$

Substituting the value of $i_{(0)} = \frac{V}{R}$, we get

$$L\left[sI(s) - \frac{V}{R}\right] + \frac{1}{sC}I(s) = 0$$

or $$sLI(s) + \frac{I(s)}{sC} = \frac{VL}{R}$$

or $$I(s) = \frac{VL/R}{sL + \frac{1}{sC}}$$

or $$I(s) = \frac{\frac{VL}{R}}{L\left(s + \frac{1}{sLC}\right)}$$

or $$I(s) = \frac{\frac{VL}{R}}{\frac{L}{s}\left(s^2 + \frac{1}{LC}\right)}$$

or $$I(s) = \frac{\frac{VLs}{RL}}{s^2 + \frac{1}{LC}}$$

or $$I(s) = \frac{\frac{V}{R}s}{s^2 + \left(\frac{1}{\sqrt{LC}}\right)^2}$$

or $$i_{(t)} = \frac{V}{R} \cdot \cos\left(\frac{1}{\sqrt{LC}}\right)t$$

Problem No 2.32 For the circuit shown in Fig. 2.39, find the current through C using Laplace transform. The switch is closed at $t = 0$ and the initial charge on the capacitor i.e., at $t = 0$ is zero. Sketch the waveform.

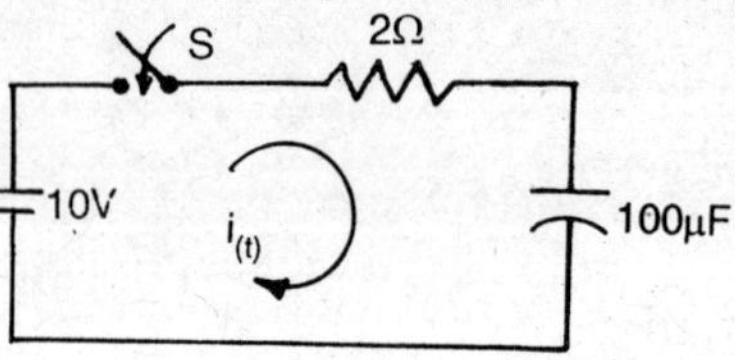

Fig. 2.39

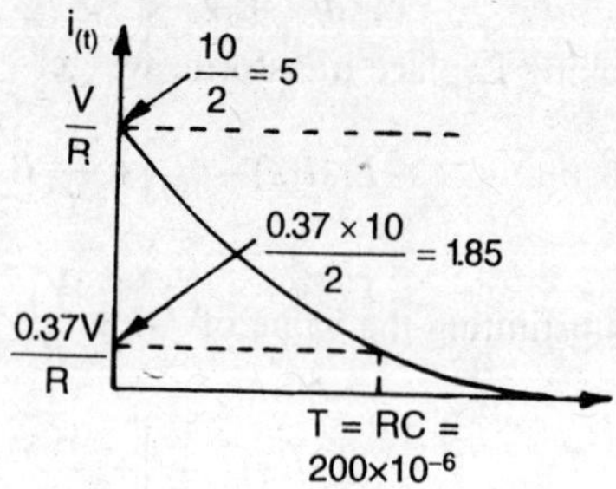

Fig. 2.39 (a)

Solution

By applying KVL, we have

$$Ri_{(t)} + \frac{1}{C}\int_{-\infty}^{t} i_{(t)}\,dt = 10$$

$$Ri_{(t)} + \frac{1}{C}\left[\int_{-\infty}^{0} \frac{dq}{dt}\,dt + \int_{0}^{t} i_{(t)}\,dt\right] = 10$$

Laplace transform gives

$$RI(s)+\frac{1}{C}\left[\frac{q(0-)}{s}+\frac{I(s)}{s}\right]=\frac{10}{s}$$

Since q (0–) = 0 (given)

Therefore,

$$I(s)\left[\frac{1}{sC}+R\right]=\frac{10}{s}$$

or $$I(s)=\frac{10/s}{R+\dfrac{1}{sC}}$$

or $$I(s)=\frac{10/s}{\dfrac{R}{s}\left[s+\dfrac{1}{RC}\right]}$$

or $$I(s)=\frac{10/R}{s+1/RC}$$

Inverse Laplace transform gives

$$i_{(t)}=\frac{10}{R}e^{-t/RC} \qquad ...(1)$$

Substituting the value of R and C, we get

$$i_{(t)}=\frac{10}{2}e^{-t/2\times100\times10^{-6}}$$

or $$i_{(t)}=5e^{-t/200\times10^{-6}}$$

From equation (1)

Since i (0) = $-\dfrac{V}{R}$, we see that particular solution for the given network is

$$i=\frac{V}{R}e^{-t/RC}\text{for } t\geq 0$$
$$= 0 \text{ for } t < 0.$$

Problem No. 2.33 In the network of Fig. 2.40, the switch is closed at t = 0 and there is no initial charge on either of the capacitors. Find the resulting current i as shown in the network. Apply the initial and final value theorems to the s domain current. Justify the result you obtained.

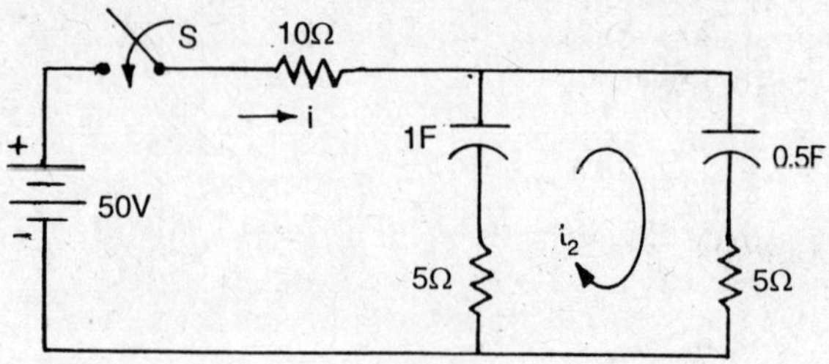

Fig. 2.40

Solution

Let the current flowing in the loop consisting of capacitors and resistance be i_2 as shown in Fig. 2.40.

When the switch is closed at $t = 0$

$$50 = 10i(t) + \frac{1}{1}\int i(t)\,dt + 5i(t) - 5i_2(t) - \frac{1}{1}\int i_2(t)\,dt$$

and $$5i_2(t) + \frac{1}{0.5}\int i_2(t)dt + 5i_2(t) + \frac{1}{1}\int i_2(t)\,dt - 5i_2(t) - \frac{1}{1}\int i_2(t)\,dt = 0$$

Taking Laplace we get

$$10I(s) + \frac{I(s)}{s} + 5I(s) - 5I_2(s) - \frac{I_2(s)}{s} = \frac{50}{s}$$

and $$5I_2(s) + \frac{I_2(s)}{0.5s} + 5I_2(s) + \frac{I_2(s)}{s} - 5I(s) - \frac{I(s)}{s} = 0$$

which gives on further simplification

$$0sI(s) + I(s) + 5sI(s) - 5sI_2(s) - I_2(s) = \frac{50}{s}$$

and $$2.5sI_2(s) + I_2(s) + 2.5sI_2(s) + 0.5I_2(s) - 2.5sI(s) - 0.5I(s) = 0$$

which further gives

$$I(s)(15s+1) - I_2(s)[5s+1] = 50 \qquad ...(1)$$

and $$I_2(s)(1.5+5s) - I(s)[2.5s+0.5] = 0 \qquad ...(2)$$

From equation (2), we get

$$I_2(s) = \frac{(2.5s+0.5)}{(1.5+5s)} I(s)$$

or $$I_2(s) = \frac{(5s+1)}{(10s+3)} I(s)$$

Substituting the value of $I_2(s)$ in equation (1), we get

$$I(s)(15s+1) - \frac{(5s+1)(5s+1)}{(10s+3)} \times I(s) = 50$$

which gives

$$I(s)\left[(15s+1)(10s+3) - (5s+1)(5s+1)\right] = 50(10s+3)$$

or $$I(s) = \frac{50(10s+3)}{125s^2 + 45s + 2}$$

$$= \frac{500(s+0.3)}{125(s^2 + 0.36s + 00.16)}$$

$$= \frac{4(s+0.3)}{s^2 + 0.36s + 0.016}$$

$$I(s) = \frac{4(s+0.3)}{(s+0.308)(s+0.052)}$$

Now

$$\frac{4(s+0.3)}{(s+0.308)(s+0.052)} = \frac{A}{(s+0.308)} + \frac{B}{(s+0.052)}$$

$$A = \left.\frac{4(s+0.3)}{(s+0.052)}\right|_{s=-0.308}$$

or $A = 0.128$

and

$$B = \left.\frac{4(s+0.3)}{(s+0.308)}\right|_{s=-0.052}$$

or $B = 3.875$

Therefore

$$I(s) = \frac{0.128}{(s+0.308)} + \frac{3.875}{(s+0.052)}$$

Therefore,

$$i_{(t)} = 0.128e^{-0.308t} + 3.875e^{-0.052t} \qquad ...(3)$$

Inital Value Theorem

$$i_{(0)} = \underset{s\to\infty}{Lt}\ s\,I(s) = \underset{s\to\infty}{Lt}\left[\frac{0.128s}{s+0.308}+\frac{3.875s}{s+0.052}\right]$$

$$= \underset{s\to\infty}{Lt}\left[\frac{0.128}{\left(1+\dfrac{0.308}{s}\right)}+\frac{3.875}{\left(1+\dfrac{0.052}{s}\right)}\right]$$

which gives

$$i_{(0)} = \frac{0.128}{1+0}+\frac{3.875}{1+0}$$

or

$$i_{(0)} = 0.128+3.875$$
$$= 4.003.$$

Final Value Theorem

$$I(\infty) = \underset{s\to 0}{Lt}\,[sI(s)] = \underset{s\to 0}{Lt}\left[\frac{0.128s}{s+0.308}+\frac{3.827s}{s+0.052}\right]$$
$$= 0$$

These results so obtained, can also be obtained if in equation (3), t is equated to zero and infinity

$$i_{(t)} = 0.128\,e^{-.308t}+3.875\,e^{-0.052t}$$

put $t = 0$

then $i_{(0)} = 0.128\,e^{-(0)}+3.875\,e^{-(0)}$

$$= 0.128+3.875$$
$$= 4.003$$

put $t = \infty$

then $i_{(\infty)} = 0.128\,e^{\infty}+3.875\,e^{\infty}$

$$= 0.$$

Problem No 2.34 In the network of Fig. 2.41, the switch S is initially closed and steady state exists. At $t = 0$ the switch S is opened.

(a) Find $i_{(t)}$ for $t > 0$

(b) Find the voltage across L_1 for $t > 0$

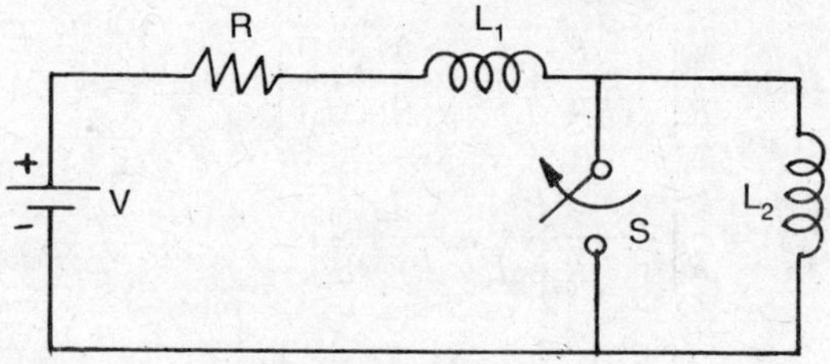

Fig. 2.41

Solution

At the time when switch is close and steady-state exists

$$i_{L_1(0)} = \frac{V}{R}, \text{(inductor } L_2 \text{ is shorted) and } i_{L_2(0)} = 0$$

At $t = 0$ when switch S is open.

$$v_{(t)} = Ri_{(t)} + \frac{L_1 di_{(t)}}{dt} + \frac{L_2 di_{(t)}}{dt}$$

Taking Laplace transform

$$\frac{V}{s} = RI(s) + L_1\left[sI(s) + i_{L1(0)}\right] + L_2\left[sI(s) - i_{L2(0)}\right]$$

or $$\frac{V}{s} = I(s)\left[R + sL_1 + sL_2\right] - L_1 i_{L1(0)}$$

or $$\frac{V}{s} = I(s)\left[R + sL_1 + sL_2\right] - \frac{L_1 V}{R}$$

or $$\frac{V}{s} + \frac{L_1 V}{R} = I(s)\left[R + sL_1 + sL_2\right]$$

or $$\frac{V}{s} + \frac{L_1 V}{R} = I(s)\left[R + sL\right]$$

where $L = L_1 + L_2$

or $$V\left[\frac{1}{s} + \frac{L_1}{R}\right] = I(s)\left[R + sL\right]$$

or $$V\left[\frac{R + L_1}{sR}\right] = I(s)\left[R + sL\right]$$

or $$I(s) = \frac{V}{R}\left[\frac{R + sL_1}{s(R + sL)}\right]$$

or $$I(s) = \frac{V}{R}\left[\frac{R}{s(R+sL)} + \frac{sL_1}{s(R+sL)}\right]$$

or $$I(s) = \frac{V}{R}\left[\frac{1}{s} - \frac{L}{R+sL} + \frac{L_1}{R+sL}\right]$$

or $$I(s) = \frac{V}{R}\left[\frac{1}{s} - \frac{L}{s+R/L} + \frac{L_1/L}{s+R/L}\right]$$

Inverse Laplace transform given

$$i_{(t)} = \frac{V}{R}\left[1 - e^{-\frac{R}{L}t} + \frac{L_1}{L}e^{-\frac{R}{L}t}\right]$$

or $$i_{(t)} = \frac{V}{R}\left[1 - e^{-\frac{R}{L}t}\left(1 - \frac{L_1}{L}\right)\right]$$

or $$i_{(t)} = \frac{V}{R}\left[1 - \frac{L-L_1}{L}e^{-\frac{R}{L}t}\right]$$

Putting $L = L_1 + L_2$, we get

or $$i_{(t)} = \frac{V}{R}\left[1 - \frac{L_1+L_2-L}{L_1+L_2}e^{-\frac{R}{L_1+L_2}t}\right]$$

or $$i_{(t)} = \frac{V}{R}\left[1 - \frac{L_2}{L_1+L_2}e^{-\frac{R}{L_1+L_2}t}\right]$$

differentiating, we get

$$\frac{di_{(t)}}{dt} = \frac{V}{R} - \frac{VL_2}{R(L_1+L_2)}e^{-\frac{R}{L_1+L_2}t}$$

$$\frac{di_{(t)}}{dt} = -\left(-\frac{R}{L_1+L_2}\right)\frac{VL_2}{R(L_1+L_2)}e^{-\frac{R}{L_1+L_2}t}$$

or $$\frac{di_{(t)}}{dt} = \frac{VL_2}{(L_1+L_2)^2}e^{-\frac{R}{L_1+L_2}t}$$

or
$$L_1 \frac{di_{(t)}}{dt} = \frac{V L_1 L_2}{(L_1 + L_2)^2} e^{-\frac{R}{L_1+L_2}t}$$

$$v_{L_1(t)} = \frac{L_1 di_{(t)}}{dt} = \frac{V L_1 L_2}{L^2} e^{-\frac{R}{L}t}$$

where $L = L_1 + L_2$

and $v_{L1(t)}$ = voltage across L_1 at $t > 0$

Problem No. 2.35 Find the current $i_{(t)}$ through the circuit shown shown in Fig. 2.42 using Laplace transform technique. Assume that the capacitor is initially uncharged and switch is initially closed at time $t = 0$.

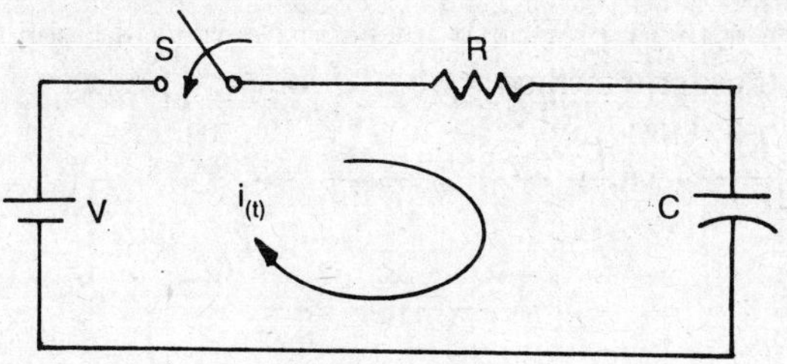

Fig. 2.42

Solution At $t = 0$, there is no charge on the capacitor. Let the switch is closed at $t = 0$. Application of Kirchoff's voltage law at the circuit yields

$$Ri + \frac{1}{C}\int_{-\infty}^{t} i\,dt = V$$

or
$$Ri + \frac{1}{C}\int_{-\infty}^{0} i\,dt + \frac{1}{C}\int_{0}^{t} i\,dt = V$$

$$RI(s) + \frac{1}{C}\left[\frac{q(0+)}{s} + \frac{I(s)}{s}\right] = \frac{V}{s}$$

Now $q\ (0+) = 0$

Therefore, the equation reduces to

$$I(s)\left[R + \frac{1}{sC}\right] = \frac{V}{s}$$

or $$I(s) = \frac{V/R}{s + \frac{1}{RC}}$$

Inverse Laplace gives, the particular solution

$$i_{(t)} = \frac{V}{R} e^{-t/RC}$$

i.e., $$i_{(t)} = \frac{V}{R} e^{-t/RC} \quad \text{for} \quad t \geq 0$$

and $i_{(t)} = 0$ for $t < 0$

Problem No 2.36 In the network shown in Fig. 2.43, the switch *K* is closed at $t = 0$ with the network previously unenergised. Find the loop current $i_1(t)$ using Laplace transform method.

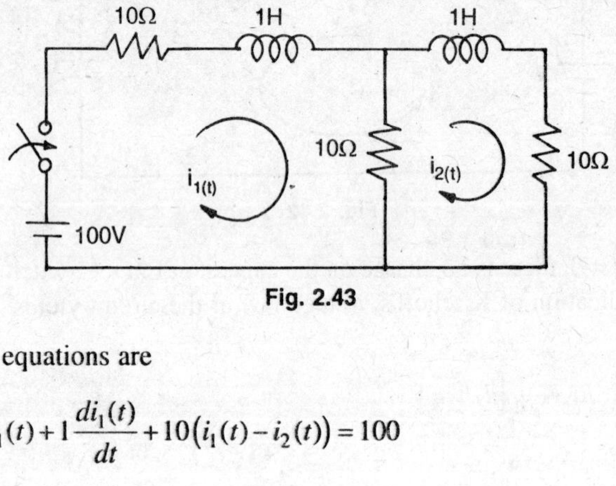

Fig. 2.43

The mesh equations are

$$10i_1(t) + 1\frac{di_1(t)}{dt} + 10(i_1(t) - i_2(t)) = 100 \quad \text{...(1)}$$

and $$1\frac{di_2(t)}{dt} + 10i_2(t) + 10(i_2(t) - i_1(t)) = 0 \quad \text{...(2)}$$

Laplace transform of the equations give

$$10I_1(s) + sI_1(s) + 10(I_1(s) - I_2(s)) = \frac{100}{s} \quad \text{...(3)}$$

and $$sI_2(s) + 10I_2(s) + 10I_2(s) - 10I_1(s) \quad \text{...(4)}$$

Simplification of the equation (3) and (4) give

$$I_1(s)(20 + s) - 10I_2(s) = \frac{100}{s} \quad \text{...(5)}$$

and $$I_2(s)(20 + s) - 10I_1(s) = 0 \quad (6)$$

Multiplying equation (5) by $\frac{(20+s)}{10}$ and adding with equation (6) gives

$$\frac{I_1(s)(20+s)^2}{10} - 10I_1(s) = \frac{100(s+20)}{s\times 10}$$

or $$I_1(s)\left[\frac{(s+20)^2-100}{10}\right] = \frac{100(s+20)}{s}$$

or $$I_1(s)\left[(s+20)^2-100\right] = \frac{100(s+20)}{s}$$

or $$I_1(s)\left[s^2+400-40s-100\right] = \frac{100(s+20)}{s}$$

or $$I_1(s) = \frac{100(s+20)}{s(s^2-40s+300)}$$

or $$I_1(s) = \frac{100(s+20)}{s(s+30)(s+10)}$$

$$I_1(s) = \frac{A}{s} + \frac{B}{s+30} + \frac{C}{s+10}$$

$$A = \left.\frac{100(s+20)}{(s+30)(s+10)}\right|_{s=0} = \frac{100\times 20}{30\times 10} = \frac{20}{3}$$

$$B = \left.\frac{100(s+20)}{s(s+10)}\right|_{s=-30} = \frac{100\times(-10)}{-30\times(-20)} = -\frac{5}{3}$$

$$C = \left.\frac{100(s+20)}{s(s+30)}\right|_{s=-10} = \frac{100\times 10}{-10\times 20} = -5$$

Therefore,

$$I_1(s) = \frac{20/3}{s} + \frac{-5/3}{s+30} + \frac{-5}{s+10}$$

Taking inverse Laplace transform gives

$$i_{1(t)} = \frac{20}{3} - \frac{5}{3}e^{-30t} - 5e^{-10t}$$

Problem No 2.37 The switch K of the network shown in Fig. 2.44 which is initially relaxed, is closed at $t = 0$. Find the current $i_2(t)$ using Laplace transform method.

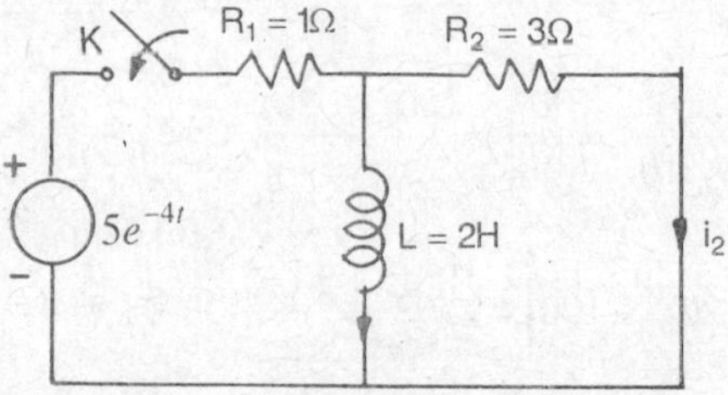

Fig. 2.44

Solution Let i_1 be the current flowing in the branch of network having inductor L = 2H. Therefore, current flowing in R_1 branch is $i_1 + i_2$.

(a) the first mesh consists of R_1, L and voltage source, and

(b) the second mesh consists of R_1, R_2 and voltage source.

Writing mesh equations, we get

$$(i_1 + i_2) \times 1 + 2\frac{di_1}{dt} = 5e^{-4t} \quad \text{...(1)}$$

$$(i_1 + i_2) \times 1 + 3i_2 = 5e^{-4t} \quad \text{...(2)}$$

Taking Laplace transform, we get

$$I_1(s) + I_2(s) + 2sI_1(s) = \frac{5}{s+4}, \text{ and} \quad \text{...(3)}$$

$$I_1(s) + I_2(s) + 3I_2(s) = \frac{5}{s+4} \quad \text{...(4)}$$

Further simplication gives

$$(1 + 2s)I_1(s) + I_2(s) = \frac{5}{s+4} \quad \text{...(5)}$$

and

$$I_1(s) + 4I_2(s) = \frac{5}{s+4} \quad \text{...(6)}$$

Multiplying equation (6) by $(1 + 2s)$ and subtracting from equation (5), we get

$$I_2(s) - 4(1+2s)I_2(s) = \frac{5}{s+4} - \frac{5(1+2s)}{s+4} \quad \text{...(7)}$$

or $$I_2(s)\big(1-4(1+2s)\big)=\frac{5s+20-5-10s}{s+4} \qquad ...(8)$$

or $$I_2(s)(1-4-8s)=\frac{15-5s}{s+4} \qquad ...(9)$$

or $$I_2(s)(-3-8s)=\frac{5(3-s)}{s+4}$$

or $$I_2(s)(8s+3)=\frac{5(s-3)}{s+4}$$

or $$I_2(s)=\frac{5(s-3)}{(s+4)(8s+3)}$$

or $$I_2(s)=\frac{\frac{5}{8}(s-3)}{(s+4)(s+3/8)}$$

or $$I_2(s)=\frac{0.625(s-3)}{(s+4)(s+3/8)}$$

or $$I_2(s)=\frac{A}{s+4}+\frac{B}{s+3/8}$$

$$A=\left.\frac{0.625(s-3)}{s+3/8}\right|_{s=-4}$$

$$=\frac{0.625(-4-3)}{-4+3/8}$$

$$=1.21$$

$$B=\left.\frac{0.625(s-3)}{(s+4)}\right|_{s=-3/8}$$

$$=\frac{0.625\left(-\frac{3}{8}-3\right)}{\left(-\frac{3}{8}+4\right)}$$

$$=-0.58$$

Therefore

$$I_2(s)=\frac{1.21}{(s+4)}+\frac{-0.58}{(s+3/8)}$$

or $$i_2(t) = 1.21e^{-4t} - 0.58e^{-3/8t}$$

Problem No 2.38 A rectangular voltage pulse of 1V magnitude and 2*s* duration as shown in Fig. 2.4 (a) is applied to initially relaxed circuit of Fig. 2.45 (b) at $t = 0$. Use Laplace transform method to find out the voltage $v_c(t)$ across the capacitor for $t > 0$ and sketch the waveform,

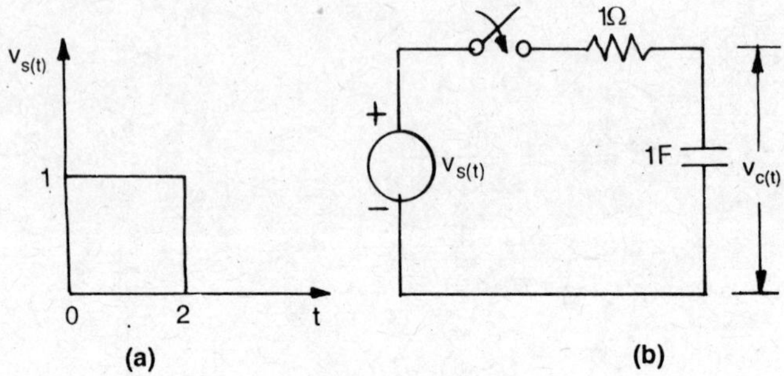

Fig. 2.45

Solution

Refer Problem No 2.17 equation (1) where equation of such type of waveform has been derived. Therefore

$$v_s(t) = 1(u(t) - u(t-a))$$

Here $a = 2$, therefore

$$v_s(t) = u(t) - u(t-2) \qquad \text{...(1)}$$

Let $i_{(t)}$ is the current flowing on closure of switch, there

$$Ri_{(t)} + \frac{1}{C}\int i_{(t)}dt = v_s(t) \qquad \text{...(2)}$$

Now

$$v_{c(t)} = \frac{1}{C}\int i_{(t)}dt \qquad \text{...(3)}$$

$$= \frac{q}{C} \quad \text{as} \quad q = Cv_{c(t)}$$

$$i_{(t)} = \frac{dq}{dt} = C\frac{dv_{c(t)}}{dt} \qquad \text{...(4)}$$

Substituting the values of R, C from the network, $i_{(t)}$ and $\frac{1}{C}\int i_{(t)}dt$ from equation (4) and (3) respectively and that of $v_{s(t)}$ from (1),

we get

$$RC\frac{dv_c(t)}{dt} + v_c(t) = v_s(t)$$

or $$1 \times 1 \times \frac{dv_c(t)}{dt} + v_c(t) = u(t) - u(t-2)$$

or $$\frac{dv_c(t)}{dt} + v_c(t) = u(t) - u(t-2) \qquad \text{...(5)}$$

Laplace transform given

$$[sV_c(s) - V_c(0+)] + V_c(s) = \frac{1}{s} - \frac{e^{-2s}}{s} \qquad \text{...(7)}$$

Now $$V_c(0+) = 0$$

Therefore,

$$V_c(s)(1+s) = \frac{1}{s} - \frac{e^{-2s}}{s}$$

as

$$V_c(s) = \frac{1}{s(1+s)} - \frac{e^{-2s}}{s(1+s)}$$

Inverse Laplace transform given

$$v_c(t) = \left(1 - e^{-t}\right)u(t) - \left(1 - e^{-(t-2)}\right)u(t-2) \qquad \text{...(8)}$$

If we study the expression of $v_c(t)$, we infer that

(a) expression $(1 - e^{-t})u(t)$ gives the value of $v_c(t)$ for $0 < t \leq T$ for +ve step input; and

(b) for the delayed negative step input $-u(t-2)$, the expression of voltage across capacitor is $-(1 - e^{-t-2})u(t-2)$

Expression $v_c(t)$ of equation (8).

Waveform of $v_c(t)$

(a) At $t = 0$, $v_c(t) = 0$

(b) At $t = 2$, $v_c(t) = 1 - e^{-2}$

(c) Between $t = 0$ and $t = 2$, $v_c(t) = 1 - e^{-t}$

(d) At $t = \infty$, $v_c(t) = 0$

(e) Between $t \Rightarrow 2$ and $t = \infty$, $v_c(t) =$ as per equation (8)

The waveform is sketched in Fig. 2.46

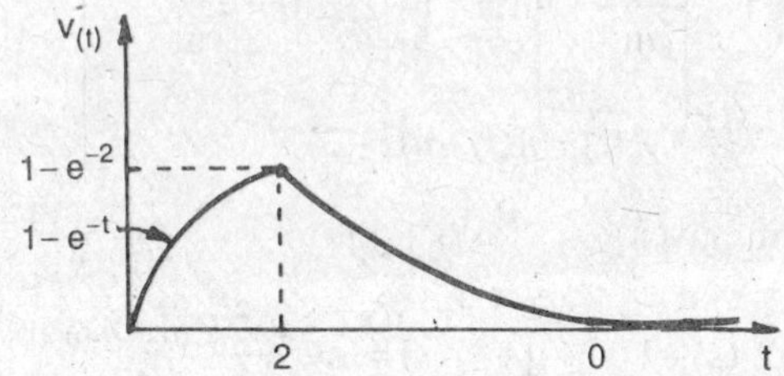

Fig. 2.46

Problem No 2.39 Determine current $i_{(t)}$ in the network shown in Fig. 2.47 and sketch the output waveform for the input voltage as shown in Fig. 2.47 with zero initial conditions.

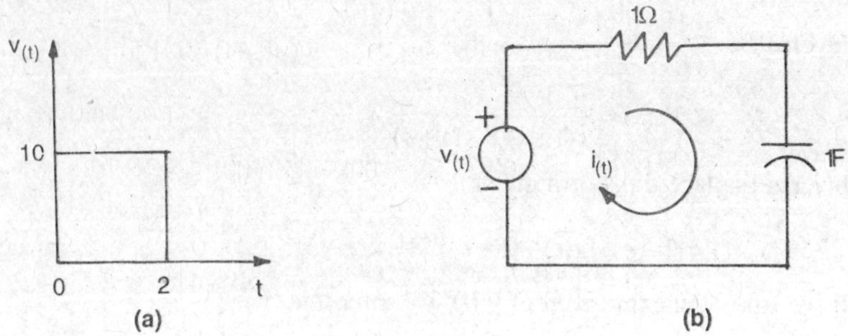

Fig. 2.47

Solution

The mesh equation is

$$Ri_{(t)} + \frac{1}{C}\int_{-\infty}^{t} i_{(t)}dt = v_{(t)}$$

$$Ri_{(t)} + \frac{1}{C}\int_{0}^{t} i_{(t)}dt + v_c(0) = v_{(t)} \quad ...(1)$$

Given

$$v_c(0) = 0;$$

$$v_{(t)} = 10[u(t) - u(t-1)];$$

$$R = 1\Omega; \text{ and}$$

$$C = 1F$$

Then equation (1) becomes

$$i_{(t)} + \int_0^t i_{(t)} dt + 0 = 10[u(t) - u(t-1)]$$

Taking Laplace transform, we get

$$I(s) + \frac{I(s)}{s} = \frac{10}{s} - \frac{10e^{-s}}{s}$$

or $$I(s) = \frac{10(1-e^{-s})}{s+1}$$

or $$I(s) = \frac{10}{s+1} - \frac{10e^{-s}}{s+1}$$

Inverse Laplace given

$$i_{(t)} = 10e^{-t}\, u(t) - 10e^{-(t-1)}\, u(t-1)$$

Waveforms are shown in Fig. 2.48

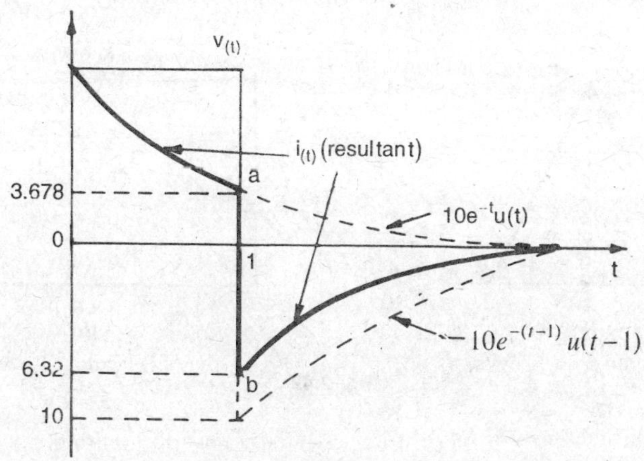

Fig. 2.48

The expression for current for various conditions of time is thus

$$i_{(t)} = 10e^{-t}\text{ ; for } 0 < t < 1 \text{ and}$$

$$i_{(t)} = -10e^{-(t-1)} \quad \text{for } t > 1$$

At $t = 0$

$$i_{(t)} = 10e^{-t}$$

$$= 10e^{-0}$$
$$= 10$$

At $t = 1$

$$i_{(t)} = 10e^{-1} = 3.678 \text{ on the +ve part of } y\text{-axis; and}$$

$$i_{(t)} = -10e^{-(t-1)}$$

$$= -10e^{-(1-1)}$$
$$= -10$$

However, the resultant curve will start at

$$= -10 + 3.678$$
$$= -6.32 \text{ on the –ve part of } y\text{-axis}$$

The resultant waveform has been shown with bold lines in Fig. 2.48.

Problem No 2.40 A triangular pulse of Fig. 2.49 (a) is applied to as input to the RL series circuit of Fig. 2.49 (b). Determine the current response $i_{(t)}$.

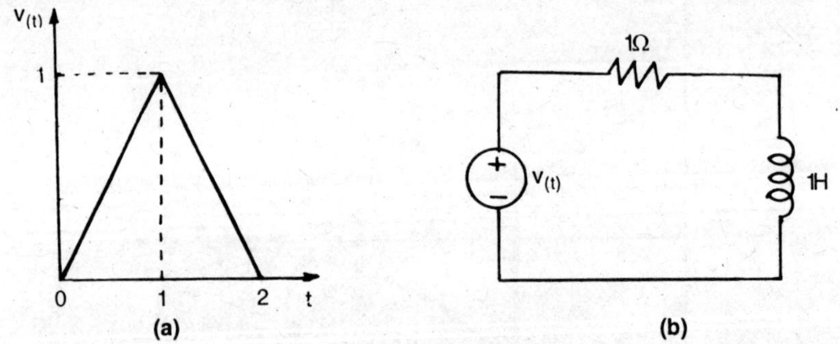

Fig. 2.49

Solution

Writing mesh equation by help of KVL

$$Ri(t) + L\frac{di(t)}{dt} = v_{(t)}$$

Taking Laplace transform, we get

$$RI(s)+sLI(s)+Li_{(0+)}=V(s)$$

Since $i_{(0+)}=0$

$$I(s)[R+sL]=V(s) \qquad ...(1)$$

Now $V(s)$ can be found as given in equation (1) of Problem No. 2.8. In this equation (1), $a=1$. Therefore,

$$V(s)=\frac{1+e^{-2s}-2e^{-s}}{s^2}$$

Therefore, equation (1) can be written as

$$I(s)[R+sL]=\frac{1+e^{-2s}-2e^{-s}}{s^2}$$

or $$I(s)=\frac{1+e^{-2s}-2e^{-s}}{s^2(R+sL)}$$

Substituting value of R and L, we get

$$I(s)=\frac{1+e^{-2s}-2e^{-s}}{s^2(s+1)}$$

Let us find out Inverse Laplace transform of $\frac{1}{s^2(s+1)}$

$$L^{-1}=\int_0^t\left(\int_0^t L^{-1}\left(\frac{1}{s+1}\right)\right)dt\,dt$$

Note: Each integration corresponds to a term $\frac{1}{s}$.

Therefore,

$$L^{-1}=\int_0^t\left(\int_0^t e^{-t}dt\right)dt=t+e^{-t}-1$$

Therefore, $i_{(t)}=L^{-1}I(s)=(t+e^{-t}-1)[u(t)-2u(t-1)+u(t-2)]$

Problem No. 2.41 A relay having a coil of inductance L and resistance $5K\Omega$ is connected in series to a battery of 50V through the switch S as shown in

Fig. 2.50. Minimum operating current of the relay in $7mA$. After closing switch S at $t = 0$, the relay operates after 0.2 second. Determine the value of the inductance and time constant. Then determine the expression of current through the relay.

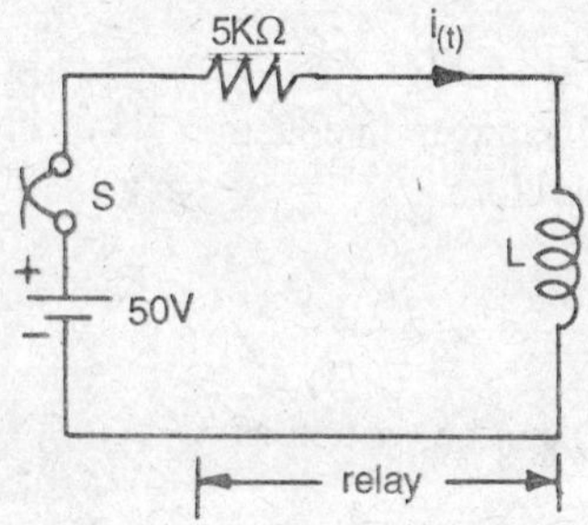

Fig. 2.50

Solution

Writing mesh equation with the help of KVL, we get

$$Ri(t) + L\frac{di(t)}{dt} = v_{(t)}$$

Taking Laplace we get

$$RI(s) + sLI(s) + Li_{(0+)} = \frac{V}{s}$$

Since $i_{(0+)} = 0$

$$I(s) = \frac{V}{sL(s + R/L)}$$

$$= \frac{V}{L}\left[\frac{A}{s} + \frac{B}{s + R/L}\right]$$

$$A = \frac{V}{L(s + R/L)}\Big|_{s=0} = \frac{V}{R}$$

$$B = \frac{V}{sL}\Big|_{s=-R/L} = -\frac{V}{R}$$

$$I(s) = \left[\frac{V/R}{s} + \frac{-V/R}{s + R/L}\right]$$

or $$i_{(t)} = L^{-1}\frac{V}{R}\left[\frac{1}{s}-\frac{1}{s+R/L}\right]$$

$$i_{(t)} = \frac{V}{R}\left(1-e^{-Rt/L}\right) \qquad ...(1)$$

Substituting $i_{(t)} = 7mA = 7\times10^{-3}\,A$

$$R = 5K\Omega = 5000\Omega$$
$$V = 50V$$
$$t = 0.2 \text{ sec.}$$

We get

$$7\times10^{-3} = \frac{50}{5000}\left(1-e^{-5000\times0.2/L}\right)$$

which gives

$$e^{-5000\times0.2/L} = 0.3$$

or

$$\frac{-1000}{L}\log_{10} e = \log_{10} 0.3$$

$$-\frac{434.3}{L} = \log_{10} 0.3$$

or $$L = \frac{-434.3}{\log_{10} 0.3}$$

$$= 831 \text{ H}$$

Time Constant $= \frac{L}{R} = \frac{831}{5000} = 0.1662$ sec.

Expression for current is derived in equation (1).

At $t = 0,\ i_{(t)} = 0$

$$t = \infty,\ i_{(t)} = \frac{V}{R} = \text{steady state value of } i_{(t)}$$

$$t = \frac{L}{R},\quad i_{(t)} = T$$

At $t = L/R$

$$i_{(t)} = \frac{V}{R}\left(1 - e^{-1}\right)$$

$$= 0.63\frac{V}{R}$$

The waveform is shown in Fig. 2.51

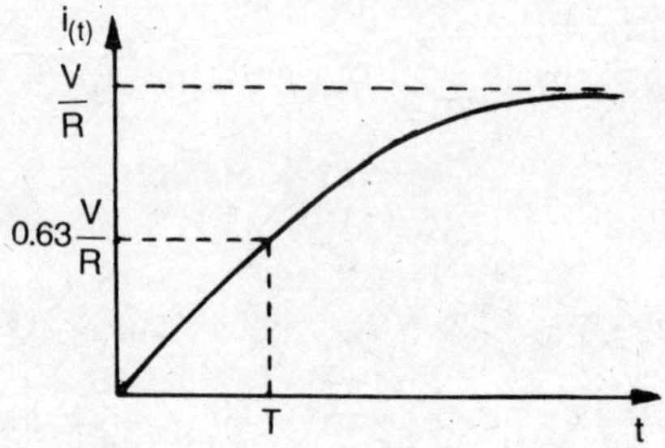

Fig. 2.51

Problem No. 2.42 The impulse response of a network is given as $(e^{-t} + e^{-2t})$. Find the transfer function. Determine the input excitation required to produce an output response as te^{-2t}.

Solution Let the response be

$$v_{(t)} = (e^{-t} + e^{-2t})u_{(t)}$$

The system transfer function

$$H(s) = \frac{V_t(s)}{V_i(s)} = L(e^{-t} + e^{-2t})u_{(t)}$$

$$= \frac{1}{s+1} + \frac{1}{s+2} = \frac{2s+3}{(s+1)(s+2)}$$

Also, in order to have response of te^{-2t} we find the Laplace transform of te^{-2t}.

$$V_t(s) = L\left[te^{-2t}\right] = \frac{1}{(s+2)^2}$$

Now $$V_i(s) = \frac{V_t(s)}{H(s)}$$

where $V_i(s)$ is the unknown excitation which has to be determined. Therefore,

$$V_i(s) = \frac{1/(s+2)^2}{2s+3/(s+1)(s+2)}$$

$$= \frac{1}{(s+2)^2} \times \frac{(s+1)(s+2)}{(2s+3)}$$

$$= \frac{(s+1)}{(2s+3)(s+2)}$$

$$= \frac{s+1}{2(s+1.5)(s+2)}$$

$$V_i(s) = \frac{A}{(s+1.5)} + \frac{B}{(s+2)}$$

$$A = \left.\frac{s+1}{2(s+2)}\right|_{s=-1.5} = \frac{-1.5+1}{2(-1.5+2)} = -0.5$$

and $$B = \left.\frac{s+1}{2(s+1.5)}\right|_{s=-2}$$

$$= \frac{-2+1}{2(-2+1.5)}$$

$$= \frac{-1}{-1}$$

$$= 1.$$

Therefore

$$V_i(s) = \frac{1}{s+2} = \frac{0.5}{s+1.5}$$

Taking inverse Laplace, we get

$$V_i(t) = \left(e^{-2t} - 0.5e^{-1.5t}\right)u_{(t)}$$

Problem No. 2.43. Tests on a certain newtork showed that the current output was $i_{(t)} = -2e^{-t} + 4e^{-3t}$ when a unit voltage was suddenly applied to the input terminals at $t = 0$. What voltage must be applied to give an output current of $i_{(t)} = 2e^{-t}$, if the network remains in the same form as the previous test?

Solution

This is similar to Problem No 2.42. A very short solution is given below

$$H(s) = \frac{I(s)}{V_i(s)}$$

$$\frac{H(s)}{s} = L\left(-2e^{-t} + 4e^{-3t}\right)$$

$$= \frac{-2}{s+1} + \frac{4}{s+3}$$

$$= \frac{-2s-6+4s+4}{(s+1)(s+3)}$$

$$= \frac{2s-2}{(s+1)(s+3)}$$

Therefore,

$$H(s) = \frac{2s(s-1)}{(s+1)(s+3)}$$

Required output current

$$I(s) = L2e^{-t} = \frac{2}{s+1}$$

$$\therefore \quad V_i(s) = \frac{I(s)}{H(s)} = \frac{2/s+1}{2s(s-1)/(s+1)(s+3)}$$

$$= \frac{2}{s+1} \times \frac{(s+1)(s+3)}{2s(s-1)}$$

or $$V_i(s) = \frac{s+3}{s(s-1)} = \frac{A}{s} + \frac{B}{s-1}$$

where

$$A = \left.\frac{s+3}{s-1}\right|_{s=0} = -3$$

$$B = \left.\frac{s+3}{s}\right|_{s=1} = 4$$

or $$V_i(s) = \frac{-3}{s} + \frac{4}{s-1}$$

or $$v_i(t) = \left(-3+4e^{t}\right)u_{(t)}$$

$$= \left(4e^{t}-3\right)u_{(t)}$$

Problem No 2.44 The current in a resistive-inductive circuit is given as $i_{(t)} = -5e^{-10t}$. Find

(a) time constant

(b) initial rate of change of current

(c) $i_{(t)}$ at t = 0.4s

(d) initial value of inductor current

Solution

(a) $\dfrac{L}{R} = T = \dfrac{1}{10} = 0.1s$

(b) $\dfrac{di(t)}{dt} = 50e^{-10t}$;

$$\left.\frac{di(t)}{dt}\right|_{t=0} = 50\,A/\text{sec}$$

(c) $i_{(t)}\big|_{t=0.4} = -5e^{-10\times0.4}$

$$= -0.1A$$

(d) From inspection of $i_{(t)}$ it is –5A.

Problem No 2.45 The unit step response of a network is given by $\left(1-e^{-bt}\right)$. Determine the impulse response $h(t)$ of the network.

Solution

We know that step response is integral of impulse response:

$$1-e^{-bt} = \int h(t)\,d(t)$$

$$\therefore \quad h(t) = \frac{d}{dt}\left(1-e^{-bt}\right)$$

$$= -e^{-bt}\times(-b)$$

$$= be^{-bt}\,.$$

Problem No 2.46 In the circuit shown in Fig. 2.52, Capacitor C_1 is initially charged to a voltage V_0. At time t = 0, switch K is closed. Obtain the expression for

(a) current and (b) change across C_1 (c) voltage across C_2 as a function of time.

If $C_1 = C_2 = 1\mu F$, $V_0 = 10V$ and $R = 10\Omega$, calculate after $t = 10\mu s$,

(a) current i (b) voltage across R (c) Charge across C_1 (d) voltage across C_2.

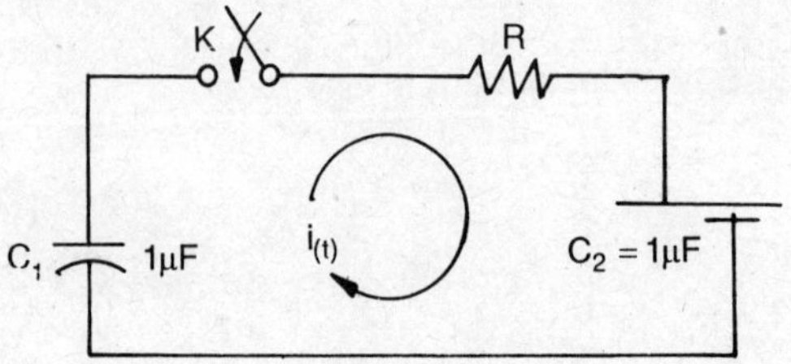

Fig. 2.52

Solution

Let $i_{(t)}$ be the mesh current flowing as shown. Writing mesh equation at instance when K is closed

$$Ri_{(t)} + \frac{1}{C_2}\int i_{(t)}dt + \frac{1}{C_1}\int i_{(t)}dt = V_0 \qquad ...(1)$$

(a) Expression for Current

Simplying equation (1), we get

$$Ri_{(t)} + \left(\frac{1}{C_2} + \frac{1}{C_1}\right)\int i_{(t)}dt = V_0$$

or

$$Ri_{(t)} + \frac{1}{C}\int i_{(t)}dt = 0 \qquad ...(2)$$

where

$$\frac{1}{C} = \frac{1}{C_1} + \frac{1}{C_2}$$

Taking Laplace transform, we get

$$RI(s) + \frac{I(s)}{C} = \frac{V_0}{s}$$

$$I(s) = \frac{V_0}{s\left(\frac{1}{sC} + R\right)}$$

or $$I(s) = \frac{V_0}{R}\left(\frac{1}{s+\frac{1}{RC}}\right)$$

or $$i_{(t)} = \frac{V}{R}e^{-t/RC}$$

Current

$$i_{(t)} = \frac{10}{10}e^{-t/10\times0.5\times10^{-6}}$$

because

$$\frac{1}{C} = \frac{1}{C_1} + \frac{1}{C_2} = \frac{1}{1} + \frac{1}{1} = 2$$

Therefore

$$C = 0.5\mu F; \text{ and}$$

given $$t = 10\mu S$$

Therefore

$$i_{(t)} = \frac{10}{10}e^{-\frac{10\times10^{-6}}{10\times0.5\times10^{-6}}}$$
$$= 0.135\ A$$

(b) Voltage across *R*

$$V_R = iR = 0.135 \times 10 = 1.35V$$

(c) Expression for Charge across C_1 The general solution is

$$q_1 = \int i_{(t)}dt$$

or $$q_1 = \int \frac{V_0}{R}e^{-t/RC}dt$$

or $$q_1 = \frac{V_0}{R}\left(\frac{e^{-t/RC}}{-1/RC}\right) + K$$

or $$q_1 = -V_0Ce^{-t/RC} + K \qquad ...(3)$$

To find particular solution; we know that at

$$t = 0, \quad q_1 = -C_1 V_0$$

Therefore, equation (3) becomes

$$-C_1 V_0 = -V_0 C e^0 + K$$

or $$-C_1 V_0 = -V_0 C + K$$

or $$K = -C_1 V_0 + V_0 C$$

Substituting in equation (3), we get

$$q_1 = -V_0 C e^{-t/RC} - C_1 V_0 + V_0 C$$

or $$q_1 = V_0 C(1 - e^{-t/RC}) - C_1 V_0 \qquad ...(4)$$

(d) Expression for charge on C_2

$$q_2 = \int i_{(t)} dt$$

or $$q_2 = \int \frac{V_0}{R} e^{-t/RC}$$

or $$q_2 = \frac{V_0}{R}\left(\frac{e^{-t/RC}}{-1/RC}\right) + K$$

or $$q_2 = -V_0 C e^{-t/RC} + K \qquad ...(5)$$

At $t = 0$, $q_2 = 0$

Therefore,

$$0 = -\frac{V_0}{R} \times RC e^{-t/RC} + K$$

or $$0 = V_0 C e^{-t/RC} + K$$

or $$K = V_0 C e^0$$

$$K = V_0 C$$

Therefore equation (5) becomes

$$q_2 = -V_0 C e^{-t/RC} + V_0 C$$

or $$q_2 = V_0 C(1 - e^{-t/RC}) \qquad ...(6)$$

Value of q_2 at $t = 10\mu S$

Substituting

$$V_0 = 10$$

$$C = 0.5 \times 10^{-6} F \qquad \left[\text{because } \frac{1}{C} = \frac{1}{C_1} + \frac{1}{C_2}\right]$$

$$R = 10\Omega$$

$$t = 10 \times 10^{-6} s$$

we get

$$q_2 = 10 \times 0.5 \times 10^{-6}\left(1 - e^{-\frac{10\times10^{-6}}{10\times0.5\times10^{-6}}}\right)$$

$$= 5 \times 10^{-6}\left(1 - e^{-2}\right)$$

$$= 5 \times 10^{-6}(1 - 0.135)$$

$$= 4.325 \times 10^{-6} C$$

$$= 4.325 \mu C$$

Charge on q_1 at t =10 μS

At $t = 0,\ Q = C_1 V_0 = 1 \times 10^{-6} \times 10 = 10 \mu C$

Therefore, at $t = 10\ \mu S$

$$q_1 = Q - q_2$$

$$= 10 - 0.4325$$

$$= 5.675\ \mu C$$

Problem No. 2.47 Determine the initial and final value of the function $f_{(t)}$ where Laplace transform is

$$F(s) = \frac{1}{s(s+a)^2}$$

Solution

(a) Initial Value

$$f_{(0+)} = \underset{s\to\infty}{Lt}\ s\,F(s)$$

$$= \underset{s\to\infty}{Lt}\ s \times \frac{1}{s(s+a)^2}$$

$$= \underset{s\to\infty}{Lt}\ \frac{1}{(s+a)^2}$$

$$= \underset{s\to\infty}{Lt}\ \frac{1}{s^2+2as+a^2}$$

$$= \underset{s\to\infty}{Lt}\ \frac{1/s^2}{1+\frac{2a}{s}+\frac{a^2}{s^2}}$$

$$= 0$$

(b) Final Value

$$f(\infty) = \underset{s\to 0}{Lt}\ s\,F(s)$$

$$= \underset{s\to 0}{Lt}\ s \times \frac{1}{s(s+a)^2}$$

$$= \underset{s\to 0}{Lt}\ \frac{1}{(s+a)^2}$$

$$= \frac{1}{(0+a)^2}$$

$$= \frac{1}{a^2}$$

Problem No 2.48 A series *RLC* circuit with $R = 2\Omega$, $L = 1$ Henry and $C = 0.5$ farad is excited with a direct voltage of $50V$ at $t = 0$. Assuming the circuit to be initially relaxed, compute the current in the circuit using Laplace transform.

Solution

When the switch is closed, by applying Kirchoff voltage law to the circuit depicted in Fig. 2.53, we have

$$2i_{(t)} + \frac{di_{(t)}}{dt} + \frac{1}{0.5}\int i_{(t)}dt = 50$$

Taking Laplace transform, we get

$$2I(s)+\left[sI(s)-i_{(0)}\right]+2I(s)+\frac{q_0}{s}=\frac{50}{s}$$

Since the initial current in the inductor and initial charge on the capacitor is zero, the above equation reduces to

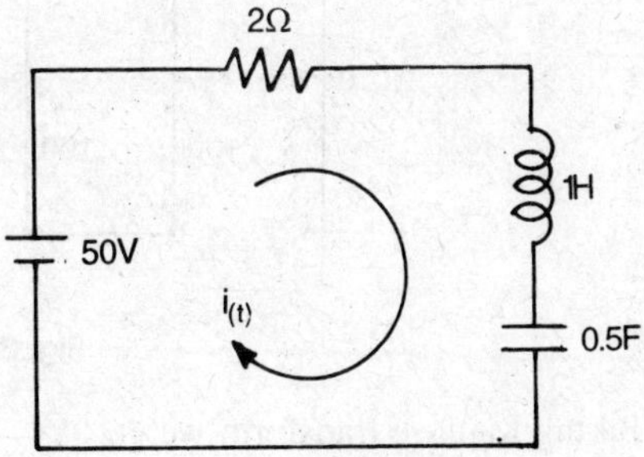

Fig. 2.53

$$2I(s)+sI(s)+\frac{2I(s)}{s}=\frac{50}{s}$$

$$I(s)\left[2+s+\frac{2}{s}\right]=\frac{50}{s}$$

$$I(s)=\frac{50}{s^2+2s+2}$$

$$I(s)=\frac{50}{s^2+2s+1+1}$$

$$=\frac{50}{(s+1)^2+1}$$

$$=50\left[\frac{1}{(s+1)^2+1^2}\right]$$

as $i_{(t)}=50e^{-t}\sin t$ $\left[\text{became } L^{-1}\dfrac{\omega}{(s+a)^2+\omega^2}=e^{-at}\sin\omega t\right]$

Problem No 2.49 A non-sinusoidal voltage of waveform as shown in Fig. 2.54 is applied across a $R-L$ series circuit with $R = 10\Omega$ and $L = 0.1$ henry. Determine the current $i_{(t)}$

Solution

Mesh equation for an $R-L$ circuit is

$$Ri(t)+L\frac{di(t)}{dt}=v_{i(t)}$$

$$10i(t)+0.1\frac{di(t)}{dt}=v_{i(t)}$$

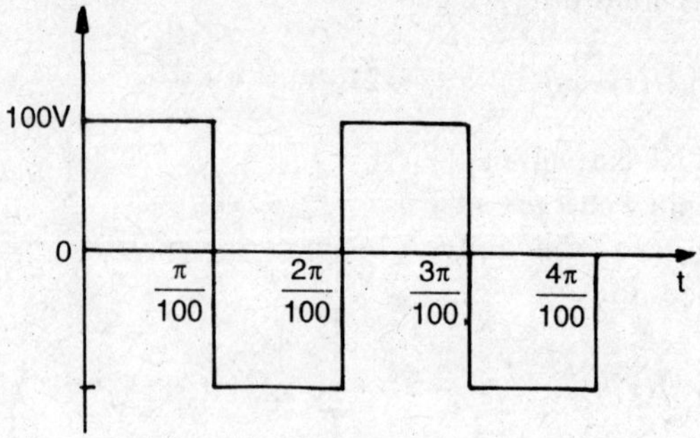

Fig. 2.54

Taking Laplace transform we get

$$10I(s) + 0.1sI(s) - i_{(0)} = V_i(s)$$

Since $i_{(0)} = 0$, therefore,

The equation of the voltage wave is

$$v_{i(t)} = 100u(t) - 200\,u(t-a) + 200\,u(t-2a) - 200\,u(t-3a)$$

$$+100\,u(t-4a)\cdot \qquad \text{where } a = \frac{\pi}{100} = \frac{314}{100} = 0.0314$$

Laplace transform is

$$V_i(s) = \frac{100}{s} - \frac{200}{s}e^{-as} + \frac{200}{s}e^{-2as} - \frac{200}{s}e^{-3as} + \frac{100}{s}e^{-4as}$$

$$= \frac{100}{s}\left[1 - 2e^{-as} + 2e^{-2as} - 2e^{-3as} + e^{-4as}\right]$$

Substituting the value of $V_i(s)$ in equation (1), we get

$$I(s) = \frac{100}{s(10+0.1s)}\left[1 - 2e^{-as} + 2e^{-2as} - 2e^{-3as} + e^{-4as}\right]$$

or

$$I(s) = \frac{1000}{s(s+100)}\left[1 - 2e^{-as} + 2e^{-2as} - 2e^{-3as} + e^{-4as}\right]$$

or

$$I(s) = 1000\left[\frac{1}{s(s+100)} - \frac{2e^{-as}}{s(s+100)} + \frac{2e^{-2as}}{s(s+100)} - \frac{2e^{-3as}}{s(s+100)} + \frac{e^{-4as}}{s(s+100)}\right]$$

or $I(s) = 1000\left[\frac{0.01}{s} - \frac{0.01}{s+100} - 2e^{-as}\left(\frac{0.01}{s} - \frac{0.01}{s+100}\right) + 2e^{-2as}\right.$

$$\left.\left(\frac{0.01}{s} - \frac{0.01}{s+100}\right) - 2e^{-3as}\left(\frac{0.01}{s} - \frac{0.01}{s+100}\right) + 2e^{-4as}\left(\frac{0.01}{s} - \frac{0.01}{s+100}\right)\right]$$

$$= 10\left[\frac{1}{s} - \frac{1}{s+100} - 2e^{-as}\left(\frac{0.01}{s} - \frac{0.01}{s+100}\right) + 2e^{-2as}\left(\frac{0.01}{s} - \frac{0.01}{s+100}\right)\right.$$

$$\left.-2e^{-3as}\left(\frac{0.01}{s} - \frac{0.01}{s+100}\right) + 2e^{-4as}\left(\frac{0.01}{s} - \frac{0.01}{s+100}\right)\right]$$

Inverse Laplace transform gives

$$i_{(t)} = 10\left[u_{(t)} - e^{-100t}u_{(t)} - 2(u(t-a) - e^{-100(t-a)}u(t-a)\right.$$

$$+2\left(u(t-2a) - e^{-100(t-2a)}u(t-2a)\right) - 2\left(u(t-3a) - e^{-100(t-3a)}u(t-3a)\right)$$

$$\left.+\left(u(t-4a) - e^{-100(t-4a)}u(t-4a)\right)\right]$$

or

$$i_{(t)} = 10\left[u_{(t)}(1 - e^{-100t}) - 2u(t-a)(1 - e^{-100(t-a)})\right.$$

$$+2u(t-2a)(1 - e^{-100(t-2a)}) - 2u(t-3a)(1 - e^{-100(t-3a)})$$

$$+u(t-4a)(1 - e^{-100(t-4a)})$$

where $a = \frac{\pi}{100} = 0.0314$

Problem No 2.50 The waveform shown in Fig. 2.55 is known as staircase

(a) Assuming that the staircase is not repeated, write an equation for it in terms of unit step functions.

(b) If this voltage is applied to an *RL* series circuit with $R = 2$ ohms and $L = 1$ henry, find an expression for the resulting current $i_{(t)}$ in the circuit.

Solution Expression of such type of waveform has been drawn in Problem No 2.14.

Therefore, $v_{(t)}$

$$v_{(t)} = u(t-2) + u(t-4) + u(t-6) + u(t-8) - 4u(t-10)$$

and
$$V(s) = \frac{e^{-2s}}{s} + \frac{e^{-4s}}{s} + \frac{e^{-6s}}{s} + \frac{e^{-8s}}{s} - \frac{e^{-10s}}{s}$$

$$= \frac{1}{s}\left[e^{-s} + e^{-4s} + e^{-6s} + e^{-8s} - 4e^{-10s}\right]$$

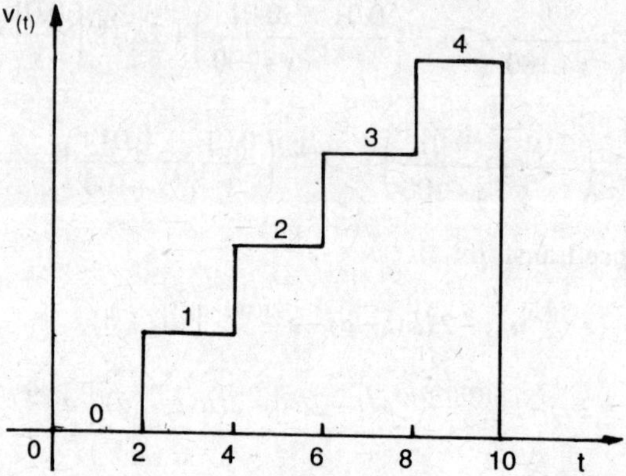

Fig. 2.55

Writing mesh equation for *RL* circuit

$$I(s) = \frac{V(s)}{(R + sL)}$$

$$= \frac{V(s)}{(2 + s)}$$

$$= \frac{V(s)}{(s + 2)}$$

Substituting value of *V*(*s*), we get

$$I(s) = \frac{1}{s(s+2)}\left[e^{-2s} + e^{-4s} + e^{-6s} + e^{-8s} - 4e^{-10s}\right]$$

$$= e^{-2s}\left[\frac{0.5}{s} - \frac{0.5}{s+2}\right] + e^{-4s}\left[\frac{0.5}{s} - \frac{0.5}{s+2}\right] + e^{-6s}\left[\frac{0.5}{s} - \frac{0.5}{s+2}\right]$$

$$+ e^{-8s}\left[\frac{0.5}{s} - \frac{0.5}{s+2}\right] - 4e^{-10s}\left[\frac{0.5}{s} - \frac{0.5}{s+2}\right]$$

Taking inverse Laplace transform, we get

$$i_{(t)} = \frac{1}{2}\Big[\{u(t-2) - e^{-2(t-2)}u(t-2)\} + \{u(t-4) - e^{-2(t-4)}u(t-4)\}$$

$$+\{u(t-6) - e^{-2(t-6)}u(t-6)\} - 4\{u(t-8) - e^{-2(t-8)}u(t-8)\}\Big]$$

or

$$i_{(t)} = \frac{1}{2}\Big[u(t-2)(1-e^{-2(t-2)}) + u(t-4)(1-e^{-2(t-4)}) + u(t-6)(1-e^{-2(t-6)})$$

$$+ 4u(t-8)(1-e^{-2(t-8)})\Big]$$

Problem No 2.51 If the capacitor is uncharged and the inductor current zero at $t = 0$ in the network shown in Fig. 2.56, show that the transform of generator current is

$$I(s) = \frac{2(s^2+s+1)}{(s^2+2s+5)(s^2+2s+2)}$$

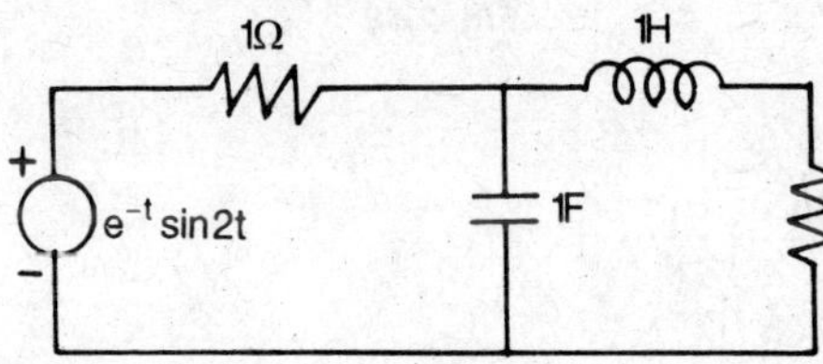

Fig. 2.56

Solution

$$v_{(t)} = e^{-t}\sin 2t$$

Therefore

$$V(s) = \frac{2}{(s+1)^2+2^2} \quad \left[\text{became} \quad Le^{-at}\sin\omega t = \frac{\omega}{(s+a)^2+\omega^2}\right]$$

Using the impedence notation, the circuit is drawn with values marked in Fig. 2.57.

Let $\quad Z_i(s) = Z_L(s) + Z_{R2}(s)$

$$= S + 1$$

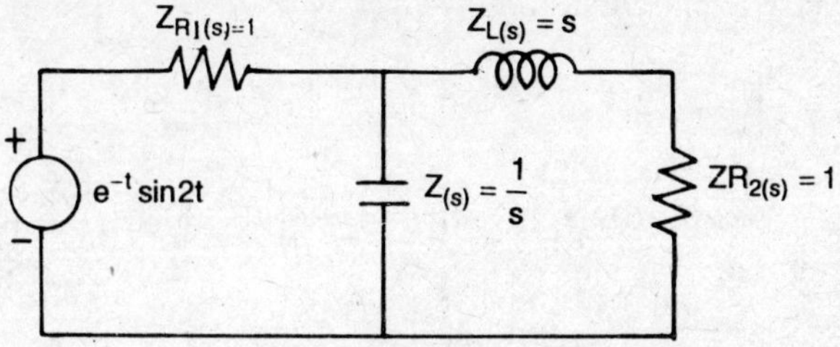

Fig. 2.57

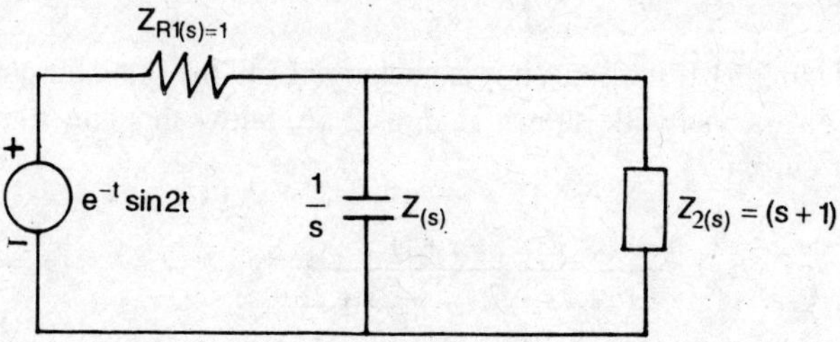

Fig. 2.58

There $\quad Y_1(s) = \dfrac{1}{s+1}$

Let $\quad Y_2(s) = Y_1(s) + Yc(s)$

$$Y_1(s) = \frac{1}{s+1} + 1 \qquad \left[\text{became } \; Yc(s) = \frac{1}{Zc(s)} = \frac{1}{1/s} = s\right]$$

$$= \frac{1+s^2+s}{s+1}$$

$$= \frac{s^2+s+1}{s+1}$$

Let $\quad Z(s) = Z_{R1}(s) + Z_{R2}(s)$

Now

$Z_{R1}(s) = 1;$ and

$$Z_2(s) = \frac{1}{Y_2(s)} = \frac{1}{s^2+s+1/s+1} = \frac{s+1}{s^2+s+1}$$

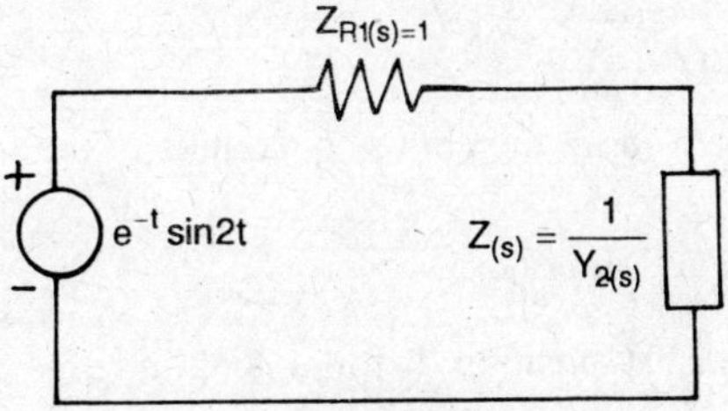

Fig. 2.59

Therefore

$$Z(s) = 1 + \frac{s+1}{s^2+s+1} = \frac{s^2+s+1+s+1}{s^2+s+1} = \frac{s^2+2s+2}{s^2+s+1}$$

Now

$$I(s) = \frac{V(s)}{Z(s)} = \frac{\dfrac{2}{(s+1)^2+2^2}}{\dfrac{s^2+2s+2}{s^2+s+1}} - \frac{\dfrac{2}{s^2+2s+1+4}}{\dfrac{s^2+2s+2}{s^2+s+1}}$$

or $$I(s) = \frac{2(s^2+s+1)}{(s^2+2s+5)(s^2+2s+2)}$$

Problem No. 2.52 In a circuit consisting of a resistance of 25Ω and inductance of 1 Henry as shown in Fig. 2.60, the switch K is in position 1 long enough and at $t = 0$ it is switched to position 2. Find the expression for the current.

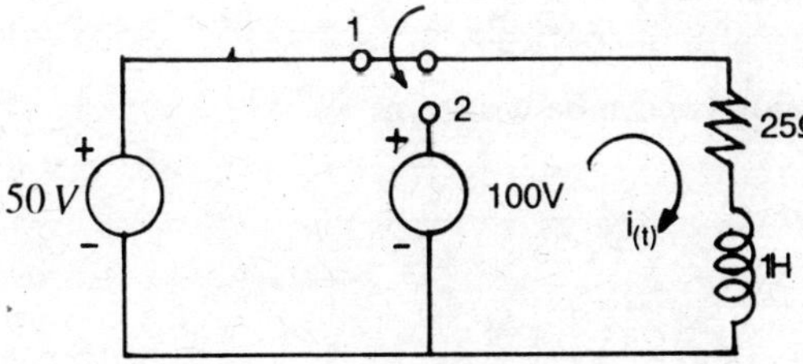

Fig. 2.60

Solution

At $t = 0$, the mesh equation is time domain is written as

$$25i_{(t)} + 1\frac{di_{(t)}}{dt} = 100v_{(t)}$$

In Laplace transform form, the above equation is of the form

$$25I(s) + sI(s) - i_{(0+)} = \frac{100}{s} \quad ...(1)$$

When the switch is at position 1 for a long time to have steady state conditions

$$i_{(0+)} = \frac{50}{25} = 2A$$

Substituting for $i_{(0+)}$ in equation (1), we get

$$25I(s) + sI(s) - 2 = \frac{100}{s}$$

$$I(s)(s+25) = \frac{100}{s} + 2$$

or
$$I(s) = \frac{100}{s(s+25)} + \frac{2}{s+25}$$

or
$$I(s) = \frac{A}{s} + \frac{B}{s+25} + \frac{2}{s+25} \quad ...(2)$$

where

$$A = \left.\frac{100}{s+25}\right|_{s=0} = \frac{100}{0+25} = 4; \text{ and}$$

$$B = \left.\frac{100}{s}\right|_{s=-25} = \frac{100}{-25} = -4$$

Therefore equation (2) can be written as

$$I(s) = \frac{4}{s} - \frac{4}{s+25} + \frac{2}{s+25}$$

$$= \frac{4}{s} - \frac{2}{s+25}$$

Inverse Laplace transform gives

$$i_{(t)} = 4 - 2e^{-25t}$$

Problem No. 2.53 In the circuit shown in Fig. 2.61, determine the current $i_{(t)}$ when the switch is at position 2. The switch is moved from position 1 to position 2 after a long time, at $t = 0$.

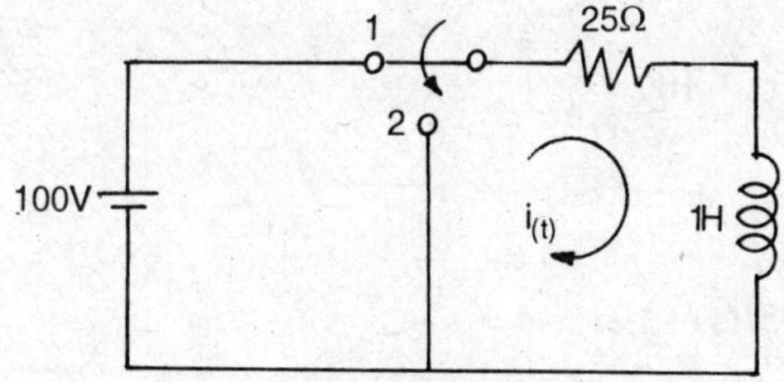

Fig. 2.61

Solution

The mesh equation is written as ($t = 0$)

$$25 i_{(t)} + 1 \frac{d i_{(t)}}{dt} = 0$$

In Laplace notation

$$25 I(s) + s I(s) - i_{(0)} = 0$$

$$i_{(0)} = \frac{100}{25} = 4A$$

Therefore,

$$25 I(s) + s I(s) - 4 = 0$$

or

$$I(s) = \frac{4}{s+25}$$

or

$$i_{(t)} = 4 e^{-25t}$$

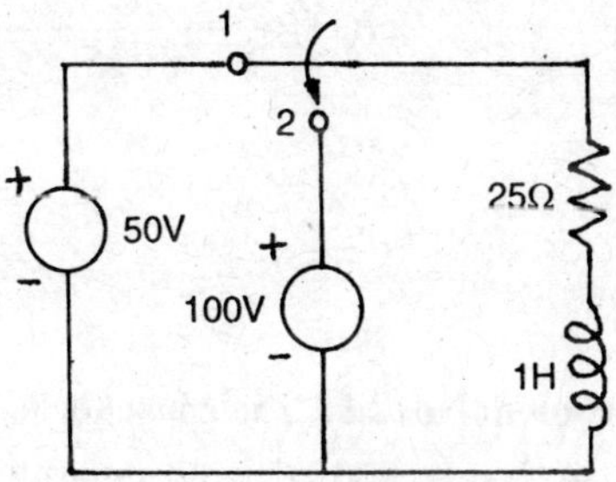

Fig. 2.62

Problem No. 2.54 If the voltage sources in the circuit are interchanged in problem No. 2.52, find the expression for the current.

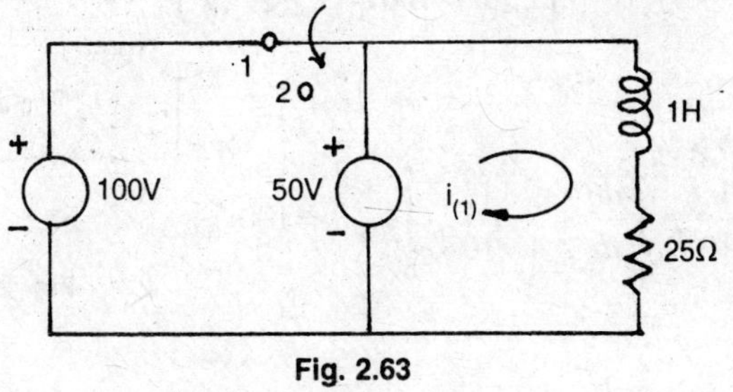

Fig. 2.63

Solution Circuit diagram with interchanged voltage sources in Fig. 2.63. Mesh equation in Laplace transfer is given as

$$25I(s)+sI(s)-i_{(0)}=\frac{50}{s}$$

$$i_{(0)}=\frac{100}{25}=4$$

Therefore

$$25I(s)+sI(s)-4=\frac{50}{s}$$

or $$I(s)(s+25)=\frac{50}{s}+4$$

or $$I(s)(s+25)=\frac{50}{s}+4$$

or $$I(s)=\frac{50}{s(s+25)}+\frac{4}{s+25}$$

or $$I(s)=\frac{2}{s}-\frac{2}{s+25}+\frac{4}{s+25}$$

$$=\frac{2}{s}+\frac{2}{s+25}$$

or $$i_{(t)}=2+2e^{-25t}$$

Problem No. 2.55 The capacitor in the circuit shown in Fig. 2.64 has initial change $q_0=800\,\mu C$ with polarity as indicated. If the switch is closed at $t=0$, obtain the current and charge for $t>0$.

Solution Mesh equation is

$$10i_{(t)}+4\times10^{-6}\int_{-\infty}^{t}i_{(t)}dt=100v_{(t)}$$

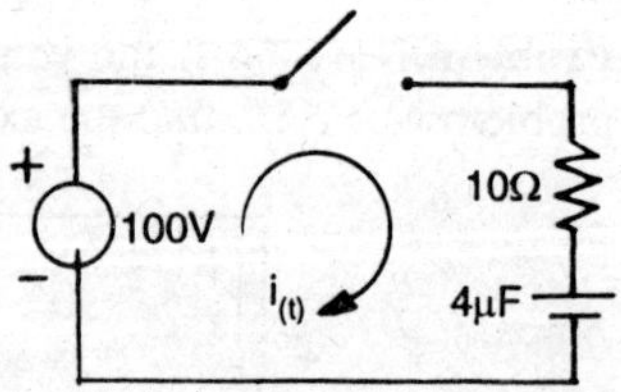

or

$$10i_{(t)}+\frac{1}{4\times10^{-6}}\int_{-\infty}^{0}\frac{dq}{dt}+\frac{1}{4\times10^{-6}}\int_{0}^{t}i_{(t)}dt=100v_{(t)}$$

Fig. 2.64

Taking Laplace transform of both the sides, we get

$$10I(s)+\frac{1}{4\times10^{-6}}\left[\frac{I(s)}{s}+\frac{q_0}{s}\right]=\frac{100}{s}$$

or $$10I(s)+\frac{1}{4\times10^{-6}}\left[\frac{I(s)}{s}+\frac{800\times10^{-6}}{s}\right]=\frac{100}{s}$$

$$10I(s)+\frac{I(s)}{4\times10^{-6}s}=\frac{100}{s}-\frac{800\times10^{-6}}{4\times10^{-6}s}$$

or $$10I(s)+\frac{I(s)}{4\times10^{-6}s}=\frac{100}{s}-\frac{200}{s}$$

or $$10I(s)+\frac{I(s)}{4\times10^{-6}s}=-\frac{100}{s}$$

or $$4\times10^{-6}s\,I(s)+I(s)=-100\times4\times10^{-6}$$

or $$I(s)=\frac{-100\times4\times10^{-6}}{\left(40\times10^{-6}s+1\right)}$$

or $$I(s)=\frac{-10\times4\times10^{-6}}{40\times10^{-6}(s+25000)}$$

or $$I(s)=\frac{-10}{s+25000}$$

or $$i_{(t)}=-10\cdot e^{-25000t} \qquad ...(1)$$

Expression for charge

Now from equation (1), we have

$$i_{(t)}=-10\cdot e^{-25000t}$$

or $$i_{(t)}=-\frac{100}{10}\cdot e^{-25000t}$$

Since $V = 100V$
and $R = 10\Omega$
We can say that

$$i_{(t)}=-\frac{V}{R}e^{-25000t}$$

or $$Ri(t) = -V\,e^{-25000t} \tag{2}$$

Now v_c the voltage across the capacitor is given by

$$v_c = V - Ri(t)$$

Substituting from equation (2), we get

$$v_{(c)} = V - \left(-Ve^{-25000t}\right)$$

or $$v_c = V\left(1 + e^{-25000t}\right)$$

Now

$$q = v_c C = VC\left(1 + e^{-25000t}\right)$$

or $$q = VC(1 + e^{-25000t})$$

or $$q = 100 \times 4 \times 10^{-6}(1 + e^{-25000t})$$

which gives

$$q = 4 \times 10^{-4}(1 + e^{-25000t})$$

Problem No. 2.56 In the *RC* circuit shown in Fig. 2.65, the switch is closed on position 1 at $t = 0$ and then moved to 2 after the passage of one time constant. Obtain the current transient for

(a) $0\ 0 < t < \tau$

(b) $t > \tau$

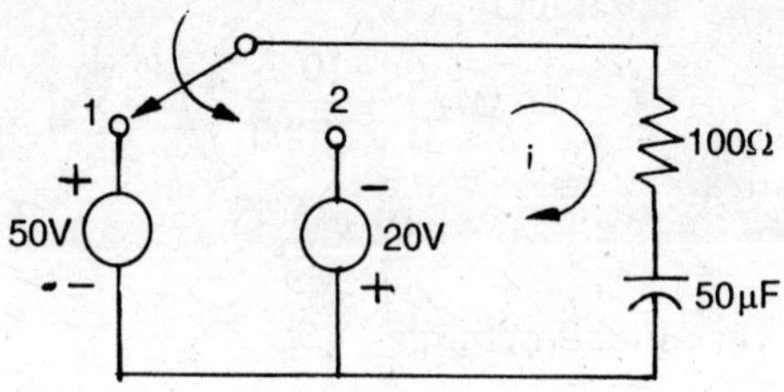

Fig. 2.65

Solution

At t = 0

Mesh equation

$$Ri_{(t)} + \frac{1}{C}\int i_{(t)}dt = 50$$

Since at $t = 0$ initial charge on $C = 0$

$$100\,I(s) + \frac{I(s)}{50 \times 10^{-6}s} = \frac{50}{s}$$

$$I(s)(100s + 0.02 \times 10^{6}) = 50$$

$$I(s) = \frac{50}{100s + 0.02 \times 10^{6}}$$

or $$I(s) = \frac{0.50}{s + 0.02 \times 10^4}$$

or $$I(s) = \frac{0.5}{s + 200}$$

$$I_{(t)} = 0.5 e^{-200t}$$

Expression for charge

$$i_{(t)} = \frac{0.5 \times 100 e^{-200t}}{100}$$

$$i_{(t)} = \frac{50}{100} e^{-200t} \quad \text{which is} = \frac{V}{R} e^{-200t}$$

Therefore

$$Ri_{(t)} = 50 e^{-200t}$$

$$v_c = V - Ri(t)$$

or $$v_c = V - 50 e^{-200t}$$

or $$v_c = 50 - 50 e^{-200t}$$

or $$v_c = 50(1 - e^{-200t})$$

or $$q = v_c C = 50C\left(1 - e^{-200t}\right)$$

or $$q = 50 \times 50 \times 10^{-6} (1 - e^{-200t})$$

or $$q = 25 \times 10^{-4} (1 - e^{-200t})$$

Since $$\tau = CR = 50 \times 10^{-6} \times 100 = 5000\,\mu s$$

$$q(\tau) = 25 \times 10^{-4}\left(1 - e^{-200 \times 5000 \times 10^{-6}}\right) = 25 \times 10^{-4}(1 - e^{-1})$$

$$= 0.25\left(1 - e^{-1}\right)\mu C$$

$$= 0.158\,\mu C$$

and $$q(\infty) = (50\ \mu F)(-20V) = -1000\,\mu C$$

Hence q for $t \geq \tau$

$$q = \left[q(\tau) - q(\infty)\right] e^{-(t-\tau)\tau} + q(\infty)$$

$$q = \left[0.25\left(1-e^{-1}\right)-(-1000)\right]e^{-(t-\tau)\tau}+(-1000)$$

$$= (0.158+1000)e^{-t/\tau}e^{1}-1000$$

$$= 1000.158e^{-t/\tau}e^{1}-1000$$

$$= 2718.71e^{-t/\tau}-1000$$

$$= 2718.71e^{-t/5000\times10^{-6}}-1000$$

$$= 2718.71e^{-200t}-1000$$

$$\frac{dq}{dt} = i(t>\tau) = 2718.71\times(-200)e^{-200t}\mu A$$

or $$i(t>\tau) = -0.544e^{-200t}A$$

Problem No. 2.57 A $2\mu F$ capacitor with initial charge $Q_0 = 1000\,\mu C$, is connected across a 100Ω resistor at $t = 0$. Calculate the time in which the transient voltage across the resistor drops from 40 to 10 volts.

Solution

The circuit diagram is shown in Fig. 2.66

In the RC circuit

$$Ri_{(t)} = v$$

and $$i_{(t)} = -\frac{Cdv}{dt}$$

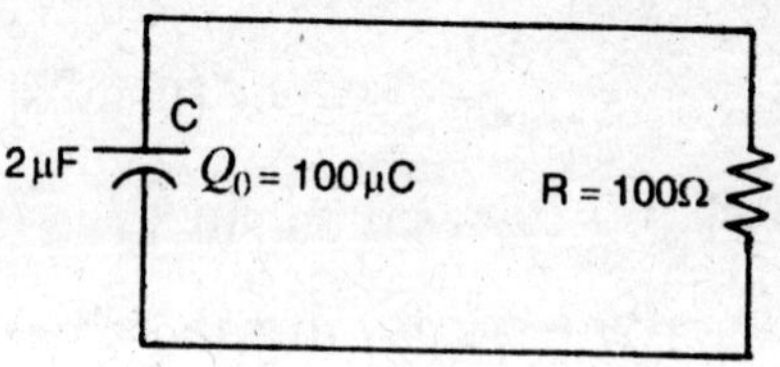

Fig. 2.66

Eliminating $i_{(t)}$ in above gives

$$\frac{dv}{dt}+\frac{1}{RC}v = 0$$

$$sV(s)+\frac{1}{RC}V(s) = 0$$

$$V(s)\left(s+\frac{1}{RC}\right) = 0$$

$$Ri_{(t)} = \frac{1}{C}\int_{-\infty}^{t} i_{(t)}dt$$

$$Ri_{(t)} = \frac{1}{C}\left[\int_{-\infty}^{0} \frac{dq}{dt}dt + \int_{0}^{t} i_{(t)}dt\right]$$

In Laplace notation

$$RI(s) = \frac{1}{C}\left[\frac{I(s)}{s} + \frac{q_0}{s}\right]$$

$$100I(s) = \frac{1}{2\times 10^{-6}}\left[\frac{I(s)}{s} + \frac{100\times 10^{-6}}{s}\right]$$

$$100\times 2\times 10^{-6}\, I(s) = \frac{I(s)}{s} + \frac{100\times 10^{-6}}{s}$$

$$100\times 2\times 10^{-6}\times s\times I(s) = I(s) + 100\times 10^{-6}$$

$$I(s)\left(100\times 2\times 10^{-6}\times s - 1\right) = 100\times 10^{-6}$$

$$I(s) = \frac{100\times 10^{-6}}{100\times 2\times 10^{-6} s - 1}$$

$$= \frac{100\times 10^{-6}}{2\times 100\times 10^{-6}\left(s - \dfrac{1}{2\times 100\times 10^{-6}}\right)}$$

$$= \frac{1/2}{(s-5000)}$$

or $\quad i_{(t)} = \frac{1}{2}e^{-5000t}$

or $\quad Ri_{(t)} = R\times \frac{1}{2}e^{-5000t}$

or $\quad v_{r(t)} = 100\times \frac{1}{2}e^{-5000t}$

or $\quad v_{r(t)} = 50e^{-5000t}$

Now the time taken to reach resistor drop from 40V to 10 V

$$40 = 50e^{-5000t}$$

which gives

$$t_{40} = \frac{\log 4/5}{-5000 \times \log e} = \frac{-0.097}{-5000 \times 0.4343} = 4.47 \times 10^{-5} \text{ sec}$$

$$= 0.000045 \text{ sec}$$

and for 10V

$$10 = 50 e^{-5000t}$$

or $$t_{10} = \frac{\log 10/50}{-5000 \times \log e} = \frac{-0.699}{-5000 \times 0.4343}$$

$$= 3.22 \times 10^{-4} \text{ sec}$$

$$= 0.000322 \text{ sec}$$

Therefore the required time is

$$= t_{10} - t_{40}$$

$$= 0.000322 - 0.000045$$

$$= 2.77 \times 10^{-4} \text{ sec}$$

or 0.277 msecs.

Problem No. 2.58 Find out the expression for current, v_c and v_R for $t > 0$ for the circuit shown in Fig. 2.67, when the switch is moved from position 1 to 2 at $t = 0$.

Solution

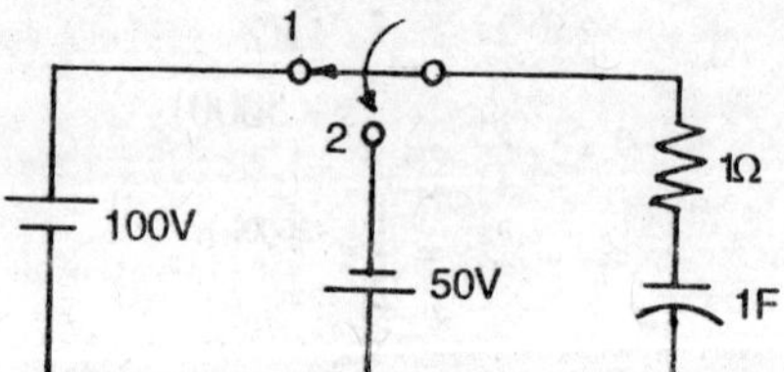

Fig. 2.67

At $t = 0$, the mesh equation is

$$Ri_{(t)} + \frac{1}{C}\int_{-\infty}^{t} i_{(t)} dt + 50 = 0$$

or $$Ri_{(t)} + \frac{1}{C}\int_{0}^{t} i_{(t)} dt + \frac{1}{C}\int_{-\infty}^{0} \frac{dq}{dt} dt = -50$$

or $$RI(s) + \frac{I(s)}{Cs} + \frac{q}{Cs} = \frac{-50}{s}$$

or $$RI(s) + \frac{I(s)}{Cs} + \frac{CV_{c(0)}}{Cs} = \frac{-50}{s}$$

or $$RI(s)+\frac{I(s)}{Cs}+\frac{V_{c(0)}}{s}=\frac{-50}{s}$$

or $$I(s)\left(R+\frac{1}{Cs}\right)+\frac{100}{s}=\frac{-50}{s}$$

or $$I(s)\left(s+\frac{1}{RC}\right)=\frac{-150}{R}$$

or $$I(s)=\frac{-150}{R(s+1/RC)}$$

or $$i_{(t)}=\frac{150}{R}e^{-t/RC} \qquad ...(1)$$

Substituting the values of *R* and *C* in equation 1, we get

$$i_{(t)}=150e^{-t} \qquad ...(2)$$

Rearranging equation (1), we get

$$Ri_{(t)}=-150e^{\;t/RC}$$

or $$v_r=-150e^{-t/RC}$$

Substituting values of R and C gives

$$v_r=-150e^{-t} \qquad ...(3)$$

Mesh equation at t = 0 can also be written as

$$v_r+v_c+50=0$$

Therefore

$$v_c=-v_r-50$$

Substituting value for v_r from equation (3), we get

$$v_c=-\left(-150e^{-t}\right)-50$$

or $$v_c=150e^{-t}-50$$

3

Two-Port Network

Problem No. 3.1 Determine the transfer impedence $Z(s)$ and the current transfer ratio of the network shown in Fig. 3.1.

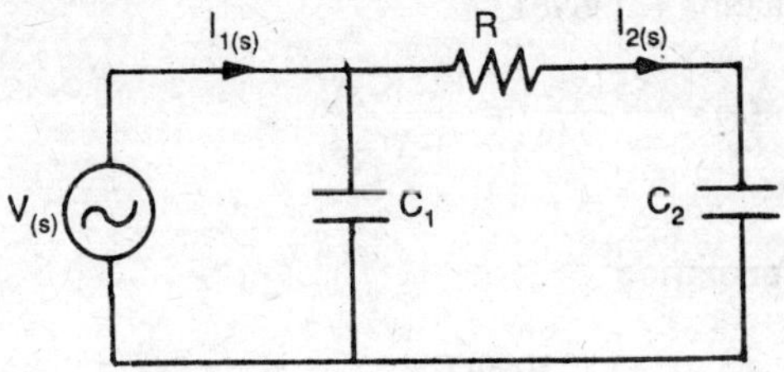

Fig. 3.1

Solution If we study the network, it is seen that there are two parallel paths

(a) one consisting of resistance R in series with C_2 and in which $I_2(s)$ is flowing, and

(b) second consisting of Capacitor C_1 in which $(I_1(s) - I_2(s))$ flows.

Across these parallel paths, is connected voltage source $V(s)$. Therefore,

$$V(s) = [I_1(s) - I_2(s)] \times \frac{1}{C_1 s}; \qquad ...(1)$$

and also

$$V(s) = I_2(s) \times \left[\frac{1}{C_2 s} + R\right]; \qquad ...(2)$$

From equation (2)

$$I_2(s) = \frac{V(s)}{R + \frac{1}{C_2 s}};$$

Substituting the value of $I_2(s)$ from equation (3), in equation (2), we get

$$V(s) = \left[I_1(s) - \frac{V(s)}{R + 1/C_2 s} \right] \frac{1}{C_1 s} \quad \text{...(4)}$$

Rearranging equation (4), we get

$$I_1(s) = V(s) C_1 s + \frac{V(s)}{R + 1/C_2 s}$$

or

$$I_1(s) = V(s) \left[C_1 s + \frac{1}{R + 1/C_2 s} \right]$$

or

$$I_1(s) = V(s) \left[C_1 s + \frac{C_2 s}{1 + RC_2 s} \right] \quad \text{...(5)}$$

Rearranging equation (2) gives

$$I_2(s) = \frac{V(s)}{R + \frac{1}{C_2 s}} = \frac{V(s) C_2 s}{1 + RC_2 s} \quad \text{...(6)}$$

Current transformation

$$\frac{I_2(s)}{I_1(s)} = \frac{\frac{V(s) C_2 s}{1 + RC_2 s}}{V_{(s)} \left[C_1 s + \frac{C_2 s}{1 + RC_2 s} \right]}$$

$$= \frac{C_2 s}{C_1 s + RC_1 C_2 s^2 + C_2 s}$$

$$= \frac{C_2}{C_1 + C_2 (1 + RC_1 s)} \quad \text{...(7)}$$

Transfer Impedence The network being a single port network having a voltage source $V(s)$ and having current $I_1(s)$, the transfer impedence is given by

$$Z(s) = \frac{V(s)}{I_1(s)} \qquad ...(8)$$

From equation (2), substituting the value of $V(s)$ in equation (7), we get

$$Z(s) = \frac{I_2(s)\left[R + \frac{1}{C_2 s}\right]}{I_1(s)} \qquad ...(9)$$

Substituting the value of $\frac{I_2(s)}{I_1(s)}$ from equation (7) in equation (9), we get

$$Z(s) = \frac{C_2}{C_1 + C_2(1 + RC_1 s)} \times \left[R + \frac{1}{C_2 s}\right]$$

$$= \frac{\frac{C_2\left[RC_2 s + 1\right]}{C_2 s}}{C_1 + C_2(1 + RC_1 s)}$$

$$= \frac{1 + RC_2 s}{s\{C_1 + C_2(1 + RC_1 s)\}}$$

Problem No. 3.2 Obtain the driving point impedence $Z(s)$ of the network shown in Fig. 3.2.

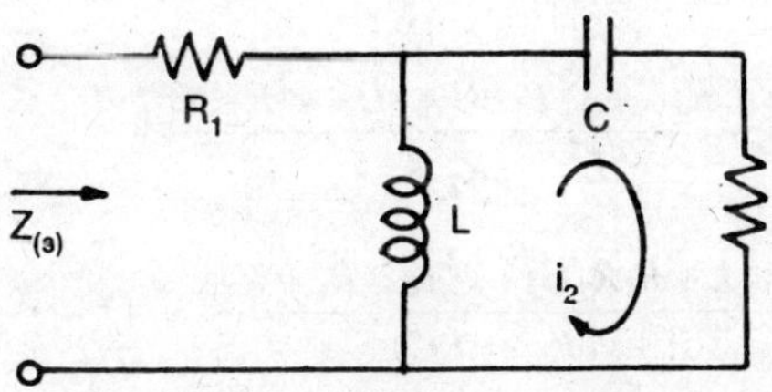

Fig. 3.2

Solution

Transform network is drawn in Fig 3.3

(a) R_2 and Capacitor C are in series, which gives

$$Z_{34} = R_2 + \frac{1}{sC} = \frac{sR_2C + 1}{sC}$$

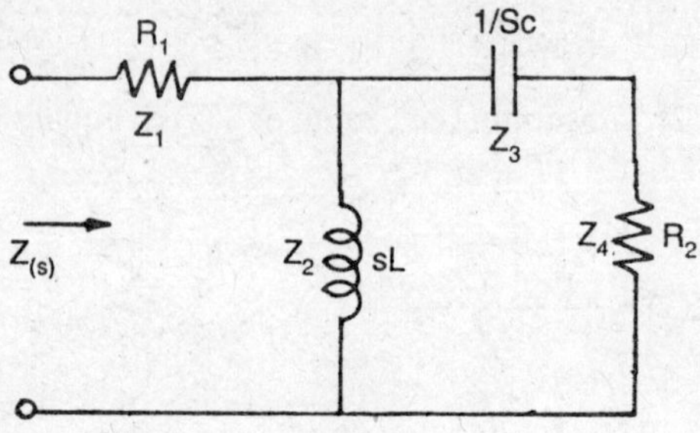

Fig. 3.3

(b) Z_2 and Z_{34} are in parallel, which gives

$$Z_{234} = \frac{Z_2 \times Z_{34}}{Z_2 + Z_{34}} = \frac{sL \times \left(\frac{sR_2C+1}{sC}\right)}{sL + \left(\frac{sR_2C+1}{sC}\right)}$$

$$= \frac{sL(1+sR_2C)}{s^2LC+sR_2C+1}$$

(c) Z_1 and Z_{234} are in series, which gives

$$Z(s) = Z_1 + Z_{234}$$

$$= R_1 + \frac{sL(1+sR_2C)}{s^2LC+sR_2C+1}$$

$$= \frac{R_1 + sR_1R_2C + s^2R_1LC + sL + s^2R_2LC}{s^2LC + sR_2 + 1}$$

$$= \frac{R_1 + s(L+R_1R_2C) + s^2LC(R_1+R_2)}{1+s(R_2+sLC)}$$

Problem No. 3.3 Determine the circuit element X in Fig. 3.4 such that the driving point impedence is 1 ohm for all Laplace transformable excitations.

Solution Assuming that element X is inductive,

$$Z(s) = \frac{1 \times \frac{1}{sC}}{1+\frac{1}{sC}} + \frac{1 \times sX}{1+sX}$$

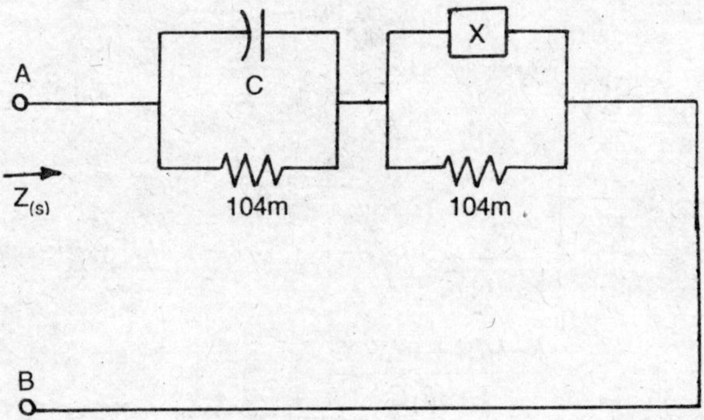

Fig. 3.4

or $$Z(s) = \frac{1}{1+sC} + \frac{sX}{1+sX}$$

or $$Z(s) = \frac{1+sX+sX+s^2CX}{(1+sC)(1+sX)}$$

or $$Z(s) = \frac{s^2CX+2sX+1}{s^2CX+sC+sX+1}$$

Putting $Z(s) = 1$, we get

$$s^2CX+sC+sX+1 = s^2CX+2sX+1$$

or $$s^2CX-s^2CX+2sX-sX-sC=0$$

or $$sX = sC$$

or $$X = C$$

Problem No. 3.4 Find the voltage transfer function of the network shown in Fig. 3.5

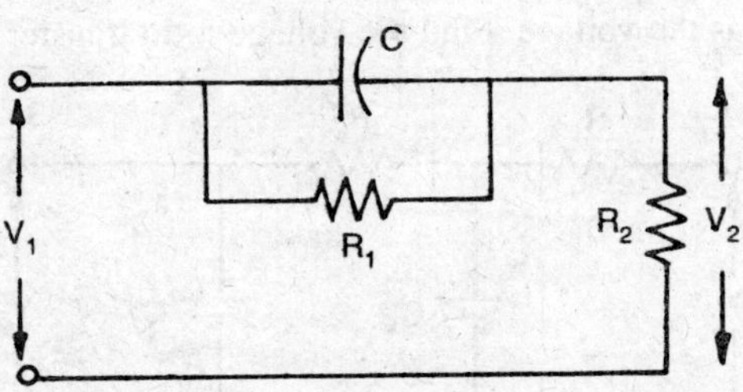

Fig. 3.5

Solution

$$Z(s) = \frac{R_1 \times \frac{1}{sC}}{R_1 + \frac{1}{sC}} + R_2$$

$$= \frac{R_1}{1 + sR_1C} + R_2$$

$$= \frac{R_1 + R_2 + sR_1R_2C}{1 + sR_1C}$$

Voltage transfer function

$$V(s) = \frac{V_2(s)}{V_1(s)} = \frac{R_2 I(s)}{Z(s)I(s)} = \frac{R_2}{Z(s)}$$

Substituting the value of $Z(s)$, we get

$$V(s) = \frac{R_2(1 + sR_1C)}{R_1 + R_2 + sR_1R_2C}$$

$$= \frac{R_1R_2C\left(s + \frac{1}{R_1C}\right)}{R_1R_2C\left(s + \frac{R_1}{R_1R_2C} + \frac{R_2}{R_1R_2C}\right)}$$

$$= \frac{s + \frac{1}{R_1C}}{s + \frac{1}{R_2C} + \frac{1}{R_1C}} = \frac{s + \frac{1}{R_1C}}{s + \frac{R_1 + R_2}{R_1R_2C}}$$

Problem No. 3.5 The two port network shown in Fig. 3.6 has V_1 as the input voltage and V_2 as the voltage. Find the voltage ratio transfer function.

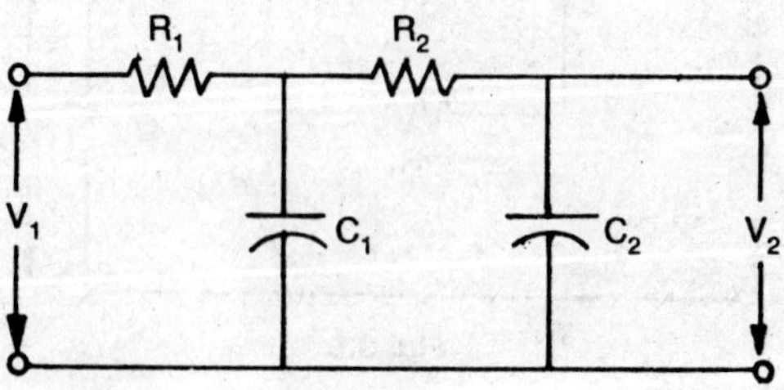

Fig. 3.6

Solution Redrawing the transformed network in s-domain (Fig. 3.7)

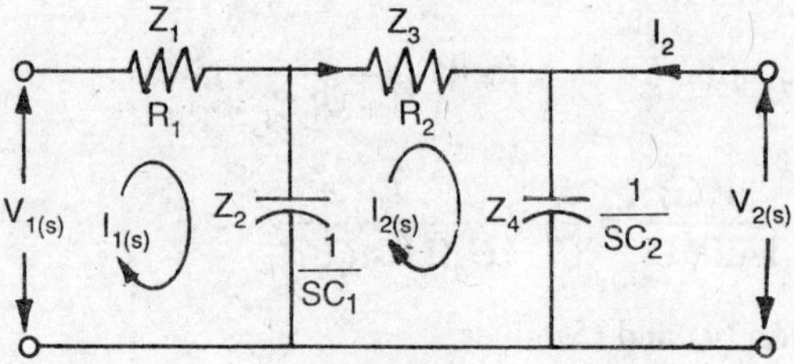

Fig. 3.7

Voltage transfer function

$$V(s)=\frac{V_2(s)}{V_1(s)}=\frac{I_2\times\frac{1}{sC_2}}{I_1\times Z(s)}=\frac{I_2/sC_2}{V_1(s)}$$

Writing mesh equation for Fig. 3.7, we get

$$V_1(s)=I_1\left[R_1+\frac{1}{sC_1}\right]-\frac{I_2}{sC_1};\text{ and} \qquad \text{...(1)}$$

$$0=I_2\left(R_2+\frac{1}{sC_2}+\frac{1}{sC_1}\right)-\frac{I_1}{sC_1} \qquad \text{...(2)}$$

Simplification of equation (1) gives,

$$V_1(s)=I_1\left(\frac{sR_1C_1+1}{sC_1}\right)-\frac{I_2}{sC_1}$$

or
$$sV_1(s)=I_1\left(\frac{1}{C_1}+sR_1\right)-\frac{I_2}{C_1}$$

or
$$sV_1(s)=I_1\left(\frac{C_1+sR_1C_1}{C_1}\right)-\frac{I_2}{C_1} \qquad \text{...(3)}$$

Simplification of equation (2), gives

$$0=I_2\left(\frac{sR_2C_2C_1+C_1+C_2}{sC_1C_2}\right)-\frac{I_1}{sC_1}$$

or $$0 = I_2\left(\frac{1}{C_1}+\frac{1}{C_2}+sR_2\right)-\frac{I_1}{C_1} \qquad ...(4)$$

Multiplying equation (3) by a factor $\frac{1}{1+sR_1C_1}$, gives

$$\frac{sV_1(s)}{1+sR_1C_1}=\frac{I_1}{C_1}-\frac{I_2}{C_2(1+sR_1C_1)} \qquad ...(5)$$

Adding equation (4) and (5) gives

$$\frac{sV_1(s)}{1+sR_1C_1}=\frac{I_2(sR_2C_2C_1+C_1+C_2)}{C_1C_2}-\frac{I_2}{C_1(1+sR_1C_1)} \qquad ...(6)$$

$$sV_1(s)=I_2\left[\frac{(1+sR_1C_1)(C_1+C_2+sR_2C_1C_2)}{C_1C_2}-\frac{1}{C_1}\right]$$

$$sV_1(s)=I_2\left[\frac{(1+sR_1C_1)(C_1+C_2+sR_2C_1C_2)-C_2}{C_1C_2}\right]$$

$$I_2=\frac{sC_1C_2V_1(s)}{(1+sR_1C_1)(C_1+C_2+sR_2C_1C_2)-C_2} \qquad ...(7)$$

Now $V(s)$, voltage transform function as found out earlier is

$$V(s)=\frac{V_2(s)}{V_1(s)}=\frac{I_2\times\frac{1}{sC_2}}{V_1(s)} \qquad ...(8)$$

Substituting value of I_2 from equation (7) in equation (8), we get

$$V(s)=\frac{\frac{sC_1C_2V_1(s)}{(1+sR_1C_1)(C_1+C_2+sR_2C_1C_2)-C_2}\times\frac{1}{sC_2}}{V_1(s)}$$

or $$V(s)=\frac{C_1}{(1+sR_1C_1)(C_1+C_2+sR_2C_1C_2)-C_2}$$

Problem No. 3.6 Calculate the driving point and transfer impedences of the network shown in Fig. 3.8.

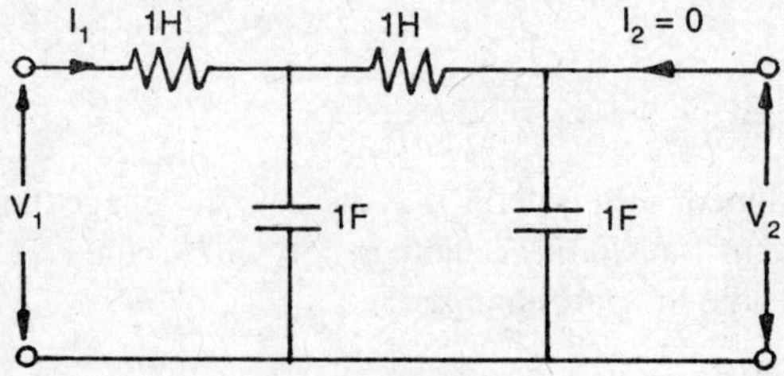

Fig. 3.8

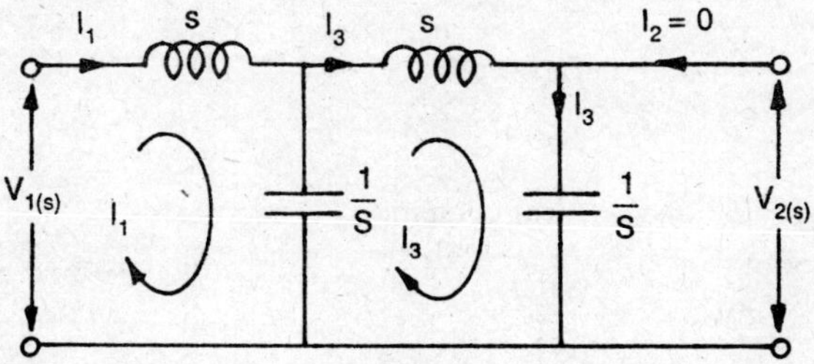

Fig. 3.9

Solution The tranformed network of Fig. 3.8 in s-domain gives a network shown in

Writing mesh equations gives

$$V_1(s) = sI_1(s) + \frac{1}{s}[I_1(s) - I_3(s)]$$

$$V_1(s) = I_1\left(s + \frac{1}{s}\right) - \frac{I_3}{s} \quad ...(1)$$

$$0 = sI_3(s) + \frac{I_3(s)}{s} + \frac{I_3(s)}{s} - \frac{I_1(s)}{s}$$

$$0 = I_3(s)\left(s + \frac{1}{s} + \frac{1}{s}\right) - \frac{I_1(s)}{s}$$

$$0 = I_3(s)(s^2 + 2) - I_1(s) \quad ...(2)$$

$$V_2(s) = \frac{I_3(s)}{s} \quad ...(3)$$

or $$I_3(s) = sV_2(s) \quad ...(4)$$

Driving point impedence = $\dfrac{V_1(s)}{I_1(s)}$ ...(5)

Transfer impedence (Z) It is defined as the ratio of voltage transform at one port to the current transform at another port. Subscript 1, 2 is affixed to notation Z to denote the port number.

Therefore, we have

$$Z_{12}(s) = \frac{V_1(s)}{I_2(s)}; \text{ and}$$

$$Z_{21}(s) = \frac{V_2(s)}{I_1(s)} \qquad ...(6)$$

Substituting value of $V_2(s)$ from equation (3), we get

$$Z_{21}(s) = \frac{I_3(s)/s}{I_1(s)} \qquad ...(6a)$$

Rewriting equations (1) and (2) as equations (7) and (8) as given below

$$V_1(s) = \frac{I_1(s)(s^2+1)}{s} - \frac{I_3(s)}{s} \qquad ...(7)$$

$$I_1(s) = I_3(s)(s^2+2) \qquad ...(8)$$

From equation (8),

$$I_3(s) = \frac{I_1(s)}{s^2+2} \qquad ...(9)$$

Substituting in equation (6a), we get

$$Z_{21}(s) = \frac{I_3(s)/s}{I_1(s)} = \frac{I_3(s)}{I_1(s)\ s(s^2+2)} = \frac{1}{s(s^2+2)}$$

Problem No. 3.7 Determine Z parameter of the network shown in Fig. 3.10.

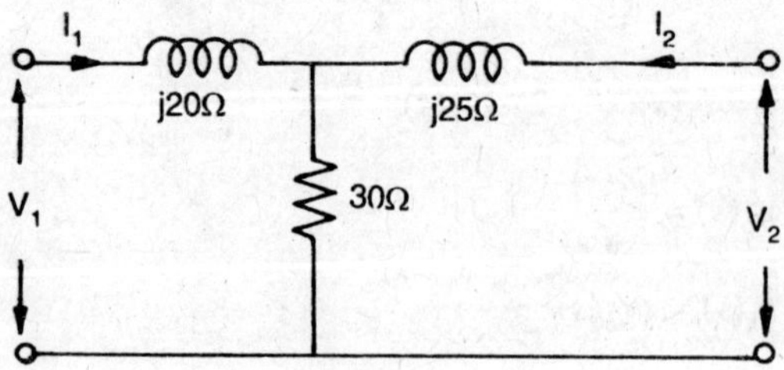

Fig. 3.10

Solution

The transform network corresponding to the given network is given in Fig. 3.11.

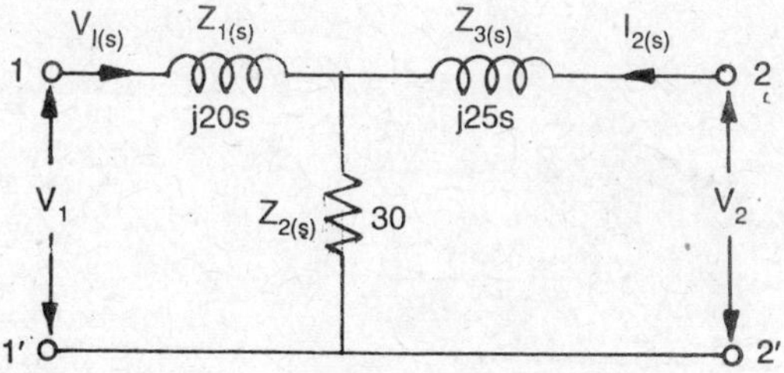

Fig. 3.11

With port 2 open circuited, we get

$$Z_{11} = \left.\frac{V_1(s)}{I_1(s)}\right|_{I_2(s)=0} = Z_1(s) + Z_2(s) = 30 + j20$$

$$Z_{21} = \left.\frac{V_2(s)}{I_1(s)}\right|_{I_2(s)=0} = \frac{I_1(s)\cdot Z_2(s)}{I_1(s)} = Z_2(s) = 30$$

With port 1 open-circuited, we get

$$Z_{12} = \left.\frac{V_1(s)}{I_2(s)}\right|_{I_1(s)=0} = \frac{I_2(s)\cdot Z_2(s)}{I_2(s)} = 30$$

$$Z_{22} = \left.\frac{V_2(s)}{I_2(s)}\right|_{I_1(s)=0} = Z_3(s) + Z_2(s) = 30 + j25$$

Therefore

$$[Z] = \begin{bmatrix} 30 + j20 & 30 \\ 30 & 30 + j25 \end{bmatrix}$$

The Z-parameter equivalent circuit is shown in Fig. 3.12.

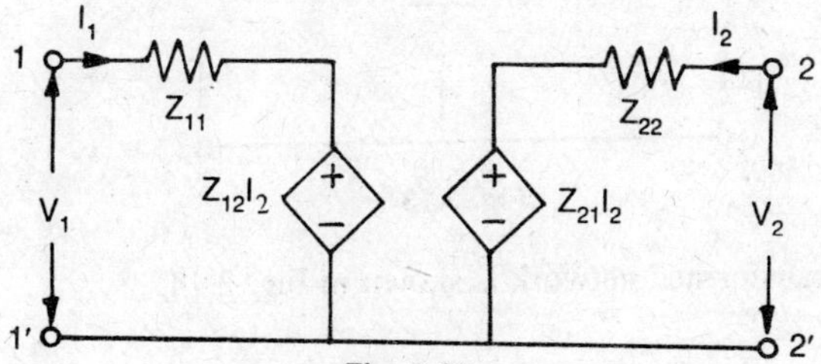

Fig. 3.12

Y-parameter

$$\Delta Z = Z_{11}Z_{22} - Z_{12}Z_{21}$$

$$= (30 + j20)(30 + j25) - 30 \times 30$$

$$= 900 + j750 + j600 - 500 - 900$$

$$= -500 + j1350$$

$$Y_{11} = \frac{Z_{22}}{\Delta Z} = \frac{30 + j25}{-500 + j1350}$$

$$= \frac{6 + j5}{-100 + j270}$$

$$= \frac{6 + j5}{-10(-10 + j27)}$$

$$Y_{12} = \frac{-Z_{12}}{\Delta Z} = \frac{-30}{-500 + j1350} = \frac{-3}{5(-10 + j27)}$$

$$Y_{21} = \frac{-Z_{21}}{\Delta Z} = \frac{-30}{-500 + j1350} = \frac{-3}{5(-10 + j27)}$$

$$Y_{22} = \frac{Z_{11}}{\Delta Z} = \frac{30}{-500 + j1350} + j20 = \frac{3 + j2}{5(-10 + j27)}$$

Problem No. 3.8 Determine the transfer function $\frac{V_0(s)}{V_i(s)}$ of the network shown in Fig. 3.13. Determine in $v_{0(t)}$, when $v_{i(t)} = 100e^{-5t}$.

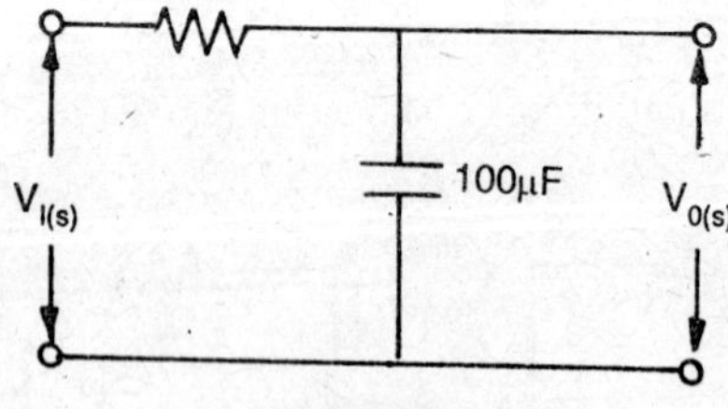

Fig. 3.13

Solution The transformed network is shown in Fig. 3.14.

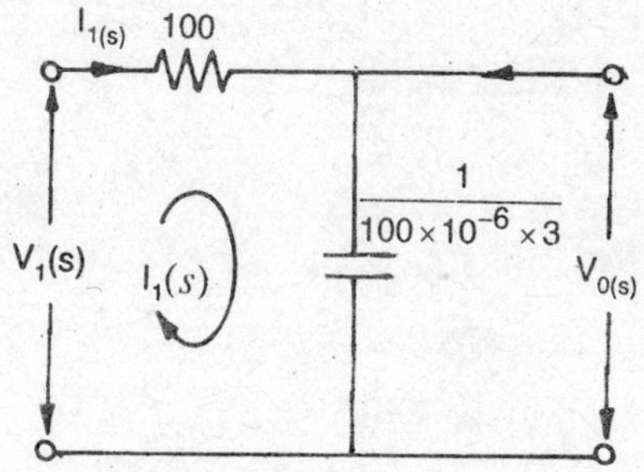

Fig. 3.14

$$\text{Transform function} = \frac{V_0(s)}{V_i(s)}$$

$$V_0(s) = \frac{1}{100 \times 10^{-6} s} I_1(s) - \frac{10^4}{s} I_1(s)$$

$$V_i(s) = \left(100 + \frac{1}{100 \times 10^{-6} s}\right) I_1(s)$$

$$= \left(100 + \frac{10^4}{s}\right) I_1(s)$$

Therefore

$$H(s) = \frac{V_0(s)}{V_1(s)} = \frac{10^4/s \times I_1(s)}{\left(100 + 10^4/s\right) I_1(s)}$$

$$= \frac{10^4}{100s + 10^4}$$

$$= \frac{100}{s+100} \qquad ...(1)$$

Now

$$V_i(t) = 100e^{-5t}$$

Therefore,

$$V_i(s) = \frac{100}{s+5}$$

From equation (1)

$$V_0(s)=\left(\frac{100}{s+100}\right)V_i(s)$$

$$=\frac{100}{(s+100)}\times\frac{100}{(s+5)}$$

$$=\frac{10000}{(s+5)(s+100)}$$

$$=\frac{A}{s+5}+\frac{B}{s+100} \quad \text{...(2)}$$

which gives

$$A=\left.\frac{10000}{s+100}\right|_{s=-5}=\frac{10000}{-5+100}=105.26$$

$$B=\left.\frac{10000}{s+5}\right|_{s=-100}=\frac{10000}{-100+5}=-105.26$$

which gives

$$V_0(s)=\frac{105.26}{s+5}+\frac{-105.26}{s+100}$$

Inverse Laplace transform gives

$$v_{0(t)}=105.26e^{-5t}-105.26e^{-100t}$$

$$=105.26(e^{-5t}-e^{-100t})$$

Problem No. 3.9 For the bridged T network shown in Fig. 3.15, evaluate the driving point admittance Y_{11} and transfer admittances Y_{12}.

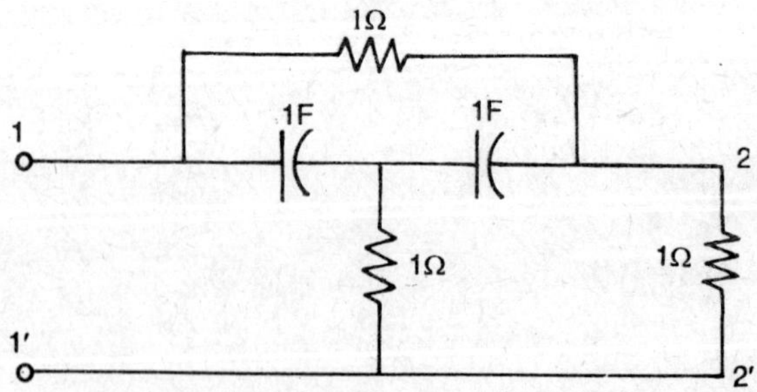

Fig. 3.15

Solution The equivalent transformed network is shown in Fig. 3.16.

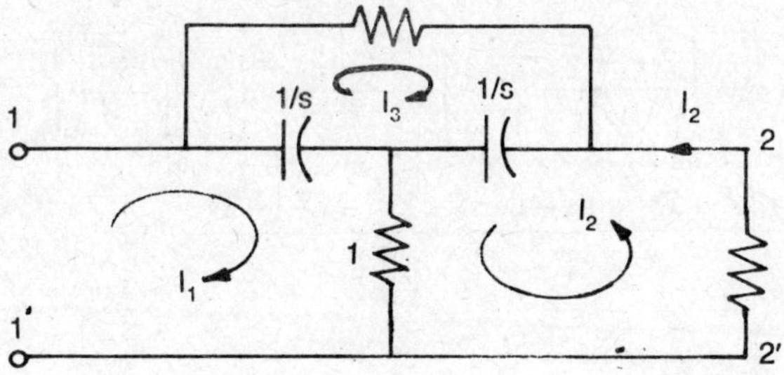

Fig. 3.16

Writing mesh equations give us

$$\left(\frac{1}{s}+1\right)I_1 + I_2 - \frac{I_3}{s} = V_1 \qquad ...(1)$$

$$I_1 + I_2\left(\frac{1}{s}+1+1\right) + \frac{I_3}{s} = 0 \qquad ...(2)$$

$$-\frac{I_1}{s} + \frac{I_2}{s} + \left(\frac{1}{s}+\frac{1}{s}+1\right)I_3 = 0 \qquad ...(3)$$

The determinant based on loop basis is

$$\Delta = \begin{vmatrix} \frac{1}{s}+1 & 1 & -\frac{1}{s} \\ 1 & \frac{1}{s}+2 & \frac{1}{s} \\ -\frac{1}{s} & \frac{1}{s} & \frac{2}{s}+1 \end{vmatrix}$$

$$\Delta = \left(\frac{1}{s}+1\right)\left[\left(\frac{1}{s}+2\right)\left(\frac{2}{s}+1\right) - \frac{1}{s^2}\right] - 1\left[1\left(\frac{2}{s}+1\right) + \frac{1}{s^2}\right]$$

$$+\left(-\frac{1}{s}\right)\left[\frac{1}{s} - \left(-\frac{1}{s} \times \left(\frac{1}{s}+2\right)\right)\right]$$

$$= \frac{1+s}{s}\left[\frac{1+2s}{s} \times \frac{2+s}{s} - \frac{1}{s^2}\right] - \left[\frac{2+s}{s} + \frac{1}{s^2}\right] - \frac{1}{s}\left[\frac{1}{s} + \frac{1+2s}{s^2}\right]$$

$$=\frac{1+s}{s}\left[\frac{(1+2s)(2+s)}{s^2}\right]-\left[\frac{(2+s)s+1}{s^2}\right]-\frac{1}{s}\left[\frac{s+1+2s}{s^2}\right]$$

$$=\frac{1+s}{s}\left[\frac{2+5s+2s^2-1}{s^2}\right]-\left[\frac{s^2+2s+1}{s^2}\right]-\frac{1}{s}\left[\frac{1+3s}{s^2}\right]$$

$$=\frac{2s^3+7s^2+6s+1-s^3-2s^2-s-1-3s}{s^3}$$

$$=\frac{s^3+5s^2+2s}{s^3}$$

$$=\frac{s^2+5s+2}{s^2}$$

Evaluation of I_1

Δ_{11} is obtained as follows. Let the three equation be

$$a_1x+b_1y+c_1z=d_1$$

$$a_2x+b_2y+c_2z=d_2$$

$$a_3x+b_3y+c_3z=d_3$$

$$\Delta x=\begin{vmatrix} d_1 & b_1 & c_1 \\ d_2 & b_2 & c_2 \\ d_3 & b_3 & c_3 \end{vmatrix}, \quad \Delta y=\begin{vmatrix} a_1 & d_1 & c_1 \\ a_2 & d_2 & c_2 \\ a_3 & d_3 & c_3 \end{vmatrix}$$

and $$\Delta z=\begin{vmatrix} a_1 & b_1 & d_1 \\ a_2 & b_2 & d_2 \\ a_3 & b_3 & d_3 \end{vmatrix}$$

$$\Delta_{11}=\begin{vmatrix} 1 & 1 & -\frac{1}{s} \\ 0 & \frac{1}{s}+2 & \frac{1}{s} \\ 0 & \frac{1}{s} & \frac{2}{s}+1 \end{vmatrix} \quad \text{and} \quad \Delta_{22}=\begin{vmatrix} \frac{1}{s}+1 & 1 & -\frac{1}{s} \\ 1 & 0 & \frac{1}{s} \\ -\frac{1}{s} & 0 & \frac{2}{s}+1 \end{vmatrix}$$

Therefore,

$$\Delta_{11}=\left(\frac{1}{s}+2\right)\left(\frac{2}{s}+1\right)-\frac{1}{s}\times\frac{1}{s}$$

$$= \frac{(1+2s)(2+s)}{s^2} - \frac{1}{s^2}$$

$$= \frac{2s^2+5s+2}{s^2} - \frac{1}{s^2}$$

$$= \frac{2s^2+5s+2-1}{s^2}$$

$$= \frac{2s^2+5s+1}{s^2}$$

and

$$\Delta_{21} = -\left[\left(\frac{2}{s}+1\right)\times 1 - \left(\frac{1}{s}\right)\left(-\frac{1}{s}\right)\right]$$

$$= -\left[\frac{2+s}{s} + \frac{1}{s^2}\right]$$

$$= -\left[\frac{2+s^2+1}{s^2}\right]$$

$$= -\frac{(s^2+2s+1)}{s^2}$$

Driving Point Admittance

$$Y_{11} = \frac{\Delta_{11}}{\Delta} = \frac{\frac{2s^2+5s+1}{s^2}}{\frac{s^2+5s+2}{s^2}} = \frac{2s^2+5s+1}{s^2+5s+2}$$

Transfer Admittance

$$Y_{12} = \frac{\Delta_{21}}{\Delta} = \frac{-\frac{s^2+2s+1}{s^2}}{\frac{s^2+5s+2}{s^2}} = -\frac{s^2+2s+1}{s^2+5s+2}$$

Problem No. 3.10 The two port network shown in Fig. 3.17 has V_1 as the input voltage and V_2 as the output voltage. Find the voltage ratio transfer function.

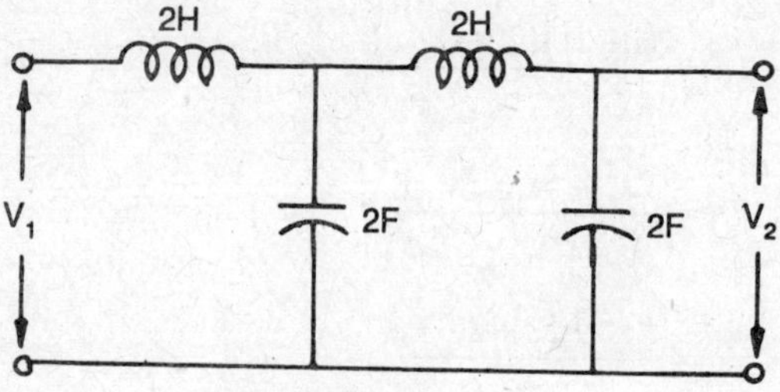

Fig. 3.17

Solution The transformed network is shown in Fig. 3.18

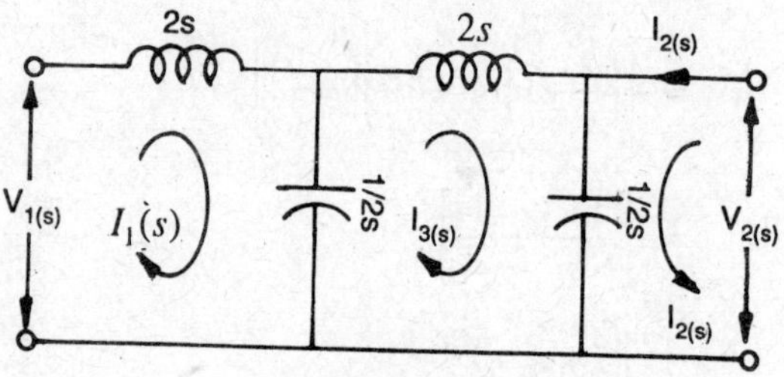

Fig. 3.18

Mesh equations are

$$V_1(s) = 2sI_1(s) + \frac{1}{2s}\left(I_1(s) - I_3(s)\right) \qquad \text{...(1)}$$

$$0 = 2sI_3(s) + \frac{1}{2s}\left(I_2(s) + I_3(s)\right) + \frac{1}{2s}\left(I_3(s) - I_1(s)\right) \qquad \text{...(2)}$$

and

$$V_2(s) = \frac{1}{2s}\left[I_2(s) + I_3(s)\right]$$

Putting $I_2(s) = 0$, we get

$$2sI_1(s) + \frac{1}{2s}I_1(s) - \frac{1}{2s}I_3(s) = V_1(s)$$

or

$$I_1(s)\left(2s + \frac{1}{2s}\right) - \frac{I_3(s)}{2s} = V_1(s)$$

or

$$\frac{\left(4s^2 + 1\right)}{2s}I_1(s) - \frac{I_3(s)}{2s} = V_1(s) \qquad \text{...(4)}$$

and

$$0 = 2sI_3(s) + \frac{I_3(s)}{2s} + \frac{I_3(s)}{2s} - \frac{I_1(s)}{2s}$$

$$\frac{\left(4s^2+2\right)}{2s} I_3(s) - \frac{I_1(s)}{2s} = 0 \qquad \text{...(5)}$$

and

$$\frac{I_3(s)}{2s} = V_2(s) \qquad \text{...(6)}$$

Multiply equation (5) by $(4s^2 + 1)$ and adding equations (4) and (5), we get

$$\frac{\left(4s^2+2\right)\left(4s^2+1\right)}{2s} I_3(s) - \frac{I_3(s)}{2s} = V_1(s)$$

or
$$I_3(s)\left[\frac{\left(16s^4+12s^2+2\right)}{2s} - \frac{1}{2s}\right] = V_1(s)$$

or
$$I_3(s)\frac{\left(16s^4+12s^2+1\right)}{2s} = V_1(s)$$

or
$$I_3(s) = \frac{2sV_1(s)}{16s^4+12s^2+1}$$

Voltage ratio Transfer function $= \frac{V_2(s)}{V_1(s)}$

$$= \frac{\frac{I_3(s)}{2s}}{V_1(s)}$$

$$= \frac{\frac{2s\,V_1(s)}{\left(16s^4+12s^2+1\right)} \times \frac{1}{2s}}{V_1(s)}$$

$$= \frac{1}{16s^4+12s^2+1}$$

Problem No. 3.11 Find the driving point impedence of the ladder network shown in Fig. 3.19.

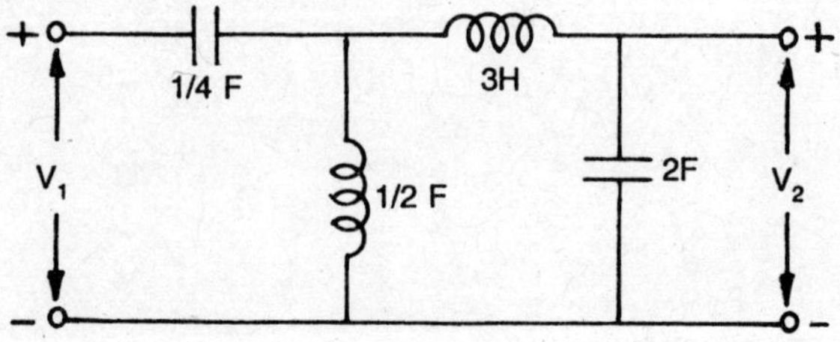

Fig. 3.19

Solution

Equivalent transformed network is shown in Fig. 3.20.

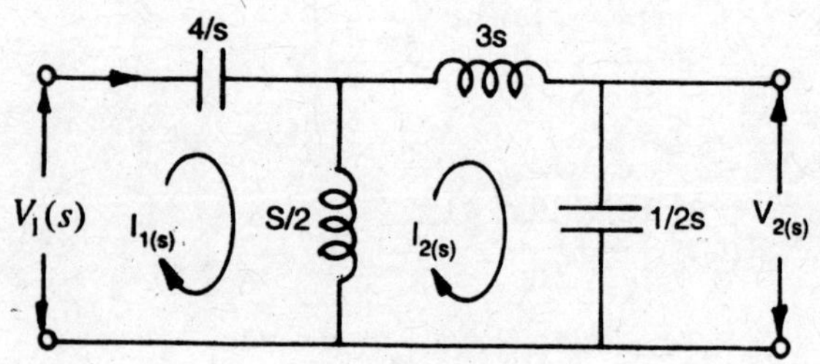

Fig. 3.20

Mesh equation gives

$$V_1(s) = \frac{4}{s} I_1(s) + \frac{s}{2} I_1(s) - \frac{s}{2} I_2(s) \qquad ...(1)$$

$$0 = 3sI_2(s) + \frac{1}{2s} I_2(s) + \frac{s}{2}\left(I_2(s) - I_1(s)\right) \qquad ...(2)$$

Simplication gives

$$V_1(s) = I_1(s)\left(\frac{4}{s} + \frac{s}{2}\right) - I_2(s)\frac{s}{2} \qquad ...(3)$$

$$0 = -\frac{s}{2} I_1(s) + I_2(s)\left(3s + \frac{1}{2s} + \frac{s}{2}\right) \qquad ...(4)$$

Equation (4) gives

$$I_2(s) = \frac{\frac{s}{2} I_1(s)}{3s + \frac{1}{2s} + \frac{s}{2}}$$

$$= \frac{\frac{s}{2} I_1(s)}{\frac{6s^2 + 1 + s^2}{2s}}$$

$$= \frac{\frac{s}{2} I_1(s)}{\frac{7s^2 + 1}{2s}}$$

$$= \frac{s^2}{7s^2 + 1} I_1(s) \qquad ...(5)$$

Substituting value of $I_2(s)$ from equation (5) gives

$$V_1(s) = I_1(s)\left(\frac{8 + s^2}{2s}\right) - \frac{s}{2} \times \frac{s^2}{7s^2 + 1} I_1(s)$$

$$= I(s)\left[\frac{s^2 + 8}{2s} - \frac{s^3}{2(7s^2 + 1)}\right]$$

$$= I_1(s)\left[\frac{(s^2 + 8)(7s^2 + 1) - s^4}{2s(7s^2 + 1)}\right]$$

$$= I_1(s)\left[\frac{7s^4 + 57s^2 + 8 - s^4}{2s(7s^2 + 1)}\right]$$

$$= I_1(s)\left[\frac{6s^4 + 57s^2 + 8}{14s^3 + 2s}\right] \qquad ...(6)$$

Driving point Impedence

$$= \frac{V_1(s)}{I_1(s)} = \frac{6s^4 + 57s^2 + 8}{14s^3 + 2s} \times \frac{I_1(s)}{I_1(s)}$$

$$= \frac{6s^4 + 57s^2 + 8}{14s^3 + 2s}$$

Problem No. 3.12 Determine the condition under which the input impedence of the network shown in Fig. 3.21 will be equal to *R*.

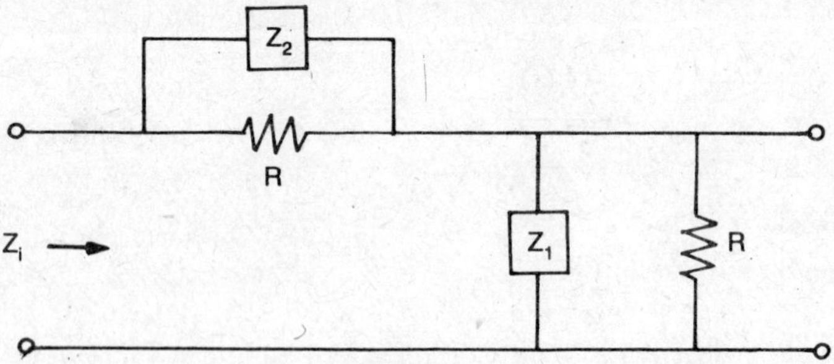

Fig. 3.21

Solution

$$Z_1 = \frac{RZ_1}{R + Z_1} + \frac{RZ_2}{R + Z_2}$$

or $$Z_1 = \frac{RZ_1(R + Z_2) + RZ_2(R + Z_1)}{(R + Z_1)(R + Z_2)}$$

or $$Z_1 = \frac{R^2Z_1 + RZ_1Z_2 + R^2Z_2 + RZ_1Z_2}{(R + Z_1)(R + Z_2)}$$

or $$Z_1 = \frac{R^2(Z_1 + Z_2) + 2RZ_1Z_2}{(R + Z_1)(R + Z_2)}$$

Putting $Z_1 = R$, we get

$$R = \frac{R^2(Z_1 + Z_2) + 2RZ_1Z_2}{(R + Z_1)(R + Z_2)}$$

or $$(R + Z_1)(R + Z_2) = R(Z_1 + Z_2) + 2RZ_1Z_2$$

or $$R^2 + RZ_2 + RZ_1 + Z_1Z_2 = RZ_1 + RZ_2 + 2Z_1Z_2$$

or $$R^2 = 2Z_1Z_2 - Z_1Z_2$$

or $$R = \sqrt{Z_1Z_2}$$

Problem No. 3.13 Determine the transfer function $\frac{V(s)}{I(s)}$ of the network shown in Fig. 3.22.

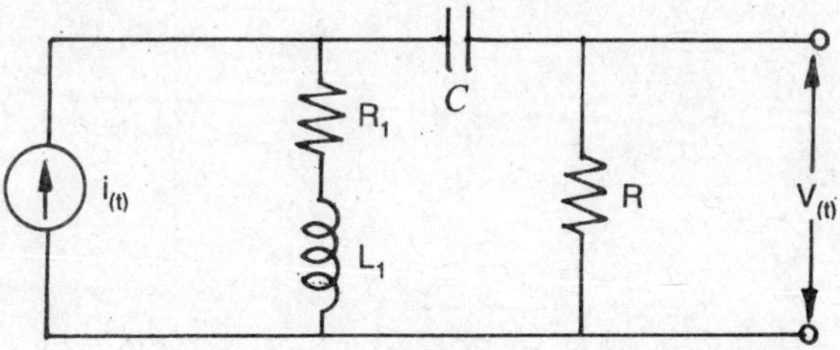

Fig. 3.22

Solution The transformed network is shown in Fig. 3.23.

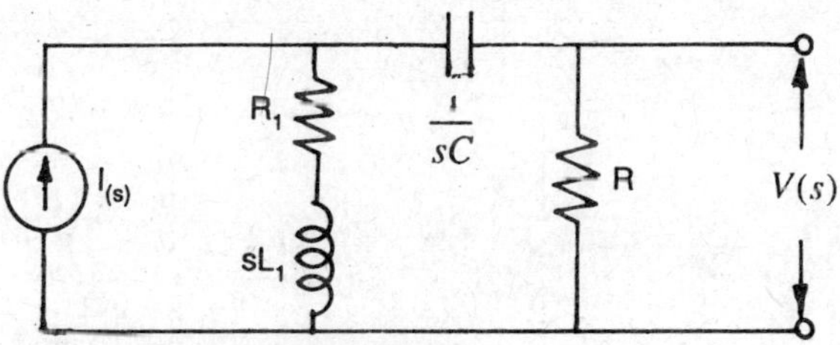

Fig. 3.23

$$Z = \frac{(R_1 + sL_1) \times \left(R + \frac{1}{sC}\right)}{(R_1 + sL_1) \times \left(R + \frac{1}{sC}\right)}$$

$$= \frac{(R_1 + sL_1) \times \left(\frac{sRC + 1}{sC}\right)}{(R_1 + sL_1) \times \left(\frac{sRC + 1}{sC}\right)}$$

$$= \frac{(R_1 + sL_1) \times \left(\frac{sRC+1}{sC}\right)}{\frac{sR_1C + s^2L_1C + sRC + 1}{sC}}$$

$$= \frac{(R_1 + sL_1)(sRC+1)}{s^2L_1C + sC(R+R_1)+1}$$

$$\frac{V(s)}{I(s)} = \frac{ZI(s)}{I(s)}$$

$$= Z$$

$$= \frac{(R_1 + sL_1)(sRC+1)}{s^2L_1C + sC(R+R_1)+1}$$

Problem No. 3.14 For the two port network shown in Fig. 3.24, determine the driving point impedence $Z_{11}(s)$, the transfer impedence $Z_{21}(s)$ and the voltage transfer ratio $G_{21}(s)$.

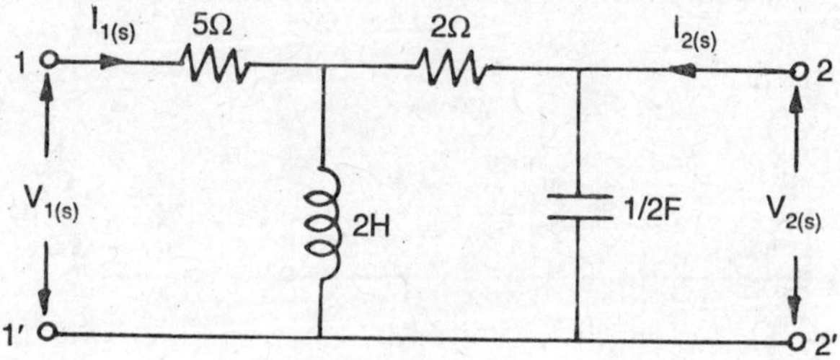

Fig. 3.24

Solution

The equivalent transformed network is shown in Fig. 3.25.

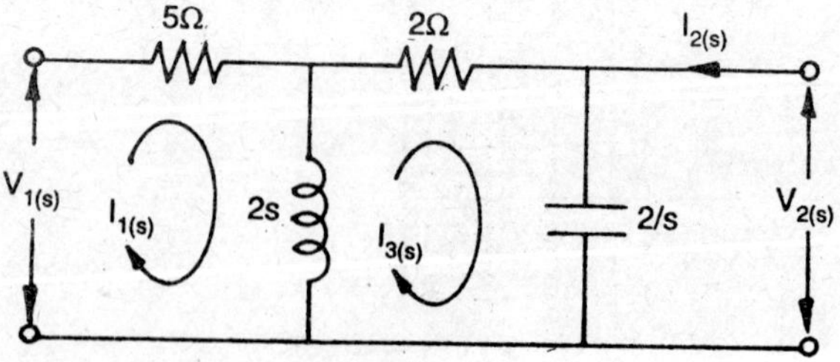

Fig. 3.25

$$Z_{11}(s) = \frac{V_1(s)}{I_1(s)} = \left[\frac{\left(2+\frac{2}{s}\right)\times 2s}{2+\frac{2}{s}+2s}+5\right]$$

$$= \frac{(2s+2)2s}{2s+2+2s^2}+5$$

$$= \frac{4s^2+4s+10s+10+10s^2}{2s^2+2s+2}$$

$$= \frac{14s^2+14s+10}{2s^2+2s+2}$$

$$= \frac{7s^2+7s+5}{s^2+s+1}$$

Writing mesh equations, give

$$V_1(s) = 5I_1(s)+2sI_1(s)-2sI_3(s)$$

or

$$V_1(s) = 7I_1(s)-2sI_3(s) \qquad ...(1)$$

and

$$0 = 2I_3(s)+\frac{2}{s}I_3(s)+2sI_3(s)-2sI_1(s)+\frac{2}{s}I_2(s)$$

or

$$0 = I_3(s)\left(2+\frac{2}{s}+2s\right)-2sI_1(s)+\frac{2}{s}I_2(s)$$

or

$$0 = \left(\frac{2s+2+2s^2}{s}\right)I_3(s)-2sI_1(s)+\frac{2}{s}I_2(s) \qquad ...(2)$$

and

$$V_2(s) = \frac{2}{s}I_2(s)+\frac{2}{s}I_3(s) \qquad ...(3)$$

From equation (2), we get

$$2sI_1(s) = \frac{2s+2+s^2}{s}I_3(s) \qquad ...(4)$$

From equation (3), we get

$$\boldsymbol{V_2(s) = \frac{2}{s}I_3(s)} \qquad ...(5)$$

From equation (4), we get

$$I_3(s) = \frac{2s^2}{2s+2+s^2} I_1(s)$$

or $$I_3(s) = \frac{s^2}{s^2+s+1} I_1(s) \quad ...(6)$$

and $$I_1(s) = \frac{s^2+s+1}{s^2} I_3(s) \quad ...(7)$$

Transfer Impedence at port 2 is

$$Z_{21}(s) = \frac{V_2(s)}{I_1(s)}$$

substituting values of $V_2(s)$ and $I_1(s)$ from equations (5) and (7) respectively, we get

$$Z_{21}(s) = \frac{\frac{2}{s} I_3(s)}{\frac{s^2+s+1}{s^2} I_3(s)}$$

$$= \frac{2s}{s^2+s+1}$$

Voltage transfer ratio $\boldsymbol{\frac{V_2(s)}{I_1(s)} = G_{21}(s)}$

Substituting value of $I_1(s)$ from equation (7) in equation (1), we get

$$V_1(s) = \frac{7(s^2+s+1)}{s^2} I_3(s) - 2sI_3(s)$$

or $$V_1(s) = \left[\frac{7s^2+7s+7-2s^3}{s^2}\right] I_3(s) \quad ...(8)$$

Therefore

$$G_{21}(s) = \frac{V_2(s)}{I_1(s)} = \frac{\frac{2}{3} I_3(s)}{\frac{7s^2+7s+7-2s^3}{s^2} I_3(s)}$$

$$= \frac{2s}{-2s^3 + 7s^2 + 7s + 7}$$

Problem No. 3.15 For the network shown in Fig. 3.26 determine

(a) $Z_{21}(s)$

(b) $Y_{21}(s)$

(c) $G_{21}(s)$

(d) $\alpha_{21}(s)$

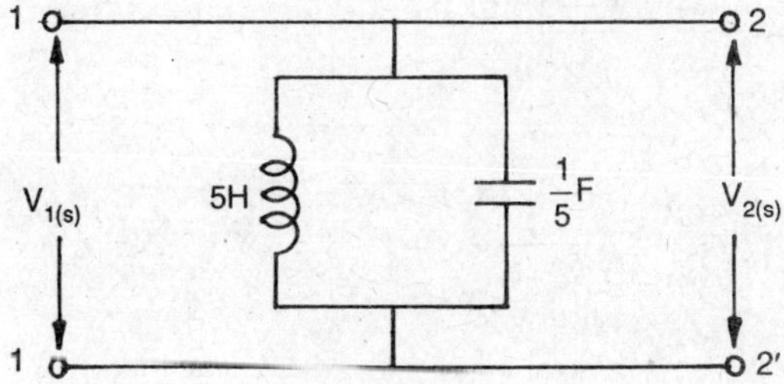

Fig. 3.26

Solution

Equivalent transformed network is shown in Fig. 3.27

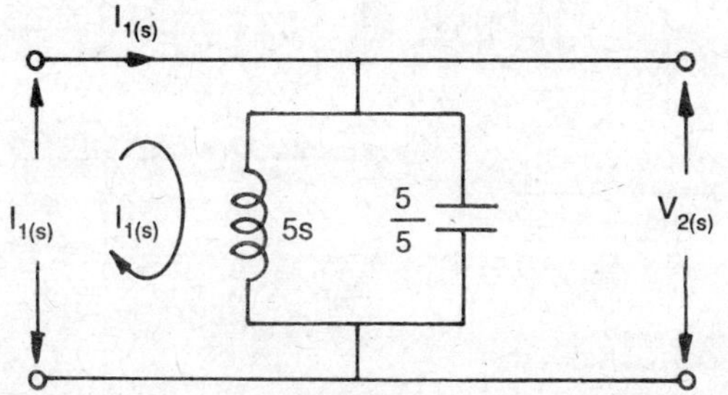

Fig. 3.27

Reducing the parallel circuit gives

$$Z = \frac{5s \times \frac{5}{s}}{5s + \frac{5}{s}} = \frac{25s}{5s^2 + 5} = \frac{5s}{s^2 + 1}$$

Mesh equation is

$$V_1(s) = \frac{5s}{s^2+1} I_1(s)$$

$$V_2(s) = \frac{5s}{s^2+1} I_1(s)$$

$$Z_{21}(s) = \frac{V_2(s)}{I_1(s)} = \frac{\frac{5s}{s^2+1} I_1(s)}{I_1(s)} = \frac{5s}{s^2+1}$$

$$Y_{21}(s) = \frac{I_1(s)}{V_2(s)} = \frac{s^2+1}{5s}$$

$$G_{21}(s) = \frac{V_2(s)}{I_1(s)} = \frac{\frac{5s}{s^2+1} I_1(s)}{\frac{5s}{s^2+1} I_1(s)} = 1$$

$$\alpha_{21}(s) = \frac{I_2(s)}{I_1(s)} = \frac{I_1(s)}{I_1(s)} = 1$$

Problem No. 3.16 Determine the driving point impedence of the network shown in Fig. 3.28 and hence plot its poles and zeroes in the s-plane.

Solution

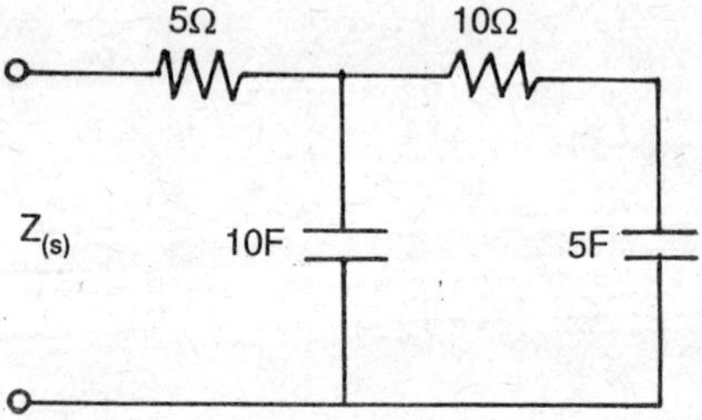

Fig. 3.28

$$Z(s) = \frac{\left(10 + \frac{1}{5s}\right) \times \frac{1}{10s}}{10 + \frac{1}{5s} + \frac{1}{10s}} + 5$$

$$= \frac{\frac{1+50s}{50s^2}}{\frac{100s+2+1}{10s}} + 5$$

$$= \frac{1+50s}{5s} \times \frac{1}{3+100s} + 5$$

$$Z(s) = \frac{50s+1}{15s+500s^2} + 5$$

$$= \frac{50s + 1 + 75s + 2500s^2}{500s^2 + 15s}$$

$$= \frac{2500s^2 + 125s + 1}{500s^2 + 15s}$$

$$= \frac{2500s^2 + 125s + 1}{s(500s + 15)}$$

$$= \frac{5s^2 + 0.25s + 0.0004}{s(s + 0.03)}$$

$$= \frac{(s + 0.05)(s + 0.0016)}{s(s + 0.03)}$$

Zeros are $s = -0.05, -0.0016$, and

Poles are $s = 0, -0.03$

Poles and zeros are shown in Fig. 3.29

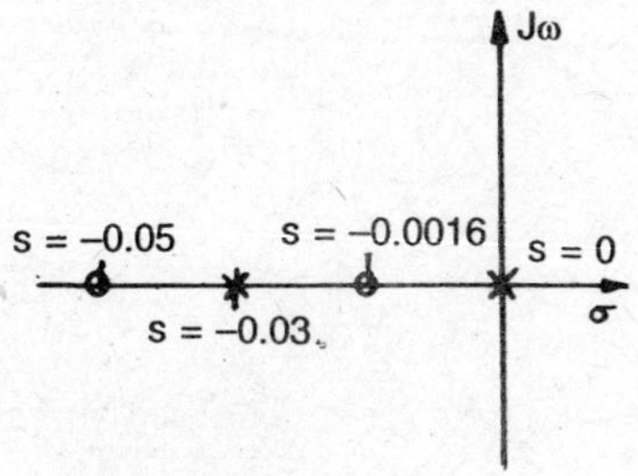

Fig. 3.29

Problem No. 3.17 For the network shown in Fig. 3.30, obtain the voltage ratio transfer function

$$G_{21}(s) = \frac{V_2(s)}{V_1(s)} \quad \text{Take L = 1H, C = 1F}$$

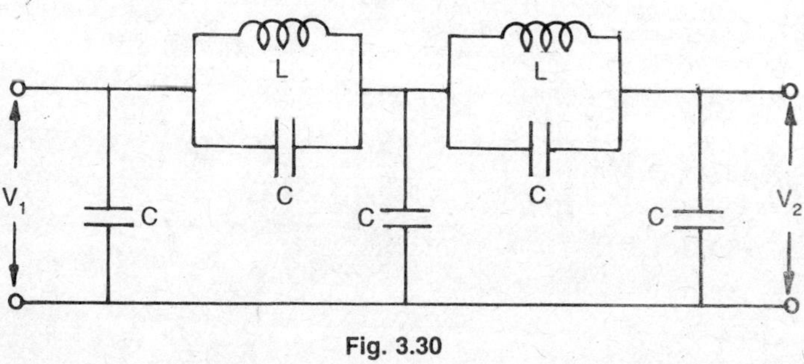

Fig. 3.30

Solution

Equivalent transformed network, after combining the parallel L and C branches is shown in Fig. 3.31.

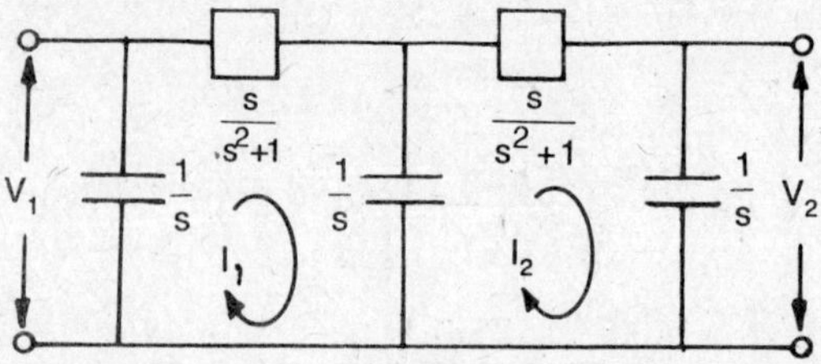

Fig. 3.31

Writing mesh equations

$$\frac{s}{s^2+1}I_1+\frac{1}{s}[I_1-I_2]+\frac{1}{s}I_1=0$$

$$\left(\frac{s}{s^2+1}+\frac{1}{s}+\frac{1}{s}\right)I_1-\frac{1}{s}I_2=0$$

$$\left(\frac{s}{s^2+1}+\frac{2}{s}\right)I_1-\frac{1}{s}I_2=0$$

or $$\left(\frac{s^2+2s^2+2}{s(s^2+1)}\right)I_1-\frac{1}{s}I_2=0$$

or $$\frac{3s^2+2}{s(s^2+1)}I_1-\frac{1}{s}I_2=0 \qquad ...(1)$$

and $$\frac{s}{s^2+1}I_2+\frac{1}{s}I_2+\frac{1}{s}I_2-\frac{1}{s}I_1=0$$

or $$\left(\frac{s}{s^2+1}+\frac{2}{s}\right)I_2-\frac{1}{s}I_1=0$$

or $$\frac{s^2+2s^2+2}{s(s^2+1)}I_2-\frac{1}{s}I_1=0$$

or $$\frac{3s^2+2}{s(s^2+1)}I_2-\frac{1}{s}I_1=0 \qquad ...(2)$$

and $$V_2=\frac{I_2}{s} \qquad ...(3)$$

and $$V_1 = \frac{I_1}{s}$$

Equation (2) gives

$$I_2 = \frac{1}{s} \times \frac{s(s^2+1)}{3s^2+2} I_1$$

or $$I_2 = \frac{s^2+1}{3s^2+2} I_1 \qquad \text{...(5)}$$

or $$I_1 = \frac{3s^2+2}{s^2+1} I_2 \qquad \text{...(6)}$$

$$G_{21} = \frac{V_2}{V_1}$$

Substituting the value of V_2 and V_1 from equations (3) and (4), we get

$$G_{21} = \frac{\frac{1}{s} I_2}{\frac{1}{s} I_1}$$

Substituting the value of I_2 from equation (5), we get

$$G_{21} = \frac{\frac{1}{s} \times \frac{s^2+1}{3s^2+2} I_1}{\frac{1}{s} I_1} = \frac{s^2+1}{3s^2+2}$$

Problem No. 3.18 Determine the driving point impedence of the network shown in Fig. 3.32 and locate the poles and zones of the network in s-plane.

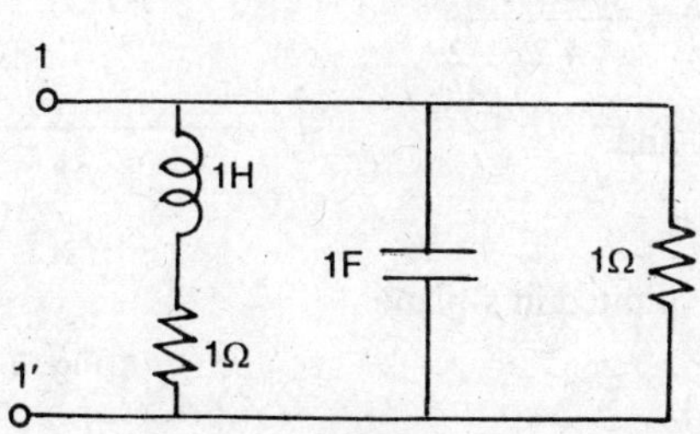

Fig. 3.32

Solution Equivalent transformed network is shown in Fig. 3.33.

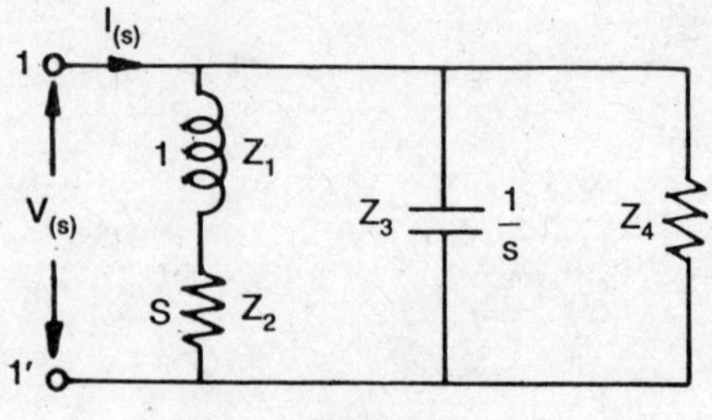

Fig. 3.33

Reducing the parallel combination one by one, we get

$$Z_{34} = \frac{1 \times \frac{1}{s}}{1 + \frac{1}{s}} = \frac{1}{s+1}$$

$$Z_{12} = 1 + s$$

$$Z(s) = Z_{1234} = \frac{\frac{1}{(s+1)} \times (1+s)}{\frac{1}{1+s} + \frac{1+s}{1}}$$

$$= \frac{1}{\frac{1+(1+s)(s+1)}{s+1}}$$

$$= \frac{s+1}{2+s+s^2+s}$$

or $$Z(s) = \frac{s+1}{s^2+2s+2}$$

Zeros are at $s = -1$ and

Poles are at $s = -1 \pm j$

Poles and zeros are plotted in s-plane in Fig. 3.34.

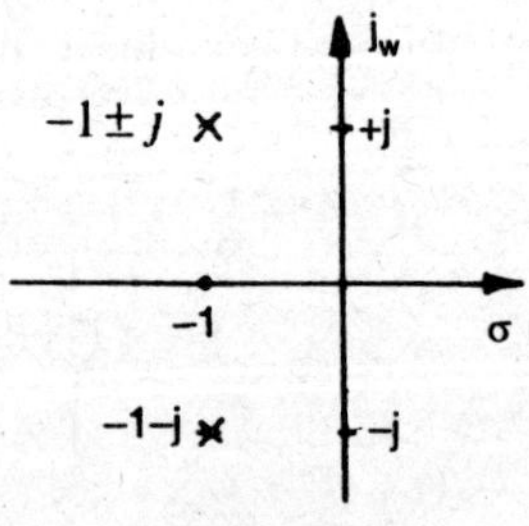

Fig. 3.34

Problem No. 3.19 For the network shown in Fig. 3.35, find the driving point impedance, $Z(s)$ and locate the poles and zeros of the impedance function in s-plane.

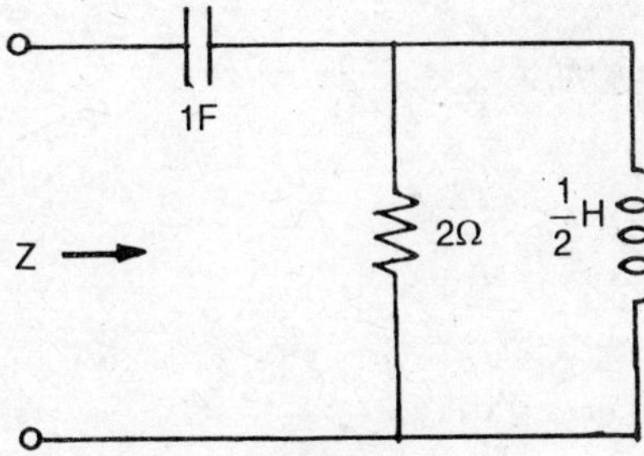

Fig. 3.35

Solution Equivalent transformed network is shown in Fig. 3.36

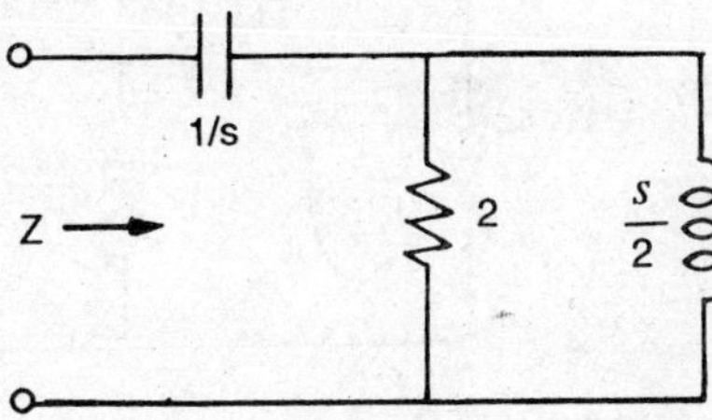

Fig. 3.36

$$Z(s) = \frac{2 \times s/2}{2 + \frac{s}{2}} = \frac{2s}{s+2}$$

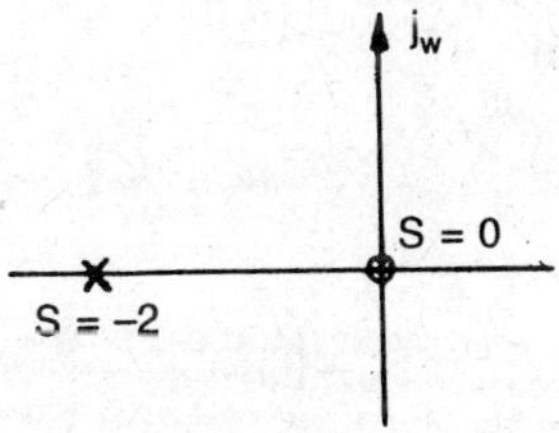

Pole is $s = -2$ and zero is $s = 0$ and are shown plotted in Fig. 3.37.

Fig. 3.37

Problem No. 3.20 For the network shown in Fig. 3.38, determine

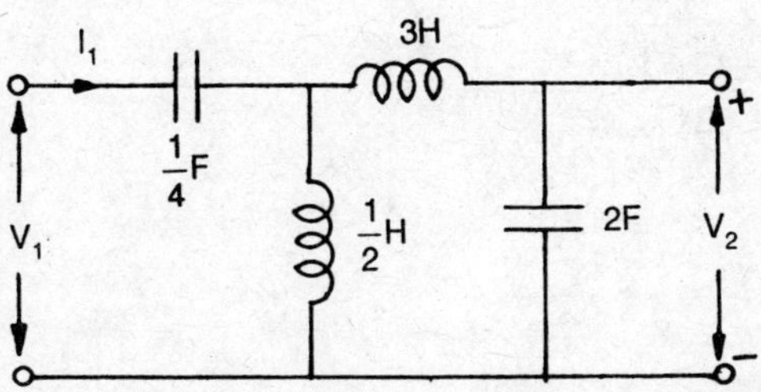

Fig. 3.38

(a) Voltage transfer ratio $\dfrac{V_2(s)}{V_1(s)}$

(b) Current transfer ratio $\dfrac{I_2(s)}{I_1(s)}$

(c) Transfer impedance $Z_{21}(s) = \dfrac{V_2(s)}{V_1(s)}$

Solution

Equivalent transformed network is shown in Fig. 3.39. Mesh equations are

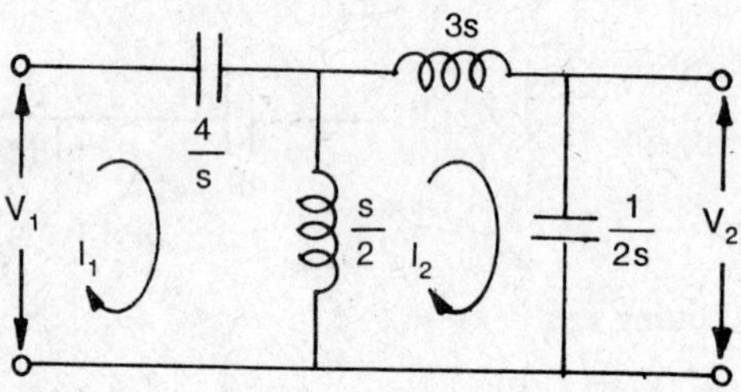

Fig. 3.39

$$\frac{4}{s}I_1 + \frac{s}{2}I_1 - \frac{s}{2}I_2 = V_1$$

$$\left(\frac{4}{s} + \frac{s}{2}\right)I_1 - \frac{s}{2}I_2 = V_1$$

$$\frac{8+s^2}{2s}I_1 - \frac{s}{2}I_2 = V_1 \qquad ...(1)$$

and

$$3sI_2 + \frac{1}{2s}I_2 + \frac{s}{2}I_2 - \frac{s}{2}I_1 = 0$$

$$\left(3s + \frac{1}{2s} + \frac{s}{2}\right)I_2 - \frac{s}{2}I_1 = 0$$

$$\frac{6s^2+1+s^2}{2s}I_2 - \frac{s}{2}I_1 = 0$$

$$\frac{7s^2+1}{2s}I_2 - \frac{s}{2}I_1 = 0 \qquad ...(2)$$

which gives

$$I_1 = \frac{2}{s} \times \frac{7s^2+1}{2s}$$

or

$$I_1 = \frac{7s^2+1}{s^2}I_2 \qquad ...(3)$$

or $$I_2 = \frac{s^2}{7s^2+1} I_1 \qquad ...(4)$$

and $$V_2 = \frac{1}{2s} I_2 \qquad ...(5)$$

Putting the value of I_1 from equation (3) in equation (1), we get

$$\frac{8+s^2}{2s} \times \frac{7s^2+1}{s^2} I_2 - \frac{s}{2} I_2 = V_1$$

or $$\left[\frac{(s^2+8)(7s^2+1)}{2s^3} - \frac{s}{2}\right] I_2 = V_1$$

or $$\left[\frac{(s^2+8)(7s^2+1)-s^4}{2s^3}\right] I_2 = V_1 \qquad ...(6)$$

Putting the value of V_2 and V_1 from equations (5) and (6) to obtain $\frac{V_2}{V_1}$, we get

$$\frac{V_2}{V_1} = \text{Voltage transformation}$$

$$= \frac{\frac{1}{2s} I_2}{\frac{(s^2+8)(7s^2+1)-s^4}{2s^3} I_2}$$

$$= \frac{s^2}{6s^4+58s^2+8} \qquad ...(7)$$

Current transfer ratio $\left(\frac{I_2}{I_1}\right)$

Putting the value of I_2 from equation (1), we get

$$\frac{I_2}{I_1} = \frac{\frac{s^2}{7s^2+1} I_1}{I_1} = \frac{s^2}{7s^2+1} \qquad ...(8)$$

Transfer impedance $\left(\dfrac{V_2(s)}{I_1(s)}\right)$

Putting the value of V_2 from equation (5), we get

$$\frac{V_{2(s)}}{I_{1(s)}} = \frac{\frac{1}{2s} I_2}{I_1} \quad \text{...(9)}$$

Putting the value of $\dfrac{I_2}{I_1}$ from equation (8), we get

$$\frac{V_2(s)}{I_1(s)} = \frac{1}{2s} \times \frac{s^2}{7s^2+1} = \frac{s}{2(7s^2+1)}$$

Problem No. 3.21 Find the driving point impedance of the circuit shown in Fig. 3.40.

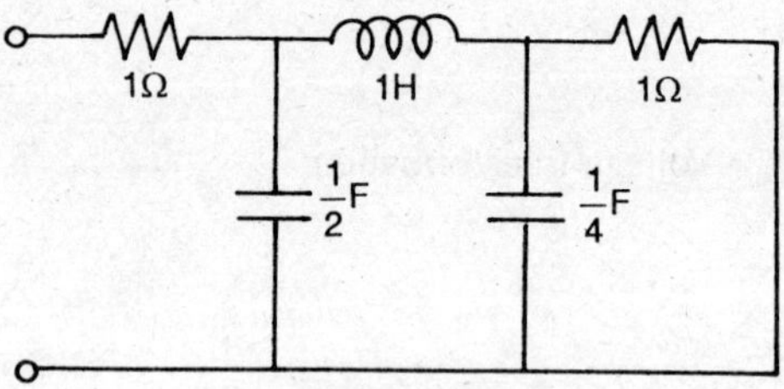

Fig. 3.40

Solution The transformed network corresponding to the given network is shown in Fig. 3.41.

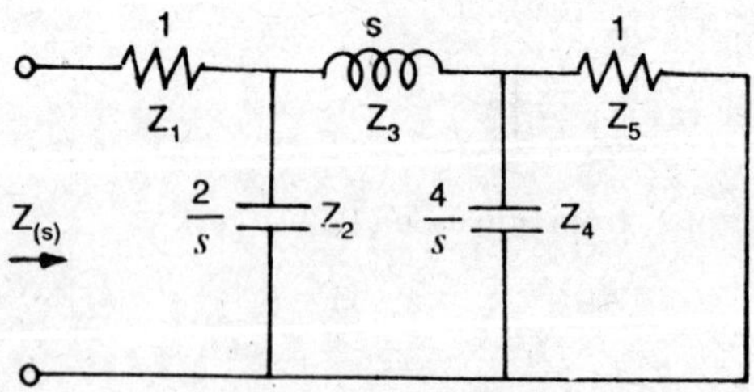

Fig. 3.41

$$Z_{45} = \frac{1 \times \frac{4}{s}}{1 + \frac{4}{s}} = \frac{4}{s+4}$$

$$Z_{345} = s + \frac{4}{s+4} = \frac{s^2+4s+4}{s+4}$$

$$Z_{2345} = \frac{\frac{2}{s} \times \frac{s^2+4s+4}{s+4}}{\frac{2}{s} + \frac{s^2+4s+4}{s+4}}$$

$$Z_{2345} = \frac{2(s^2+4s+4)}{2s+8+s^3+4s^2+4s} = \frac{2(s^2+4s+4)}{s^3+4s^2+6s+8}$$

$$Z(s) = Z_{12345} = 1 + \frac{2(s^2+4s+4)}{s^3+4s^2+6s+8}$$

$$= \frac{s^3+4s^2+6s+8+2s^2+8s+8}{s^3+4s^2+6s+8}$$

$$= \frac{s^3+6s^2+14s+16}{s^3 \mathbin{\text{ı}} 4s^2+6s+8}$$

Problem 3.22 Find the open-circuit and short-circuit parameters of the two terminal pair network shown in Fig. 3.42

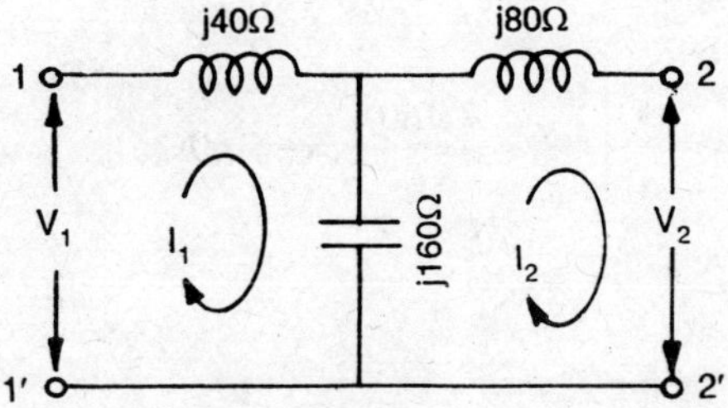

Fig. 3.42

Solution

Open Circuit Parameter

(a) $Z_{11} = \left.\frac{V_1}{I_1}\right|_{I_2=0}$

Writing mesh equation with $I_2 = 0$

$$V_1 = I_1(j40 - j160) = -j120I_1$$

Therefore,

$$Z_{11} = \frac{-j120I_1}{I_2} = -j120\Omega$$

(b) $Z_{12} = \left.\frac{V_1}{I_2}\right|_{I_1=0}$

Writing mesh equation with $I_1 = 0$, we get

$$V_1 = -j160I_2$$

Therefore,

$$Z_{12} = \left.\frac{V_1}{I_2}\right|_{I_1=0} = \frac{-j160I_2}{I_2} = -j160\Omega$$

(c) $Z_{21} = \left.\frac{V_2}{I_1}\right|_{I_2=0}$

Mesh equation with $I_2 = 0$, we gives

$$V_2 = -j160I_1$$

Therefore,

$$Z_{21} = \left.\frac{V_2}{I_1}\right|_{I_2=0} = \frac{-j160I_1}{I_1} = -j160\Omega$$

(d) $Z_{22} = \left.\frac{V_2}{I_2}\right|_{I_1=0}$

Mesh equation with $I_1 = 0$, gives

$$V_2 = (j80 - j160)I_2 = -j80I_2$$

Therefore,

$$Z_{22} = \left.\frac{V_2}{I_2}\right|_{I_1=0} = \frac{-j80I_2}{I_2} = -j80\Omega$$

Short-Circuit Parameter

$$\Delta Z = Z_{11}\,Z_{22} - Z_{12}\,Z_{21}$$
$$= (-j120)(-j80) - (-j160)(-j160)$$
$$= -9600 - (-25600)$$
$$= 16000.$$

$$Y_{11} = \frac{Z_{22}}{\Delta Z} = \frac{-j80}{16000} = \frac{-j}{200}\ \text{mho}$$

$$Y_{12} = \frac{-Z_{12}}{\Delta Z} = \frac{-(-j160)}{16000} = \frac{j}{100}\ \text{mho}$$

$$Y_{21} = \frac{-Z_{21}}{\Delta Z} = \frac{-(-j160)}{16000} = \frac{j}{100}\ \text{mho}$$

$$Y_{22} = \frac{-Z_{11}}{\Delta Z} = \frac{-(-j160)}{16000} = \frac{-j}{133.3}\ \text{mho}$$

Note Use Table 3 for conversion of Z-parameters with Y-parameters.

Problem No. 3.23 Determine h-parameters with the following data

(a) Output short circuited

$$V_1 = 25V,\ I_1 = 1A,\ I_2 = 2A$$

(b) With input terminals Open-circuited

$$V_1 = 10V,\ V_2 = 50V,\ I_2 = 2A$$

Solution

$$h_{11} = \left.\frac{V_1}{I_1}\right|_{V_2=0} \quad \text{with the output short-circuited}$$

$$= \frac{25}{1} = 25$$

$$h_{12} = \left.\frac{V_1}{V_2}\right|_{I_1=0} \quad \text{with the input open-circuited}$$

$$= \frac{10}{50} = 0.2$$

$$h_{21} = \left.\frac{I_2}{I_1}\right|_{V_2=0} \quad \text{with output short-circuited}$$

$$= \frac{2}{1}$$

$$= 2$$

$$h_{22} = \left.\frac{I_2}{V_2}\right|_{I_1=0} \quad \text{with input open-circuited}$$

$$= \frac{2}{50} = 0.04$$

In matrix form

$$\begin{bmatrix} V_1 \\ I_2 \end{bmatrix} = \begin{bmatrix} h_{11} & h_{12} \\ h_{21} & h_{22} \end{bmatrix} \begin{bmatrix} I_1 \\ V_2 \end{bmatrix}$$

$$\begin{bmatrix} V_1 \\ I_2 \end{bmatrix} = \begin{bmatrix} 25 & 0.2 \\ 2 & 0.04 \end{bmatrix} \begin{bmatrix} I_1 \\ V_2 \end{bmatrix}$$

Alternatively

An h-parameter equivalent network based on the given data is drawn in Fig. 3.43

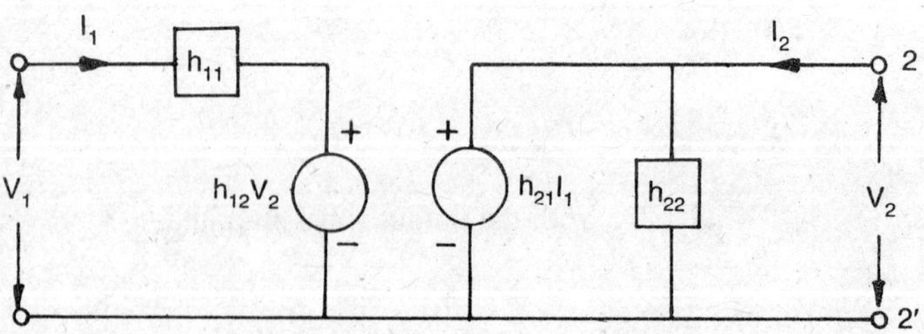

Fig. 3.43

Applying Kirchoffs law, we get

$$V_1 = h_{11}I_1 + h_{12}V_2 \quad ...(1)$$

$$I_2 = h_{21}I_1 + h_{22}V_2 \quad ...(2)$$

(a) when output shorted i.e. $V_2 = 0$, substitution of values given in the problem in equations (1) and (2), gives

$$25 = h_{11} \times 1 + h_{12} \times 0 = h_{11}$$

and $$2 = h_{22} \times 1 = h_{22}$$

(b) when the input is open circuited i.e., $I_1 = 0$, substitution of values given in the problem in equations (1) and (2) gives

$$10 = h_{11} \times 0 + h_{12} \times 50 = 50h_{12}$$

or $$h_{12} = \frac{10}{50} = 0.2$$

and $$2 = h_{11} \times 0 + h_{22} \times 50 = 50h_{22}$$

or $$h_{22} = \frac{2}{50} = 0.04$$

Problem No. 3.24 Two identical sections of the network shown in Fig. 3.43, (a) are connected is parallel. Obtain the Y-parameter of the resulting network and verify the result by direct calculations.

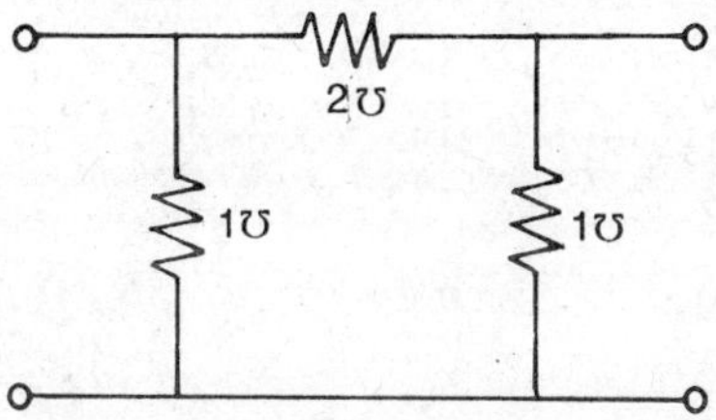

Fig. 3.43 (a)

Solution Let the two networks be N_a and N_b. We will calculate Y-parameter of one network. The Y-parameters of the other network will be same. Both the Y-parameters will then be added to obtain the result.

Network N_a (Fig. 3.44)

The node equations are

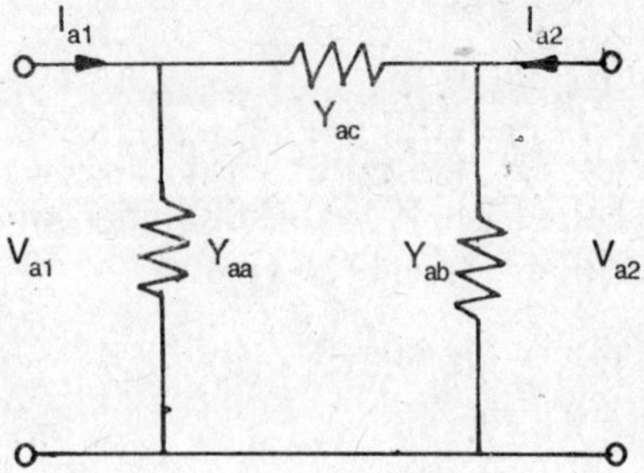

Fig. 3.44

$$I_{a1} = V_{a1}Y_{aa} + (V_{a1} - V_{a2})Y_{ac} = (Y_{aa} + Y_{ac})V_{a1} - Y_{ac}V_{a2} \qquad ...(1)$$

and

$$I_{a2} = V_{a2}Y_{ab} + (V_{a2} - V_{a1})Y_{ac} = -Y_{ac}V_{a1} + (Y_{ab} + Y_{ac})V_{a2} \qquad ...(2)$$

The terminal characteristics are given by

$$I_1 = Y_{11}V_1 + Y_{12}V_2\text{; and} \qquad ...(3)$$

$$I_2 = Y_{21}V_1 + Y_{22}V_2 \qquad ...(4)$$

Comparing equations (1) and (2) with (3) and (4) respectively, we get

$$Y_{11a} = Y_{11} = \left.\frac{I_1}{V_1}\right|_{V_2=0} = \left.\frac{I_{a1}}{V_{a1}}\right|_{V_{a2}=0} = Yaa + Yac = 1 + 2 = 3 \text{ mho}$$

$$Y_{22a} = Y_{22} = \left.\frac{I_2}{V_2}\right|_{V_1=0} = \left.\frac{I_{a2}}{V_{a2}}\right|_{V_{a1}=0} = Yab + Yac = 1 + 2 = 3 \text{ mho}$$

$$Y_{12a} = Y_{12} = \left.\frac{I_1}{V_1}\right|_{V_2=0} = \left.\frac{I_{a1}}{V_{a1}}\right|_{V_{a2}=0} = -Yac = -2 \text{ mho}$$

$$Y_{21a} = Y_{21} = \left.\frac{I_2}{V_1}\right|_{V_2=0} = \left.\frac{I_{a2}}{V_{a1}}\right|_{V_{a2}=0} = -Yac = -2 \text{ mho}$$

Similarly for network N_b which is identical to network N_a

$$Y_{11b} = 3 \text{ mho}$$

$$Y_{22b} = 3 \text{ mho}$$

$$Y_{12b} = Y_{21b} = -2 \text{ mho}$$

Since N_a and N_b are combined to be connected in parallel, the Y-parameters of the complete network are

$$Y_{11} = Y_{11a} + Y_{11b} = 3+3 = 6 \text{ mho}$$

$$Y_{22} = Y_{22a} + Y_{22b} = 3+3 = 6 \text{ mho}$$

$$Y_{12} = Y_{12a} + Y_{12b} = -2-2 = -4 \text{ mho}$$

$$Y_{21} = Y_{21a} + Y_{21b} = -2-2 = -4 \text{ mho}$$

Verification

Fig. 3.45 shows network N_a and identical network N_b in parallel

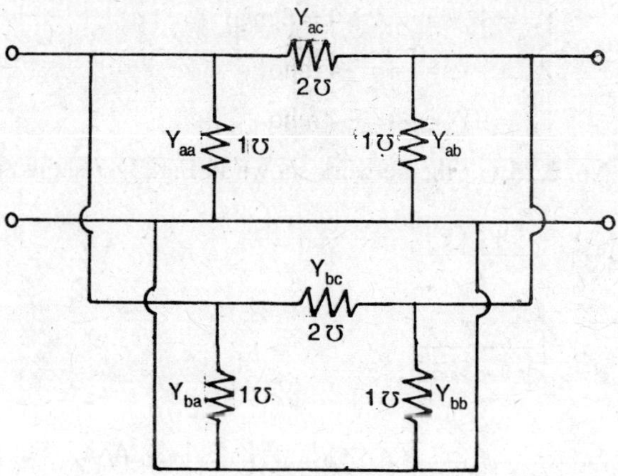

Fig. 3.45

The resultant equivalent network after combining the parallel branches, i.e.,

$$Y_c = Y_{ac} + Y_{bc} = 2+2 = 4 \text{ mho}$$

$$Y_a = Y_{aa} + Y_{ba} = 1+1 = 2 \text{ mho}$$

$$Y_a = Y_{aa} + Y_{ba} = 1+1 = 2 \text{ mho}$$

is shown in Fig. 3.46

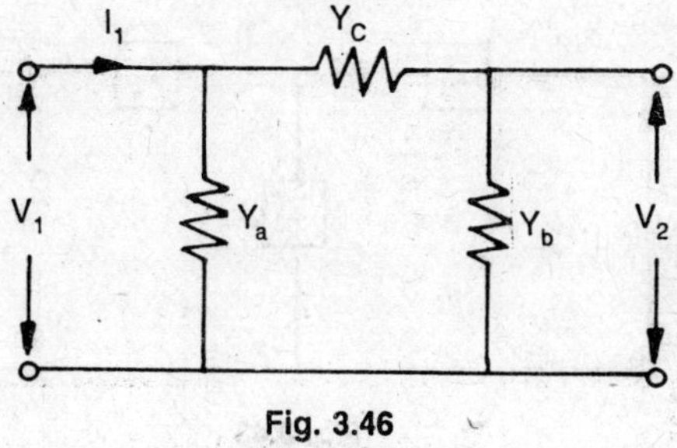

Fig. 3.46

The node equations are

$$I_1 = V_1 Y_a + (V_1 - V_2) Y_c = (Y_a + Y_c) V_1 - Y_c V_2 \quad ...(5)$$

and

$$I_2 = V_2 Y_b + (V_2 - V_1) Y_c = -Y_c V_1 + (Y_b + Y_c) V_2 \quad ...(6)$$

Comparing equations (5) and (6) with (1) and (2) respectively, we get

$$Y_{11} = Y_a + Y_c = 2 + 4 = 6 \text{ mho}$$

$$Y_{22} = Y_b + Y_c = 2 + 4 = 6 \text{ mho}$$

$$Y_{12} = -Y_c = -4 = -4 \text{ mho}$$

$$Y_{21} = -Y_c = -4 = -4 \text{ mho}$$

Problem No. 3.25 For the network shown in Fig. 3.47 show that

$$Z_{11}(s) = \frac{s^2 + 2s + 1}{s}$$

and $$Y_{11}(s) = \frac{3s}{3s^2 + 5s + 3}$$

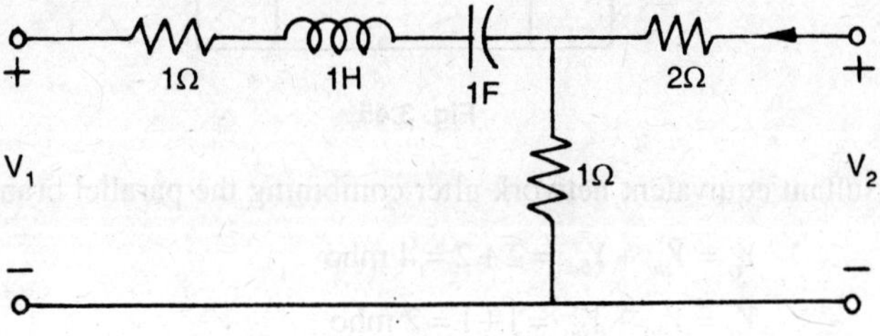

Fig. 3.47

Solution The equivalent transformed network is shown in Fig. 3.48.

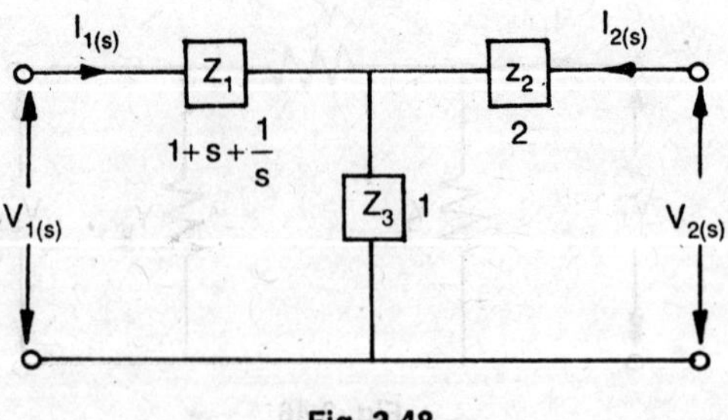

Fig. 3.48

$Z_{11}(s)$ Writing the mesh equation with $I_2(s) = 0$, we get

$$V_1(s) = \left(1 + s + \frac{1}{s}\right) I_1(s) + I_1(s) \text{ ; and}$$

$$= I_1(s)\left(1 + s + \frac{1}{s} + 1\right)$$

$$= I_1(s)\left(\frac{s + s^2 + 1 + s}{s}\right)$$

$$= I_1(s)\left(\frac{s^2 + 2s + 1}{s}\right)$$

Now

$$Z_{11}(s) = \left.\frac{V_1(s)}{I_1(s)}\right|_{I_2(s)=0} = \frac{\frac{\left(s^2 + 2s + 1\right)}{s} I_1(s)}{I_1(s)}$$

$$= \frac{\left(s^2 + 2s + 1\right)}{s}$$

Y-parameter $Y_{11}(s)$ Let us find first, other Z-parameters.

Writing node equations for Fig. 3.48, we get

$$V_1 = Z_1 I_1 + Z_3(I_1 + I_2) = (Z_1 + Z_3) I_1 + Z_3 I_2 \qquad ...(1)$$

$$V_2 = Z_2 I_2 + Z_3(I_1 + I_2) = Z_3 I_1 + (Z_2 + Z_3) I_2 \qquad ...(2)$$

The Z-parameters are given by

$$V_1 = Z_{11} I_1 + Z_{12} I_2 \qquad ...(3)$$

$$V_{21} = Z_{21} I_1 + Z_{22} I_2 \qquad ...(4)$$

Comparing equations (1) and (2), with (3) and (4) respectively, we get

$$Z_{11} = \left.\frac{V_1}{I_1}\right|_{I_2=0} = Z_1 + Z_3 = \left(1 + s + \frac{1}{s}\right) + 1$$

$$= \frac{s + s^2 + 1 + s}{s}$$

$$= \frac{s^2 + 2s + 1}{s} \qquad \text{(Already found out)}$$

$$Z_{12} = Z_{21} = Z_3 = 1$$

$$Z_{22} = Z_2 + Z_3 = 2 + 1 = 3$$

Now

$$\Delta Z = Z_{11}Z_{22} - Z_{12}Z_{21}$$

$$= \left(\frac{s^2 + 2s + 1}{s}\right) \times 3 - 1 \times 1$$

$$= \frac{3(s^2 + 2s + 1)}{s} - 1$$

$$= \frac{3(s^2 + 2s + 1) - s}{s}$$

$$= \frac{3s^2 + 5s + 3}{s}$$

Now

$$Y_{11}(s) = \frac{Z_{22}}{\Delta Z} = \frac{3}{\dfrac{3s^2 + 5s + 3}{s}}$$

$$= \frac{3s}{3s^2 + 5s + 3}$$

Problem No. 3.26 For the network shown in Fig. 3.49, where the Z-parameters are

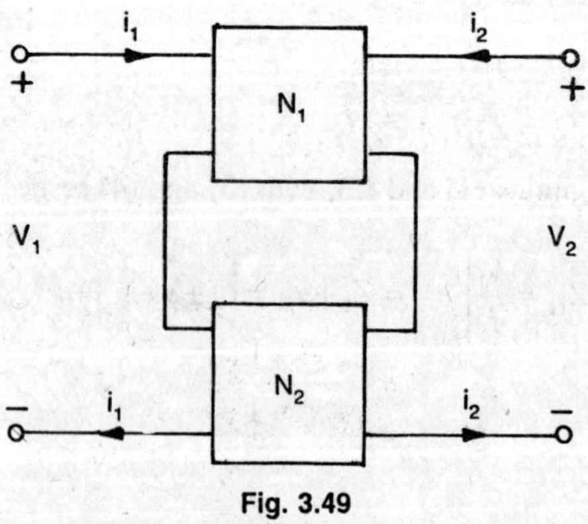

Fig. 3.49

$$Z_{n1} = \begin{bmatrix} 1 & 3 \\ 3 & 4 \end{bmatrix}, \quad Z_{n2} = \begin{bmatrix} 5 & 7 \\ 7 & 8 \end{bmatrix}, \text{ show that } \begin{bmatrix} v_1 \\ v_2 \end{bmatrix} = \begin{bmatrix} 6 & 10 \\ 10 & 12 \end{bmatrix} \begin{bmatrix} i_1 \\ i_2 \end{bmatrix}$$

Solution Fig. 3.49 depicts series connection of two port-networks N_1 and N_2. Fig. 3.49 is redrawn as Fig. 3.50, showing various input and output parameters of individual networks N_1 and N_2.

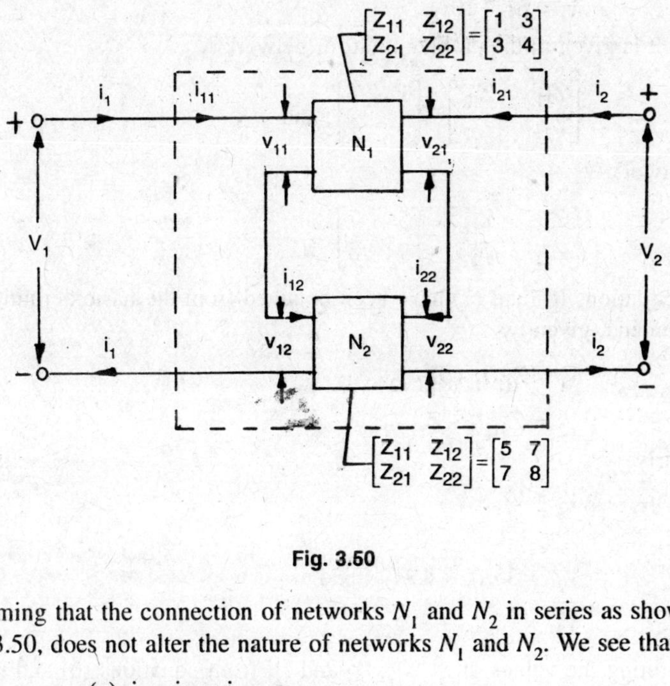

Fig. 3.50

Assuming that the connection of networks N_1 and N_2 in series as shown in Fig. 3.50, does not alter the nature of networks N_1 and N_2. We see that

(a) $i_1 = i_{11} = i_{12}$ ...(1)

(b) $i_2 = i_{21} = i_{22}$ ...(2)

(c) $V_1 = V_{11} + V_{12}$ ...(3)

(d) $V_2 = V_{21} + V_{22}$ ...(4)

Now

$V_1 = V_{11} + V_{12}$ ...(5)

$V_{11} = 1 \times i_{11} + 3 \times i_{21}$; and ...(6)

$V_{12} = 5 \times i_{12} + 7 \times i_{22}$ ...(7)

Substituting the values of i_{11}, i_{12}, i_{21} and i_{22} from equations (1) and (2) in equations (6) and (7), we get

$$V_{11} = i_1 + 3i_2 \text{ ; and} \quad \text{...(8)}$$

$$V_{12} = 5i_1 + 7i_2 \quad \text{...(9)}$$

Substituting values of V_{11} and V_{12} from equations (8) and (9) in equations (3), we get

$$V_1 = (1+5)i_1 + (3+7)i_2$$

or

$$V_1 = 6i_1 + 10i_2 \quad \text{...(10)}$$

Note: It is given in the problem that for network N_1

$$\begin{bmatrix} Z_{11} & Z_{12} \\ Z_{21} & Z_{22} \end{bmatrix} = \begin{bmatrix} 1 & 3 \\ 3 & 4 \end{bmatrix} \text{; and}$$

for network N_2

$$\begin{bmatrix} Z_{11} & Z_{12} \\ Z_{21} & Z_{22} \end{bmatrix} = \begin{bmatrix} 5 & 7 \\ 7 & 8 \end{bmatrix}$$

Also equations (6) and (7) have been obtained from the basic definition of Z-parameter, given by

$$V_1 = Z_{11}\, I_1 + Z_{22}\, I_2 \quad \text{...(11)}$$

$$V_2 = Z_{21}\, I_1 + Z_{22}\, I_2 \quad \text{...(12)}$$

Similarly

$$V_2 = V_{21} + V_{22} \quad \text{...(13)}$$

Now

$$V_{21} = 3 \times i_{11} + 4 \times i_{21} \text{ ; and} \quad \text{...(14)}$$

$$V_{22} = 7 \times i_{12} + 8 \times i_{22} \quad \text{...(15)}$$

Substituting the values of i_{11}, i_{12}, i_{21} and i_{22} from equations (6) and (7) in equations (14) and (15), we get

$$V_{21} = 3i_1 + 4i_2 \quad \text{...(15a)}$$

$$V_{22} = 7i_1 + 8i_2$$

Substituting values of V_{21} and V_{22} from equation (14) in equation (13), we get

$$V_2 = (3+7)i_1 + (4+8)i_2$$

or

$$V_2 = 10i_1 + 12i_2 \quad \text{...(16)}$$

Comparing equations (10) and (16) with equations (11) and (12), we can write

$$\begin{bmatrix} V_1 \\ V_2 \end{bmatrix} = \begin{bmatrix} 6 & 10 \\ 10 & 12 \end{bmatrix} = \begin{bmatrix} i_1 \\ i_2 \end{bmatrix}$$

Problem No. 3.27 Show that h-parameters will not exist for a two port network when $Z_{22} = 0$.

Solution For a two-port network, the Z-parameters equations are

$$V_1 = Z_{11} I_1 + Z_{12} I_2 \qquad ...(1)$$

$$V_2 = Z_{21} I_1 + Z_{22} I_2 \qquad ...(2)$$

Rewriting equation (2), we get

$$I_2 = \left(\frac{-Z_{21}}{Z_{22}}\right) I_1 + \left(\frac{1}{Z_{22}}\right) V_2 \qquad ...(3)$$

or $$I_2 = h_{22} I_1 + h_{22} V_2 \qquad ...(4)$$

Substituting value of I_2 from equation (3) in (1), we get

$$V_1 = Z_{11} I_1 + Z_{12} I_2$$

$$= Z_{11} I_1 + Z_{12} \left\{ \left(\frac{-Z_{21}}{Z_{22}}\right) I_1 + \frac{1}{Z_{22}} V_2 \right\}$$

$$= Z_{11} I_1 - \frac{Z_{12} Z_{21}}{Z_{22}} I_1 + \frac{Z_{12}}{Z_{22}} V_2$$

$$= Z_{11} Z_{22} I_1 - Z_{12} Z_{21} I_1 + \frac{Z_{12}}{Z_{22}} V_2$$

$$= \frac{(Z_{11} Z_{22} - Z_{12} Z_{21}) I_1}{Z_{22}} + \frac{Z_{12}}{Z_{22}} V_2$$

$$= \frac{\Delta Z}{Z_{22}} I_1 + \frac{Z_{12}}{Z_{22}} V_2 \quad (\Delta Z = Z_{11} Z_{22} - Z_{12} Z_{21})$$

or $$V_1 = h_{11} I_1 + h_{12} V_2 \qquad ...(5)$$

From equations (4) and (5) it is seen that

$$h_{11} = \frac{\Delta Z}{Z_{22}}; \quad h_{12} = \frac{Z_{12}}{Z_{22}};$$

$$h_{21} = \frac{-Z_{21}}{Z_{22}}; \text{ and}$$

$$h_{22} = \frac{1}{Z_{22}}$$

Therefore, if $Z_{22} = 0$, all the h-parameter will not exist.

Problem No. 3.28 Show that when two port networks N_1 and N_2 are connected in parallel, the equivalent Y-parameter of the combined network is

$$Y_{eq} = Y_{N1} + Y_{N2}$$

Solution In Fig. 3.51, networks N_1 and N_2 are shown connected in parallel

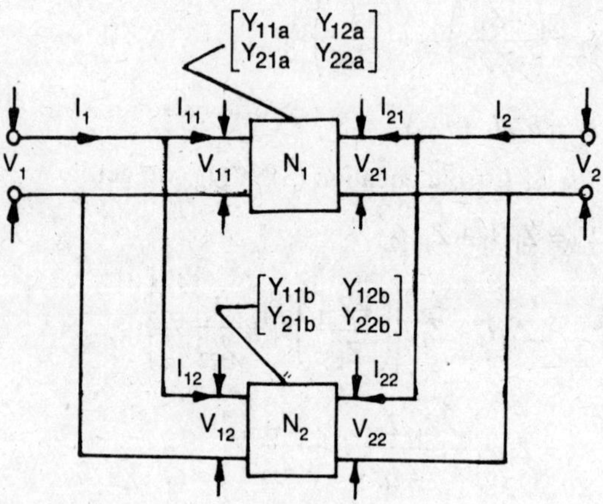

Fig. 3.51

From Fig. 3.51, it is seen that for **Network 'N_1'**

$$I_{11} = Y_{11a}V_{11} + Y_{12a}V_{21}; \text{ and} \quad ...(1)$$

$$I_{21} = Y_{21a}V_{11} + Y_{22a}V_{21} \quad ...(2)$$

Similarly for **Network 'N_2'**

$$I_{12} = Y_{11b}V_{12} + Y_{12b}V_{22}; \text{ and} \quad ...(3)$$

$$I_{22} = Y_{21b}V_{12} + Y_{22b}V_{22} \quad ...(4)$$

Also

$$V_1 = V_{11} = V_{12}; \text{ and} \quad ...(5)$$

$$V_2 = V_{21} = V_{22} \quad ...(6)$$

and
Also,

$$I_1 = I_{11} + I_{12}\text{; and} \qquad ...(7)$$

$$I_2 = I_{21} + I_{22} \qquad ...(8)$$

Equation (7) after substitution of values from equations (1) and (3), gives

$$I_1 = I_{11} + I_{12}$$

$$= Y_{11a}V_{11} + Y_{12a}V_{21} + Y_{11b}V_{12} + Y_{12b}V_{22}$$

$$= Y_{21a}V_1 + Y_{12a}V_2 + Y_{11b}V_1 + Y_{12b}V_2$$

or

$$I_1 = (Y_{11a} + Y_{11b})V_1 + (Y_{12a} + Y_{12b})V_2$$

or

$$I_1 = Y_{11}V_1 + Y_{12}V_2 \qquad ...(9)$$

Equation (8) after substitution of values from equation (2) and (4), gives

$$I_2 = I_{21} + I_{22}$$

$$= Y_{21a}V_{11} + Y_{22a}V_{21} + Y_{21b}V_{12} + Y_{22b}V_{22}$$

$$= Y_{21a}aV_1 + Y_{22a}V_2 + Y_{21b}V_1 + Y_{22b}V_2$$

$$= (Y_{21a} + Y_{21b})V_1 + (Y_{22a} + Y_{22b})V_2$$

or

$$I_2 = Y_{21}V_1 + Y_{22}V_2 \qquad ...(10)$$

The combined Y-parameter is given by

$$\begin{bmatrix} I_1 \\ I_2 \end{bmatrix} = \begin{bmatrix} Y_{11} & Y_{12} \\ Y_{21} & Y_{22} \end{bmatrix} \begin{bmatrix} V_1 \\ V_2 \end{bmatrix}$$

where

$$\left.\begin{aligned} Y_{11} &= Y_{11a} + Y_{11b} \\ Y_{12} &= Y_{12a} + Y_{12b} \\ Y_{21} &= Y_{21a} + Y_{21b} \\ Y_{22} &= Y_{22a} + Y_{22b} \end{aligned}\right\} \qquad ...(11)$$

or

$$\begin{bmatrix} I_1 \\ I_2 \end{bmatrix} = \begin{bmatrix} Y_{11a} & Y_{12a} \\ Y_{21a} & Y_{22a} \end{bmatrix} \begin{bmatrix} V_1 \\ V_2 \end{bmatrix} + \begin{bmatrix} Y_{11b} & Y_{12b} \\ Y_{21b} & Y_{22b} \end{bmatrix} \begin{bmatrix} V_1 \\ V_2 \end{bmatrix}$$

or Equation (11), shows that

$$Y_{eq} = Y_{N1} + Y_{N2}$$

Problem No. 3.29 Determine Z, h and Y-parameters of the network shown in Fig. 3.52.

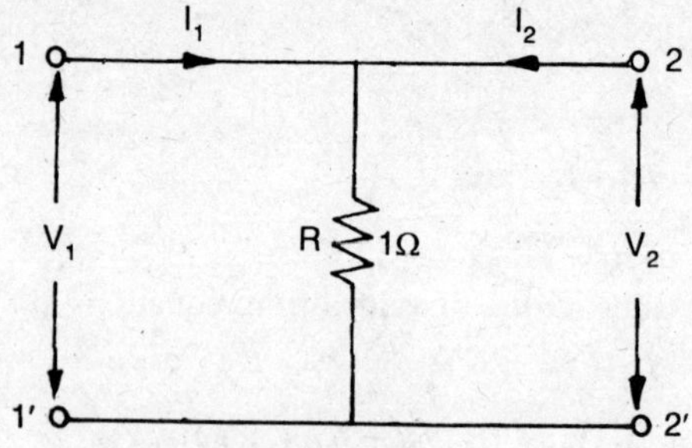

Fig. 3.52

Solution

(a) **Z-parameter**

(i) $Z_{11} = \left.\frac{V_1}{I_1}\right|_{I_2=0} = \frac{I_1 R}{I_1} = R = 1\Omega$

$Z_{22} = \left.\frac{V_2}{I_2}\right|_{I_1=0} = \frac{I_2 R}{I_2} = R = 1\Omega$

$Z_{12} = \left.\frac{V_1}{I_2}\right|_{I_1=0} = \frac{I_2 R}{I_2} = R = 1\Omega$

$Z_{21} = \left.\frac{V_2}{I_1}\right|_{I_2=0} = \frac{I_1 R}{I_1} = R = 1\Omega$

Therefore, Z-parameter are

$$Z = \begin{bmatrix} 1 & 1 \\ 1 & 1 \end{bmatrix}$$

$$\Delta Z = Z_{11}Z_{22} - Z_{12}Z_{21} \quad ...(1)$$

$$= 1 \times 1 - 1 \times 1$$
$$= 0$$

h-parameter

$$h_{11} = \frac{\Delta Z}{Z_{22}} = \frac{0}{1} = 0$$

$$h_{12} = \frac{Z_{12}}{Z_{22}} = \frac{1}{1} = 1$$

$$h_{21} = \frac{-Z_{21}}{Z_{22}} = \frac{-1}{1} = -1$$

$$h_{22} = \frac{1}{Z_{22}} = \frac{1}{1} = 1$$

Therefore,

$$h = \begin{bmatrix} 0 & 1 \\ -1 & 1 \end{bmatrix}$$

Y-parameters

$$Y_{11} = \frac{Z_{22}}{\Delta Z}, \quad Y_{12} = \frac{-Z_{12}}{\Delta Z}, \quad Y_{21} = \frac{-Z_{21}}{\Delta Z}$$

and $$Y_{22} = \frac{Z_{11}}{\Delta Z}$$

Since $\Delta Z = 0$, parameter will not exist.

Problem No. 3.30 Determine V_2/I in terms of R_1, R_2 and Z-parameters of network N of Fig. 3.53 when

$$Z_N = \begin{bmatrix} Z_{11} & Z_{12} \\ Z_{21} & Z_{22} \end{bmatrix}$$

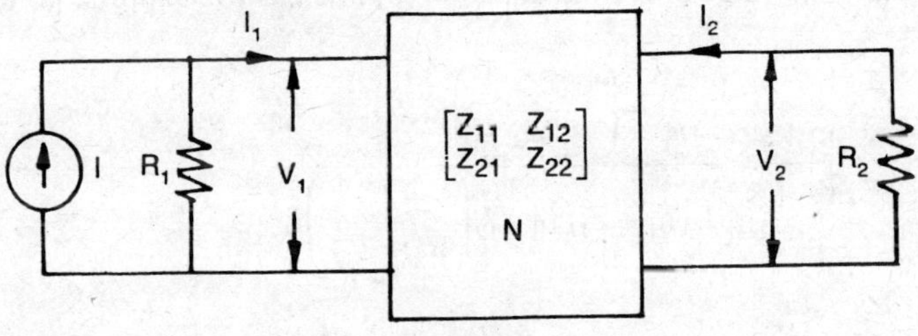

Fig. 3.53

Solution

We know that

$$V_1 = Z_{11}I_1 + Z_{12}I_2; \qquad ...(1)$$

$$V_2 = Z_{21}I_1 + Z_{22}I_2; \qquad ...(2)$$

Also

$$V_2 = -I_2R_2 \qquad ...(3)$$

and

$$V_1 = (I - I_1) R_1 \quad \text{...(4)}$$

or

$$I = I_1 + \frac{V_1}{R_1} \quad \text{...(5)}$$

Substituting the value of V_2 from equation (3) in equation (2), we get

$$-I_2 R_2 = Z_{21} I_1 + Z_{22} I_2$$

or

$$I_2 = \frac{-Z_{21}}{R_2 + Z_{22}} I_1 \quad \text{...(6)}$$

Substituting the value of V_1 from equation (1) in equation (5), we get

$$I = I_1 + \frac{Z_{11} I_1 + Z_{12} I_2}{R_1} \quad \text{...(7)}$$

Now, we have to find

$$\frac{V_2}{I} = \frac{-I_2 R_2}{I_1 + \dfrac{Z_{11} I_1 + Z_{12} I_2}{R_1}}$$

$$= \frac{-I_2 R_1 R_2}{I_1 R_1 + Z_{11} I_1 + Z_{12} I_2}$$

Substituting the value of I_2 from equation (6) in the above expression, we get

$$\frac{V_2}{I} = \frac{-\left(\dfrac{-Z_{21}}{R_2 + Z_{22}}\right) R_1 R_2 I_1}{I_1 (R_1 + Z_{11}) + Z_{12}\left(\dfrac{-Z_{21}}{R_2 + Z_{22}}\right) I_1}$$

or

$$\frac{V_2}{I} = \frac{R_1 R_2 Z_{21}}{(R_1 + Z_{11})(R_2 + Z_{22}) - Z_{12} Z_{21}}$$

or

$$\frac{V_2}{I} = \frac{R_1 R_2 Z_{21}}{R_1 R_2 + R_1 Z_{22} + R_2 Z_{11} + Z_{11} Z_{22} - Z_{12} Z_{21}}$$

or

$$\frac{V_2}{I} = \frac{R_1 R_2 Z_{21}}{R_1 R_2 + R_1 Z_{22} + R_2 Z_{11} + \Delta Z}$$

Problem No. 3.31 The network N shown in Fig. 3.54 may be described by the Z-parameters. Show that with port 2 open

$$G_{12} = \frac{Z_{21}}{Z_{11}}$$

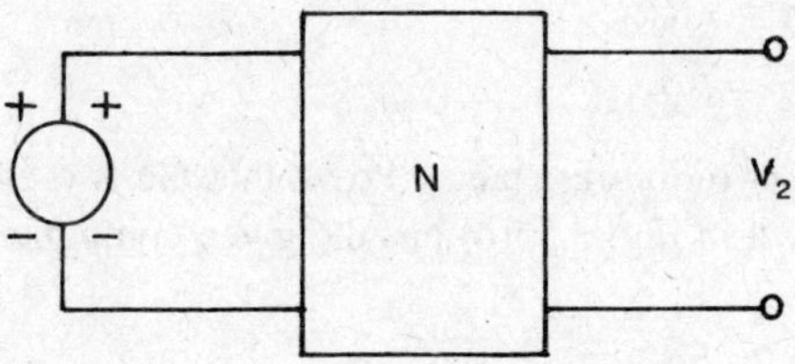

Fig. 3.54

Solution

The Z-parameter equations are

$$V_1 = Z_{11}I_1 + Z_{12}I_2 \quad ...(1)$$

$$V_2 = Z_{21}I_1 + Z_{22}I_2 \quad ...(2)$$

We have to prove that

$$G_{12} = \frac{Z_{21}}{Z_{11}} \qquad \text{i.e., } G_{12} = \frac{V_2}{V_1}$$

$$Z_{21} = \left.\frac{V_2}{I_1}\right|_{I_2=0}$$

and

$$Z_{11} = \left.\frac{V_1}{I_1}\right|_{I_2-0}$$

$I_2 = 0$ means port 2 open circuited and this is the requirement in this problem. Therefore, putting $I_2 = 0$, equation (1) and (2) give

$$V_1 = Z_{11}I_1, \text{ i.e., } Z_{11} = V_1/I_1 \quad ...(3)$$

and

$$V_2 = Z_{21}I_1, \text{ i.e., } Z_{21} = V_2/I_1 \quad ...(4)$$

Substituting in G_{12}, gives

$$G_{12} = \frac{V_2/I_1}{V_1/I_1} = \frac{V_2}{V_1} \text{ and hence proved.}$$

Problem No. 3.32 The element represented in the network in Fig. 3.55(a) is a gyrator which is described by the set of equation

$$v_1 = -R_0 i_2$$

$$v_2 = R_0 i_1$$

where R_0 is the gyration resistance. Prove that the two elements equivalent network as shown in Fig. 3.55(b) has the given component value.

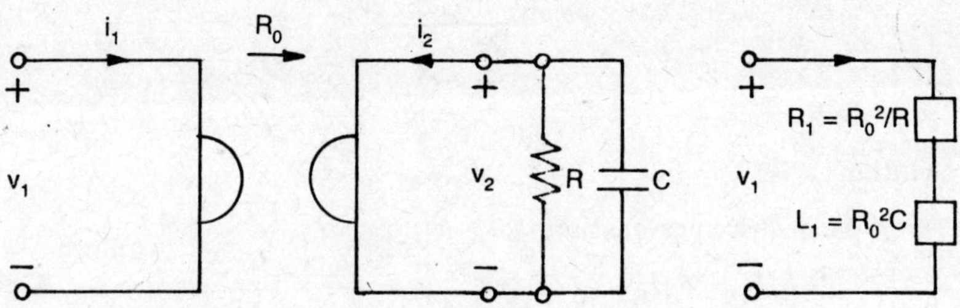

Fig. 3.55

Solution It is seen that gyrator has been terminated in a resistor and a capacitor

$$v_1 = -R_0 i_2$$

$$= -R_0\left[\frac{-v_2}{R} - C\frac{dv_2}{dt}\right]$$

$$= -R_0\left[\frac{-R_0 i_1}{R} - C\frac{d(R_0 i_1)}{dt}\right]$$

$$= \frac{R_0^2}{R} i_1 + R_0^2 C\frac{di_1}{dt}$$

Thus at the input terminals, the $v - i$ relationship is that of

(a) an inductor with inductance equal to $R_0^2 C$; and

(b) a resistance equal to $\frac{R_0^2}{R}$.

Problem No. 3.33 Find Z and Y parameter for the resistance network shown in Fig. 3.56.

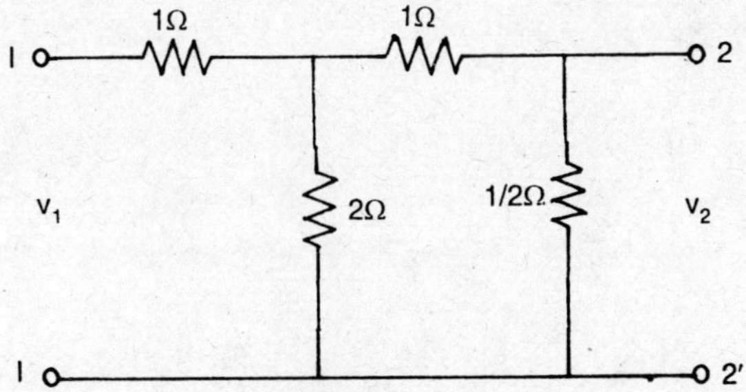

Fig. 3.56

Z-parameter

(a) $Z_{11} = \left.\frac{V_1}{I_1}\right|_{I_2=0}$

The equivalent impedance of the circuit looking from the terminals 1 – 1′ is

$$Z_{eq} = 1 + \frac{2 \times 1.5}{2 + 1.5} = \frac{65}{35} = \frac{13}{7}\Omega$$

(b) $Z_{21} = \left.\frac{V_2}{I_1}\right|_{I_2=0}$

Consider Fig. 3.57

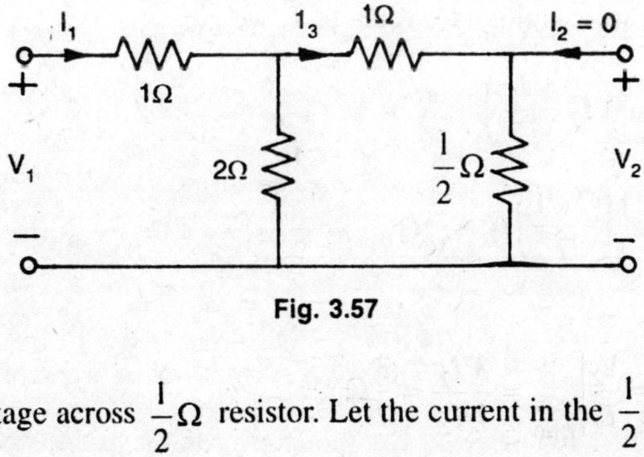

Fig. 3.57

V_2 is the voltage across $\frac{1}{2}\Omega$ resistor. Let the current in the $\frac{1}{2}\Omega$ resistor be I_3.

Then

$$I_3 = I_1 \times \frac{2}{2+1+\frac{1}{2}} = \frac{2I_1}{3.5}$$

Now

$$V_2 = \frac{1}{2} \times I_3 = \frac{1}{2} \times \frac{2}{1} \times \frac{I_1}{3.5} = \frac{I_1}{3.5}$$

Therefore,

$$Z_{21} = \left.\frac{V_2}{I_1}\right|_{I_2=0} = \frac{I_1/3.5}{I_1} = \frac{1}{3.5} = \frac{10}{35} = \frac{2}{7}\,\Omega$$

(c) $Z_{22} = \left.\frac{V_2}{I_2}\right|_{I_1=0}$

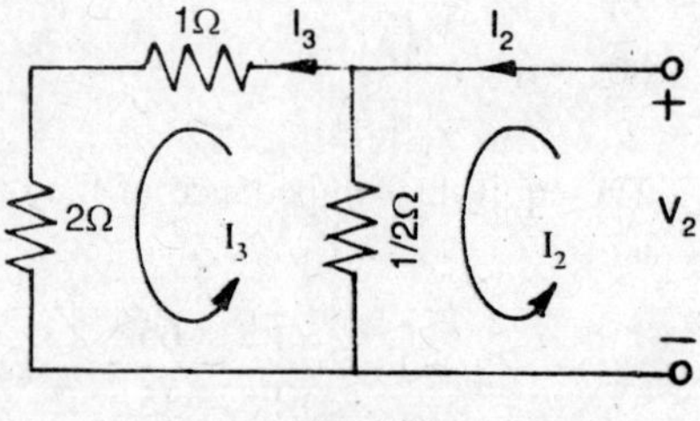

Fig. 3.58

Writing mesh equation for Fig. 3.58, we get

$$V_2 = \frac{1}{2}(I_2 - I_3) \qquad ...(1)$$

$$0 = I_3 + 2I_3 + \frac{1}{2}(I_3 - I_2) \qquad ...(2)$$

or

$$3.5I_3 = I_2$$

$$I_3 = \frac{I_2}{2 \times 3.5} = \frac{I_2}{7} \qquad ...(3)$$

Substituting in equation (1), value of I_3 we get

$$V_2 = \frac{1}{2}\left(I_2 - \frac{I_2}{7}\right)$$

or $$V_2 = \frac{1}{2}\left(\frac{7-1}{7}\right)I_2 = \frac{3}{7}I_2$$

Therefore,

$$Z_{22} = \left.\frac{V_2}{I_2}\right|_{I_1=0} = \frac{3}{7}\frac{I_2}{I_2} = \frac{3}{7}\,\Omega$$

(d) $Z_{12} = \left.\frac{V_1}{I_2}\right|_{I_1=0}$

$V_1 = 2 \times I_3$ (from equation (3))

$= \frac{2}{7} I_2$

Therefore,

$$Z_{12} = \left.\frac{V_1}{I_2}\right|_{I_1=0}$$

$$= \frac{2}{7}\frac{I_2}{I_2}$$

$$= \frac{2}{7}\Omega$$

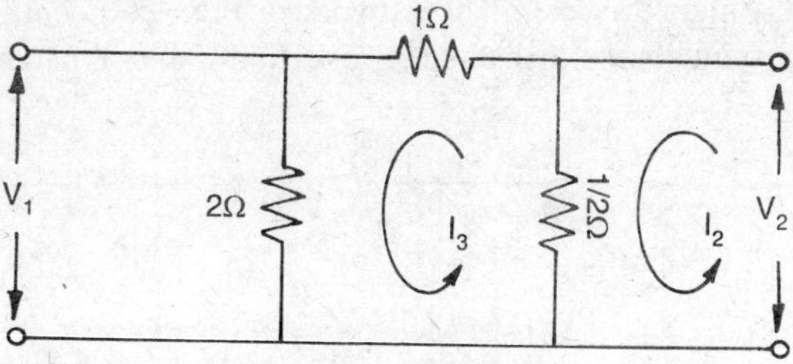

Fig. 3.59

$$Z = \begin{bmatrix} \frac{13}{7} & \frac{2}{7} \\ \frac{2}{7} & \frac{3}{7} \end{bmatrix}$$

$$\Delta Z = Z_{11}Z_{22} - Z_{12}Z_{21}$$

$$= \frac{13}{7} \times \frac{3}{7} - \frac{2}{7} \times \frac{2}{7}$$

or $$\Delta Z = \frac{39}{49} - \frac{4}{49}$$

$$= \frac{35}{49} = \frac{5}{7}$$

From conversion Table (4)

$$Y_{11} = \frac{Z_{22}}{\Delta Z} = \frac{3/7}{5/7} = \frac{3}{5}$$

$$Y_{12} = \frac{-Z_{12}}{\Delta Z} = \frac{-2/7}{5/7} = \frac{-2}{5}$$

$$Y_{21} = \frac{-Z_{21}}{\Delta Z} = \frac{-2/7}{5/7} = \frac{-2}{5}$$

$$Y_{22} = \frac{Z_{11}}{\Delta Z} = \frac{13/7}{5/7} = \frac{13}{5}$$

Therefore

$$Y = \begin{bmatrix} \frac{3}{5} & -\frac{2}{5} \\ -\frac{2}{5} & \frac{13}{5} \end{bmatrix}$$

Problem No. 3.34 The network of Fig. 3.60 contains a current-controlled current source. For this network, find Z and Y parameters.

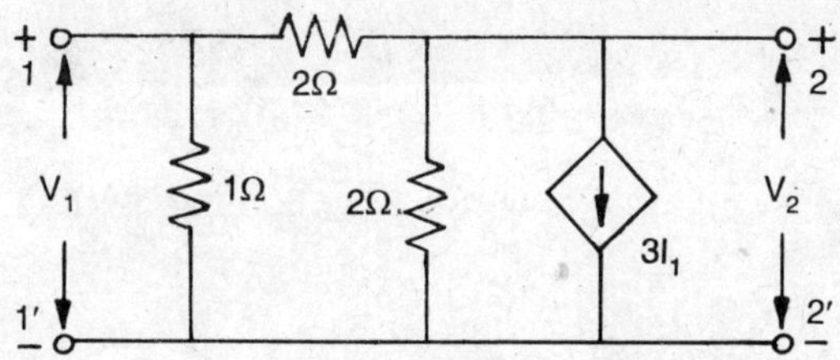

Fig. 3.60

Solution

Converting resistive π-network into *T*, we get equivalent Fig. 3.61.

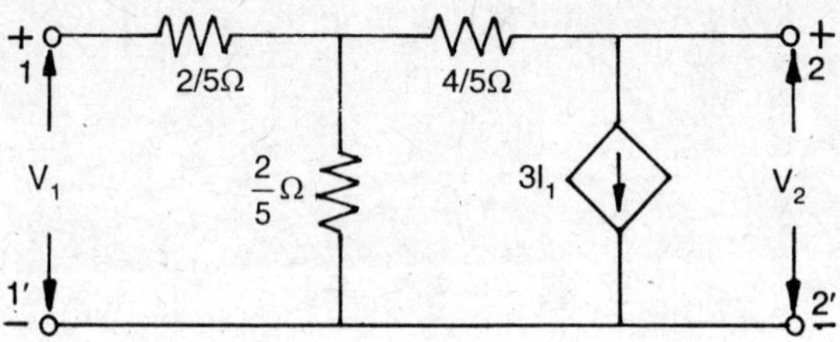

Fig. 3.61

Z-parameters

Open-circuiting V_2 and applying V_1 at terminals 1-1′ we get a circuit shown in Fig. 3.62.

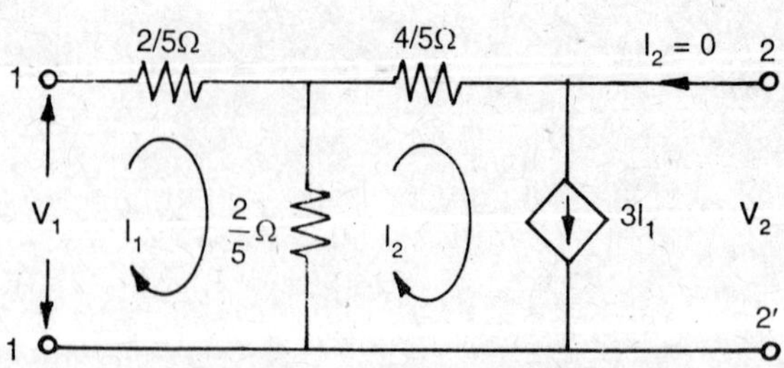

Fig. 3.62

Using KVL equations

$$V_1 = \frac{2}{5}I_1 + \frac{2}{5}I_1 - \frac{2}{5}I_2'$$

or
$$V_1 = \frac{4}{5}I_1 + \frac{2}{5}I_2' \qquad ...(1)$$

since port 2-2′ is open-circuited, $I_2 = 0$ and therefore,

$$I_2' = 3I_1 \qquad ...(2)$$

Substituting value of $I_2' = 3I_1$ from equation (2) in equation (1), we get

$$V_1 = \frac{4}{5}I_1 - \frac{2}{5} \times 3I_1$$

or
$$V_1 = \frac{4}{5}I_1 - \frac{6}{5}I_1$$

or
$$V_1 = -\frac{2}{5}I_1$$

Therefore,

$$Z_{11} = \left.\frac{V_1}{I_1}\right|_{I_2=0} = \frac{\frac{-2}{5}I_1}{I_1} = \frac{-2}{5} = -0.4\Omega$$

Also,

$$V_2 = \frac{2}{5}(I_1 - I_2') - \frac{4}{5}I_2'$$

or

$$V_2 = \frac{2}{5}I_1 - \frac{6}{5}I_2'$$

but $I_2' = 3I_1$ (from equation (2))

Therefore,

$$V_2 = \frac{2}{5}I_1 - \frac{6}{5} \times 3I_1$$

or
$$V_2 = \frac{2}{5}I_1 - \frac{18}{5}I_1$$

or
$$V_2 = \frac{-16}{5}I_1$$

Now

$$Z_{21} = \left.\frac{V_2}{I_1}\right|_{I_2=0} = \frac{-\frac{16}{5}I_1}{I_1} = -\frac{16}{5} = -3.2$$

Now let us open circuit port 1-1′ (i.e. $I_1 = 0$) apply V_2 at 2-2′ as shown in Fig. 3.63

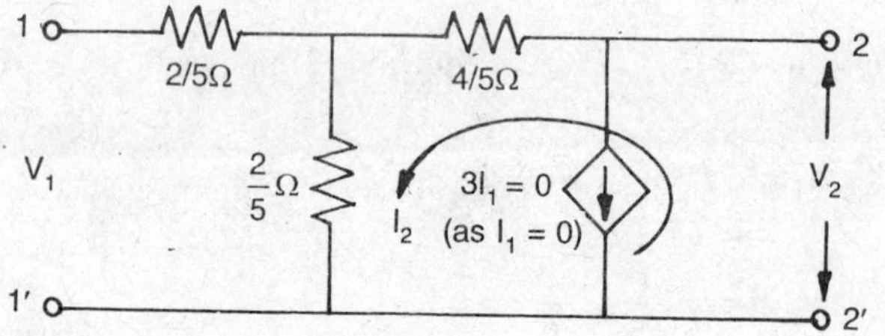

Fig. 3.63

From the circuit given in Fig. 3.63

$$I_2 = \frac{V_2}{\frac{4}{5}+\frac{2}{5}} = \frac{5}{6}V_2 \qquad \text{...(3)}$$

$$V_1 = \frac{2}{5}I_2 \qquad (\text{since } I_1 = 0) \qquad \text{...(4)}$$

Substituting value of I_2 from equation (3) in (4), we get

$$V_1 = \frac{2}{5}\times\frac{5}{6}V_2 = \frac{10}{30}V_2 = \frac{1}{3}V_2 \qquad \text{...(5)}$$

Now

$$Z_{12} = \left.\frac{V_1}{I_2}\right|_{I_1=0} = \frac{\frac{1}{3}V_2}{\frac{5}{6}V_2} = \frac{2}{5} = 0.4$$

and

$$Z_{22} = \left.\frac{V_2}{I_2}\right|_{I_1=0} = \frac{V_2}{\frac{5}{6}V_2} = \frac{6}{5} = 1.2$$

Therefore

$$Z = \begin{bmatrix} Z_{11} & Z_{12} \\ Z_{21} & Z_{22} \end{bmatrix} = \begin{bmatrix} -0.4 & 0.4 \\ -3.2 & 1.2 \end{bmatrix}$$

Y-Parameter

$$\begin{aligned} \Delta Z &= Z_{11}Z_{22} - Z_{12}Z_{21} \\ &= -0.4 \times 12 - 0.4(-32) \\ &= -0.48 + 1.28 \\ &= 0.8 \end{aligned}$$

From Conversion Table 4, we get

$$Y_{11} = \frac{Z_{22}}{\Delta Z} = \frac{1.2}{0.8} = 1.5 \text{ mho}$$

$$Y_{12} = \frac{-Z_{12}}{\Delta Z} = -\frac{0.4}{0.8} = -0.5 \text{ mho}$$

$$Y_{21} = -\frac{Z_{21}}{\Delta Z} = -\frac{(-3.2)}{0.8} = 4 \text{ mho}$$

$$Y_{22} = \frac{Z_{11}}{\Delta Z} = \frac{-0.4}{0.8} = -0.5 \text{ mho}$$

Therefore

$$Y = \begin{bmatrix} Y_{11} & Y_{12} \\ Y_{21} & Y_{22} \end{bmatrix} = \begin{bmatrix} 1.5 & -0.5 \\ 4 & -0.5 \end{bmatrix}$$

Problem No. 3.35 Find Z and Y-parameters for the resistive network containing a controlled source shown in Fig. 3.64

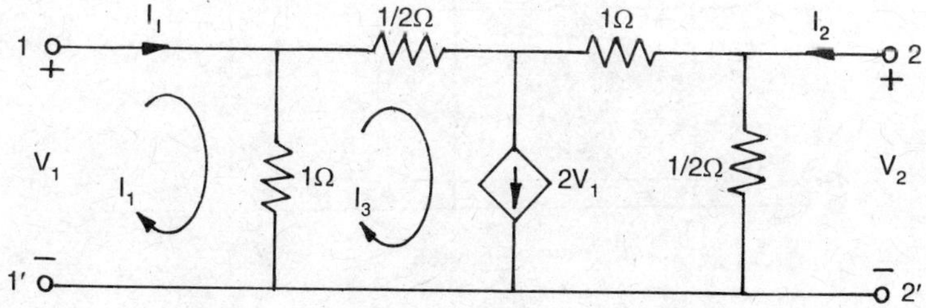

Fig. 3.64

Solution

Open-circuit port 2-2′ and with $I_2 = 0$

(a) $Z_{11} = \left.\dfrac{V_1}{I_1}\right|_{I_2=0}$

$$V_1 = 1 \times (I_1 - I_3)$$

or $\quad V_1 = I_1 - I_3 \quad$ ...(1)

Also $\quad I_3 = 2V_1 \quad$ ...(2)

which gives

$$V_1 = I_1 - 2V_1$$

or $\quad 3V_1 = I_1$

or $\quad \dfrac{V_1}{I_1} = \dfrac{1}{3} \quad$ ...(3)

Now

$$Z_{11} = \left.\frac{V_1}{I_1}\right|_{I_2=0} = \frac{1}{3}\Omega$$

(b) $Z_{21} = \left.\dfrac{V_2}{I_1}\right|_{I_2=0}$

Mesh equations as applied to Fig. 3.65, gives

$$V_1 = 0.5(I_3 + I_4) + I_4 + 0.5I_4$$

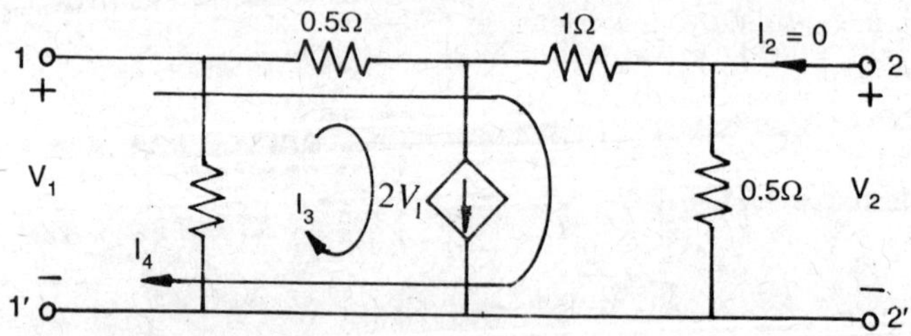

Fig. 3.65

or $\quad V_1 = 2I_4 + 0.5I_3 \quad$ ...(4)

Also $\quad \left.\begin{array}{l} V_2 = 0.5I_4; \quad \text{and} \\ I_3 = 2V_1 \end{array}\right\} \quad$ ...(5)

Substituting values of I_4 and I_3 from equation (5) in (4), we get

$$V_1 = \frac{2 \times V_2}{0.5} + 0.5(2V_1)$$

or $$4V_2 = 0$$

or $$V_2 = 0$$

Therefore,

$$Z_{21} = \left.\frac{V_2}{I_1}\right|_{I_2=0} = \frac{0}{I_1} = 0$$

Now open-circuiting port 1-1′ and making $I_1 = 0$, as shown in Fig. 3.66, we see that

(c) $$\mathbf{Z_{12}} = \left.\frac{\mathbf{V_1}}{\mathbf{I_2}}\right|_{I_1=0}$$

From the network of Fig. 3.66, we get

$$\left.\begin{aligned} I_3' &= 2V_1; \quad \text{and} \\ I_4' &= \frac{V_1}{1} = V_1 \end{aligned}\right\} \qquad ...(6)$$

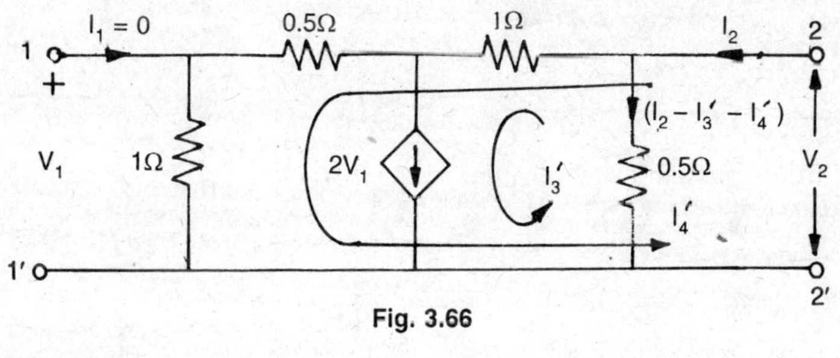

Fig. 3.66

Mesh equations give

$$(I_2 - I_3' - I_4') \times 0.5 = V_2 \qquad ...(7)$$

and

$$I_4'(1 + 0.5 + 1) + I_3' \times 1 = V_2$$

or

$$2.5I_4' + I_3' = V_2 \qquad ...(8)$$

Equating equations (7) and (8), gives

$$0.5(I_2 - I_3' - I_4') = 2.5I_4' + I_3'$$

or $$I_2 = \frac{3I_4' + 1.5I_3'}{0.5}$$

or $$I_2 = 3I_3' + 6I_4' \qquad ...(9)$$

Putting the values of I_3' and I_4' from equations (6) and (7) in (9), we get

$$I_2 = 6V_1 + 6V_1 = 12V_1$$

Therefore

$$Z_{12} = \left.\frac{V_1}{I_2}\right|_{I_1=0} = \frac{V_1}{12V_1} = \frac{1}{12}\Omega \qquad ...(10)$$

(d) $\boldsymbol{Z_{22} = \left.\frac{V_2}{I_2}\right|_{I_1=0}}$

Rewriting equation (8)

$$I_3' + 2.5I_4' = V_2$$

substituting values of I_3' and I_4' from equation (6) and (7) in the above equation, gives

$$2V_1 + 2.5V_1 = V_2$$

$$\frac{V_2}{V_1} = 4.5 \qquad ...(11)$$

But, from equation (10) which is

$$\frac{V_1}{I_2} = \frac{1}{12}$$

or $$V_1 = \frac{I_2}{12} \qquad ...(12)$$

substituting value of V_1 from equation (12) in (11), give

$$\frac{V_2}{I_2/12} = 4.5$$

or

$$\frac{V_2}{I_2} = \frac{4.5}{12} = \frac{3}{8}\Omega$$

But, since

$$\frac{V_2}{I_2} = Z_{22},\ \text{therefore,}$$

$$Z_{22} = \left.\frac{V_2}{I_2}\right|_{I_1=0} = \frac{3}{8}\,\Omega$$

Y-parameters

$$\Delta Z = Z_{11}Z_{22} - Z_{12}Z_{21}$$

$$= \frac{1}{3}\times\frac{3}{8} - \frac{1}{12}\times 0$$

$$= \frac{1}{8}\,\Omega$$

Therefore,

$$Y_{11} = \frac{Z_{22}}{\Delta Z} = \frac{3/8}{1/8} = 3 \text{ mho}$$

$$Y_{21} = \frac{-Z_{21}}{\Delta_Z} = \frac{0}{1/8} = 0$$

$$Y_{12} = \frac{-Z_{12}}{\Delta_Z} = \frac{-1/12}{1/8} = -\frac{2}{3} \text{ mho}$$

$$Y_{22} = \frac{Z_{11}}{\Delta_Z} = \frac{1/3}{1/8} = \frac{8}{3} \text{ mho}$$

Problem No. 3.36 Fig. 3.67 shows a resistive network containing a single controlled source. For this network, find Z and Y-parameters.

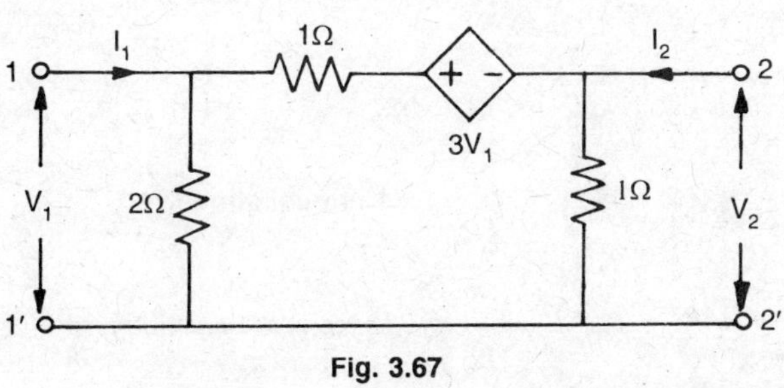

Fig. 3.67

Solution

(1) Open Circuiting Port 2-2′ ($I_2 = 0$)

This results into Fig. 3.68

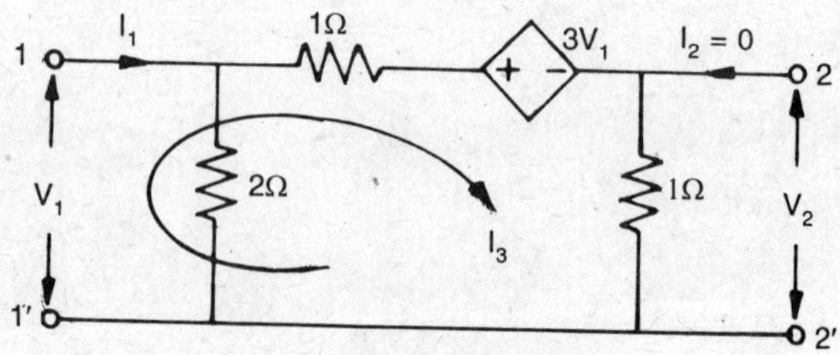

Fig. 3.68

(a) $Z_{11} = \left.\dfrac{V_1}{I_1}\right|_{I_2=0}$

$$V_1 = 2(I_1 - I_3) \quad \text{...(1)}$$

Also

$$1 \times I_3 + 3V_1 + 1 \times I_3 = V_1$$

or $$2I_3 + 3V_1 = V_1 \quad \text{...(2)}$$

$$I_3 = -V_1 \quad \text{...(2a)}$$

and also

$$1 \times I_3 = V_2$$

i.e. $$I_3 = V_2 \quad \text{...(3)}$$

and

$$(I_1 - I_3) \times 2 = V_1 \qquad \text{(from equation (1))}$$

or $$I_1 - I_3 = \frac{V_1}{2}$$

or $$I_1 + V_1 = \frac{V_1}{2} \qquad \text{(from equation (2a))}$$

or $$I_1 = -\frac{V_1}{2}$$

or $$\frac{V_1}{I_1} = -2 \quad \text{...(4)}$$

i.e.

$$Z_{11} = \left.\frac{V_1}{I_1}\right|_{I_2=0} = -2 \qquad ...(5)$$

(b) $\mathbf{Z_{21}} = \left.\dfrac{V_2}{I_1}\right|_{I_2=0}$

Now equating equations (1) and (2)

$$2I_3 + 3V_1 = 2(I_1 - I_3)$$

or $$4I_3 = 2I_1 - 3V_1 \qquad ...(6)$$

Since $I_3 = V_2$, from equation (3), equation (6) becomes

$$4V_2 = 2I_1 - 3V_1 \qquad ...(7)$$

Since $4V_2 = 2I_1 - 3V_1$, from equation (4), equation (7) becomes

$$4V_2 = 2I_1 - 3(-2I_1)$$

or $$4V_2 = 2I_1 + 6I_1$$

or $$4V_2 = 8I_1$$

or $$\frac{V_2}{I_1} = \frac{8}{4} = 2\Omega \qquad ...(8)$$

Since

$$Z_{21} = \left.\frac{V_2}{I_1}\right|_{I_2=0}$$, Therefore, from equation (8)

$$Z_{21} = \frac{V_2}{I_1} = 2\Omega$$

Open-circuiting port 1-1′ and applying V_2 at 2-2′ and putting $I_1 = 0$, we find out

(c) $\mathbf{Z_{12}} = \left.\dfrac{V_1}{I_2}\right|_{I_1=0}$

Fig. 3.69 gives the following mesh equations

$$V_1 = 2I_4 \qquad ...(9)$$

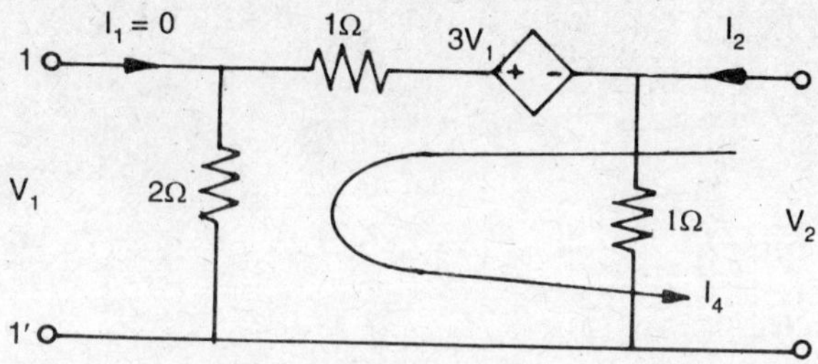

Fig. 3.69

$$(1+2)I_4 - 3V_1 = V_2$$

or

$$3I_4 - 3V_1 = V_2 \qquad ...(10)$$

Also,

$$V_1 = 2 \times I_4 \qquad ...(11)$$

or

$$I_4 = \frac{V_1}{2} \qquad ...(12)$$

Substitution in equation (10), $I_4 = \frac{V_1}{2}$, gives

$$3 \times \frac{V_1}{2} - 3V_1 = V_2$$

or

$$V_2 = -\frac{3}{2}V_1 \qquad ...(13)$$

Also

$$(I_2 - I_4) \times 1 = V_2 \qquad ...(14)$$

substituting $I_4 = \frac{V_1}{2}$ from equation (12) in (14), gives

$$I_2 - \frac{V_1}{2} = V_2 \qquad ...(15)$$

substituting $V_2 = -\frac{3}{2}V_1$ from equation (13) in equation (15), gives

$$I_2 - \frac{V_1}{2} = -\frac{3}{2}V_1$$

or

$$I_2 = -V_1 \qquad ...(16)$$

or $$\frac{V_1}{I_2} = -1 \qquad \text{...(16a)}$$

Since $Z_{12} = \left.\frac{V_1}{I_2}\right|_{I_1=0}$, therefore, from equation (16a), we get

$$Z_{12} = \frac{V_1}{I_2} = -1\Omega \qquad \text{...(17)}$$

Rewriting equation (16)

$$I_2 = -V_1 \qquad \text{...(18)}$$

substituting value of V_1 from equation (13) in (18), we get

$$I_2 = \frac{2}{3} V_2$$

or $$\frac{V_2}{I_2} = \frac{3}{2} \qquad \text{...(19)}$$

Since

(d) $Z_{22} = \left.\frac{V_2}{I_2}\right|_{I_1=0}$, therefore, from equation (19), we get

$$Z_{22} = \frac{V_2}{I_2} = \frac{3}{2}\Omega \qquad \text{...(20)}$$

Y-parameters

$$\Delta Z = Z_{11}Z_{22} - Z_{21}Z_{21}$$

$$= -2 \times \frac{3}{2} - 2 \times -1$$

$$= -1$$

Therefore,

$$Y_{11} = \frac{Z_{22}}{\Delta Z} = \frac{3/2}{-1} = -\frac{3}{2} \text{ mho}$$

$$Y_{22} = \frac{Z_{11}}{\Delta Z} = \frac{-2}{-1} = 2 \text{ mho}$$

$$Y_{12} = \frac{-Z_{12}}{\Delta Z} = \frac{-1}{-1} = -1 \text{ mho}$$

$$Y_{21} = \frac{-Z_{21}}{\Delta Z} = \frac{-2}{-1} = 2\Omega \text{ mho}$$

Problem No. 3.37 Find the Z and Y-parameters for the RC ladder network shown in Fig. 3.70

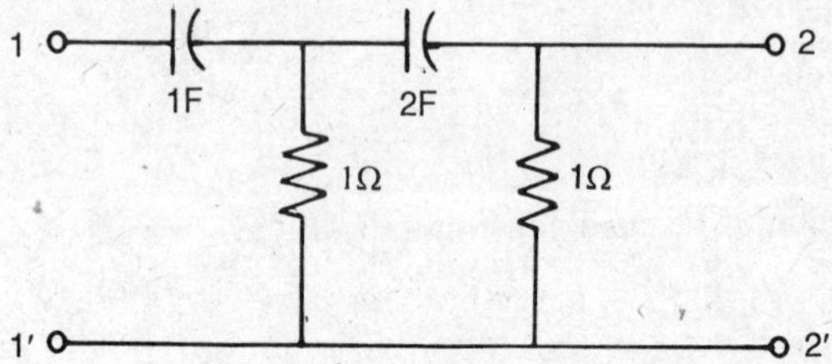

Fig. 3.70

Solution

Transformed network is shown in Fig. 3.71

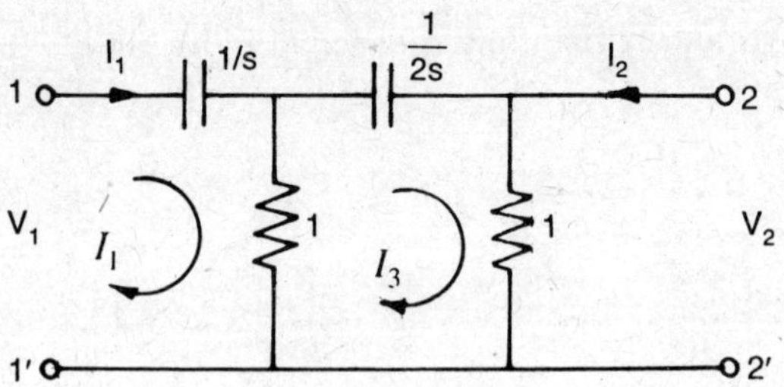

Fig. 3.71

with port 2-2′ open circuited and $I_2 = 0$

$$V_2 = I_3 \times 1 = I_3 \quad \text{...(1)}$$

$$I_3\left(1+1+\frac{1}{2s}\right) = I_1 \quad \text{...(2)}$$

$$\left(1+\frac{1}{s}\right)I_1 - I_3 = V_1 \quad \text{...(3)}$$

From equation (2), we get

$$I_3 = \frac{2s}{1+4s} I_1 \qquad ...(4)$$

From equation (3), we get

$$\left(\frac{s+1}{s}\right) I_1 - I_3 = V_1 \qquad ...(5)$$

substituting value of I_3 from equation (4) in (5), we get

$$\left(\frac{s+1}{s}\right) I_1 - \left(\frac{2s}{1+4s}\right) I_1 = V_1$$

$$I_1\left(\frac{4s^2+5s+1}{3s(1+4s)}\right) = V_1$$

or $$Z_{11} = \left.\frac{V_1}{I_1}\right|_{I_2=0} = \frac{2s^2+5s+1}{s(1+4s)} \qquad ...(6)$$

Also

$$V_2 = I_3 \qquad \text{(from equation (1))}$$

or $$V_2 = \frac{2s}{1+4s} I_1 \qquad \text{(from equation (4))}$$

or $$Z_{21} = \left.\frac{V_2}{I_1}\right|_{I_2=0} = \frac{2s}{1+4s} \qquad ...(7)$$

Open circuiting port 1-1′ and putting $I_1 = 0$ as shown in Fig. 3.72, we get

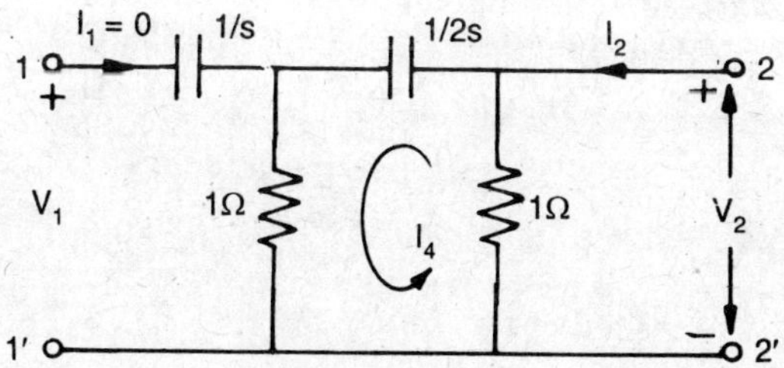

Fig. 3.72

From Fig. 3.72

$$V_1 = I_4 \qquad ...(8)$$

$$\left(1+1+\frac{1}{2s}\right)I_4 = I_2 \qquad ...(9)$$

$$V_2 = I_2 - I_4 \qquad ...(10)$$

From equation (9), we get

$$I_4 = \frac{2s}{4s+1} I_2 \qquad ...(11)$$

From equation (10), we get

$$V_2 = I_2 - \frac{2s}{4s+1} I_2, \text{ (from equation 11)} \qquad ...(12)$$

$$\frac{V_2}{I_2} = Z_{22} = \left(1 - \frac{2s}{4s+1}\right)$$

or

$$Z_{22} = \left.\frac{V_2}{I_2}\right|_{I_1=0} = \frac{2s+1}{4s+1} \qquad ...(13)$$

Also

$$V_1 = I_4 = I_2\left(\frac{2s}{4s+1}\right) \qquad ...(14)$$

(by putting the value of I_4 from equation (11) in equation (8))

or

$$\frac{V_1}{I_2} = \frac{2s}{4s+1}$$

i.e.

$$Z_{12} = \left.\frac{V_1}{I_2}\right|_{I_1=0} = \frac{2s}{4s+1} \qquad ...(15)$$

Y-parameter

$$\Delta Z = Z_{11}Z_{22} - Z_{12}Z_{21}$$

$$= \frac{4s^2+5s+1}{4s+1} \times \frac{2s+1}{4s+1} - \frac{2s}{4s+1} \times \frac{2s}{4s+1}$$

$$= \frac{(4s^2+5s+1)(2s+1) - 4s^2}{(4s+1)(4s+1)}$$

$$= \frac{8s^3 + 4s^2 + 10s^2 + 5s + 2s + 1 - 4s^2}{(4s+1)^2} = \frac{8s^3 + 10s^2 + 7s + 1}{(4s+1)^2}$$

$$Y_{11} = \frac{Z_{22}}{\Delta Z} = \frac{\dfrac{2s+1}{4s+1}}{\dfrac{8s^3 + 10s^2 + 7s + 1}{(4s+1)^2}} = \frac{(2s+1)(4s+1)}{8s^3 + 10s^2 + 7s + 1}$$

$$Y_{22} = \frac{Z_{11}}{\Delta Z} = \frac{\dfrac{4s^2 + 5s + 1}{(4s+1)}}{\dfrac{8s^3 + 10s^2 + 7s + 1}{(4s+1)^2}} = \frac{(4s^2 + 5s + 1)(4s+1)}{8s^3 + 10s^2 + 7s + 1}$$

$$Y_{12} = \frac{-Z_{12}}{\Delta Z} = \frac{\dfrac{-2s}{4s+1}}{\dfrac{8s^3 + 10s^2 + 7s + 1}{(4s+1)^2}} = \frac{-2s(4s+1)}{8s^3 + 10s^2 + 7s + 1}$$

$$Y_{21} = \frac{-Z_{21}}{\Delta_Z} = \frac{-2s/4s+1}{8s^3 + 10s^2 + 7s + 1/(4s+1)^2} = \frac{-2s(4s+1)}{8s^3 + 10s^2 + 7s + 1}$$

Problem No. 3.38 The network shown in Fig. 3.73 consists of a resistive T-network and resistive π-network connected in parallel. For the element values given, determine the Y-parameters.

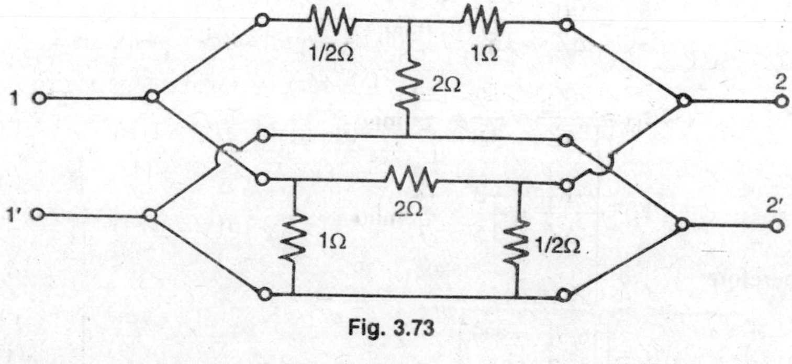

Fig. 3.73

Solution T-network is shown in Fig. 3.74,

Applying KVL around the two loops in Fig. 3.74, with loop currents I_1 and I_2 to obtain

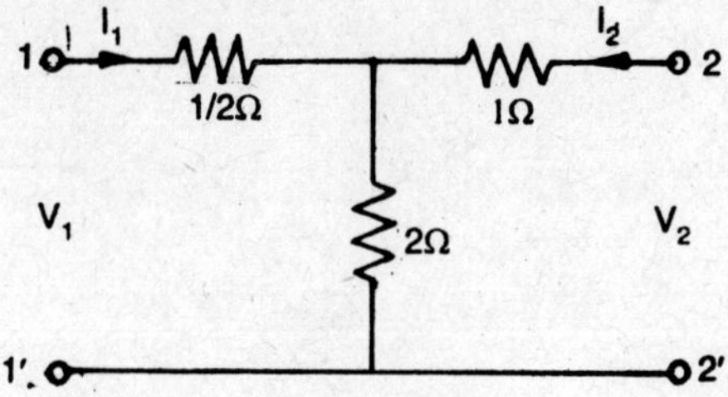

Fig. 3.74

$$V_1 = \frac{1}{2} I_1 + 2(I_1 + I_2) = 2.5I_1 + 2I_2$$

$$V_2 = 1 \times I_2 + 2(I_1 + I_2) = 2I_1 + 3I_2$$

$$Z = \begin{bmatrix} 2.5 & 2 \\ 2 & 3 \end{bmatrix}$$

$$\Delta Z = Z_{11}Z_{22} - Z_{12}Z_{21}$$

$$= 2.5 \times 3 - 2 \times 2$$

$$= 7.5 - 4$$

$$= 3.5$$

$$Y_{11} = \frac{Z_{22}}{\Delta Z} = \frac{3}{3.5} = \frac{6}{7} \text{ mho}$$

$$Y_{22} = \frac{Z_{11}}{\Delta Z} = \frac{2.5}{3.5} = \frac{5}{7} \text{ mho}$$

$$Y_{12} = \frac{-Z_{12}}{\Delta Z} = \frac{-2}{3.5} = \frac{-4}{7} \text{ mho}$$

$$Y_{21} = \frac{-Z_{21}}{\Delta Z} = \frac{-2}{3.5} = \frac{-4}{7} \text{ mho}$$

Therefore

$$Y = \begin{bmatrix} 6/7 & -4/7 \\ -4/7 & 5/7 \end{bmatrix} \qquad ...(1)$$

π-network is shown in Fig. 3.75

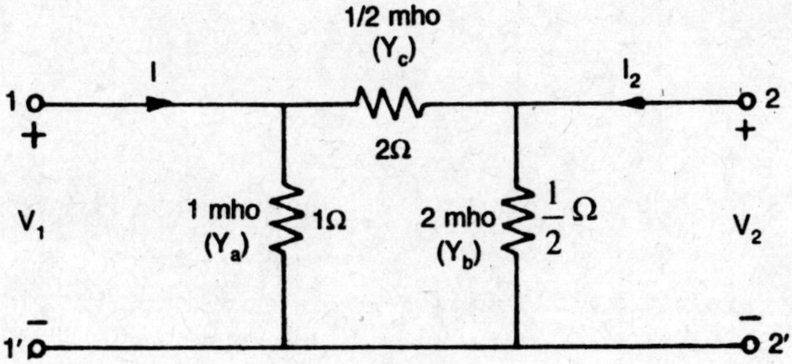

Fig. 3.75

The node equations are

$$I_1 = (Y_a + Y_c)V_1 - Y_c V_2$$

$$I_2 = -Y_c V_1 + (Y_b + Y_c)V_2$$

From Fig. 3.75

$$\left.\begin{aligned} I_1 &= \left(1+\frac{1}{2}\right)V_1 - \frac{1}{2}V_2 = \frac{3}{2}V_1 - \frac{1}{2}V_2 \\ &\text{and} \\ I_2 &= -\frac{1}{2}V_1 + \left(2+\frac{1}{2}\right)V_2 = -\frac{1}{2}V_1 + \frac{5}{2}V_2 \end{aligned}\right\} \quad ...(2)$$

Comparing equation (2) with equation (3) given below

$$\left.\begin{aligned} I_1 &= Y_{11}V_1 + Y_{12}V_2 \\ I_2 &= Y_{21}V_1 + Y_{22}V_2 \end{aligned}\right\} \quad ...(3)$$

We get

$$Y_{11} = \frac{3}{2} \text{ mho}$$

$$Y_{12} = -\frac{1}{2} \text{ mho}$$

$$Y_{21} = -\frac{1}{2} \text{ mho}$$

$$Y_{22} = \frac{5}{2} \text{ mho}$$

or

$$Y = \begin{bmatrix} \frac{3}{2} & \frac{-1}{2} \\ -\frac{1}{2} & \frac{5}{2} \end{bmatrix} \qquad ...(5)$$

The total Y-parameters of the network is sum of equations (1) and (5)

$$Y_{total} = \begin{bmatrix} \frac{6}{7} & -\frac{4}{7} \\ -\frac{4}{7} & \frac{5}{7} \end{bmatrix} + \begin{bmatrix} \frac{3}{2} & -\frac{1}{2} \\ -\frac{1}{2} & \frac{5}{2} \end{bmatrix}$$

$$= \begin{bmatrix} \frac{6}{7}+\frac{3}{2} & \frac{-4}{7}-\frac{1}{2} \\ -\frac{4}{7}-\frac{1}{2} & \frac{5}{7}+\frac{5}{2} \end{bmatrix}$$

$$= \begin{bmatrix} \frac{33}{14} & -\frac{15}{14} \\ -\frac{15}{14} & \frac{45}{14} \end{bmatrix}$$

Problem No. 3.39 Determine h_{11} and h_{21} of the network shown in Fig. 3.76

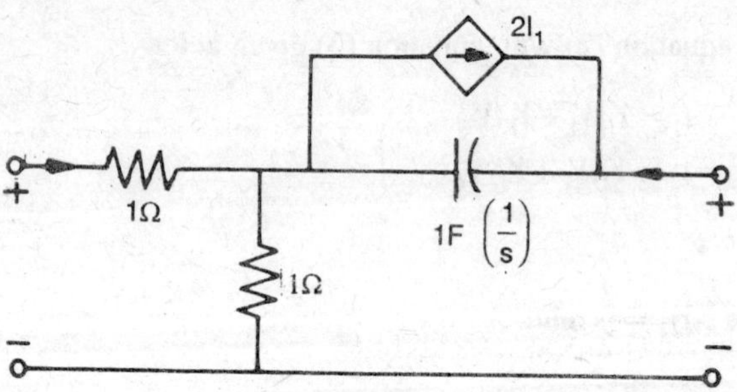

Fig. 3.76

Solution Putting $V_2 = 0$ as shown in Fig. 3.77, we proceed to determine h_{21}
Applying KCL at node 'a'

$$I_1 = 2I_1 + s \times 1 \times V_a + \frac{V_a}{1} = 2I_1 + V_a(s+1) \qquad ...(1)$$

but

$$V_a = V_1 - I_1 \times 1 = V_1 - I_1 \qquad ...(2)$$

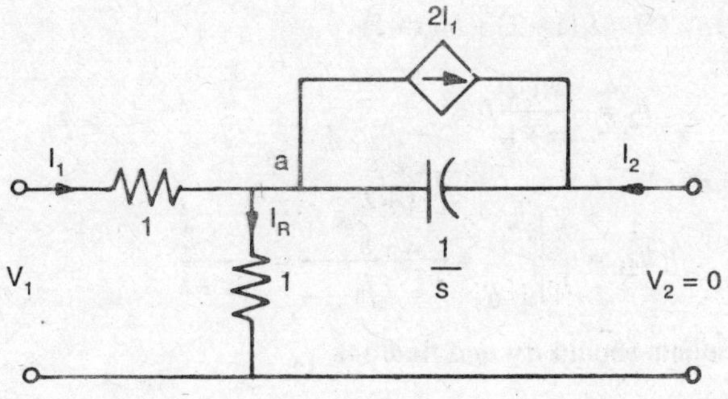

Fig. 3.77

Therefore, by putting value of V_a from equation (2) in (1) we get

$$I_1 = 2I_1 + (s+1)(V_1 - I_1)$$

or $$I_1 = 2I_1 + sV_1 - sI_1 + V_1 - I_1$$

or $$I_1 = I_1 - sI_1 + sV_1 + V_1$$

or $$V_1(s+1) = sI_1$$

or $$\frac{V_1}{I_1} = \frac{s}{s+1}$$

which means

$$h_{11} = \left.\frac{V_1}{I_1}\right|_{V_2=0} = \frac{s}{s+1}$$

Again applying KCL at node 'a'.

$$I_1 = 2I_1 + s \times 1 \times V_a + V_a$$

or $$I_1 = 2I_1 + sV_a + 1 \times I_R$$

or $$I_1 = 2I_1 + sV_a + I_R \quad ...(3)$$

Now $$\left.\begin{aligned} I_R &= I_1 + I_2 \quad \text{and} \\ V_a &= 1 \times I_R = I_R \end{aligned}\right\} \quad ...(4)$$

substituting in equation (3), value of I_R and V_a from equations (4), we get

$$I_1 = 2I_1 + sI_R + I_R$$

or $$I_1 = 2I_1 + s(I_1 + I_2) + (I_1 + I_2)$$

or $$I_1 = 2I_1 + sI_1 + sI_2 + I_1 + I_2$$

or $$I_1 = 3I_1 + sI_1 + sI_2 + I_2$$

or $$0 = I_1(s+2) + I_2(s+1)$$

or $$I_2 = -\frac{s+2}{s+1} I_1 \qquad ...(5)$$

$$h_{21} = \left.\frac{I_2}{I_1}\right|_{V_2=0} = -\frac{\frac{s+2}{s+1} \times I_1}{I_1} = -\frac{s+2}{s+1}$$

Note: Students should try and find out

$$h_{22} = \frac{s}{1+s} \quad \text{and} \quad h_{12} = \frac{s}{1+s}$$

where

$$h_{22} = \left.\frac{I_2}{V_2}\right|_{I_1=0} \quad \text{and} \quad h_{12} = \left.\frac{V_1}{V_2}\right|_{I_1=0}$$

Problem No. 3.40 Show that for the network shown in Fig. 3.78, the following equation holds

$$\begin{bmatrix} V_2 \\ I_2 \end{bmatrix} = \begin{bmatrix} 1.5 & 6.5 \\ 0.25 & 1.25 \end{bmatrix} \begin{bmatrix} V_1 \\ -I_1 \end{bmatrix}$$

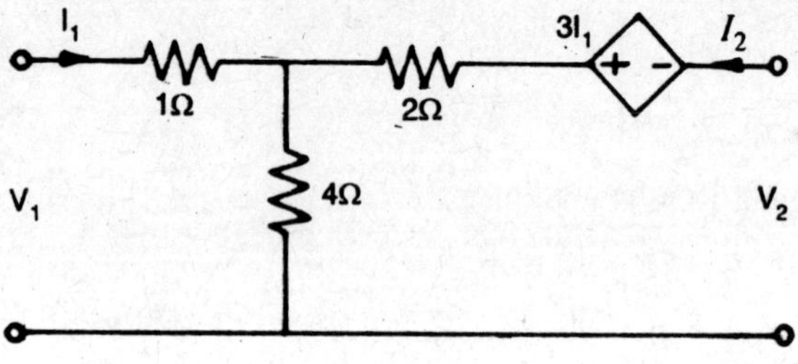

Fig. 3.78

Solution The given equation shows description of inverse transmission parameters of a network

With Sending end Open-Circuited

$$A' = \left.\frac{V_2}{V_1}\right|_{I_1=0}$$

writing mesh equation gives

$$\left.\begin{aligned} V_1 &= 1\times I_1 + 4(I_1 + I_2) \quad \text{and} \\ V_2 &= -3I_1 + 2I_2 + 4I_2 \end{aligned}\right\} \qquad \text{...(1)}$$

since $I_1 = 0$, above equations reduce to

$$\left.\begin{aligned} V_1 &= 4I_2; \quad \text{and} \\ V_2 &= 6I_2 \end{aligned}\right\} \qquad \text{...(2)}$$

Therefore,

$$A' = \left.\frac{V_2}{V_1}\right|_{I_1=0} = \frac{6I_2}{4I_2} = 1.5$$

$$C' = \left.\frac{I_2}{V_1}\right|_{I_1=0} = \frac{I_2}{4I_2} = 0.25$$

With sending end short-circuited the network is shown in Fig. 3.79

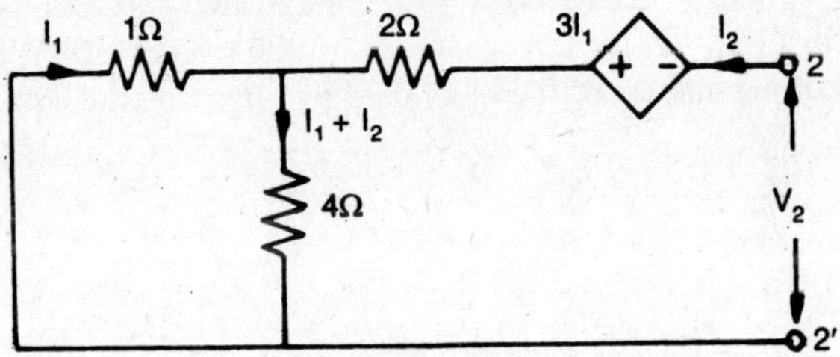

Fig. 3.79

Mesh equations give

$$1\times I_1 + 4(I_1 + I_2) = 0$$

or $$I_2 = -\frac{5}{4}I_1 \qquad \text{...(3)}$$

and $$V_2 = -3I_1 + 2I_2 + 4(I_1 + I_2)$$

or $$V_2 = I_1 + 6I_2 \qquad \text{...(4)}$$

substituting value of I_2 from equation (3) in (4), we get

$$V_2 = I_1 + 6\times\left(-\frac{5}{4}\right)I_1$$

$$= -\frac{13}{2}I_1 \qquad \text{...(5)}$$

Now

$$B' = \left.\frac{V_2}{-I_1}\right|_{V_1=0} = \frac{-\frac{13}{2}I_1}{-I_2} = 6.5$$

$$D' = \left.\frac{I_2}{-I_1}\right|_{V_1=0} = \frac{-\frac{5}{4}I_1}{-I_1} = 1.25$$

In matrix form

$$\begin{bmatrix} V_2 \\ I_2 \end{bmatrix} = \begin{bmatrix} A' & B' \\ C' & D' \end{bmatrix} \begin{bmatrix} V_1 \\ -I_1 \end{bmatrix} \quad \text{...(6)}$$

or after substituting values in equation (6), we get

$$\begin{bmatrix} V_2 \\ I_2 \end{bmatrix} = \begin{bmatrix} 1.5 & 6.5 \\ 0.25 & 1.25 \end{bmatrix} \begin{bmatrix} V_1 \\ -I_1 \end{bmatrix}$$

Problem No. 3.41 For the cascade connection of two ports depicted in Fig. 3.80, show that the transfer impedance Z_{12} of the overall circuit is given in terms of the Z-parameters of the individual two ports by the equation

$$Z_{12} = \frac{Z_{12a}Z_{12b}}{Z_{11b} + Z_{22a}}.$$

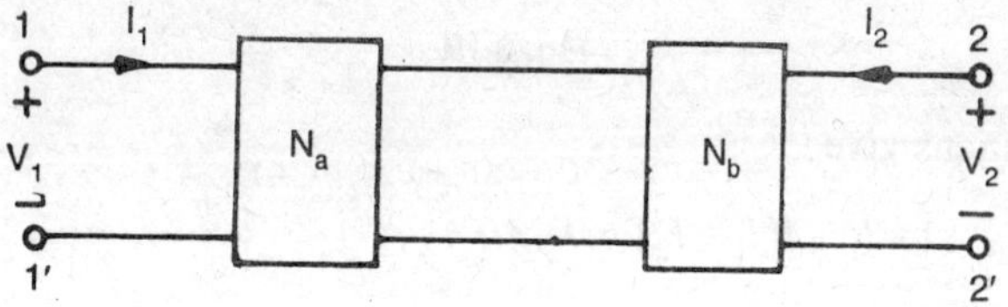

Fig. 3.80

In addition show that the short circuit admittance Y_{12} is given by

$$Y_{12} = \frac{-Y_{12a}Y_{12b}}{Y_{11b} + Y_{22a}}$$

Solution Z-parameter equations of network are described by

$$V_1 = Z_{11}I_1 + Z_{12}I_2 \quad \text{...(1a)}$$

$$V_2 = Z_{21}I_1 + Z_{22}I_2 \quad \text{...(1b)}$$

Network 'N_a'

Considering I_1 as the input current at port 1 and I_2' as the input to the network at port 2 as shown in Fig. 3.81, the

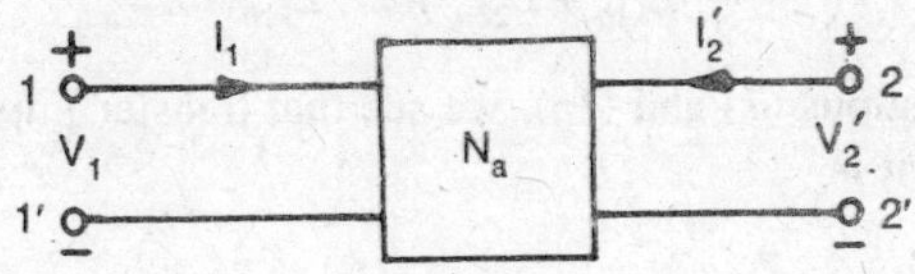

Fig. 3.81

Z-parameter equations are described as

$$V_1 = Z_{11a}I_1 + Z_{12a}I_2' \quad ...(2)$$

$$V_2' = Z_{21a}I_1 + Z_{22a}I_2' \quad ...(3)$$

Network 'N_b'

Input current I_2' at port 2 of network N_a becomes the input current at port 1 of network N_b but with a negative sign attached to it as shown in Fig. 3.82.

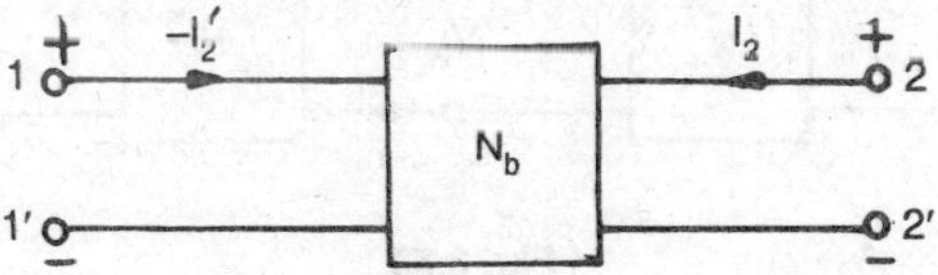

Fig. 3.82

Z-parameter equations for network N_b are

$$V_2' = -Z_{11b}I_2' + Z_{12b}I_2 \quad ...(4)$$

$$V_2 = -Z_{21b}I_2' + Z_{22b}I_2 \quad ...(5)$$

Equating equations (3) and (4), we get

$$I_2' = \frac{Z_{12b}I_2 - Z_{12a}I_1}{Z_{11b} + Z_{22a}} \quad ...(6)$$

substituting value of I_2' from equation (6) in (2), we get

$$V_1 = Z_{11a}I_1 + Z_{12a}\left(\frac{Z_{12b}I_2 - Z_{12a}I_1}{Z_{11b} + Z_{22a}}\right)$$

or $$V_1 = Z_{11a}I_1 - \frac{Z_{12a}Z_{12a}I_1}{Z_{11b}+Z_{22a}} + \frac{Z_{12a}Z_{12b}I_2}{Z_{11b}+Z_{22a}}$$

or $$V_1 = \left(Z_{11a} - \frac{Z_{12a}Z_{12a}}{Z_{11b}+Z_{22a}}\right)I_1 + \frac{Z_{12a}Z_{12b}}{Z_{11b}+Z_{22a}}I_2 \qquad ...(7)$$

Comparing equations (7) abd (1a), we see that transfer impedance (Z_{12}) of the overall circuit is

$$Z_{12} = \frac{Z_{12a}Z_{12b}}{Z_{11b}+Z_{22a}}$$

Short-Circuit Admittance Y_{12} Equations (8) defines the short-circuit admittance parameters as

$$I_1 = Y_{11}V_1 + Y_{12}V_2 \qquad ...(8a)$$

$$I_2 = Y_{21}V_1 + Y_{22}V_2 \qquad ...(8b)$$

Equations (9) and (10) describe the short-circuit admittance parameters of the networks N_a and N_b shown in Fig. 3.83

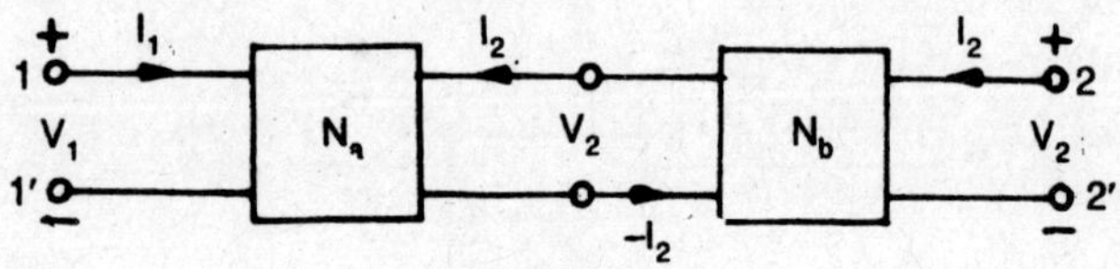

Fig. 3.83

For network 'N_a'

$$I_1 = Y_{11a}V_1 + Y_{12a}V_2' \qquad ...(9a)$$

$$I_2' = Y_{21a}V_1 + Y_{22a}V_2' \qquad ...(9b)$$

For network 'N_b'

$$-I_2' = Y_{11b}V_2' + Y_{12b}V_2 \qquad ...(10a)$$

$$I_2 = Y_{21b}V_2' + Y_{22b}V_2 \qquad ...(10b)$$

Equating equations (9b) and (10a), we get

$$Y_{21a}V_1 + Y_{22a}V_2' = -(Y_{11b}V_2' + Y_{12b}V_2)$$

$$V_2'(Y_{22a} + Y_{11b}) = -Y_{21a}V_1 - Y_{12b}V_2$$

$$V_2' = -\frac{Y_{21a}V_1 + Y_{12b}V_2}{Y_{22a} + Y_{11b}} \qquad ...(11)$$

Substituting the value of V_2' from equation (11) in (9a), we get

$$I_1 = Y_{11a}V_1 - \frac{(Y_{21a}V_1 + Y_{12b}V_2)Y_{12a}}{Y_{22a} + Y_{11b}} \qquad ...(12)$$

or

$$I_1 = \left(Y_{11a} - \frac{Y_{21a}Y_{12a}}{Y_{22a} + Y_{11b}}\right)V_1 + \left(\frac{-Y_{12a}Y_{12b}}{Y_{22a} + Y_{11b}}\right)V_2 \qquad ...(13)$$

Comparing equation (13) with (8a), we get short-circuit admittance

$$Y_{12} = -\frac{Y_{12a}Y_{12b}}{Y_{11b} + Y_{22a}}$$

Problem No. 3.42 Prove that for a passive reciprocal network $AD - BC = 1$, where A, B, C, D are the elements of transmission matrix

Solution Let us consider a two port network shown in Fig. 3.84

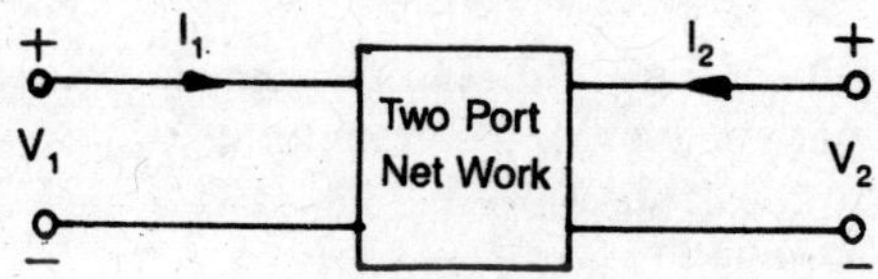

Fig. 3.84

The transmission or ABCD-para-meter equations of the considered network can be written as

$$V_1 = AV_2 + B(-I_2) \qquad ...(1)$$

$$I_1 = CV_2 + D(-I_2) \qquad ...(2)$$

$$\begin{bmatrix} V_1 \\ I_1 \end{bmatrix} = \begin{bmatrix} A & B \\ C & D \end{bmatrix}\begin{bmatrix} V_2 \\ -I_2 \end{bmatrix}$$

or

Determitant of transmission matrix is given by

$$-AD + BC \qquad ...(3)$$

substituting the impedance parameters in equation (3), we get

$$BC - AD = \frac{\Delta Z}{Z_{21}} \times \frac{1}{Z_{21}} - \frac{Z_{11}}{Z_{21}}\frac{Z_{22}}{Z_{21}}$$

$$= \frac{\Delta Z}{(Z_{21})^2} - \frac{Z_{11}Z_{22}}{(Z_{21})^2}$$

$$= \frac{Z_{11}Z_{22} - Z_{12}Z_{21} - Z_{11}Z_{22}}{(Z_{21})^2}$$

$$= -\frac{Z_{12}Z_{21}}{(Z_{21})^2}$$

$$= -\frac{Z_{12}}{Z_{21}} \qquad ...(4)$$

For a reciprocal network

$$Z_{12} = Z_{21} \qquad ...(5)$$

substituting condition of equation (5) in (4), we get

$$BC - AD = -1$$

or $$AD - BC = 1$$

Problem No. 3.43 Find the Z and Y-parameters of the transformer (non-ideal) shown in the Fig. 3.85. Determine T and π-equivalent circuits for the transformer in term of L_1, L_2 and M.

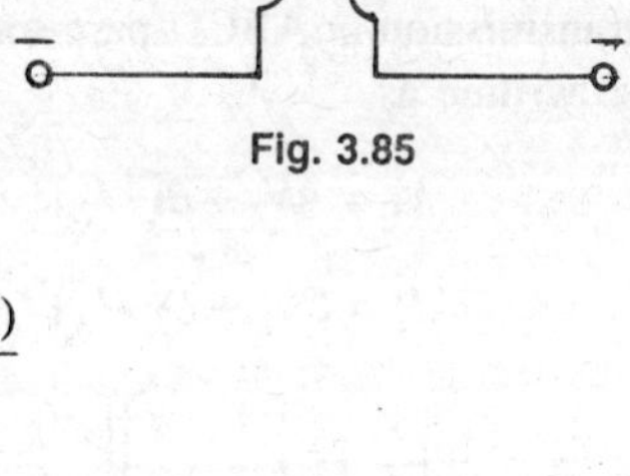

Fig. 3.85

Solution

$$v_1(t) = L_1 \frac{di_1(t)}{dt} + M \frac{di_2(t)}{dt}$$

and $$v_2(t) = L_2 \frac{di_2(t)}{dt} + M \frac{di_1(t)}{dt}$$

In Laplace transform form

$$V_1(s) = sL_1 I_1(s) + sMI_2(s)$$

and $$V_2(s) = sL_2 I_2(s) + sMI_1(s)$$

The Z-parameter matrix is

$$Z = \begin{bmatrix} sL_1 & sM \\ sM & sL_2 \end{bmatrix}$$

$$\Delta Z = Z_{11}Z_{22} - Z_{12}Z_{21}$$
$$= sL_1 sM - sM\, sL_2$$
$$= s^2 L_1 M - s^2 L_2 M$$
$$= s^2 M(L_1 - L_2)$$

Equivalent T-circuit representation is shown in Fig. 3.86

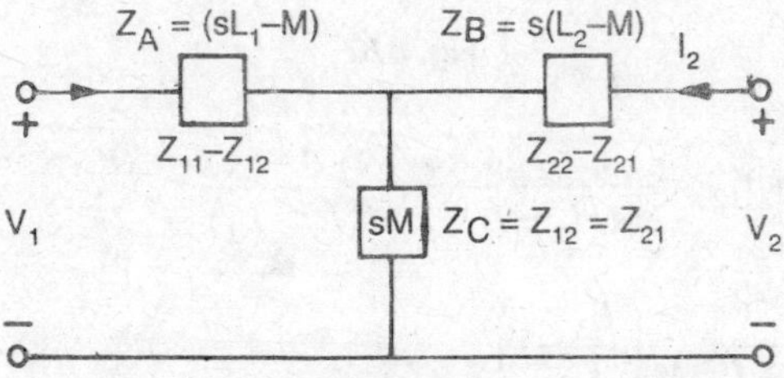

Fig. 3.86

In Fig. 3.86

$$Z_A = Z_{11} - Z_{12} = sL_1 - sM = s(L_1 - M)$$

$$Z_B = Z_{22} - Z_{21} = sL_2 - sM = s(L_2 - M)$$

$$Z_c = Z_{12} = Z_{21} = sM$$

Y-parameter

$$Y_{11} = \frac{Z_{22}}{\Delta Z} = \frac{sL_2}{s^2 M(L_1 - L_2)} = \frac{L_2}{sM(L_1 - L_2)}$$

$$Y_{12} = \frac{-Z_{12}}{\Delta Z} = \frac{-sM}{s^2 M(L_1 - L_2)} = \frac{-1}{M(L_1 - L_2)}$$

$$Y_{21} = \frac{-Z_{21}}{\Delta Z} = \frac{-sM}{s^2 M(L_1 - L_2)} = \frac{-1}{M(L_1 - L_2)}$$

$$Y_{22} = \frac{Z_{11}}{\Delta Z} = \frac{sL_1}{s^2 M(L_1 - L_2)} = \frac{L_1}{sM(L_1 - L_2)}$$

π-equivalent network representation is shown in Fig. 3.87

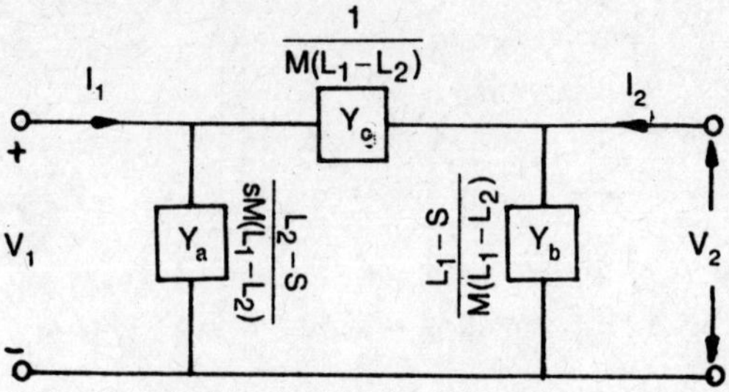

Fig. 3.87

$$Y_a = Y_{11} + Y_{12} = \frac{L_2}{sM(L_1 - L_2)} - \frac{1}{M(L_1 - L_2)}$$

$$= \frac{1}{(L_1 - L_2)M}\left(\frac{L_2}{s} - 1\right)$$

$$= \frac{L_2 - s}{sM(L_1 - L_2)}$$

$$Y_b = Y_{22} + Y_{12}$$

$$= \frac{L_1}{sM(L_1 - L_2)} - \frac{1}{M(L_1 - L_2)}$$

$$= \frac{1}{M(L_1 - L_2)}\left[\frac{L_1}{s} - 1\right]$$

$$= \frac{L_1 - s}{M(L_1 - L_2)}$$

$$Y_c = -Y_{12} = -Y_{21}$$

$$= \frac{1}{M(L_1 - L_2)}$$

Problem No. 3.44 The circuit in part (a) of Fig. 3.88 is to be described by an equivalent input circuit shown in Fig. 3.88(b). Determine Z_{eq} in Fig. 3.88(b) as a function of the elements and voltages in Fig. 3.88(a). What will be the value of Z_{eq}, if R_2 is infinite with the output open-circuited and also if R_2 i zero, with output short-circuited

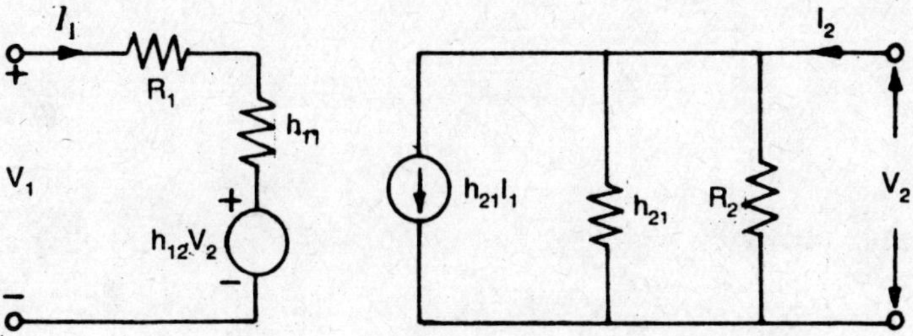

Fig. 3.88 (a)

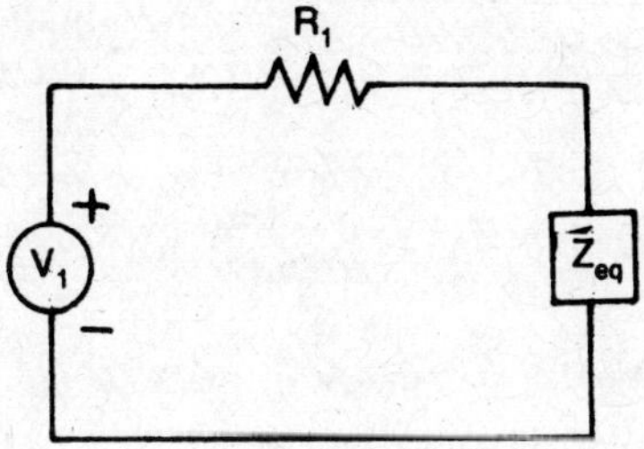

Fig. 3.88 (b)

Solution

Here

$$V_1 = (R_1 + h_{11})I_1 + h_{12}V_2 \qquad ...(1)$$

$$I_2 = h_{21}I_1 + h_{22}V_2 \qquad ...(2)$$

$$V_2 = -I_2R_2 \qquad ...(3)$$

Putting value of V_2 from equation (3) in equation (2), we get

$$I_2 = \left(\frac{h_{21}}{1 + h_{22}R_2}\right)I_1 \qquad ...(4)$$

and by putting the value of I_2 from equation (4) in equation (3), we get

$$V_2 = -\frac{h_{21}R_2I_1}{1 + h_{22}R_2} \qquad ...(5)$$

Substituting value of V_2 from equation (5) in equation (1), we get

$$V_1 = (R_1 + h_{11})I_1 - \frac{h_{12}h_{21}R_2I_1}{1 + h_{22}R_2}$$

Therefore,

$$\frac{V_1}{I_1} = R_1 + h_{11} - \frac{h_{12}h_{21}R_2}{1+h_{22}R_2}$$

$$= R_1 + Z_{eq}.$$

where

$$Z_{eq} = h_{11} - \frac{h_{12}h_{21}R_2}{1+h_{22}R_2}$$

If R_2 is infinite, with the output open-circuited, then

$$Z_{eq} = h_{11} - \frac{h_{12}h_{21}}{1/R_2 + h_{22}} = \frac{h_{11}h_{22} - h_{12}h_{21}}{h_{22}}$$

and if $R_2 = 0$, with the output short-circuited

$$Z_{eq} = h_{11} - 0$$

$$= h_{11}$$

Problem No. 3.45 Find T-parameter of the configuration shown in Fig. 3.89.

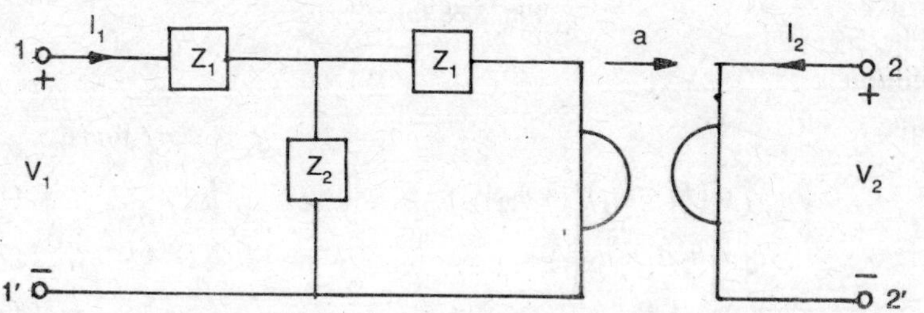

Fig. 3.89

Solution

Fig. 3.89 shows a T-network in tandem with a gyrater.

T-parameter of T-network

Refer Fig. 3.90, when a-a′ is open

$$A = \left.\frac{V_1}{V_2'}\right|_{I_2'=0}$$

$$V_1 = (Z_1 + Z_2)I_1$$

and

$$V_2' = Z_2 I_1$$

Therefore, $$A = \frac{(Z_1 + Z_2)I_1}{Z_2 I_1} = \frac{Z_1 + Z_2}{Z_2}$$

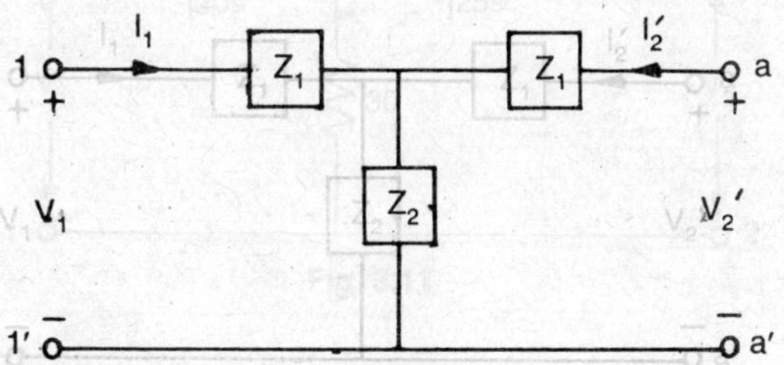

Fig. 3.90

$$C = \left.\frac{I_1}{V_2'}\right|_{I_2'=0} = \frac{I_1}{Z_2 I_1} = \frac{1}{Z_2}$$

when aa′ is short-circuited as shown in Fig. 3.91

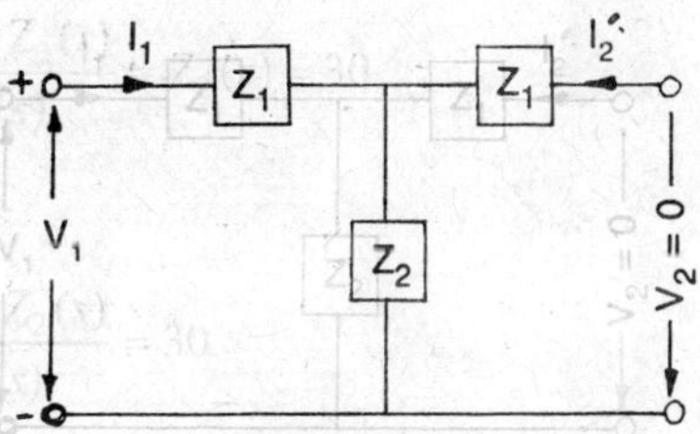

Fig. 3.91

$$V_2' = 0$$

$$B = \left.\frac{-V_1}{I_2'}\right|_{V_2'=0}$$

Writing mesh equations give

$$V_1 = (Z_1 + Z_2)I_1 + Z_2 I_2' \quad \text{...(1)}$$

$$0 = (Z_1 + Z_2)I_2' + Z_2 I_1 \quad \text{...(2)}$$

From equation (2)

$$I_1 = -\frac{Z_1 + Z_2}{Z_2} I_2' \quad \text{...(3)}$$

and

$$I_2' = -\frac{Z_2}{Z_1 + Z_2} I_1 \quad \text{...(4)}$$

substituting value of I_1 from equation (3) in equation (1), we get

$$V_1 = -\frac{(Z_1 + Z_2)(Z_1 + Z_2)}{Z_2} I_2' + Z_2 I_2'$$

$$V_1 = \left(\frac{Z_2^2 - (Z_1 + Z_2)^2}{Z_2} \right) I_2'$$

$$B = -\left.\frac{V_1}{I_2'}\right|_{V_2'=0}$$

$$= -\left(\frac{Z_2^2 - (Z_1 + Z_2)^2}{Z_2 I_2'} \right)_{I_2'} = -\frac{Z_2^2 - (Z_1 + Z_2)^2}{Z_2} = \frac{Z_1^2 + 2Z_1Z_2}{Z_2}$$

$$D = \left.\frac{-I_1}{I_2}\right|_{V_2'=0}$$

$$= -\left(-\frac{Z_1 + Z_2}{Z_2} \right) \frac{I_2'}{I_2'} = \frac{Z_1 + Z_2}{Z_2}$$

Therefore, transmission matrix is

$$= \begin{bmatrix} \frac{Z_1 + Z_2}{Z_2} & \frac{Z_1^2 + 2Z_1Z_2}{Z_2} \\ \frac{1}{Z_2} & \frac{Z_1 + Z_2}{Z_2} \end{bmatrix}$$

For the gyrator

$$V_2' = a(-I_2); \text{ and}$$

$$I_2' = \frac{1}{a} V_2$$

Transmission matrix is thus

$$= \begin{bmatrix} 0 & a \\ 1/a & 0 \end{bmatrix}$$

The overall transmission matrix is obtained by product of the transmission matrix and is

$$\begin{bmatrix} A & B \\ C & D \end{bmatrix} = \begin{bmatrix} \dfrac{Z_1 + Z_2}{Z_2} & \dfrac{Z_1^2 + 2Z_1Z_2}{Z_2} \\ \dfrac{1}{Z_2} & \dfrac{Z_1 + Z_2}{Z_2} \end{bmatrix} \begin{bmatrix} 0 & a \\ 1/a & 0 \end{bmatrix}$$

$$= \begin{bmatrix} \dfrac{Z_1^2 + 2Z_1Z_2}{aZ_2} & \dfrac{a(Z_1 + Z_2)}{Z_2} \\ \dfrac{Z_1 + Z_2}{aZ_2} & \dfrac{a}{Z_2} \end{bmatrix}$$

Problem No. 3.46 In Fig. 3.92 is shown two identical transformers. Determine the Z-parameters of the network

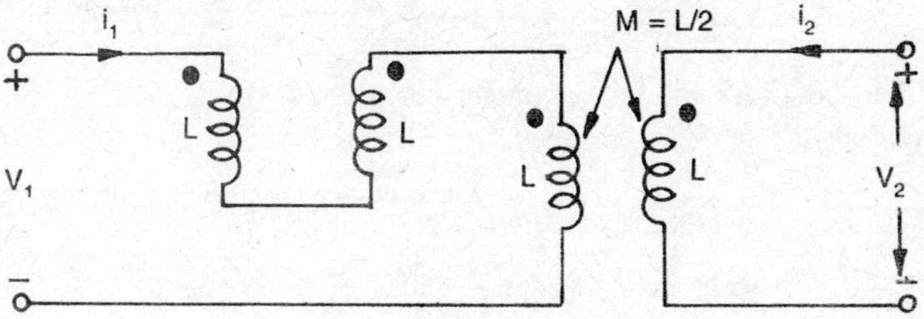

Fig. 3.92

Solution

Mesh equations are

$$V_1 = \frac{L di_1(t)}{dt} + \frac{L di_1(t)}{dt} + \frac{M di_2(t)}{dt}$$

and $$V_2 = \frac{L di_2(t)}{dt} + \frac{M di_1(t)}{dt}$$

In Laplace transform notation

$$V_1(s) = 2sL\, I_1(s) + sM\, I_2(s)$$

and $$V_2(s) = sM\, I_1(s) + sL\, I_2(s)$$

Putting $M = L/2$, we get

$$V_1(s) = 2sL\, I_1(s) + \frac{sL}{2} I_2(s) \qquad ...(1)$$

and $$V_2(s) = \frac{sL}{2} I_1(s) + sL\, I_2(s) \qquad ...(2)$$

Z-parameters are defined by

$$V_1(s) = Z_{11}I_1(s) + Z_{12}I_2(s) \quad \text{...(3)}$$

and
$$V_2(s) = Z_{21}I_1(s) + Z_{22}I_2(s) \quad \text{...(4)}$$

Comparing equation (1) and (3) with (3) and (4) we get

$$Z_{11} = 2sL, \qquad Z_{21} = \frac{sL}{2}$$

$$Z_{12} = \frac{sL}{2}, \qquad Z_{22} = sL$$

Therefore,

$$Z = \begin{bmatrix} 2sL & \frac{sL}{2} \\ \frac{sL}{2} & sL \end{bmatrix}$$

Problem No. 3.47 Find Z-parameters of the network shown in Fig. 3.93, where coils are perfectly coupled.

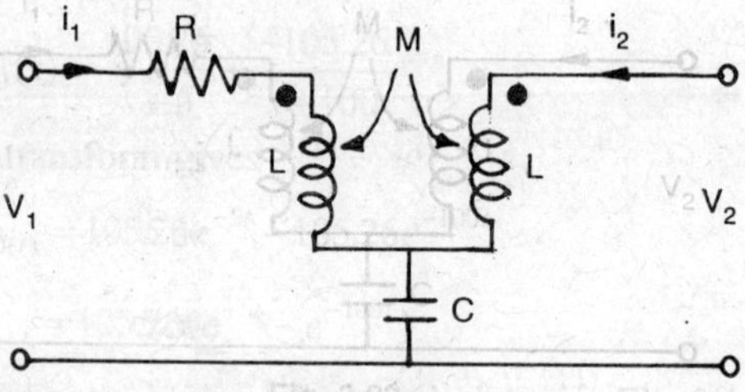

Fig. 3.93

Solution

Since coils are perfectly coupled, then
$K = 1$

$$K = \frac{M}{\sqrt{L_1 L_2}}$$

$$1 = \frac{M}{\sqrt{L_1 L_2}}$$

Therefore,

$$M = L$$

Mesh equations are

$$v_1(t) = Ri_1(t) + L\frac{di_1(t)}{dt} + M\frac{di_2(t)}{dt} + \frac{1}{C}\left(\frac{di_1(t)}{dt} + \frac{di_2(t)}{dt}\right)$$

or

$$V_i(s) = RI_1(s) + sLI_1(s) + \frac{I_1(s)}{sC} + sMI_2(s) + \frac{I_2(s)}{sC}$$

or

$$V_1(s) = \left(R + sL + \frac{1}{sC}\right)I_1(s) + \left(sM + \frac{1}{sC}\right)I_2(s) \qquad ...(1)$$

and

$$V_2(s) = L\frac{di_2(t)}{dt} + \frac{Mdi_1(t)}{dt} + \frac{1}{C}\left(\frac{di_1(t)}{dt} + \frac{di_2(t)}{dt}\right)$$

or

$$V_2(s) = sLI_2(s) + sMI_1(s) + \frac{I_1(s)}{sC} + \frac{I_2(s)}{sC}$$

or

$$V_2(s) = \left(sM + \frac{1}{sC}\right)I_1(s) + \left(sL + \frac{1}{sC}\right)I_2(s) \qquad ...(2)$$

Putting $M = 1$ in equations (1) and (2), we get

$$V_1(s) = \left(R + sL + \frac{1}{sC}\right)I_1(s) + \left(s + \frac{1}{s}\right)I_2(s) \qquad ...(3)$$

and

$$V_2(s) = \left(s + \frac{1}{s}\right)I_1(s) + \left(sL + \frac{1}{sC}\right)I_2(s) \qquad ...(4)$$

Z-parameter equations are defined by

$$V_1 = Z_{11}I_1 + Z_{12}I_2 \qquad ...(5)$$

$$V_2 = Z_{21}I_1 + Z_{22}I_2 \qquad ...(6)$$

Comparing equations (3) and (4) with (5) and (6), we get

$$Z = \begin{bmatrix} Z_{11} & Z_{12} \\ Z_{21} & Z_{22} \end{bmatrix} = \begin{bmatrix} R + sL + \frac{1}{sC} & s + \frac{1}{s} \\ s + \frac{1}{s} & sL + \frac{1}{sC} \end{bmatrix}$$

Problem No. 3.48 Find the T-parameter of the configuration shown in Fig. 3.94.

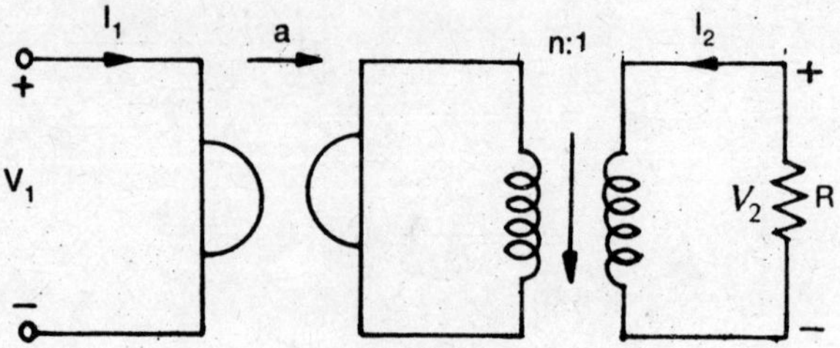

Fig. 3.94

Solution

T-parameter of the gyrator has been found in Problem No. 3.32 and 3.45 and is

$$T_G = \begin{bmatrix} 0 & a \\ \frac{1}{a} & 0 \end{bmatrix} \qquad ...(1)$$

$$V_2 = -RI_2;$$

$$I_2 = -nI_1;$$

or
$$V_2 = -R(-nI_1)$$

$$= nRI_1$$

or
$$I_1 = \frac{1}{nR} V_2$$

and

$$V_1 = nV_2$$

$$= n(-RI_2)$$

$$= nR(-I_2) \qquad ...(2)$$

From equations (1) and (2), we get

$$\begin{bmatrix} V_1 \\ I_1 \end{bmatrix} = \begin{bmatrix} 0 & nR \\ \frac{1}{nR} & 0 \end{bmatrix} \begin{bmatrix} V_2 \\ -I_2 \end{bmatrix}$$

Therefore

$$T_{trans} = \begin{bmatrix} 0 & nR \\ \dfrac{1}{nR} & 0 \end{bmatrix}$$

$$T_{total} = T_G \times T_{trans}$$

$$= \begin{bmatrix} 0 & a \\ \dfrac{1}{a} & 0 \end{bmatrix} \begin{bmatrix} 0 & nR \\ \dfrac{1}{nR} & 0 \end{bmatrix}$$

$$= \begin{bmatrix} \dfrac{a}{nR} & 0 \\ 0 & \dfrac{nR}{a} \end{bmatrix}$$

Problem No. 3.49 Find the Z-parameter of the circuit shown in Fig. 3.96

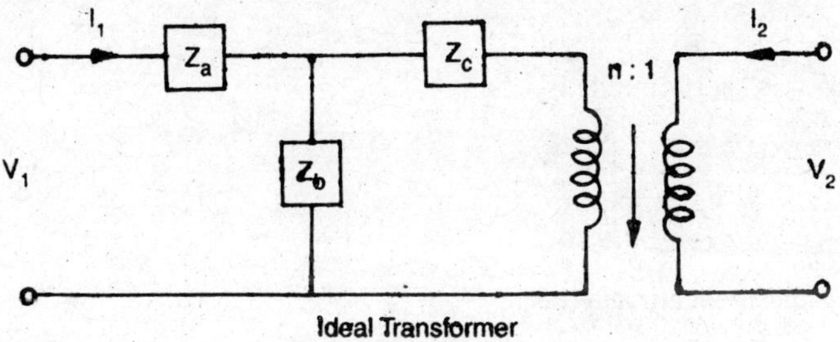

Fig. 3.96

Solution Fig. 3.97 shows a T-network connected in tandem with an ideal transformer. We will find the transmission parameter of the combination and then convert to Z-parameter.

T-network (Fig. 3.97)

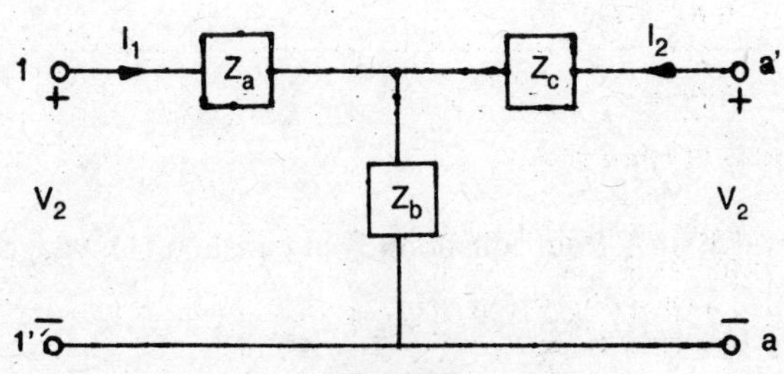

Fig. 3.97

Let a-a′ is open

$$A = \left.\frac{V_1}{V_2'}\right|_{I_2'=0}$$

$$V_1 = (Z_a + Z_b)I_1$$

and

$$V_2' = Z_b I_1$$

Therefore

$$A = \left.\frac{V_1}{V_2'}\right|_{I_2'=0} = \frac{(Z_a + Z_b)I_1}{Z_b I_1} = \frac{Z_a + Z_b}{Z_b}$$

$$C = \left.\frac{I_1}{V_2'}\right|_{I_2'=0} = \frac{I_1}{Z_b I_1} = \frac{1}{Z_b}$$

when a-a′ is short circuited, $V_2' = 0$

$$B = \left.\frac{-V_1}{I_2'}\right|_{V_2'=0}$$

Writing mesh equations give

$$V_1 = (Z_a + Z_b)I_1 + Z_b I_2' \quad \text{...(1)}$$

and

$$0 = (Z_b + Z_c)I_2' + Z_b I_1 \quad \text{...(2)}$$

From equation (2), we get

$$I_1 = -\frac{(Z_b + Z_c)I_2'}{Z_b} \quad \text{...(3)}$$

and

$$I_2' = \frac{-Z_b}{Z_b + Z_c} I_1 \quad \text{...(4)}$$

Substituting value of I_1 from equation (3) in equation (1), we get

$$V_1 = \frac{-(Z_a + Z_b)(Z_b + Z_c)}{Z_b} I_2' + Z_b I_2'$$

or $$V_1 = \left(\frac{Zb^2 - (Z_a + Z_b)(Z_b + Z_c)}{Z_b} \right) I_2'$$

$$B = \left. \frac{-V_1}{I_2'} \right|_{V_2'=0}$$

$$= -\frac{\left(Z_b^2 - (Z_a + Z_b)(Z_b + Z_c)\right)}{Z_b \times I_2'} I_2'$$

$$= \frac{-\left(Z_b^2 - (Z_a + Z_b)(Z_b + Z_c)\right)}{Z_b}$$

$$D = \left. \frac{-I_1}{I_2'} \right|_{V_2'=0}$$

$$= -\left\{ -\frac{(Z_b + Z_c)}{Z_b} \right\} \frac{I_2'}{I_2'}$$

$$= \frac{Z_b + Z_c}{Z_b}$$

Therefore Transmission-parameter matrix is

$$= \begin{bmatrix} \dfrac{Z_a + Z_b}{Z_b} & \dfrac{Z_b^2 - (Z_b + Z_c)(Z_a + Z_b)}{Z_b} \\ \dfrac{1}{Z_b} & \dfrac{Z_b + Z_c}{Z_b} \end{bmatrix}$$

or $$= \begin{bmatrix} \dfrac{Z_a + Z_b}{Z_b} & Z_a Z_b + Z_b Z_c + Z_a Z_c \\ \dfrac{1}{Z_b} & \dfrac{Z_b + Z_c}{Z_b} \end{bmatrix}$$

Ideal Transformer

Refer Fig. 3.98

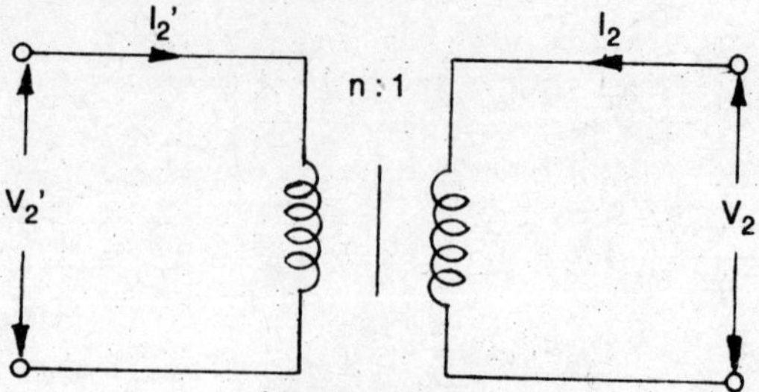

Ideal Transformer

Fig. 3.98

$$V_2' = nV_2$$

$$I_2' = \frac{1}{n}(-I_2)$$

In matrix form

$$\begin{bmatrix} V_2' \\ I_2' \end{bmatrix} = \begin{bmatrix} n & 0 \\ 0 & \frac{1}{n} \end{bmatrix} \begin{bmatrix} V_2 \\ -I_2 \end{bmatrix}$$

Therefore transmission matrix of an ideal transformer is

$$= \begin{bmatrix} n & 0 \\ 0 & \frac{1}{n} \end{bmatrix} \qquad ...(6)$$

Transmission matrix of the combination is obtained by multiplying equations (5) and (6)

$$\begin{bmatrix} A & B \\ C & D \end{bmatrix} = \begin{bmatrix} \frac{Z_a + Z_b}{Z_b} & \frac{Z_aZ_b + Z_bZ_c + Z_cZ_a}{Z_b} \\ \frac{1}{Z_b} & \frac{Z_b + Z_c}{Z_b} \end{bmatrix} \begin{bmatrix} n & 0 \\ 0 & \frac{1}{n} \end{bmatrix}$$

$$= \begin{bmatrix} \frac{n(Z_a + Z_b)}{Z_b} & \frac{Z_aZ_b + Z_bZ_c + Z_cZ_a}{nZ_b} \\ \frac{n}{Z_b} & \frac{Z_b + Z_c}{nZ_b} \end{bmatrix}$$

$$\Delta T = AD - BC$$

$$= \frac{n(Z_a + Z_b)}{Z_b} \times \frac{(Z_b + Z_c)}{nZ_b} - \frac{n(Z_aZ_b + Z_bZ_c + Z_cZ_a)}{nZ_b \times Z_b}$$

$$= \frac{(Z_a + Z_b)(Z_b + Z_c)}{Z_b^2} - \frac{Z_aZ_b + Z_bZ_c + Z_cZ_a}{Z_b^2}$$

$$= \frac{Z_aZ_b + Z_aZ_c + Z_bZ_c + Z_b^2 - Z_aZ_b - Z_bZ_c - Z_cZ_a}{Z_b^2} = 1.$$

$$Z_{11} = \frac{A}{C} = \frac{n\dfrac{(Z_a + Z_b)}{Z_b}}{\dfrac{n}{Z_b}} = Z_a + Z_b$$

$$Z_{12} = \frac{\Delta T}{C} = \frac{1}{n/Z_b} = \frac{Z_b}{n}$$

$$Z_{21} = \frac{1}{C} = \frac{1}{n/Z_b} = \frac{Z_b}{n}$$

$$Z_{22} = \frac{D}{C} = \frac{\dfrac{Z_b + Z_c}{nZ_b}}{\dfrac{n}{Z_b}} = \frac{Z_b + Z_c}{n^2}$$

$$Z = \begin{bmatrix} Z_a + Z_b & \dfrac{Z_b}{n} \\ \dfrac{Z_b}{n} & \dfrac{Z_b + Z_c}{n^2} \end{bmatrix}$$

Problem No. 3.50 Determine Z-parameters of the configuration shown in Fig. 3.99 (a).

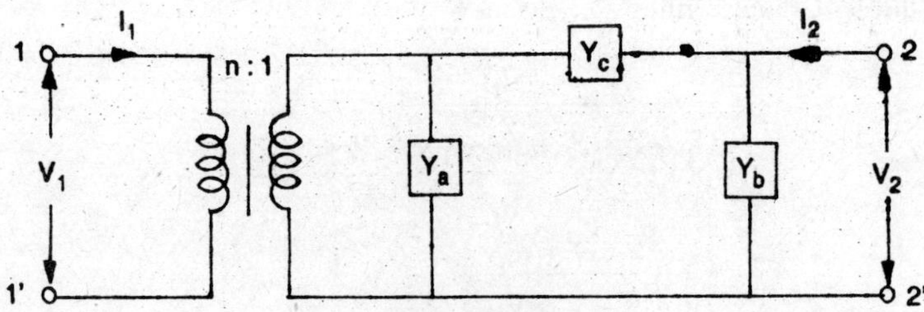

Ideal Transformer

Fig. 3.99 (a)

Solution

Ideal transference Transmission-parameters has been found in Problem No. 3.49 and are

$$\begin{bmatrix} A & B \\ C & D \end{bmatrix} = \begin{bmatrix} n & 0 \\ 0 & \frac{1}{n} \end{bmatrix} \qquad ...(1)$$

Pie-network

Let port 2-2′ is open and set $I_2 = 0$

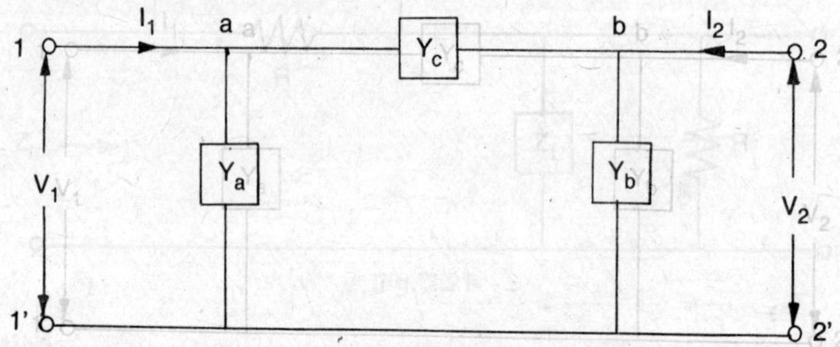

Fig. 3.99 (b)

Node 'a'

$$(Y_c + Y_a)V_1 + (-Y_c)V_2 = I_1 \qquad ...(1)$$

Node 'b'

$$(-Y_c)V_1 + (Y_c + Y_b)V_2 = 0 \qquad ...(2)$$

Solving by Cramer's rule

$$V_1 = \frac{(Y_c + Y_b)I_1}{(Y_c + Y_a)(Y_c + Y_b) - \frac{1}{Y_c^2}}$$

$$V_2 = \frac{Y_c I_1}{(Y_c + Y_a)(Y_c + Y_b) - \frac{1}{Y_c^2}}$$

Thus

$$A = \left.\frac{I_1}{V_2}\right|_{I_2=0} = \frac{Y_c + Y_b}{Y_c} = \frac{Y_b + Y_c}{Y_c}$$

$$C = \left.\frac{I_1}{V_2}\right|_{I_2=0} = \frac{Y_A Y_B + Y_B Y_C + Y_A Y_C}{Y_c}$$

Similarly, for B and D, set $V_2 = 0$ and short-circuit, port 2-2′. Then,

$$I_1 = (Y_c + Y_a) V_1$$

and $$I_2 = -Y_c V_1$$

or $$V_1 = \frac{-I_2}{Y_c}$$

Thus,

$$B = \left.\frac{V_1}{-I_2}\right|_{V_2=0} = \frac{-I_2/Y_c}{-I_2} = \frac{1}{Y_c}$$

$$D = \frac{I_1}{-I_2} = \frac{(Y_c + Y_a)V_1}{-(-Y_c V_1)} = \frac{Y_a + Y_c}{Y_c}$$

Therefore, transmission-parameter in matrix form are

$$\begin{bmatrix} A & B \\ C & D \end{bmatrix} = \begin{bmatrix} \dfrac{Y_B + Y_C}{Y_C} & \dfrac{1}{Y_C} \\ \dfrac{Y_A Y_B + Y_B Y_C + Y_C Y_A}{Y_C} & \dfrac{Y_a + Y_c}{Y_C} \end{bmatrix}$$

$$\Delta T = AD - BC$$

$$= \frac{Y_B + Y_C}{Y_C} \times \frac{Y_a + Y_c}{Y_C} - \frac{1}{Y_C} \times \frac{Y_A Y_B + Y_B Y_C + Y_C Y_A}{Y_C}$$

$$= \frac{Y_c^{\,2}}{Y_c^{\,2}}$$

$$= 1$$

Z-parameters

$$Z_{11} = \frac{A}{C} = \frac{n\dfrac{(Y_B + Y_C)}{Y_C}}{\dfrac{Y_A Y_B + Y_B Y_C + Y_c Y_a}{nY_c}} = n^2 \frac{(Y_B + Y_C)}{Y_A Y_B + Y_B Y_C + Y_C Y_A}$$

$$Z_{12} = \frac{\Delta T}{C} = \frac{1}{\dfrac{Y_AY_B + Y_BY_C + Y_cY_a}{nY_c}} = \frac{nY_C}{Y_AY_B + Y_BY_C + Y_CY_A}$$

$$Z_{21} = \frac{1}{C} = \frac{1}{\dfrac{Y_AY_B + Y_BY_C + Y_CY_A}{nY_c}} = \frac{nY_C}{Y_AY_B + Y_BY_C + Y_CY_A}$$

$$Z_{22} = \frac{D}{C} = \frac{\dfrac{Y_A + Y_C}{nY_C}}{\dfrac{Y_AY_B + Y_BY_C + Y_CY_A}{nY_c}} = \frac{Y_A + Y_C}{Y_AY_B + Y_BY_C + Y_CY_A}$$

Therefore

$$Z = \begin{bmatrix} \dfrac{n^2(Y_B + Y_C)}{Y_AY_B + Y_BY_C + Y_CY_A} & \dfrac{nY_C}{Y_AY_B + Y_BY_C + Y_CY_A} \\ \dfrac{nY_c}{Y_AY_B + Y_BY_C + Y_CY_A} & \dfrac{Y_A + Y_C}{Y_AY_B + Y_BY_C + Y_CY_A} \end{bmatrix}$$

Transmission parameters of the combination is

$$\begin{bmatrix} A & B \\ C & D \end{bmatrix} = \begin{bmatrix} n & 0 \\ 0 & \dfrac{1}{n} \end{bmatrix} \begin{bmatrix} \dfrac{Y_B + Y_C}{Y_C} & \dfrac{1}{Y_C} \\ \dfrac{Y_AY_B + Y_BY_C + Y_CY_A}{Y_C} & \dfrac{Y_A + Y_C}{Y_C} \end{bmatrix}$$

$$= \begin{bmatrix} \dfrac{n(Y_B + Y_C)}{Y_C} & \dfrac{n}{Y_C} \\ \dfrac{Y_AY_B + Y_BY_C + Y_CY_A}{nY_C} & \dfrac{Y_A + Y_C}{nY_C} \end{bmatrix}$$

$$\Delta T = AD - BC$$

$$= \frac{n(Y_B + Y_C)}{Y_C} \times \frac{(Y_A + Y_C)}{nY_C} - \frac{n}{Y_C} \times \frac{Y_AY_B + Y_BY_C + Y_CY_A}{nY_C}$$

$$= \frac{Y_C^2}{Y_C^2}$$

$$= 1.$$

Problem No. 3.51 Determine the image impedances of the network shown in Fig. 3.100.

Solution

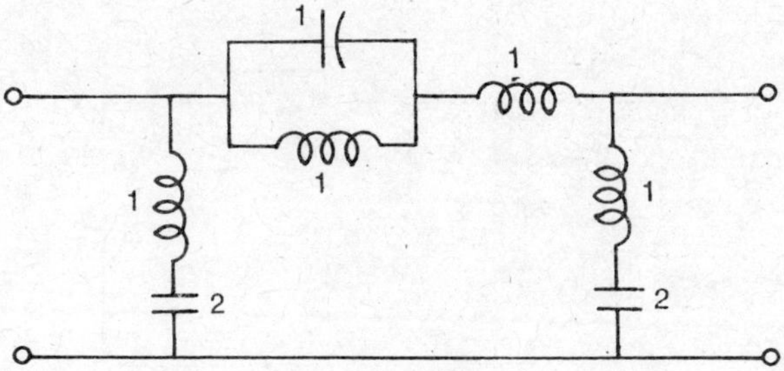

Fig. 3.100

Solution

Considering that the units of the elements of the circuit is in ohms, the equivalent circuit is shown in Fig. 3.101. We have

$$V_1 = AV_2 - BI_2 \quad ...(1)$$

$$I_1 = CV_2 - DI_2 \quad ...(2)$$

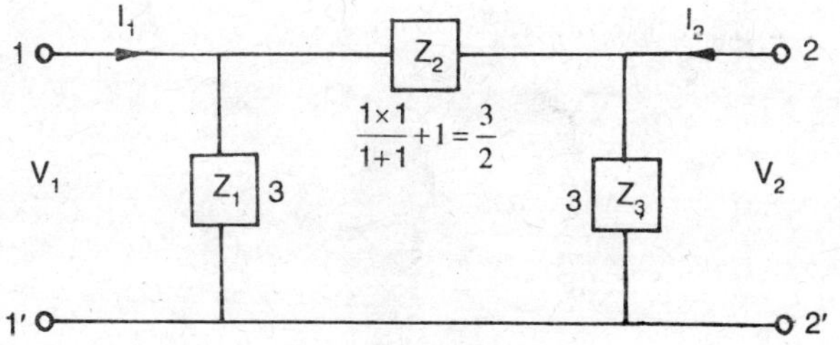

Fig. 3.101

With port 2-2′ short-circuited, with V_1 across $a-a'$, $V_2 = 0$

$$V_1 = AX_0 - BI_2$$

or
$$-I_2 = \frac{V_1}{2} \quad ...(3)$$

and

$$I_1 = CX_0 - DI_2$$

or
$$I_1 = -DI_2$$

or $$D = -\frac{I_1}{I_2} \qquad \text{...(4)}$$

On short-circuiting port 2-2′, the circuit reduces to Fig. 3.102

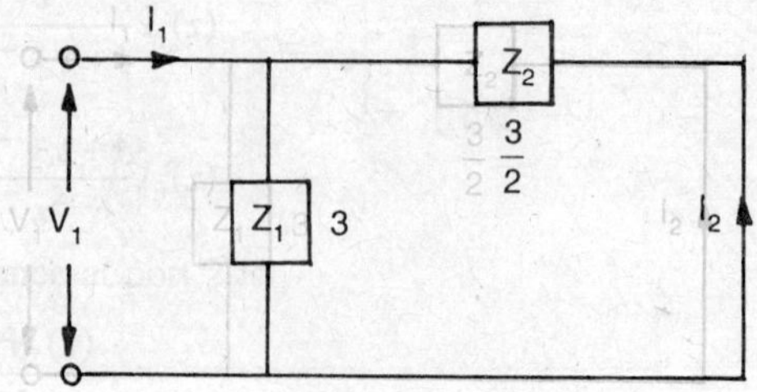

Fig. 3.102

Therefore,

$$V_1 = -I_2 \times Z_2$$

$$= \frac{-3}{2} I_2$$

or $$I_2 = \frac{-2}{3} V_1 \qquad \text{...(5)}$$

$$B = \frac{-V_1}{I_2}\bigg|_{V_2=0} = -\frac{-V_1}{\frac{-2}{3}V_1} = \frac{3}{2}\,\Omega$$

Also

$$I_1 = \frac{V_1}{3} + \frac{V_1}{3/2}$$

$$= \frac{V_1}{3} + \frac{2V_1}{3}$$

$$= \frac{3V_1}{3}$$

$$= V_1$$

But since $V_1 = \frac{-3}{2} I_2$ (equation 5), we get

$$I_1 = \frac{-3}{2} I_2 .$$

Hence,

$$D = \frac{-I_1}{I_2}\bigg|_{V_2=0} = \frac{-\left(-\frac{3}{2}I_2\right)}{I_2} = \frac{3}{2}\,\Omega$$

Alternatively

$$V_1 = 3(I_1 + I_2) \qquad \text{...(6)}$$

$$0 = \frac{3}{2}I_2 + 3(I_1 + I_2) \qquad \text{...(7)}$$

Equation (7) gives

$$0 = \frac{3}{2}I_2 + 3I_2 + 3I_1$$

or $$0 = \frac{9}{2}I_2 + 3I_1$$

or $$I_1 = \frac{-9}{2 \times 3}I_2$$

or $$I_1 = \frac{-3}{2}I_2.$$

hence $$D = \frac{-I_1}{I_2} = \frac{-(-3/2)I_2}{I_2} = \frac{3}{2}\,\Omega$$

With port 2-2′ open circuited, we get $I_2 = 0$. Let the currents be as shown in Fig. 3.103

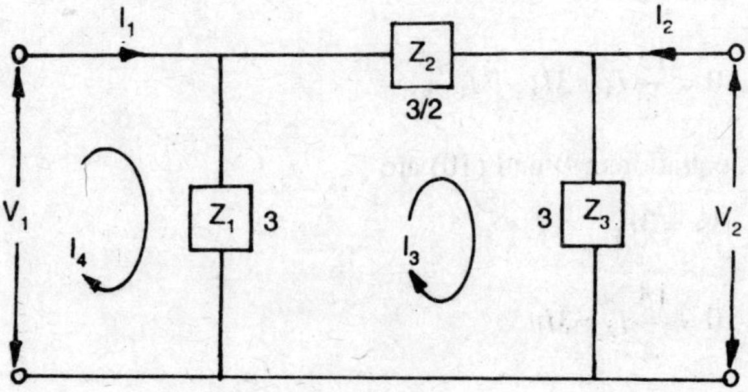

Fig. 3.103

Then,

$$I_4 = \frac{V_1}{Z_1} = \frac{V_1}{3}$$

$$I_3 = \frac{V_1}{3/2+3} = \frac{2V_1}{9}$$

$$I_1 = I_3 + I_4 = \frac{V_1}{3} + \frac{2V_1}{9} = \frac{5V_1}{9}$$

$$V_2 = I_3 \times 3 = \frac{2V_1}{9} \times 3 = \frac{2}{3} V_1$$

Hence,

$$A = \left.\frac{V_1}{V_2}\right|_{I_2=0} = \frac{V_1}{2/3\, V_1} = \frac{3}{2}\Omega$$

$$C = \left.\frac{I_1}{V_2}\right|_{I_2=0} = \frac{5/9\, V_1}{2/3\, V_1} = \frac{5}{6}\Omega$$

Alternatively

$$V_1 = 3I_1 - 3I_3; \qquad ...(8)$$

$$V_2 = 3I_2 + 3I_3; \qquad ...(9)$$

and $$0 = \frac{3}{2} I_3 + 3I_3 + 3I_2 + 3I_3 - 3I_1$$

or $$0 = \left(\frac{3}{2} + 3 + 3\right) I_3 + 3I_2 - 3I_1$$

or $$0 = \frac{15}{2} I_3 + 3I_2 - 3I_1 \qquad ...(10)$$

Since, $I_2 = 0$, equations (9) and (10) are

$$V_2 = 3I_3 \qquad ...(11)$$

and $$0 = \frac{15}{2} I_3 - 3I_1 \qquad ...(12)$$

or $$I_3 = \frac{3 \times 2}{15} I_1$$

or $$I_3 = \frac{2}{5} I_1 \qquad \text{...(13)}$$

Putting in equation (8), value of I_3, we get

$$V_1 = 3I_1 - \frac{3 \times 2}{5} I_1$$

or $$V_1 = 3I_1 - \frac{6}{5} I_1$$

or $$V_1 = \frac{9}{5} I_1 \qquad \text{...(14)}$$

Also, from equation (11)

$$V_2 = 3I_3$$

or $$V_2 = \frac{3 \times 2}{5} I_1 = \frac{6}{5} I_1 \qquad \text{...(15)}$$

Hence,

$$A - \left.\frac{V_1}{V_2}\right|_{I_2=0} = \frac{9/5 I_1}{6/5 I_1} = \frac{3}{2} \Omega.$$

and $$C = \left.\frac{I_1}{V_2}\right|_{I_2=0} = \frac{I_1}{6/5 I_1} = \frac{5}{6} \Omega.$$

Image Parameters

$$Z_{I1} = \sqrt{\frac{AB}{CD}} = \sqrt{\frac{3/2 \times 3/2}{5/6 \times 3/2}} = \sqrt{\frac{3}{2} \times \frac{6}{5}} = \sqrt{\frac{9}{5}} = 1.34\,\Omega$$

$$Z_{I2} = \sqrt{\frac{BD}{AC}} = \sqrt{\frac{3/2 \times 3/2}{5/6 \times 3/2}} = \sqrt{\frac{9}{5}} = 1.34\,\Omega$$

Note: Since $A = D$, network is symetrical and since $Z_{I1} = Z_{I2}$, the network is symetrical reciprocal. The image impedances are also called the **characteristic** or **iterative** impedance.

$$\phi = \tanh^{-1} \sqrt{\frac{BC}{AD}} = \sqrt{\frac{3/2 \times 5/6}{3/2 \times 3/2}} = \sqrt{\frac{5/6}{3/2}} = 0.963.$$

4

Representation: Networks and Systems

4.1 Representation of a System

Systems are generally described by a set of differential equations. the interrelationships of the equations are depicted pictorially with the help of block diagrams and signal flow graphs.

4.1.1 Block Diagram Block diagram is a pictorial representation of a system. A simple open-loop system having one input and one output is shown in Fig. 4.1. R(s) is the signal into the block representing the input signal and C(s) is the signal going out of the block representing the output signal. The block represents the transfer function G(s). (R)s and C(s) are the Laplace transform of the input and output signals.

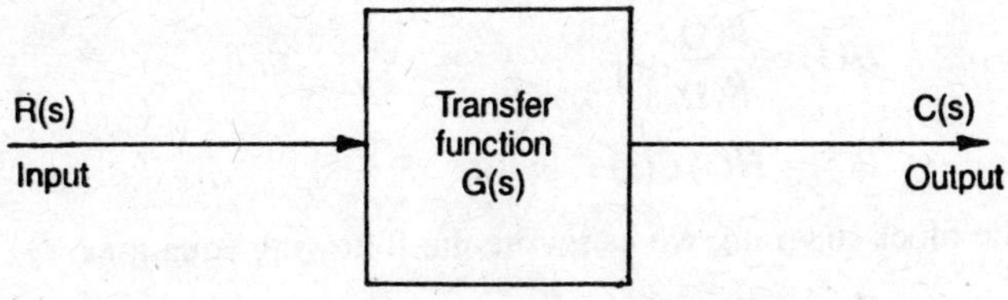

Fig. 4.1

The transfer function relating the input and output is thus given by

$$G(s) = \frac{C(s)}{R(s)} \qquad ...(4.1)$$

4.1.2 Block Diagram of a Closed-Loop System. Let us consider a closed-loop system with a negative feedback represented by a block diagram as shown in Fig. 4.2

The various notations used in the block diagram are explained below:

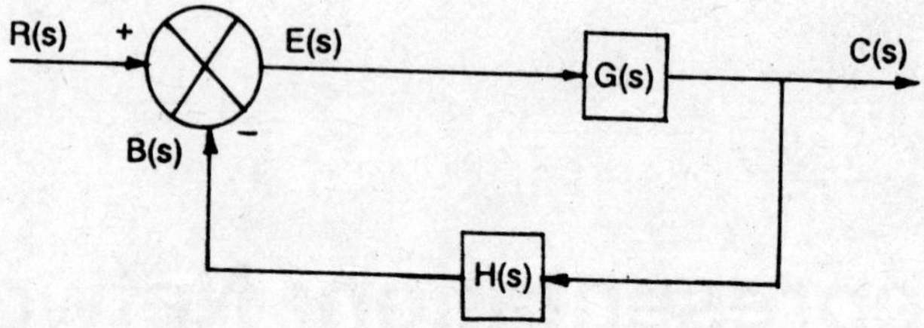

Fig. 4.2

R(s) : Input signal

C(s) : Output signal

E(s) : Actuating signal

B(s) : Feedback signal

G(s) : Forward path transfer function

H(s) : Feedback path transfer function

T(s) : Transfer function of closed-loop

From definition of transfer function, we can write

$$G(s) = \frac{C(s)}{E(s)}, \text{ or}$$

$$C(s) = G(s)\,E(s) \qquad ...(4.2)$$

$$T(s) = \frac{C(s)}{R(s)}, \text{ and} \qquad ...(4.3)$$

$$H(s) = \frac{B(s)}{R(s)}, \text{ or}$$

$$B(s) = H(s)\,C(s) \qquad ...(4.4)$$

From the block diagram, we can write the following equations

$$C(s) = G(s)\,E(s) \qquad ...(4.5)$$

$$E(s) = R(s) - B(s), \text{ or} \qquad ...(4.6)$$

$$E(s) = R(s) - H(s)\,C(s) \qquad ...(4.7)$$

Substituting the value of E(s) in equation (4.5), we get

$$C(s) = G(s)[R(s) - H(s)\,C(s)]$$

$$C(s) + G(s)\,H(s)\,C(s) = G(s)\,R(s)$$

$$C(s)[1 + G(s)\,H(s)] = G(s)\,R(s)$$

or

$$\frac{C(s)}{R(s)} = \frac{G(s)}{1 + G(s)\,H(s)} \qquad \text{...(4.8)}$$

where $\dfrac{C(s)}{R(s)} = T(s)$

Equation (4.8) can be represented by a block diagram as shown in Fig. 4.3

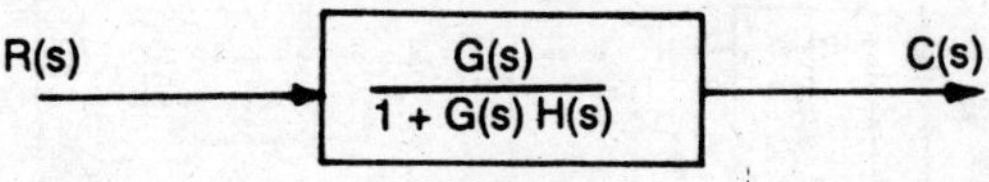

Fig. 4.3

The loop transfer function or open loop transfer function of a closed-loop system is defined as

$$\frac{B(s)}{E(s)} = G(s)\,H(s) \qquad \text{...(4.9)}$$

This is derived as given below

$$\frac{B(s)}{E(s)} = \frac{B(s)}{C(s)} \times \frac{C(s)}{E(s)}$$

$$= H(s)\,G(s)$$

$$= G(s)\,H(s) \qquad \text{...(4.10)}$$

In many cases where $|G(s)\,H(s)| >> 1$, the closed-loop transfer function

$$\frac{C(s)}{R(s)} = \frac{1}{H(s)} \qquad \text{...(4.11)}$$

This is called the approximate transfer function and is the inverse of feedback transfer function. The significance of this concept lies in the fact that for cases where $|G(s)\,H(s)| >> 1$, the closed-loop approximate transfer function is independent of forward path transfer function. It only depends upon the feedback transfer function H(s). In open-loop transfer function, the transfer function is dependent only on forward path transfer function G(s) which is assembled with power elements. In an open-loop system, in order to achieve the desired objective, the power elements constituting G(s) have to be selected very carefully. On the other hand in a closed-loop system, transfer function being dependent on H(s), consists of measuring elements and hence require careful selection

Table 4.1 Block Diagram Reduction Rules

SL. No.	Operation.	Block Diagram.	Equivalent Block Diagram.	Equation.
1.	Combining blocks in cascade	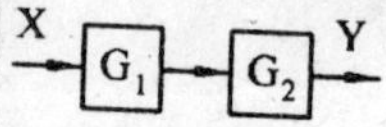	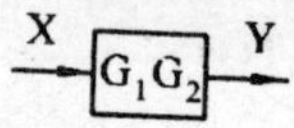	$Y = (G_1 G_2) X$
2.	Combining blocks in parallel; or eliminating a forward loop.	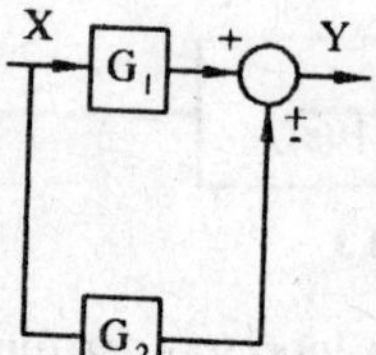	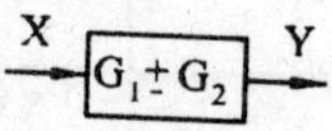	$Y = G_1 X \pm G_2 X$
3.	Removing a block from a forward path.		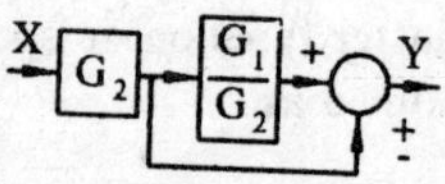	$Y = G_1 X \pm G_2 X$
4.	Eliminating a feedback loop.	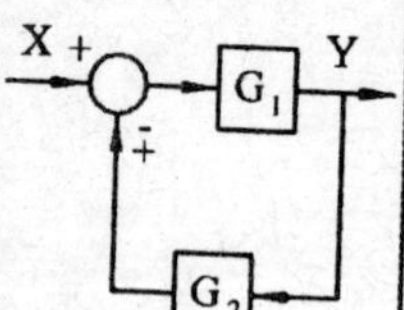	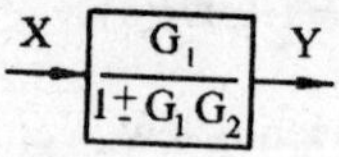	$Y = G_1 (X \mp G_2 Y)$
5.	Removing a block from a feedback loop.		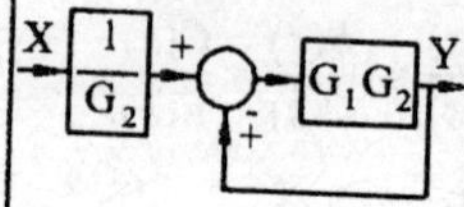	$Y = G_1 (X \mp G_2 Y)$
6.	Rearranging summing points.	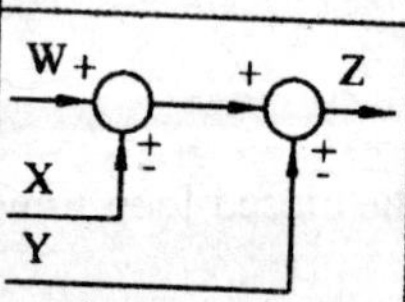	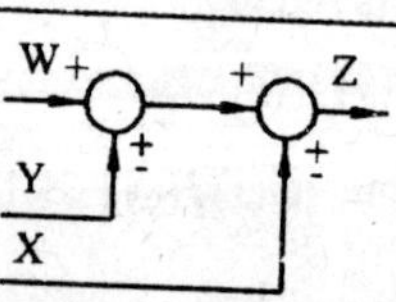	$Z = W \pm X \pm Y$
7.	Rearranging summing points.	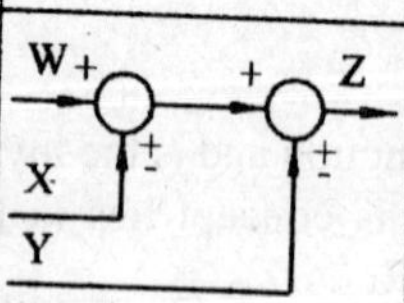	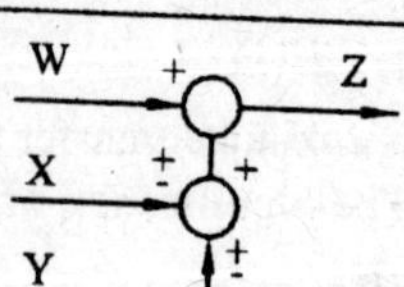	$Z = W \pm X \pm Y$

(Contd.)

8.	Moving a summing point beyond a block.	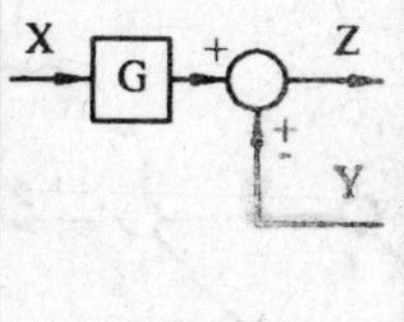	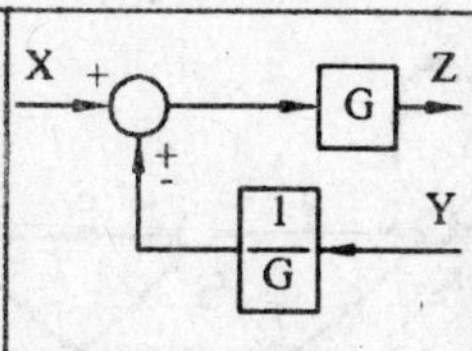	Z = G X ± Y
9.	Moving a summing point beyond a block.	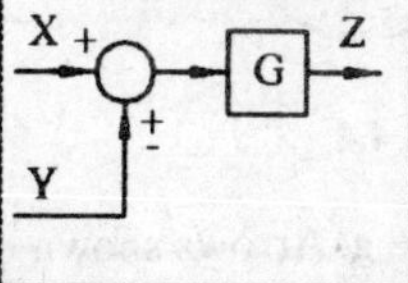	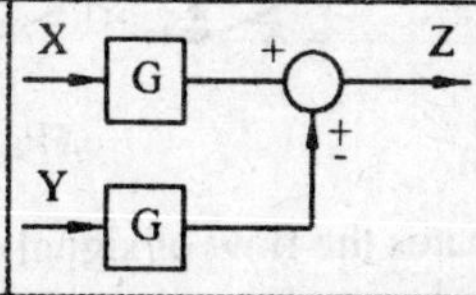	Z = G [X ± Y]
10.	Moving a takeoff point ahead of a block.	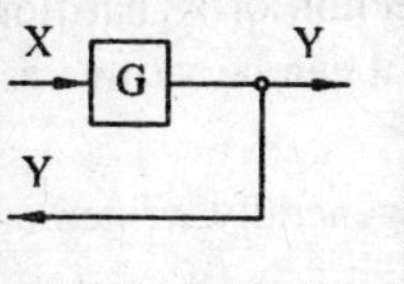	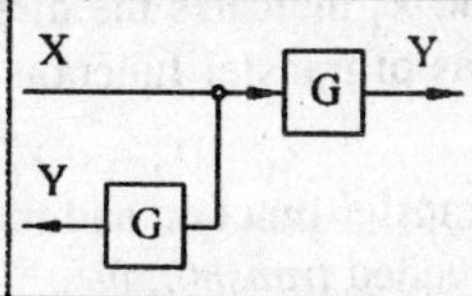	Y = G X
11.	Moving a takeoff point beyond a block.	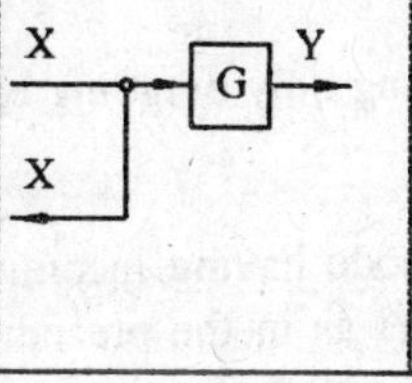	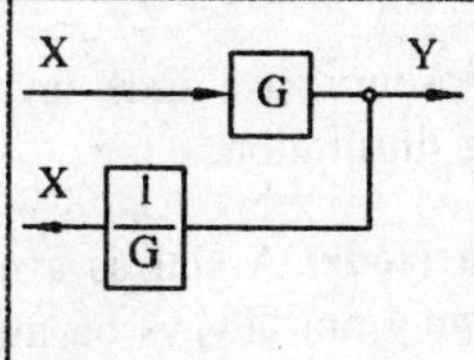	Y = G X

Any complex system can be conveniently represented by a block diagram and all complex representations of block diagrams can be reduced to a very simple diagram by the use of block diagram reduction algebra. These rules have been illustrated in Table 4.1

4.1.3 Signal Flow Graph. Any linear control system can be described by a set of linear equations having the form

$$Y_i = \sum_{i=1}^{n} a_{ij} Y_i \qquad \text{...(4.12)}$$

where $\qquad i = 1, 2, \ldots\ldots, n$

A signal flow graph represents a set of equations of this type by means of branches and nodes. Each variable is represented by a node and branches connect the variables. Before proceeding further let us understand commonly used terms in the signal-flow diagrams. A signal-flow diagram is shown in Fig. 4.4.

Node: Nodes are the variables or signal of the system e.g. x_1, x_2, x_3, x_4, x_5, x_6.

Branch: It is line joining or connecting two nodes.

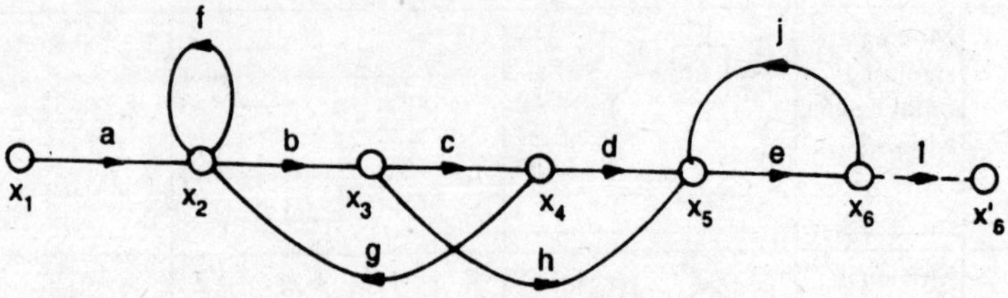

Fig. 4.4

Arrow: It indicates the flow of signal e.g. Arrows shown on branch connecting nodes x_1 and x_2 indicates the direction of signal flow from node x_1 to node x_2. In terms of transfer function it can be written as

$$x_2 = a\, x_1,$$

where a is the transfer function and is generally written on the top of branch itself. It is•also called *transmittance.*

Source Node: A source is a node having only outgoing branches such as x_1 in the preceding illustration.

Sink or Output Node: A sink is a node having incoming branches only. This condition may not always be met as in the preceding illustration. In such cases an additional branch with unit transmittance as shown by the dotted line is introduced in order to meet the specified condition. In the preceding illustration since x_6 node was not meeting the required condition, an outgoing branch from x_6 has been introduced terminating at x_6' having unit gain. Now x_6' meets the required condition. Therefore x_6' is the sink node.

$$x_6' = 1 \times x_6$$
$$= x_6$$

Forward Path: Forward paths are paths which originate from a source node and terminate at a sink node and along which no node is encountered more than once. In the preceding illustration there are two forward paths, namely

$x_1 - x_2 - x_3 - x_4 - x_5 - x_6 - x_6'$ having gain '*abcde*', and

$x_1 - x_2 - x_3 - x_5 - x_6'$ having gain '*abhe*'.

Path Gain: It is the product of the gains of all branches along the path as explained in the preceding defination.

Loop: Loop is a closed path which originates from a node and terminates at the same node along which no intermediate path is traversed twice e.g.

$x_2 - x_3 - x_4 - x_2$ having gain '*bcg*',

$x_5 - x_6 - x_5$ having gain '*cj*', and

$x_2 - x_2$ having gain '*f*'.

Table 4.2: Signal-Flow Graph Algebra

SL. No.	Operation.	Original.	Equivalent Equation or Diagram.
1.	Addition.	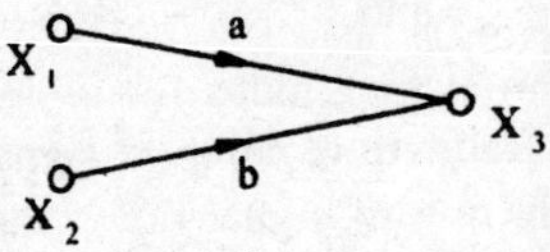 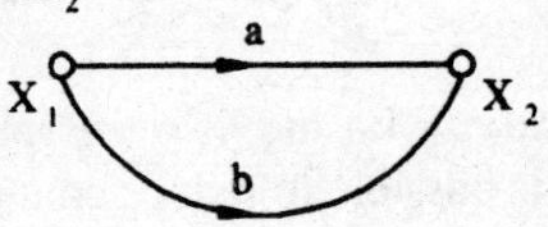	$X_3 = a x_1 + b x_2$ 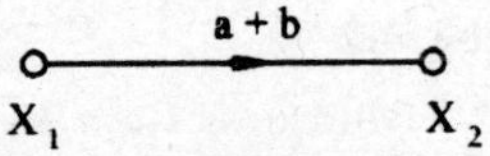 $X_2 = (a+b) x_1$
2.	Multipli - cation.	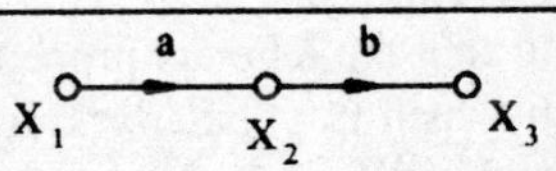	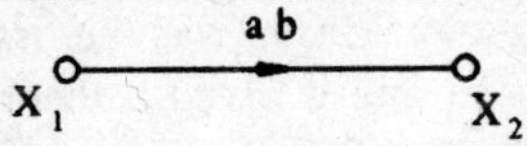 $X_2 = a b x_1$
3.	Feedback loop elimination.	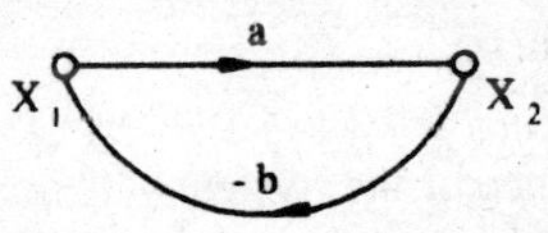 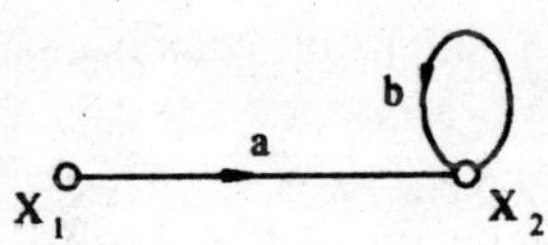	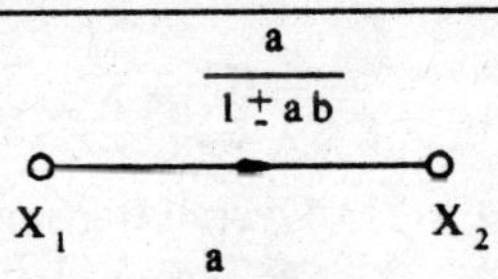 $x_2 = \frac{a}{1 \pm ab} x_1$ 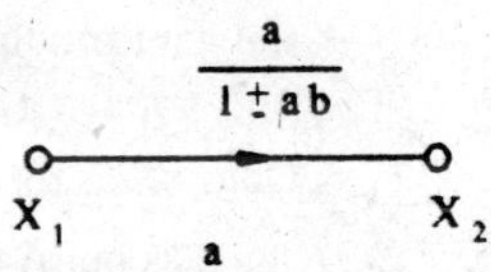 $x_2 = \frac{a}{1 \pm b} x_1$
4.	Misc.	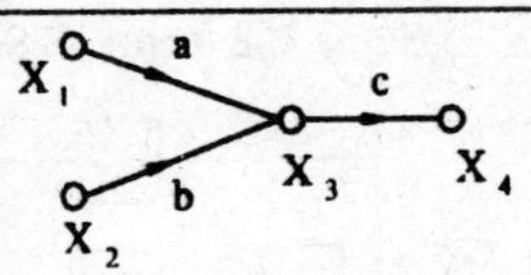 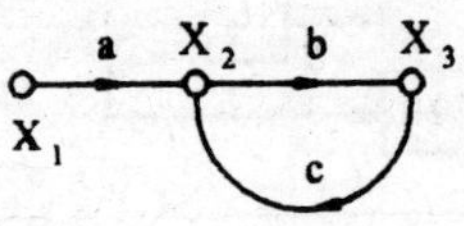	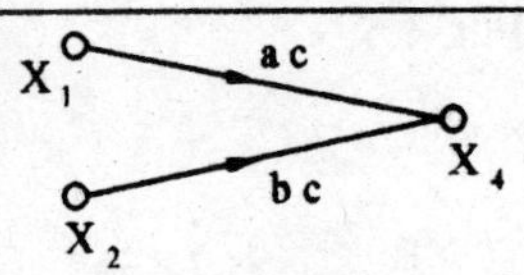 $X_4 = b c x_2 + a c x_1 = c(b x_2 + a x_1)$ 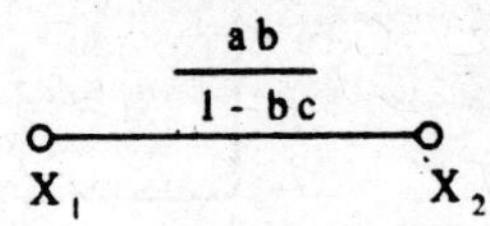

Self or Feedback Loop: It is a path originating from a node and terminating at the same node without encountering any other node, e.g. $x_2 - x_2$ is a self loop having transmittance *f*.

Non-touching Loops: Non-touching loops are those loops having no path/ branches in common and in addition without any common node. In the preceding illustration

$x_2 - x_3 - x_4 - x_2$ and $x_5 - x_6 - x_5$, and

$x_2 - x_2$ and $x_5 - x_6 - x_5$, are the two

possible combinations of non-touching loops.

4.1.4 Signal-Flow Diagram Reduction. complex signal flow diagrams of a system can be simpled by the use of signal-flow graph algebra shown in Table 4.2.

4.1.5 Manson's Gain Formula: With the help of repeated applications of signal-flow graph algebra, it is possible to reduce complex and complicated graph to one source and sink nodes. However, This is time consuming and tedious. Mason was the first to spot the inherent property of the signal-flow graph and evolved a theorem which is popularly known as Mason's gain formula. This, by mere inspection of the signal-flow graph gives the desired answer. The general expression for signal flow graph as given by Mason is

$$T = \frac{1}{\Delta} \sum P_K \Delta_K, \text{ where}$$

T = Overall transmittance of the System

$\Delta =$ 1 – (sum of all individual loops) + (sum of the gain product of all possible combination of two non-touching loops) – (sum of the gain product of all possible combination of there non-touching loops) + (...) – (...) + ...

P_k = Gain of k_{th} forward path.

Δ_k = Same as but Δ formed by loops not touching the k_{th} forward path.

Problem No. 4.1 Obtain the transfer function for the block diagram shown in Fig. 4.5

(Pune University, B.Sc. Engg. Elect. 1989)

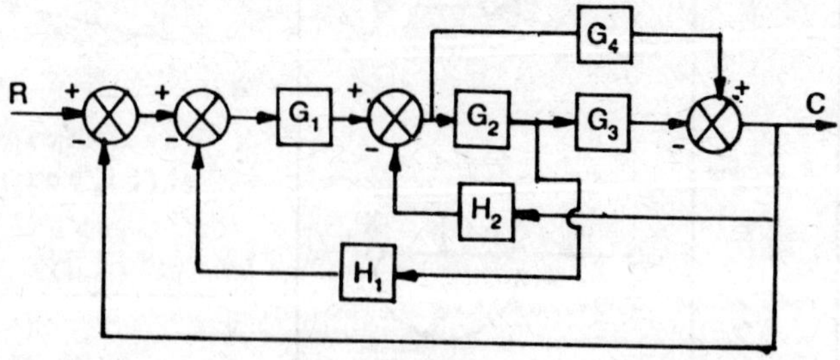

Fig. 4.5

Solution

Moving the take off point between blocks G_2 and G_3 ahead of G_2

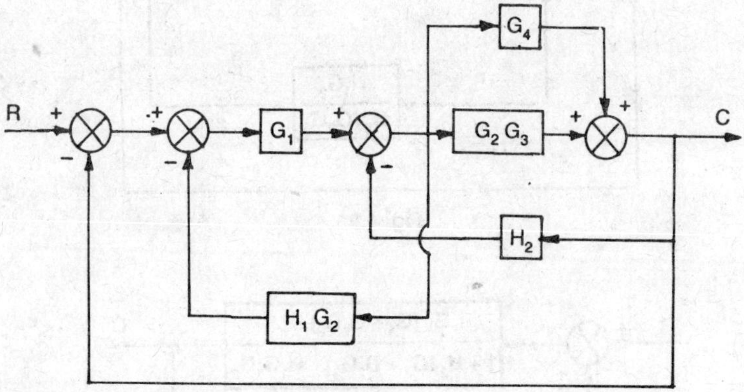

Fig. 4.6

Eliminating forward path G_4

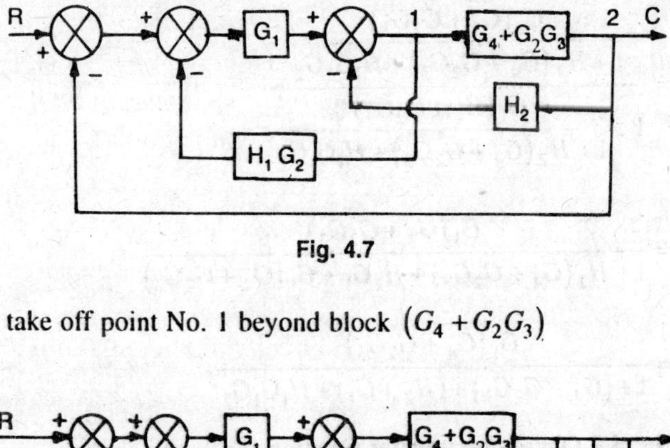

Fig. 4.7

Moving take off point No. 1 beyond block $(G_4 + G_2G_3)$

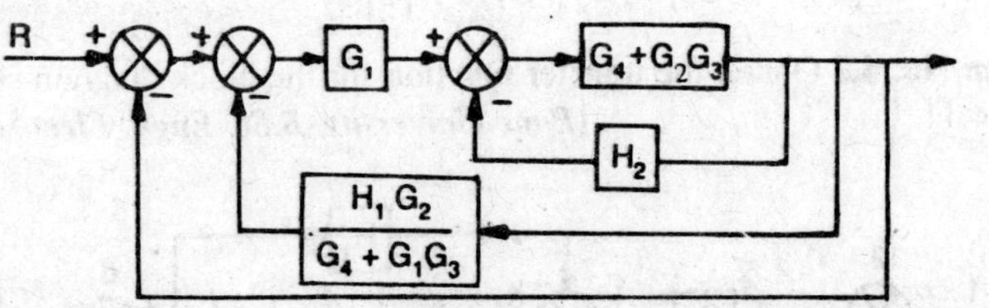

Fig. 4.8

Eliminating the feed back loops

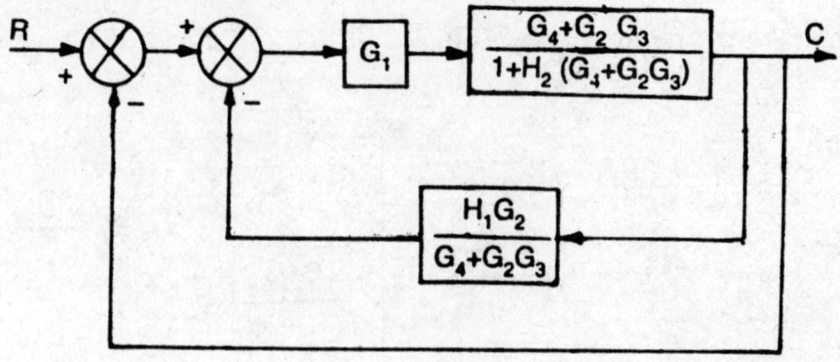

Fig. 4.9

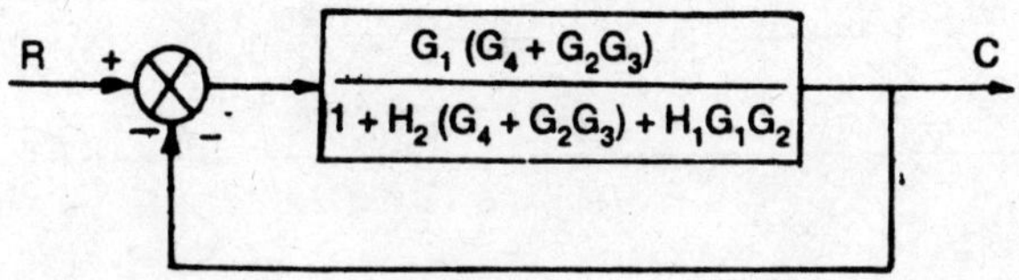

Fig. 4.10

$$\frac{C}{R}=\frac{\dfrac{G_1(G_4+G_2G_3)}{1+H_2(G_4+G_2G_3)+H_1G_1G_2}}{1+\dfrac{G_1(G_4+G_2G_3)}{1+H_2(G_4+G_2G_3)+H_1G_1G_2}}$$

$$\frac{C}{R}=\frac{G_1(G_4+G_2G_3)}{1+H_2(G_4+G_2G_3)+H_2G_1+G_1(G_4+G_2G_3)}$$

$$\frac{C}{R}=\frac{G_1(G_4+G_2G_3)}{1+(G_4+G_2G_3)+(H_2+G_1)+H_1G_1G_2}$$ **Ans.**

Problem No. 4.2 Obtain the transfer function for the block diagram shown in Fig. 4.11 (*Pune University, B.Sc. Engg. Elect. 1988*)

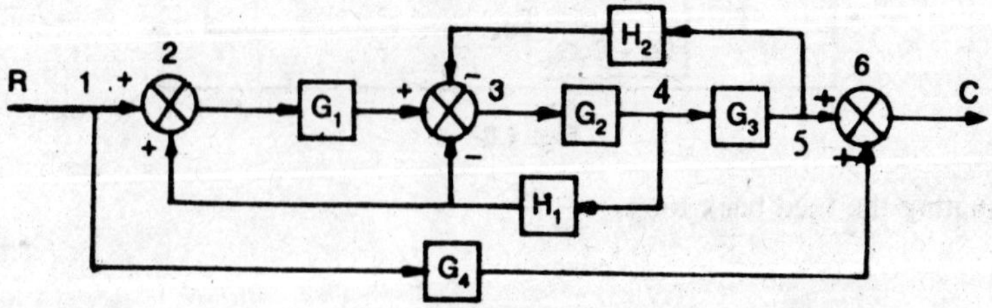

Fig. 4.11

Solution

Moving take-off point No. 4 beyond block G_3

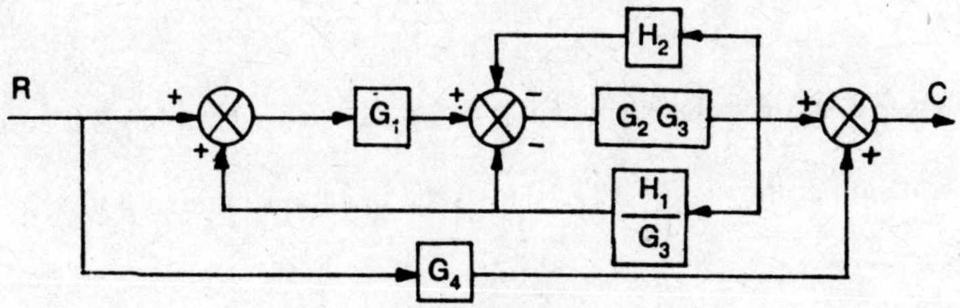

Fig. 4.12

Eliminating feedback having transfer function H_2

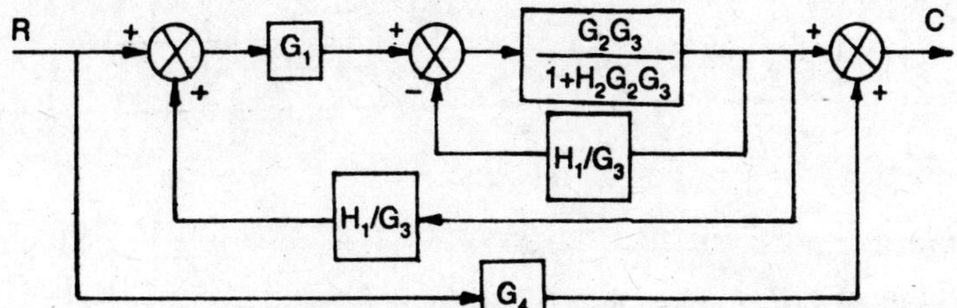

Fig. 4.13

Eliminating feedback loop

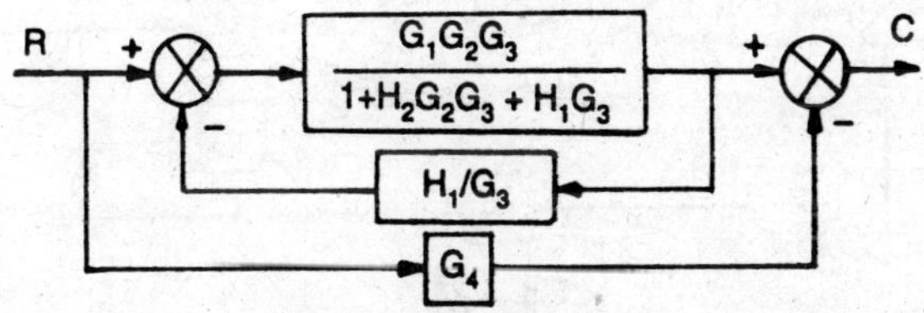

Fig. 4.14

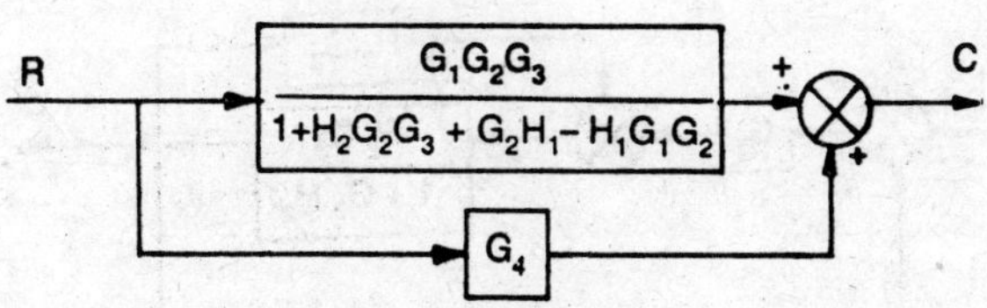

Fig. 4.15

Eliminating the forward path loop

$$\frac{C}{R} = G_4 + \frac{G_1 G_2 G_3}{1 + H_2 G_2 G_3 + G_2 H_1 (1 - G_1)}$$ **Ans.**

Problem No. 4.3 Evaluate $\frac{C}{R_1}$ and $\frac{C}{R_2}$ for a system whose block diagram representation is shown in Fig. 4.16 R_1 is the input to summing point No. 1.

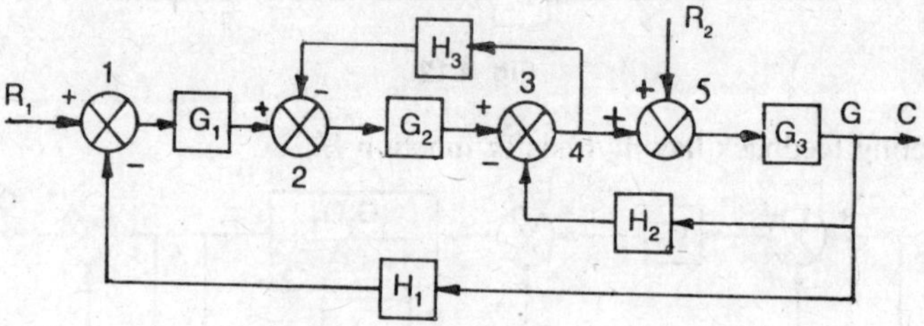

Fig. 4.16

Solution

Evaluation of C/R_1 Assume $R_2 = 0$. Therefore summing point No. 5 can be removed. Shift take off point No. 4 beyond block G_3

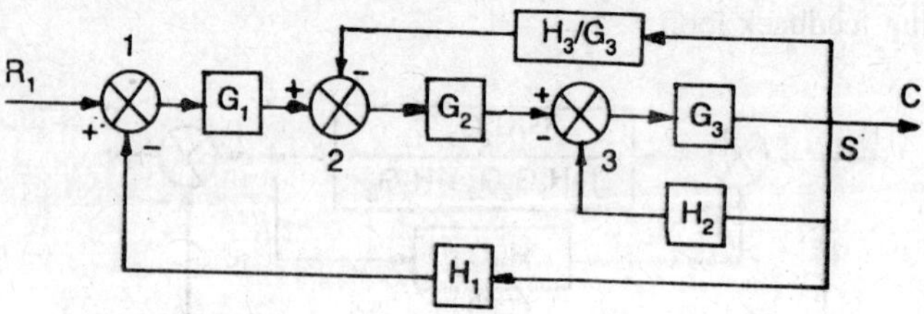

Fig. 4.17

Eliminate the feedback loop between points 3 and 6

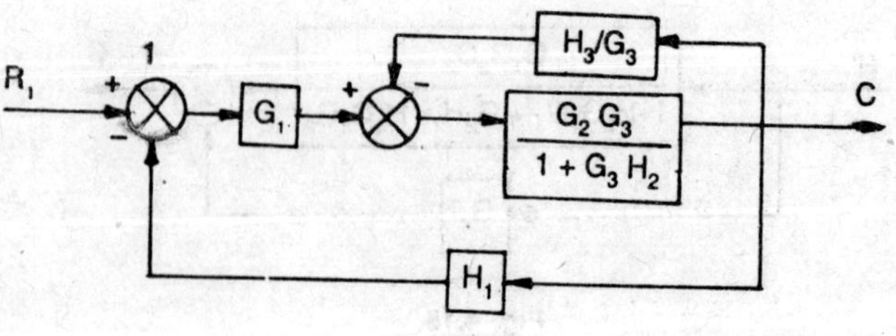

Fig. 4.18

Eliminating the feed back loop again

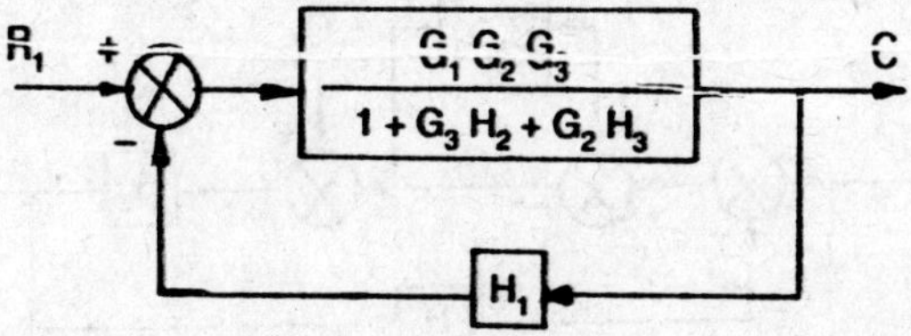

Fig. 4.19

$$\frac{C}{R_1} = \frac{G_1G_2G_3}{1+G_3H_2+H_3G_2+G_1G_2G_3H_1}$$ **Ans.**

Evaluation of $\frac{C}{R_2}$

Assume $R_1 = 0$. Thus summing point No. 1 can be removed

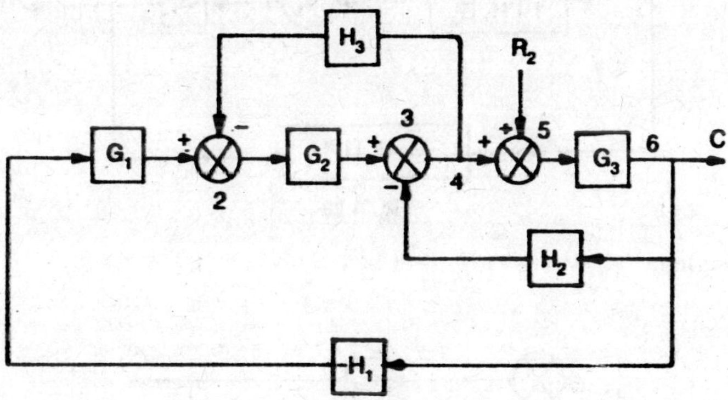

Fig. 4.20

Shifting the summing point No. 2 and rearranging beyond G_2

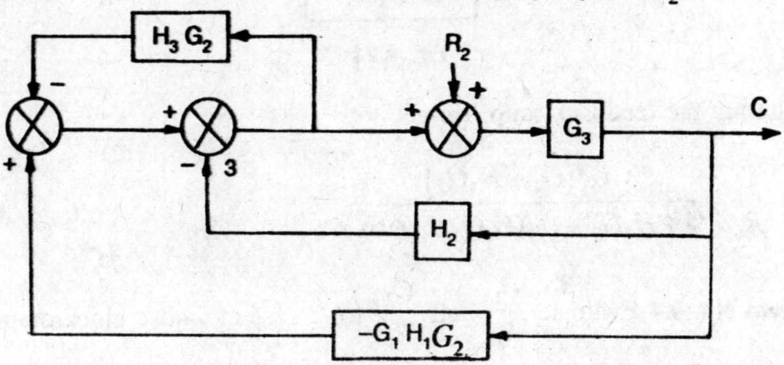

Fig. 4.21

Rearranging, we get

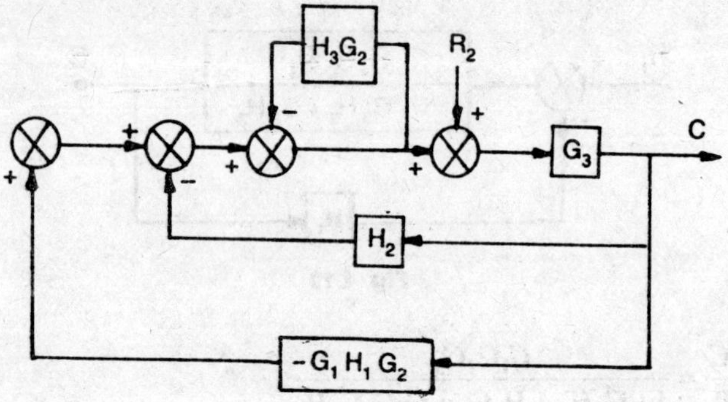

Fig. 4.22

Rearranging and eliminating the feedback loop

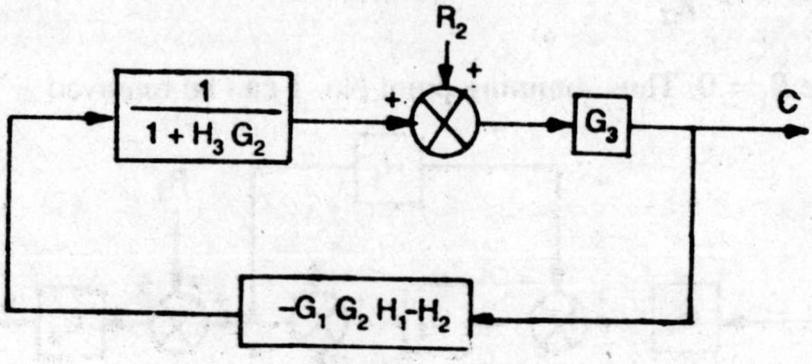

Fig. 4.23

Rearranging,

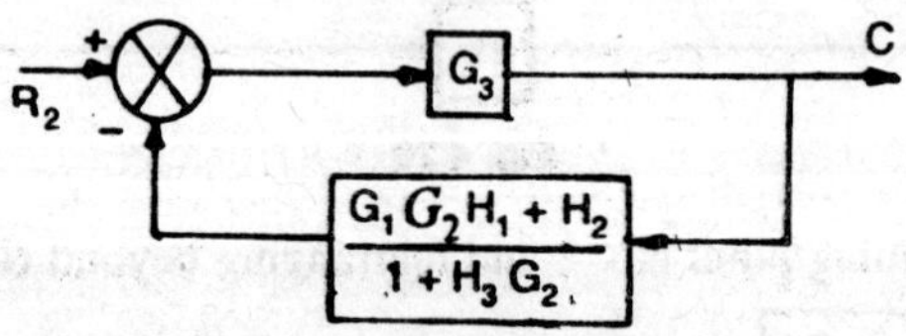

Fig. 4.24

Eliminating the feedback loop, we get

$$\frac{C}{R_2} = \frac{G_3(G_4 + H_3G_2)}{1 + H_3G_2 + G_3(G_1G_2H_1 + H_2)}$$ **Ans.**

Problem No. 4.4 Evaluate $\frac{C_1}{R_1}$ and $\frac{C_2}{R_1}$ for a system whose block diagram representation is shown in Fig. 4.25.

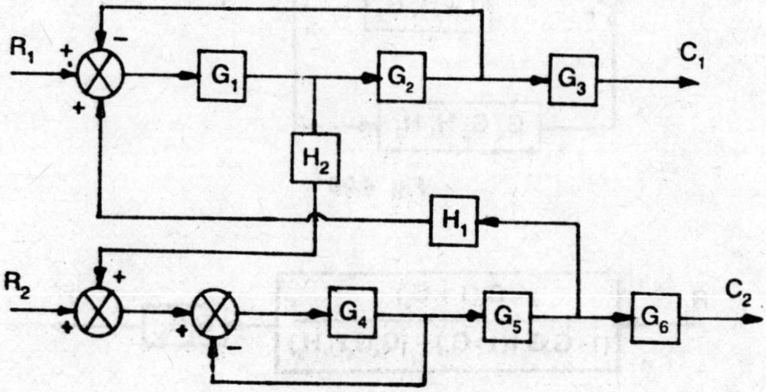

Fig. 4.25

Solution

Assuming $R_2 = 0$

Evaluation of $\frac{C_1}{R_1}$

Assuming $C_2 = 0$ **and rearranging, we get**

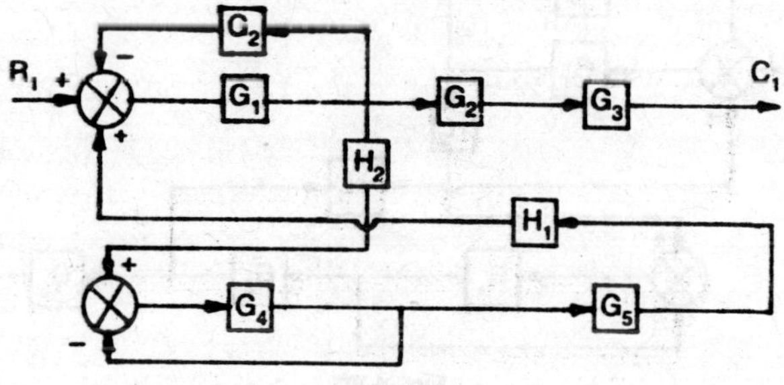

Fig. 4.26

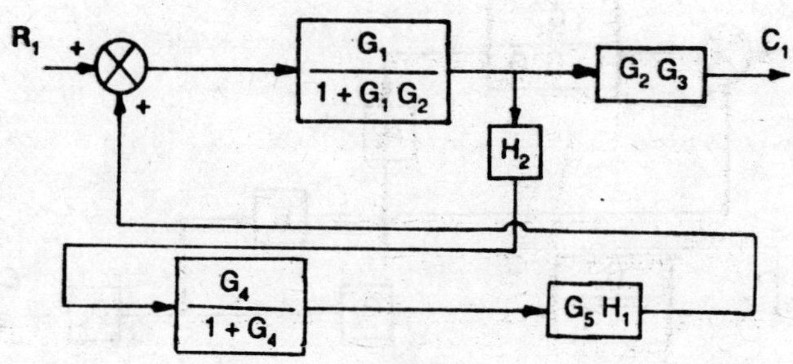

Fig. 4.27

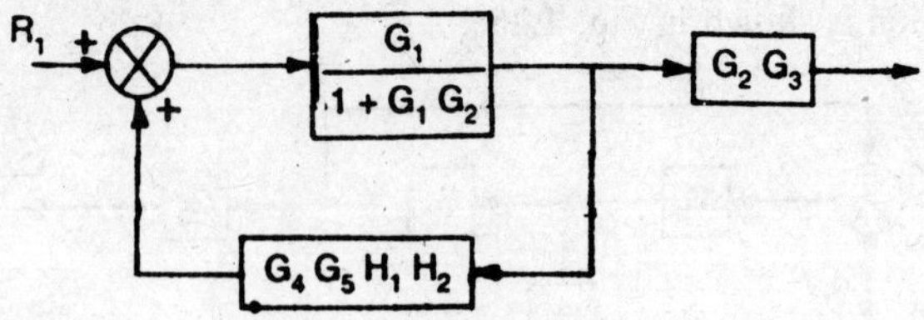

Fig. 4.28

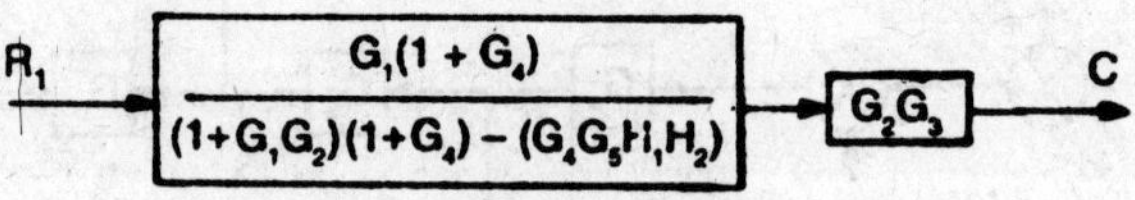

Fig. 4.29

$$\therefore \quad \frac{C_1}{R_1} = \frac{G_1G_2G_3(1+G_4)}{(1+G_1G_2)(1+G_4)-G_4G_5H_1H_2} \qquad \textbf{Ans.}$$

Evaluation of $\dfrac{C_2}{R_1}$ **Assuming** $C_1 = 0$ **and rearranging**

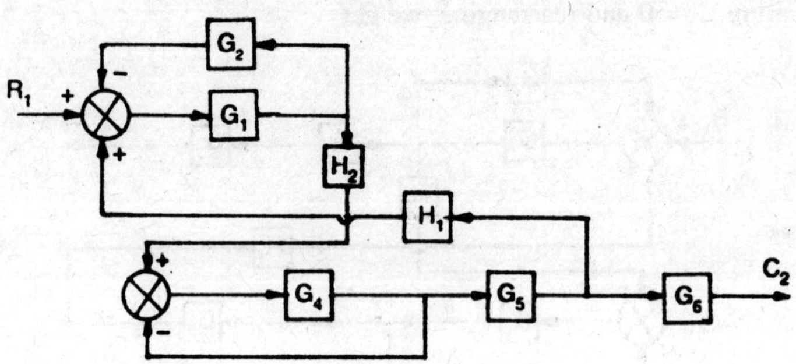

Fig. 4.30

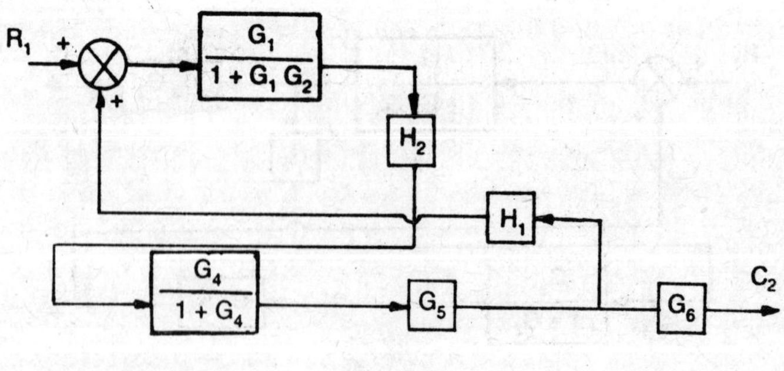

Fig. 4.31

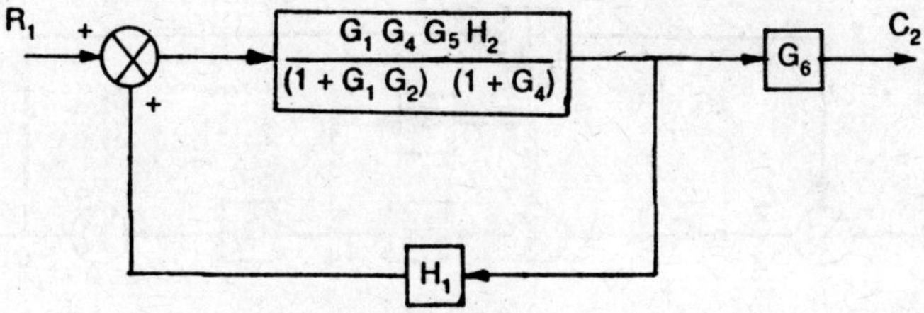

Fig. 4.32

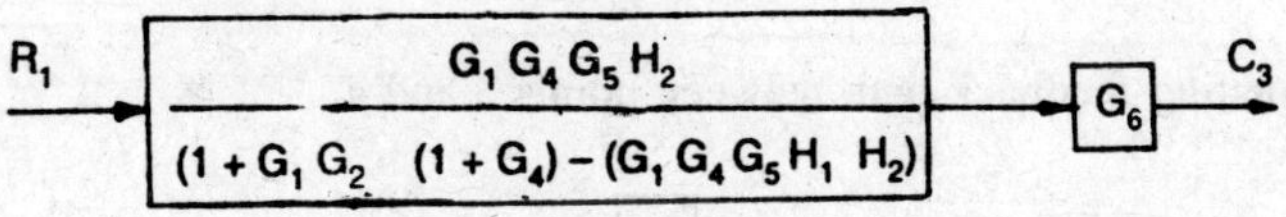

Fig. 4.33

$$\therefore \quad \frac{C_2}{R_1} = \frac{G_1G_4G_5G_6H_2}{(1+G_1G_2)(1+G_4)-(G_1G_4G_5H_1H_2)} \qquad \textbf{Ans.}$$

Problem No. 4.5 Evaluate $\dfrac{C_2}{R_1}$ for the system whose block diagram representation is shown in Fig. 4.34.

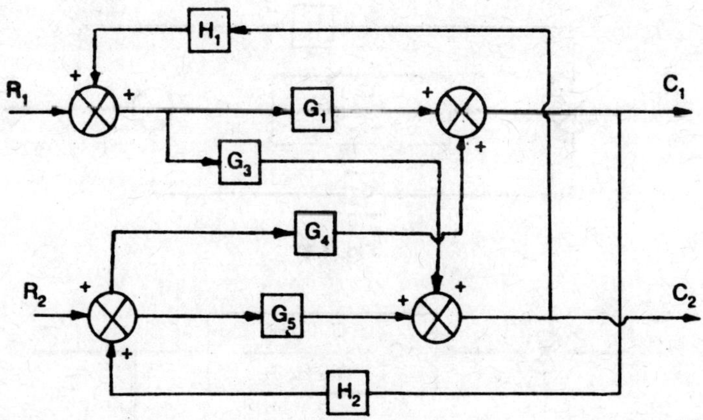

Fig. 4.34

Solution

Assuming $R_2 = 0$ and $C_1 = 0$. Rearranging, we get

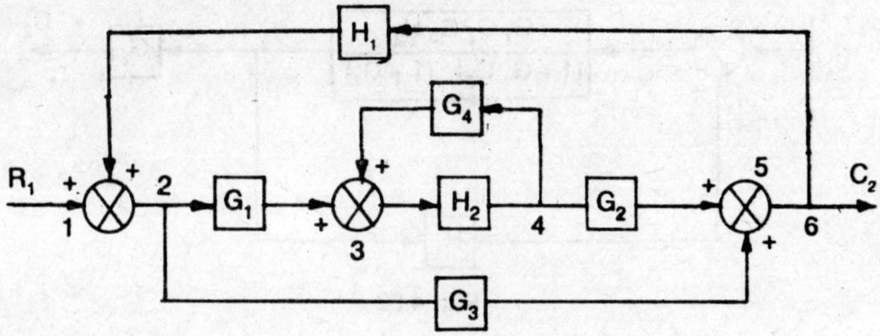

Fig. 4.35

Eliminating feedback path between points 3 and 4

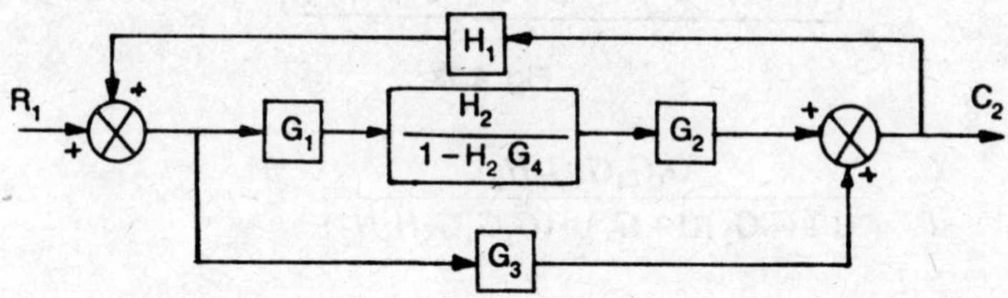

Fig. 4.36

Further simplification gives

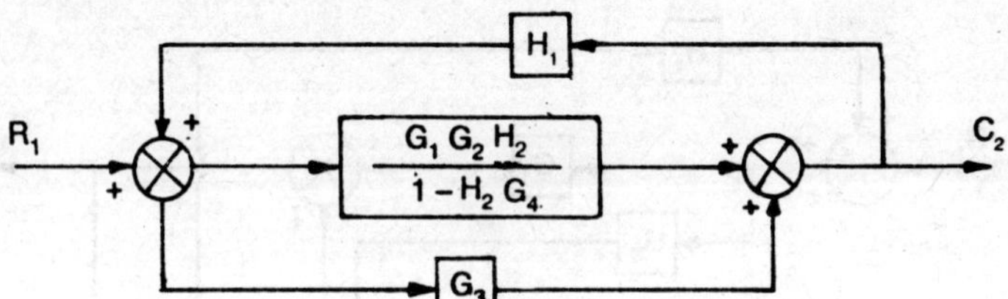

Fig. 4.37

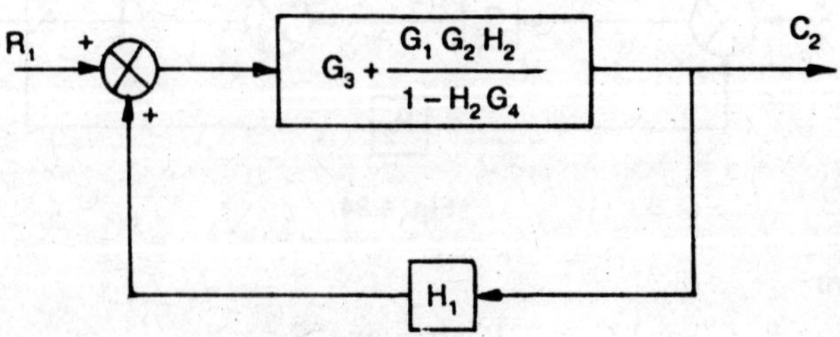

Fig. 4.38

$$\frac{C_2}{R_1} = \frac{G_3 + \dfrac{G_1G_2H_2}{1-H_2G_4}}{1-\left(G_3 + \dfrac{G_1G_2H_2}{1-H_2G_4}\right)H_1}$$

$$= \frac{G_3 - H_2G_3G_4 + G_1G_2H_2}{(1-H_2G_4 - G_3H_1 + H_1H_2G_4G_3 - G_1G_2H_1H_2)}$$

$$= \frac{G_3(1-H_2G_4)+G_1G_2H_2}{(1-G_4H_2)+H_1\left[G_3(H_2G_4-1)-G_1G_2H_2\right]}$$ **Ans.**

Problem No. 4.6 Find the transfer function for the system whose block diagram representation is shown in Fig. 4.39. **(Pune University)**

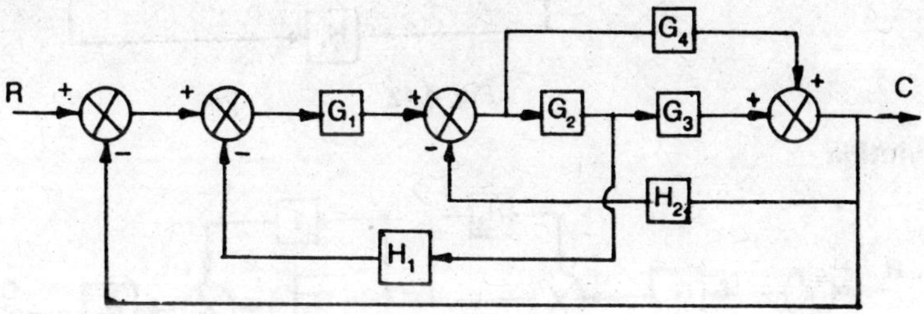

Fig. 4.39

Solution

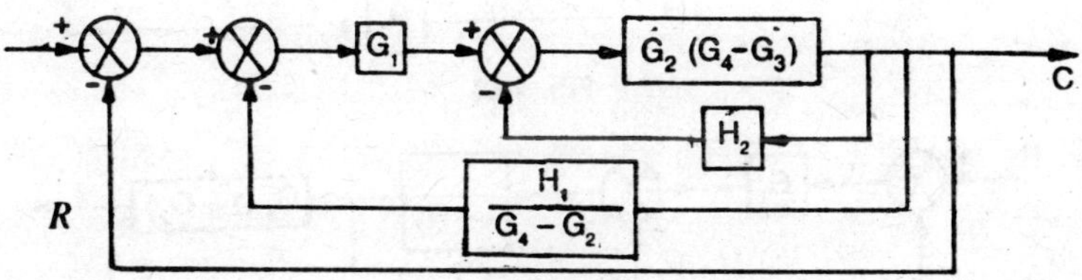

Fig. 4.40

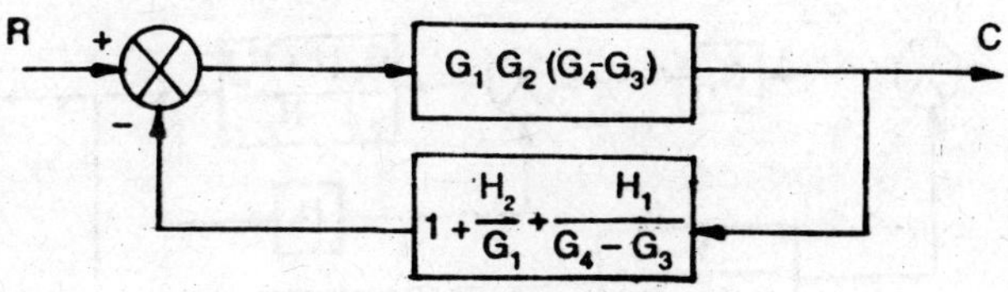

Fig. 4.41

$$\frac{C}{R} = \frac{G_1G_2(G_4 - G_3)}{1 + G_1G_2(G_4 - G_3)\left[1 + \frac{H_2}{G_1} + \frac{H_1}{G_4 - G_3}\right]}$$

$$= \frac{G_1G_2(G_4 - G_3)}{1 + H_2G_2(G_4 - G_3) + G_1G_2H_1 + G_1G_2(G_4 - G_3)}$$ **Ans.**

Problem No. 4.7 Find transfer function for the block diagram shown in Fig. 4.42.

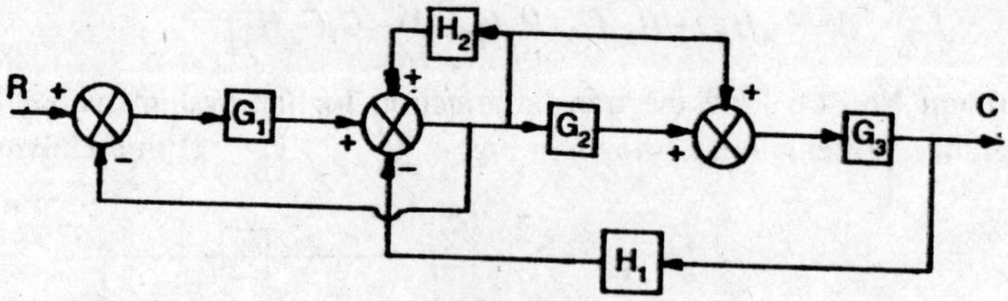

Fig. 4.42

Solution

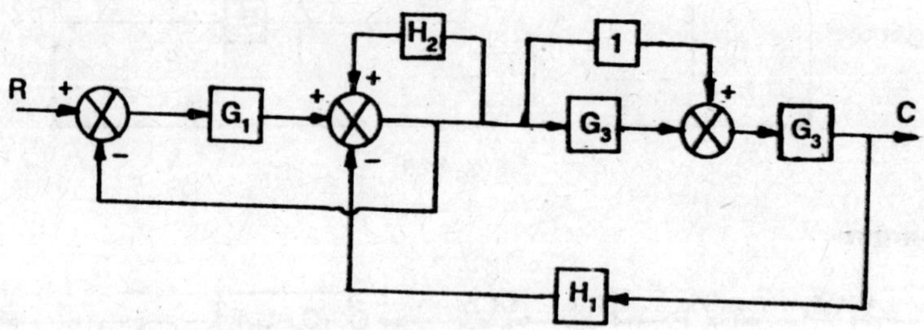

Fig. 4.43

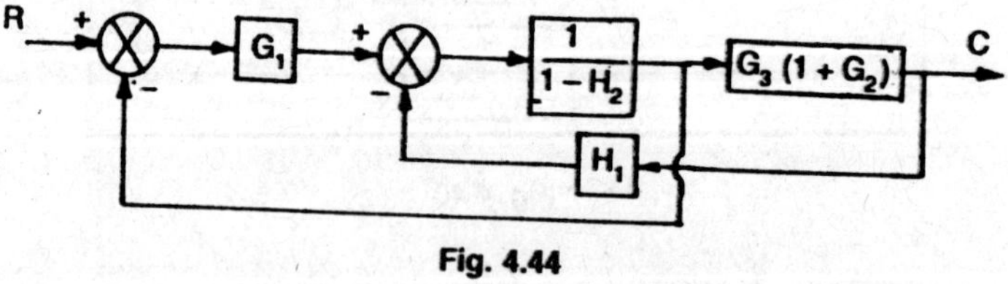

Fig. 4.44

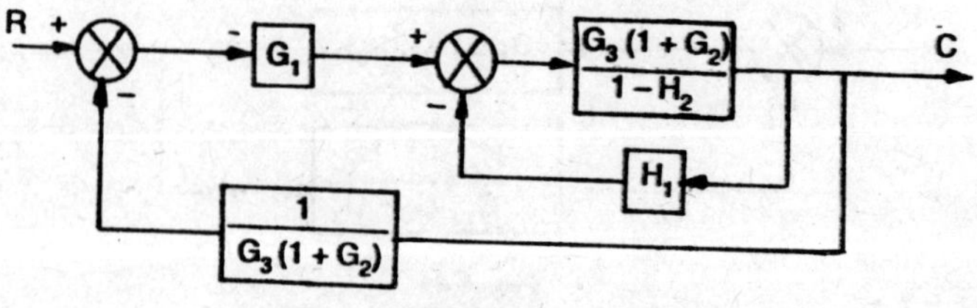

Fig. 4.45

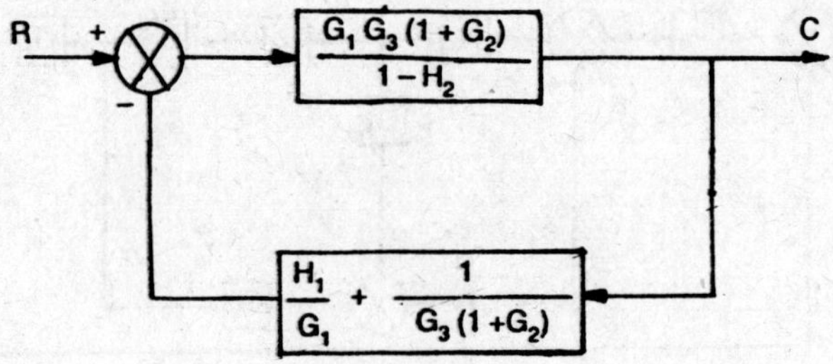

Fig. 4.46

$$\frac{C}{R}=\frac{\dfrac{G_1G_3(1+G_2)}{1-H_2}}{1+\dfrac{G_1G_3(1+G_2)}{1-H_2}\left[\dfrac{H_1}{G_1}+\dfrac{1}{G_3(1+G_2)}\right]}$$

$$\therefore \quad \frac{C}{R}=\frac{G_1G_3(1+G_2)}{(1-H_2)+H_1G_3(1+G_2)+G_1} \qquad \textbf{Ans.}$$

Problem No. 4.8 Find the closed loop transfer function of the system shown in Fig. 4.47.

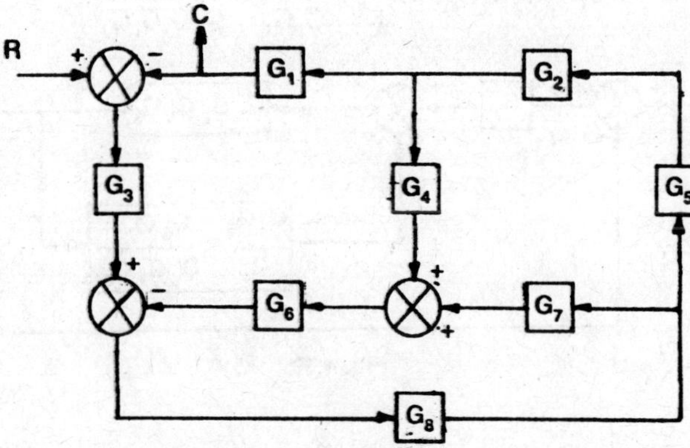

Fig. 4.47

Solution

Redrawing the block diagram of Fig. 4.47.

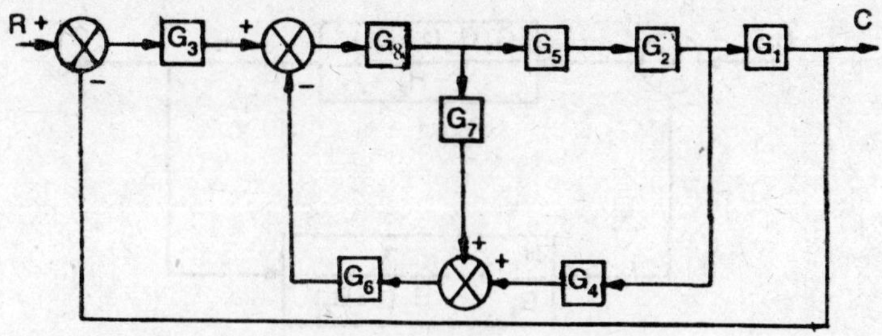

Fig. 4.48

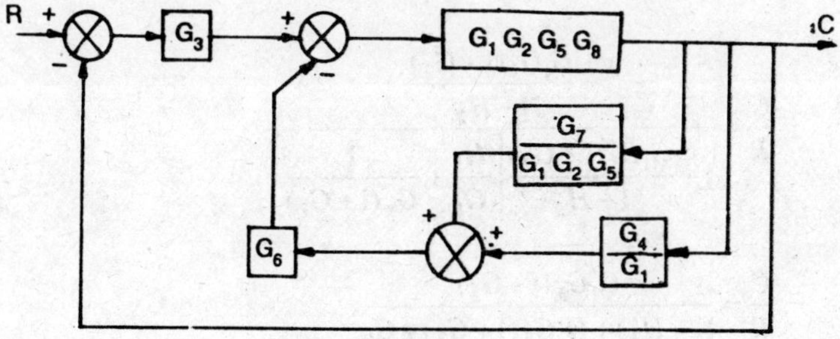

Fig. 4.49

$$\frac{C}{R} = \frac{G_1G_2G_3G_5G_8}{1+G_1G_2G_3G_5G_8 - G_2\left(1+\dfrac{G_4G_6}{G_1G_3}+\dfrac{G_6G_7}{G_3G_5G_1G_2}\right)}$$

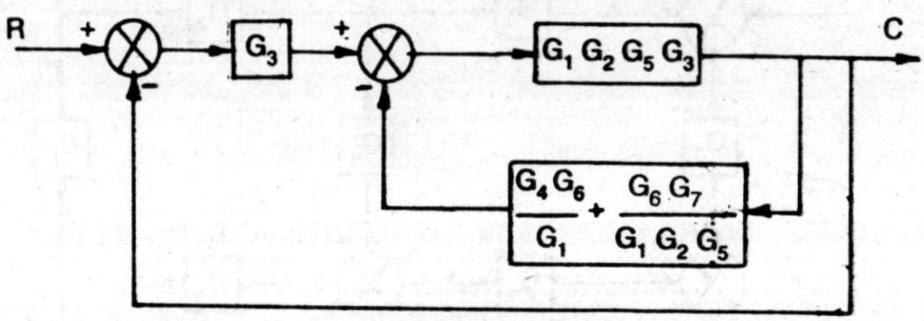

Fig. 4.50

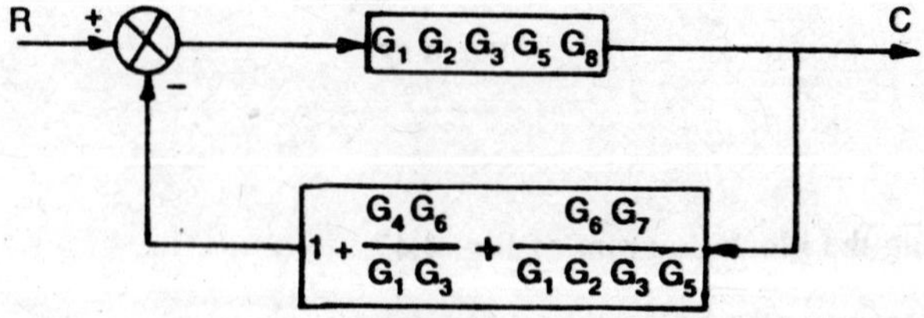

Fig. 4.51

$$\frac{C}{R}=\frac{G_1G_2G_3G_5G_8}{1+G_1G_2G_3G_5G_8+G_4G_6G_8+G_6G_7G_8}$$ **Ans.**

Problem No. 4.9 Find closed loop transfer function of system shown in Fig. 4.52.

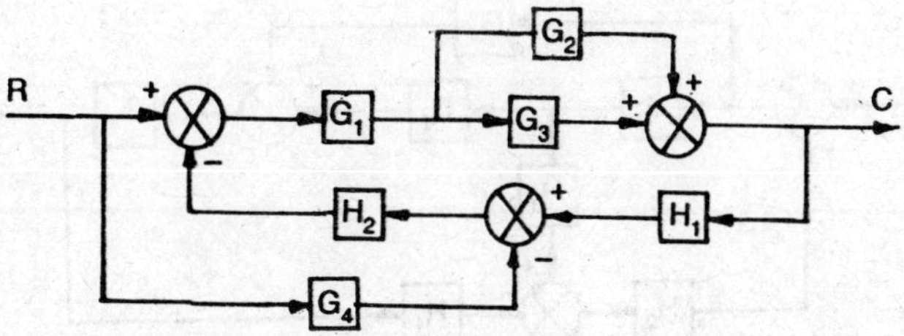

Fig. 4.52

Solution

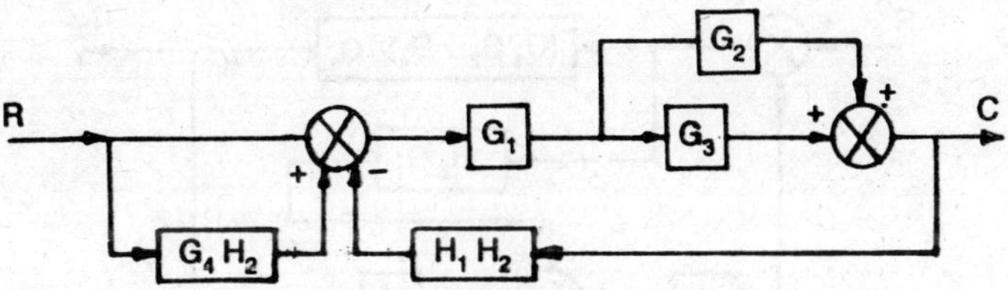

Fig. 4.53

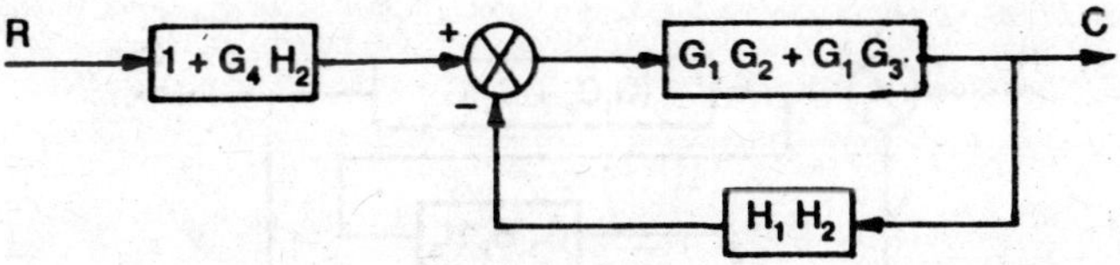

Fig. 4.54

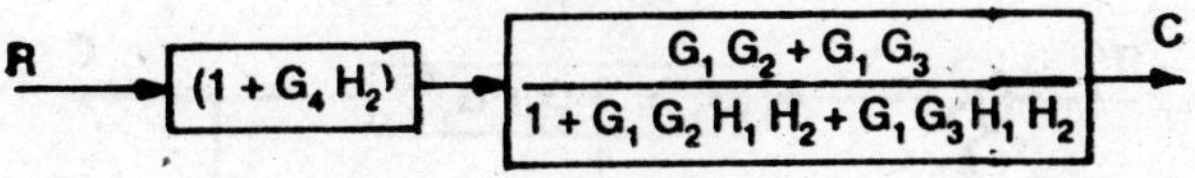

Fig. 4.55

$$\therefore \quad \frac{C}{R}=\frac{(1+G_4H_2)(G_1G_2+G_1G_3)}{1+G_1G_2H_1H_2+G_1G_3H_1H_2}$$

or $$\frac{C}{R}=\frac{G_1(G_2+G_3)(1+G_4H_2)}{1+G_1H_1H_2(G_2+G_3)}$$ **Ans.**

Problem No. 4.10 Find the closed loop transfer function of the system shown in Fig. 4.56.

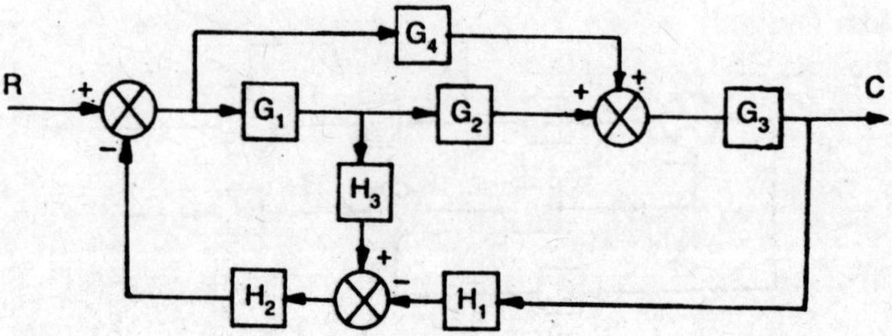

Fig. 4.56

Solution

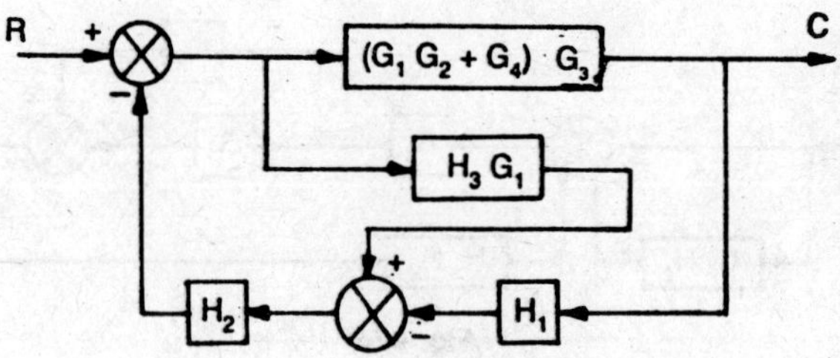

Fig. 4.57

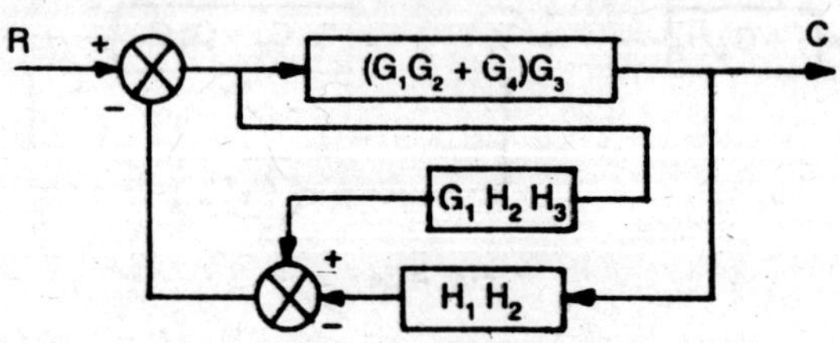

Fig. 4.58

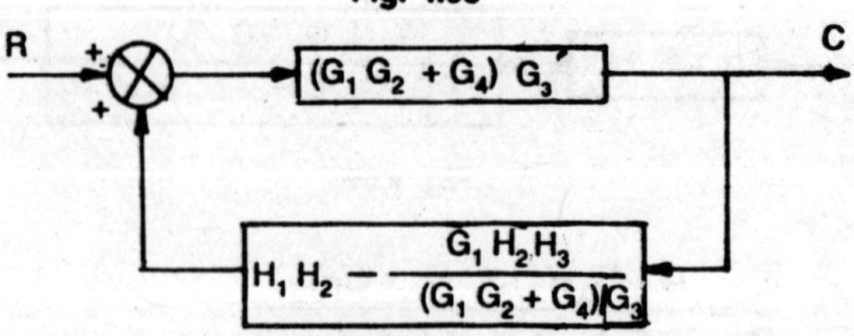

Fig. 4.59

$$\frac{C}{R} = \frac{G}{1-GH}$$

$$= \frac{(G_1G_2 + G_4)G_3}{1 + G_1H_2(H_3 - G_2G_3H_1) - G_3G_4H_1H_2} \qquad \textbf{Ans.}$$

Problem No. 4.11 Reduce the block diagram in Fig. 4.60 to its simplest possible from and hence obtain its closed loop transfer function.

(Pune University)

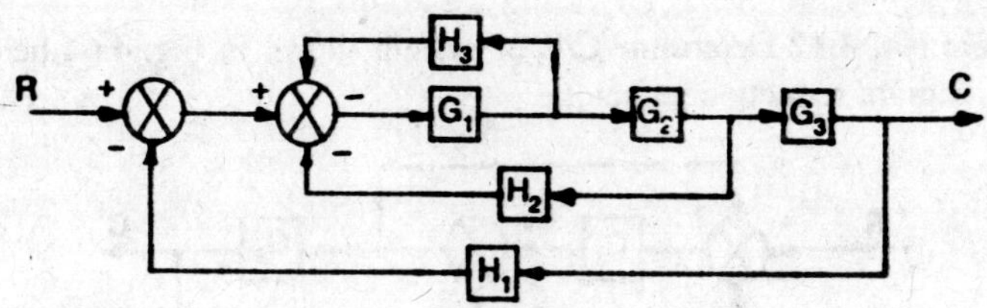

Fig. 4.60

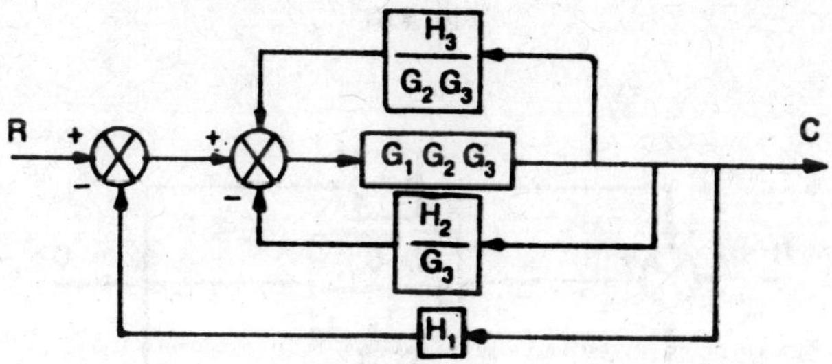

Fig. 4.61

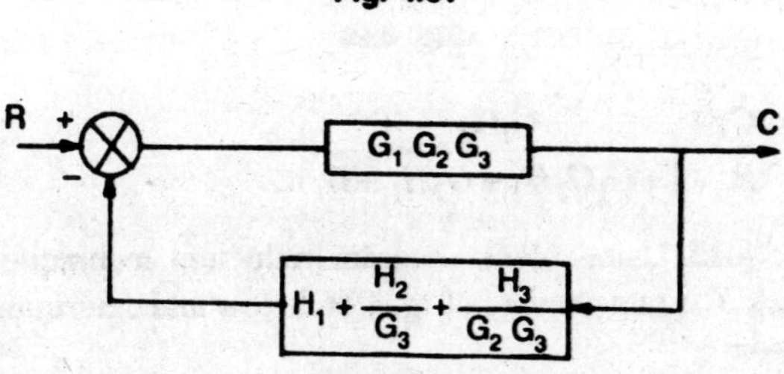

Fig. 4.62

$$\frac{C}{R} = \frac{G_1G_2G_3}{1 + G_1G_2G_3\left(H_1 + \dfrac{H_2}{G_3} + \dfrac{H_3}{G_2G_3}\right)}$$

or
$$= \frac{G_1G_2G_3}{1+G_1G_2G_3H_1+G_1G_2H_2+G_1H_3}$$

The simplest block diagram is shown in Fig. 4.63

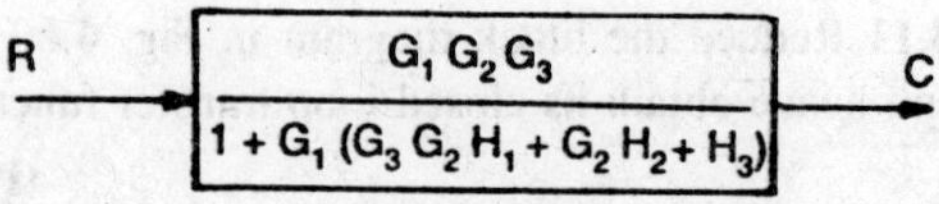

Fig. 4.63

Problem No. 4.12 Determine C/R of system shown in Fig. 4.64 below by block diagram reduction technique. **(AMIE)**

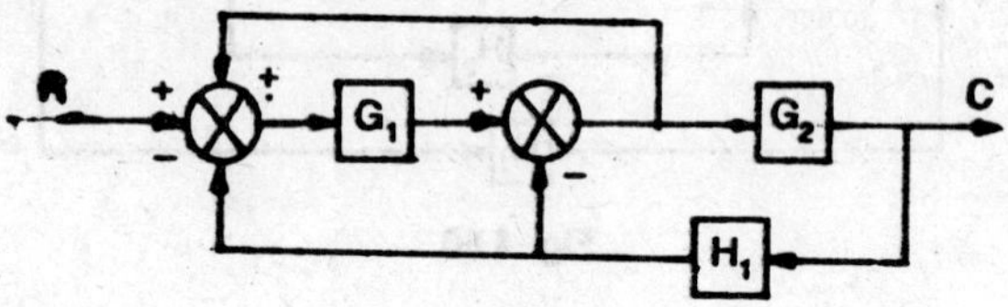

Fig. 4.64

Solution

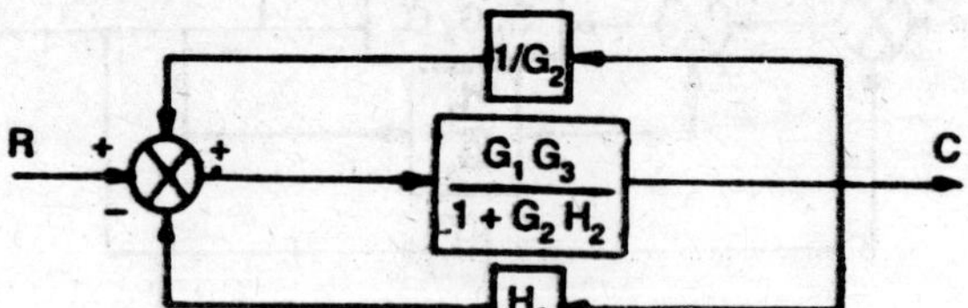

Fig. 4.65

$$\frac{C}{R} = \frac{G_1G_2}{1+G_1G_2H_1+G_2H_1-G_1}$$ **Ans.**

Problem No. 4.13 Using block diagram reduction technique reduce the following block diagram shown in Fig. 4.66 below and determine the overall transfer function. **(AMIE)**

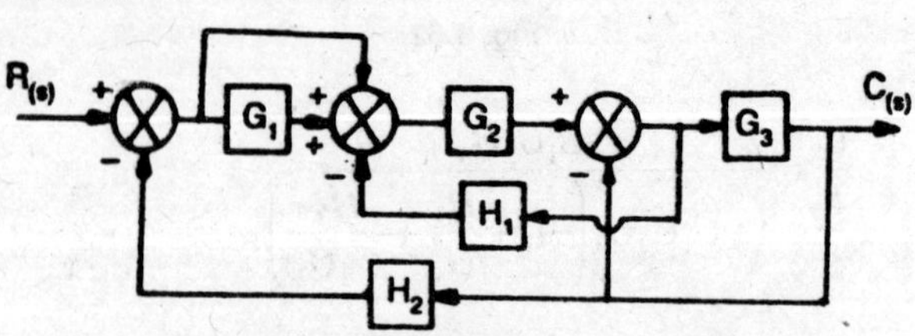

Fig. 4.66

Solution

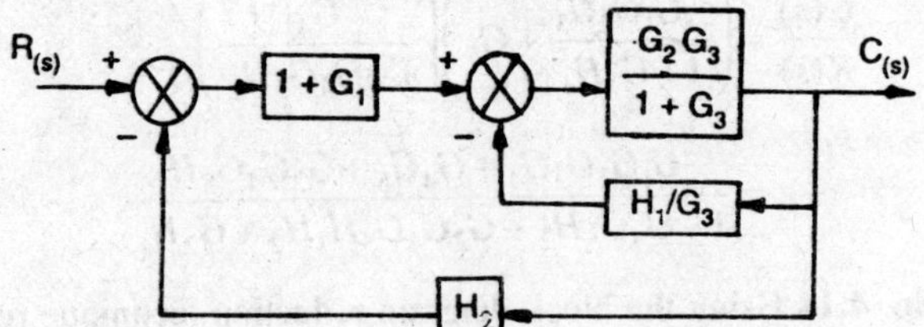

Fig. 4.67

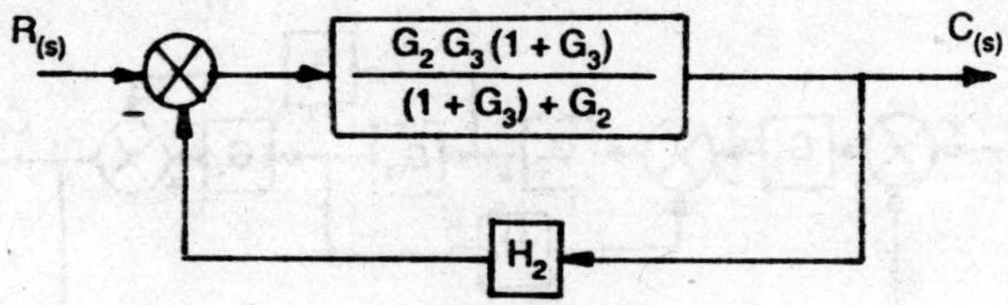

Fig. 4.68

$$\therefore \quad \frac{C(s)}{R(s)} = \frac{G}{1+GH}$$

$$= \frac{G_1G_2G_3 + G_2G_3}{1 + G_3 + G_2H_1 + G_2G_3H_2 + G_1G_2G_3H_2} \quad \textbf{Ans.}$$

Problem No. 4.13 Using block diagram reduction technique find the transfer function for the block diagram shown in Fig. 4.69. **(AMIE)**

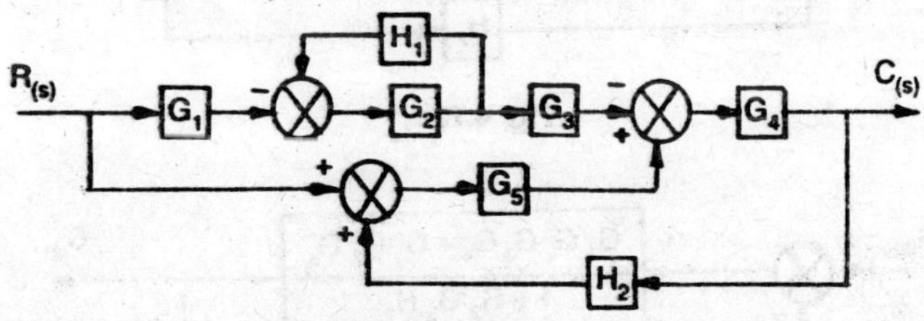

Fig. 4.69

Solution

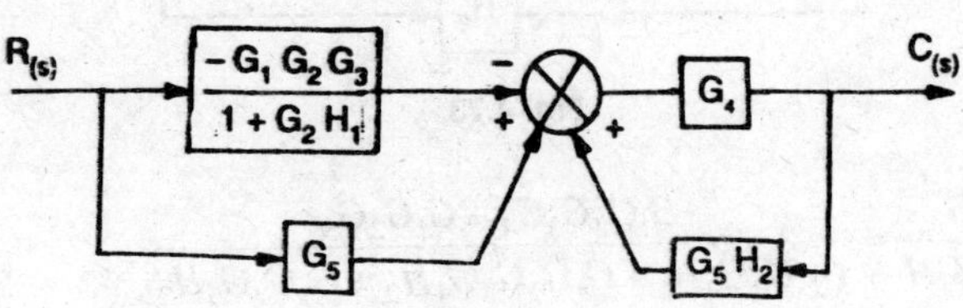

Fig. 4.70

$$\therefore \quad \frac{C(s)}{R(s)} = \left(\frac{-G_1G_2G_3}{1+G_2H_1} + G_5\right)\left(\frac{G_4}{1-G_4G_5H_2}\right)$$

$$= \frac{-G_1G_2G_3G_4 + G_4G_5 + G_2G_4G_5H_1}{1 - G_4G_5H_2 - G_2G_4G_5H_1H_2 + G_2H_1} \qquad \textbf{Ans.}$$

Problem No. 4.14 Using the block diagram reduction technique reduce the system shown in Fig. 4.71 to simplest possible form and find the transfer function.

(Pune University)

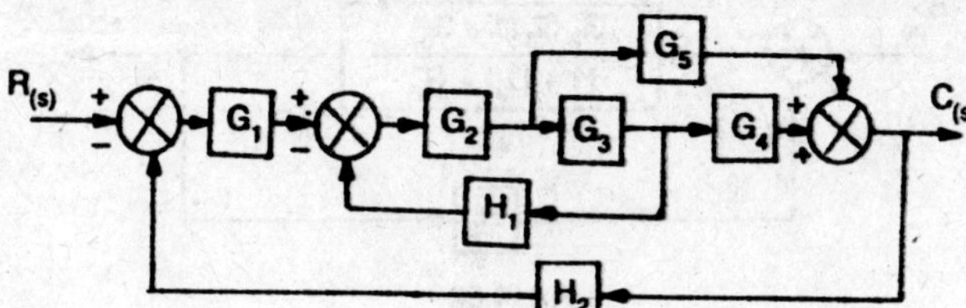

Fig. 4.71

Solution

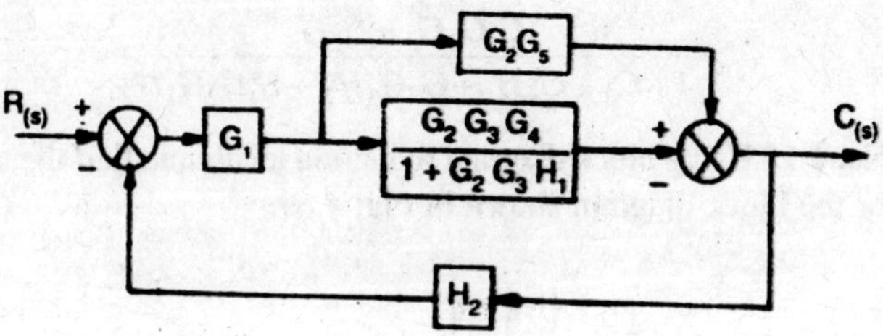

Fig. 4.72

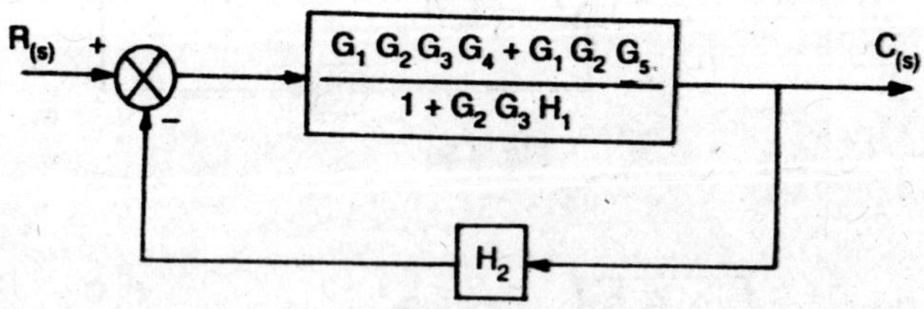

Fig. 4.73

$$\frac{C(s)}{R(s)} = \frac{G}{1+GH} = \frac{G_1G_2G_3G_4 + G_1G_2G_5}{1 + G_2G_3H_1 + G_1G_2G_3G_4H_2 + G_1G_2G_5H_2} \qquad \textbf{Ans.}$$

Problem No. 4.15 Detrmine the ratio $\frac{C(s)}{R(s)}$ for the multiple-loop system shown in Fig. 4.74. **(Pune University)**

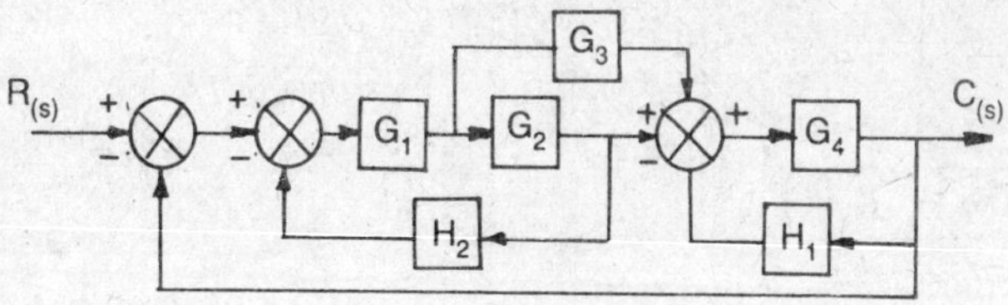

Fig. 4.74

Solution

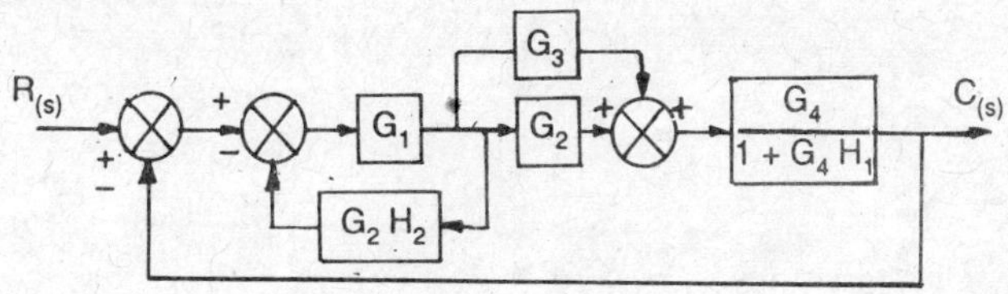

Fig. 4.75

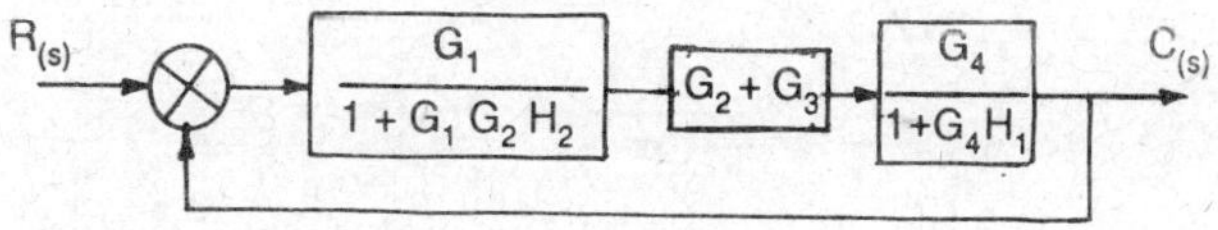

Fig. 4.76

$$\frac{C(s)}{R(s)} = \frac{G}{1+GH}$$

$$= \frac{G_1G_2G_4 + G_1G_3G_4}{1+G_1G_2H_2 + G_4H_1 + G_1G_2G_4H_1H_2 + G_1G_2G_4 + G_1G_3G_4}$$ **Ans.**

Problem No. 4.16 Draw signal flow graph for the following equations

(a) $y_2 = a_1 \frac{dy_1}{dt}$ (b) $y_3 = \frac{d^2y_2}{dt^2} + \frac{dy_1}{dt} - y_1$

(c) $\frac{d^2y}{dx^2} + \frac{2}{3}\frac{dy}{dx} + \frac{11}{2}y = x$

Solution

(a) $y_2 = a_1 \dfrac{dy_1}{dt}$

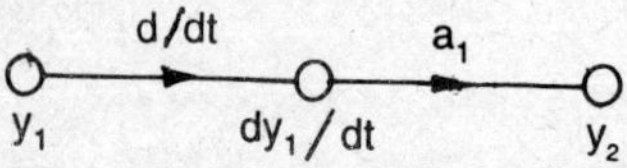

Fig. 4.77

(b) $y_3 = \dfrac{d^2 y_2}{dt^2} + \dfrac{dy_1}{dt} - y_1$

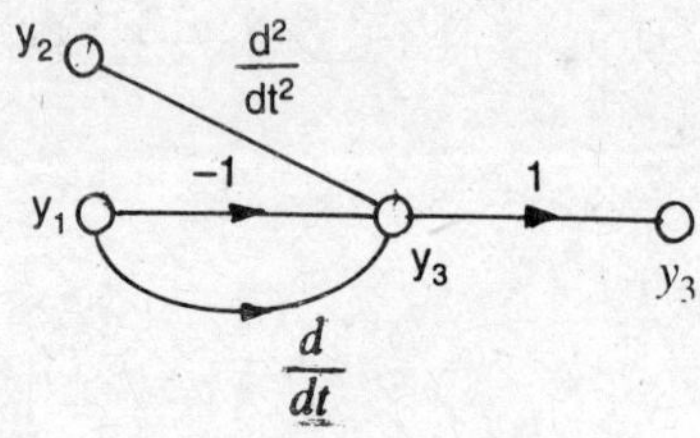

Fig. 4.78

(c) $\dfrac{d^2 y_2}{dx^2} + \dfrac{2}{3}\dfrac{dy}{dx} + \dfrac{11}{2} y = x$

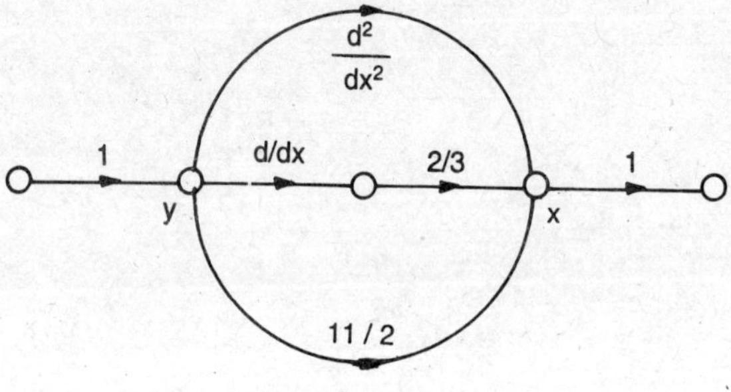

Fig. 4.79

Alternatively

$$\frac{d^2 y}{dx^2} + \frac{2}{3}\frac{dy}{dx} + \frac{11}{2} y = x$$

or
$$s^2 Y(s) + \frac{2}{3} Y(s) + \frac{11}{2} Y(s) = X(s)$$

$$\frac{Y(s)}{X(s)} = \frac{1}{s^2 + \frac{2}{3}s + \frac{11}{2}}$$

$$= \frac{1/s^2}{1 + \frac{2}{3s} + \frac{11}{2s^2}}$$

$$= \frac{1/s^2}{1 - \left(-\frac{2}{3s} - \frac{11}{2s^2}\right)}$$

Comparing with Mason's gain formula, there is

(a) One forward path with gain of $1\ s^2$

(b) two feed back loops with gain of

$$\frac{-2}{3s}, \frac{-11}{2s^2}$$

The signal flow graph is shown in Fig. 4.80.

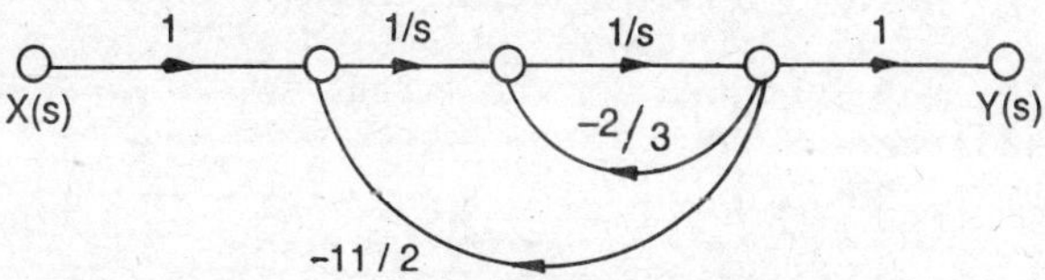

Fig. 4.80

Problem No. 4.17 Obtain the closed loop transfer function for the system shown in Fig. 4.81.

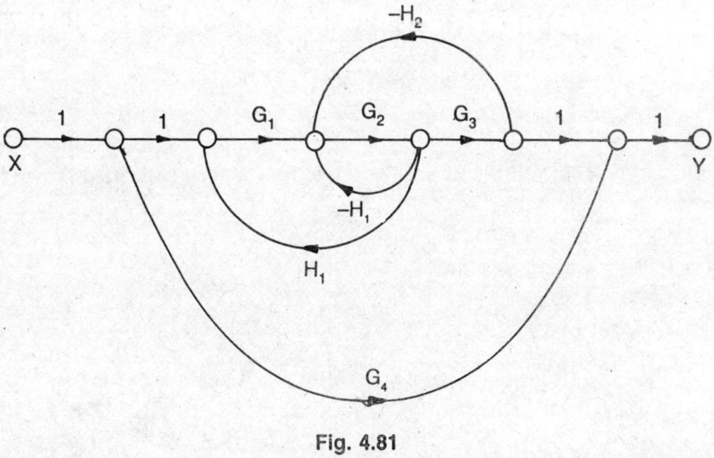

Fig. 4.81

Solution

This problem is similar to Problem No. 4.2.

Forward Paths There are two forward paths

$$P_1 = G_1G_2G_3 \qquad P_2 = G_1$$

Loops There are three feed back loops

$$L_1 = -G_2H_1$$

$$L_2 = G_1G_2H_1$$

$$L_3 = -G_2G_3H_2$$

$$\Delta = 1-(L_1+L_2+L_3)$$

$$= 1+G_2H_1 - G_1G_2H_1 + G_2G_3H_2$$

$$\Delta_1 = 1 \qquad \Delta_2 = \Delta$$

$$\frac{Y}{X} = \frac{P_1\Delta_1 + P_2\Delta_2}{\Delta}$$

$$= \frac{G_1G_2G_3 + G_4(1+G_2H_1 - G_1G_2H_1 + G_2G_3H_2)}{1+G_2H_1 - G_1G_2H_1 + G_2G_3H_2}$$ **Ans.**

Problem No. 4.18 Find *C/R* for the system whose block diagram representation is shown in Fig. 4.82 by signal flow graph technique.

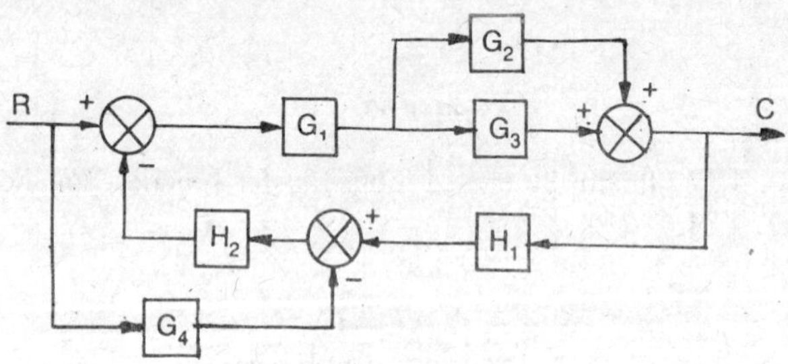

Fig. 4.82

Solution

The signal flow graph for the given block diagram representation is shown in Fig. 4.83.

Forward Paths

$$P_1 = G_1G_3 \qquad \therefore \Delta_1 = 1$$

$$P_2 = G_1G_2 \qquad \therefore \Delta_2 = 1$$

$$P_3 = G_1 G_3 G_4 H_2 \qquad \therefore \Delta_3 = 1$$

$$P_4 = G_1 G_2 G_4 H_2 \qquad \therefore \Delta_4 = 1$$

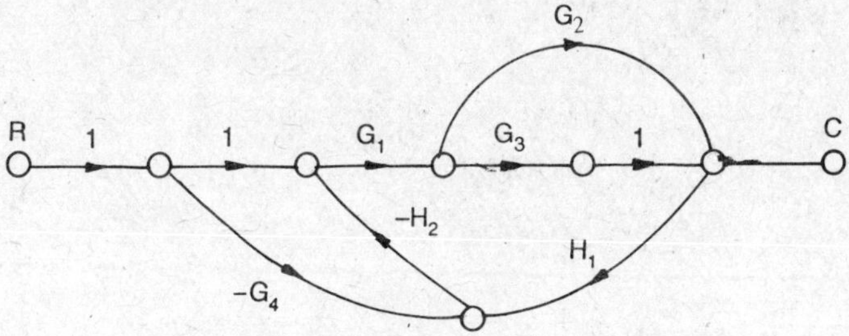

Fig. 4.83

Loops

$$L_1 = -G_1 G_3 H_1 H_2$$

$$L_2 = -G_1 G_2 H_1 H_2$$

$$\Delta = 1 - (L_1 + L_2)$$

$$= 1 + G_1 G_2 H_1 H_2 + G_1 G_3 H_1 H_2$$

$$T = \frac{P_1\Delta_1 + P_2\Delta_2 + P_3\Delta_3 + P_4\Delta_4}{\Delta}$$

or $$T = \frac{G_1G_3 + G_1G_2 + G_1G_3G_4G_2 + G_1G_2G_4H_2}{1 + G_1G_3H_1H_2 + G_1G_2H_1H_2}$$

or $$T = \frac{G_1(G_3 + G_2)(1 + G_4H_2)}{1 + G_1H_1H_2(G_3 + G_2)}$$ **Ans.**

Problem No. 4.19 Find $\frac{C}{R}$ for the system shown in Fig. 4.84.

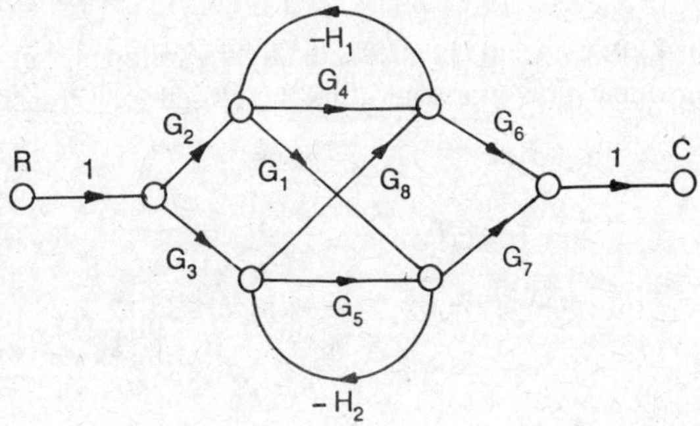

Fig. 4.84

Solution

Forward Paths

$$P_1 = G_2G_4G_6$$
$$P_2 = G_3G_5G_7$$
$$P_3 = G_2G_1G_7$$
$$P_4 = G_3G_8G_6$$
$$P_5 = -G_2G_1H_2G_8G_6$$
$$P_6 = -G_3G_8H_1G_1G_7$$

Loops

$$L_1 = -G_4H_1$$
$$L_2 = -G_5H_2$$
$$L_3 = G_1H_2G_8H_1$$

Nontouching Loops There is one pair having gain product

$$= G_4H_1G_5H_2$$
$$\Delta = 1 + G_4H_1 + G_5H_2 - G_1H_2G_8H_1 + G_4H_1G_5H_2$$
$$\Delta_1 = 1 + G_5H_2$$
$$\Delta_2 = 1 + G_4H_1$$
$$\Delta_3 = \Delta_4 = \Delta_5 = \Delta_6 = 1$$
$$T = \frac{P_1\Delta_1 + P_2\Delta_2 + P_3\Delta_3 + P_4\Delta_4 + P_5\Delta_5 + P_6\Delta_6}{\Delta}$$
$$= \frac{\begin{array}{c} G_2G_4G_6(1+G_5H_2) + G_3G_5G_7(1+G_4H_1) \\ + G_2G_1G_7 + G_3G_8G_6 \\ -G_1G_2G_6G_8H_2 - G_1G_3G_7G_8H_1 \end{array}}{1 + G_4H_1 + G_5H_2 + G_4G_5H_1H_2 - G_1G_8H_1H_2} \quad \textbf{Ans.}$$

Problem No. 4.20 A system is represented by the following set of equations. Construct the signal flow graph and find the closed loop transfer function.

$$x = x_1 + t_3u$$
$$\dot{x}_1 = q_1x_1 + x_2 + t_2u$$
$$\dot{x}_2 = -q_2x_1 + t_1u$$

Solution

Consider x = output and u = input. Writing the given equations in Laplace domain.

$$X = X_1 + t_3 U \qquad ...(1)$$

$$sX_1 = q_1 X_1 + X_2 + t_2 U$$

or $$X_1 = \frac{X_2}{s+q_1} + \frac{t_2}{s+q_1} U \qquad ...(2)$$

$$sX_2 = q_2 X_1 + t_1 U$$

or $$X_2 = \frac{-q_2}{s} X_1 + \frac{t_1}{s} U$$

The signal flow graph is shown in Fig. 4.85.

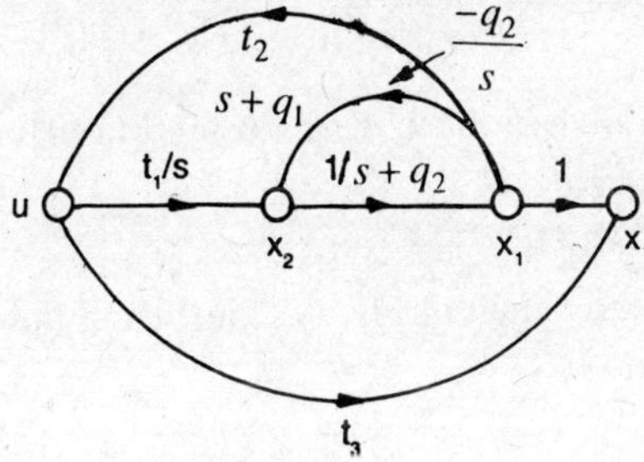

Fig. 4.85

Forward Paths

$$P_1 = \frac{t_1}{s(s+q_1)}$$

$$P_2 = \frac{t_2}{s+q_1}$$

$$P_3 = t_3$$

Loops

$$L_1 = \frac{q_2}{s(s+q_1)}$$

$$\Delta = 1 + \frac{q_2}{s(s+q_1)}$$

$$\Delta_1 = \Delta_2 = 1$$

$$\Delta_3 = 1 + \frac{q_2}{s(s+q_1)} = \frac{s^2 + q_1 s + q_2}{s(s+q_1)}$$

$$T = \frac{P_1\Delta_1 + P_2\Delta_2 + P_3\Delta_3}{\Delta}$$

$$= \frac{\dfrac{t_1}{s(s+q_1)} + \dfrac{t_2}{(s+q_1)} + \dfrac{t_3(s^2 + q_1 s + q_2)}{s(s+q_1)}}{1 + \dfrac{q_2}{s(s+q_1)}}$$

or $$T = \frac{t_1 + t_2 s + t_3(s^2 + q_1 s + q_2)}{s^2 + q_1 s + q_2}$$ **Ans.**

Problem No. 4.21 Solve problem 4.1 by using Mason's Gain formula.

Solution

Converting the block diagram of Fig. 4.5 into signal flow graph, we get the network as shown in Fig. 4.86.

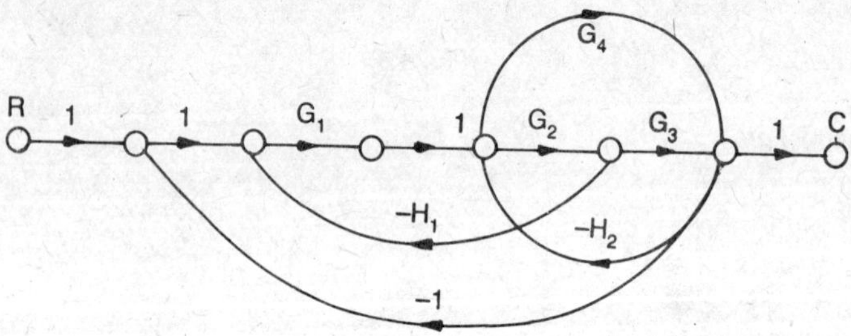

Fig. 4.86

Forward Paths

$$P_1 = G_1 G_2 G_3$$
$$P_2 = G_1 G_4$$

Loops

$$L_1 = -G_1 G_2 H_1$$

$$L_2 = -G_2G_3H_2$$
$$L_3 = -G_1G_2G_3$$
$$L_4 = -G_4H_2$$
$$L_5 = -G_1G_4$$
$$\Delta = 1 + G_1G_2H_1 + G_2G_3H_2 + G_1G_2G_3 + G_4H_2 + G_1G_4$$
$$\Delta_1 = \Delta_2 = 1$$

$$T = \frac{G_1G_2G_3 + G_1G_4}{1 + G_1G_2H_1 + G_2G_3H_2 + G_1G_2G_3 + G_4H_2 + G_1G_4}$$

Ans.

Problem No. 4.22 Solve problem 4.5 using Mason's gain formula.

Solution

Converting the block diagram of Fig. 4.34 into signal flow graph, the network diagram obtained is shown in Fig. 4.87.

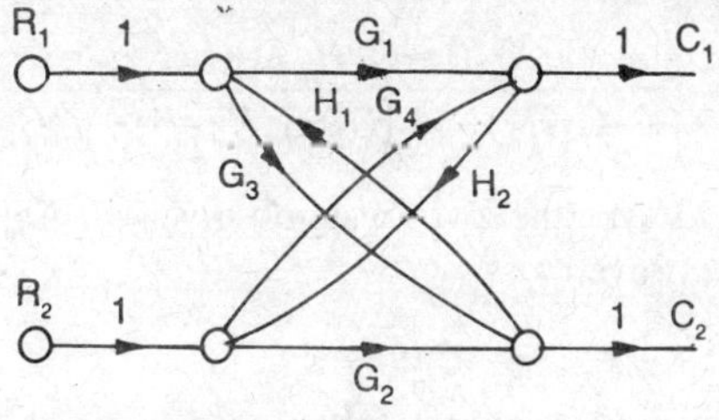

Fig. 4.87

This can be redrawn as shown in Fig. 4.88. Assuming $C_1 = 0$ and $R_2 = 0$.

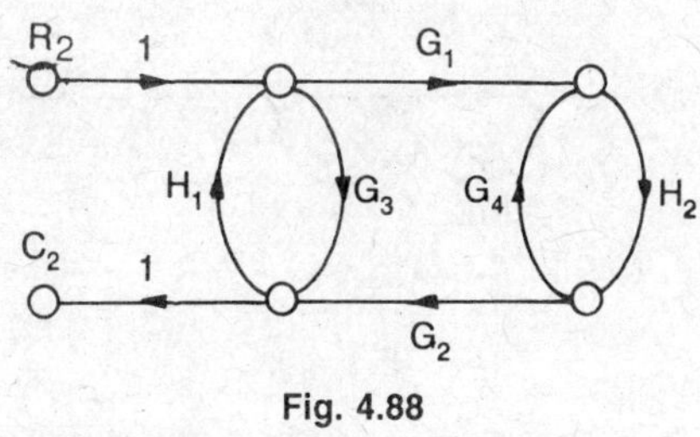

Fig. 4.88

Forward Paths

$$P_1 = G_3$$

$$P_2 = G_1 G_2 H_2$$

Loops

$$L_1 = G_3 H_1$$
$$L_2 = G_4 H_2$$
$$L_3 = G_1 H_2 G_2 H_1$$

Non-touching Loops

$$\Delta = 1 - (G_3 H_1 + G_4 H_2 + G_1 G_2 H_1 H_2) + G_3 G_4 H_1 H_2$$
$$\Delta_1 = 1 - G_4 H_2$$
$$\Delta_2 = 1$$

$$T = \frac{P_1 \Delta_1 + P_2 \Delta_2}{\Delta}$$

$$\therefore \quad T = \frac{G_3(1 - G_4 H_2) + G_1 G_2 H_2}{1 - G_3 H_1 - G_4 H_2 + G_1 G_2 H_1 H_2 + G_3 G_4 H_1 H_2}$$

$$T = \frac{G_3(1 - G_4 H_2) + G_1 G_2 H_2}{(1 - G_4 H_2) + H_1 [G_3(G_4 H_2 - 1) - G_1 G_2 H_2]}$$ **Ans.**

Problem No. 4.23 Are the two systems given in Fig. 4.89 (a) and (b) equivalent. If not, prove.

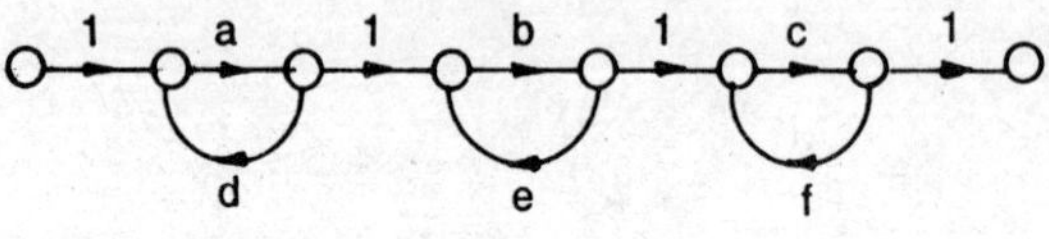

(a)

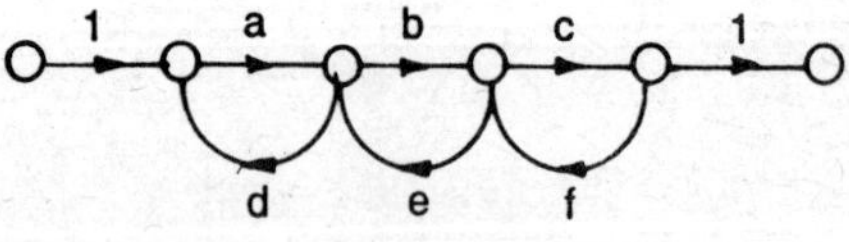

(b)

Fig. 4.89

The systems are not equivalent as the transfer function for Fig. 4.89 (a) is

$$T = \frac{abc}{1 - (ad + be + cf) + cadbe + (adef + becf) - (adbecf)}$$

and the transfer function for Fig. 4.15 (b) is

$$T = \frac{abc}{1-(ad+be+cf)+(adcf)}$$

Problem No. 4.24 Find the transfer function for the system shown in Fig. 4.90.

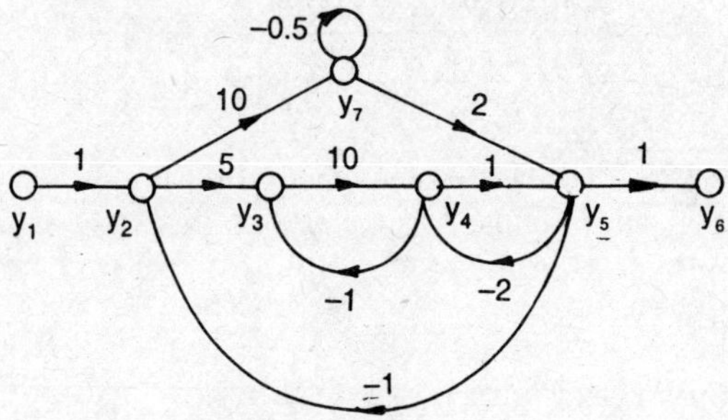

Fig. 4.89

Solution

Forward Paths

$$P_1 = y_1y_2y_3y_4y_5y_6 = 1\times5\times10\times1\times1 = 50$$

$$P_2 = y_1y_2y_7y_5y_6 = 1\times10\times2\times1 = 20$$

Loops

$$L_1 = y_2y_3y_4y_5y_2 = 5\times10\times1\times-1 = -50$$

$$L_2 = y_3y_4y_3 = 10\times-1 = -10$$

$$L_3 = y_4y_5y_4 = 1\times-2 = -2$$

$$L_4 = y_7y_7 = -0.5$$

$$L_5 = y_2y_7y_5y_2 = 10\times2\times-1 = -20$$

Non-touching loops

(1) L_2 and $L_4 = -10\times-0.5 = 5$

(2) L_3 and $L_4 = -2\times-0.5 = 1$

(3) L_1 and $L_4 = -50\times-0.5 = 25$

$$\Delta = 1-(-50-10-2-0.5-20)+(5+1+25) = 114.5$$

$$\Delta_1 = 1-(L_4) = 1-(-0.5) = 1.5$$

$$\Delta_2 = 1-(L_2) = 1-(-10) = 11$$

$$T = \frac{P_1\Delta_1 + P_2\Delta_2}{\Delta}$$

$$= \frac{50\times 1.5 + 20\times 11}{114.5}$$

$$= 2.576$$ **Ans.**

Problem No. 4.25 Find the transfer function for the system shown in Fig. 4.91 by Masons gain formula. Assume $N = 0$. **(AMIE, Pune University)**

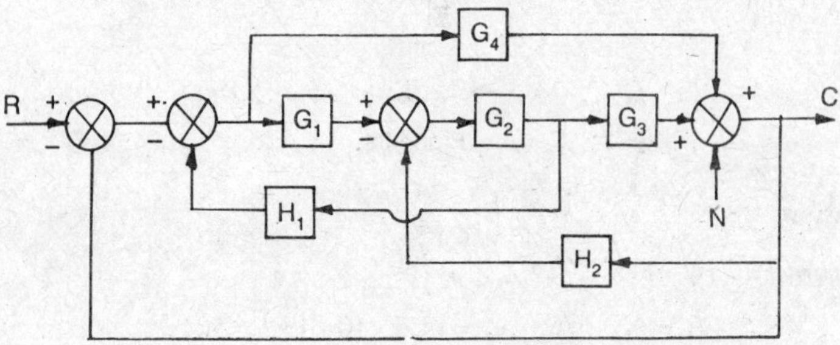

Fig. 4.91

Solution

The signal flow graph for the system is shown in Fig. 4.92.

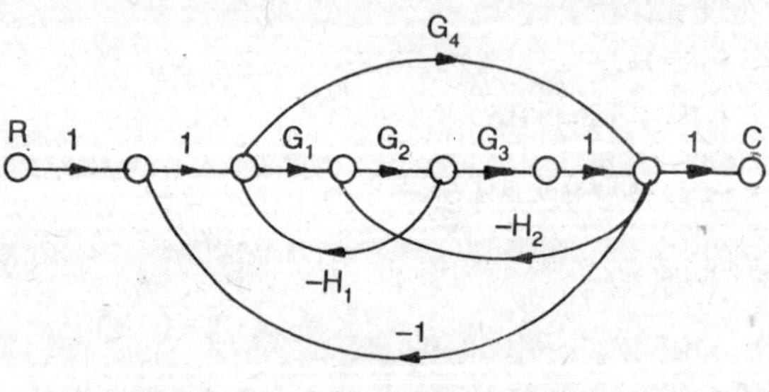

Fig. 4.92

Forward Paths

$$P_1 = G_1G_2G_3$$

$$P_2 = G_4$$

Loops

$$L_1 = -G_1G_2H_1$$
$$L_2 = -G_1G_2G_3$$
$$L_3 = -G_2G_3H_2$$
$$L_4 = -G_4$$
$$L_5 = G_4H_2G_2H_1$$
$$\Delta = 1 + G_1G_2H_1 + G_1G_2G_3 + G_2G_3H_2 + G_4 - G_2G_4H_1H_2$$
$$\Delta_1 = \Delta_2 = 1$$

$$\therefore \quad T = \frac{G_1G_2G_3 + G_4}{1 + G_1G_2H_1 + G_1G_2G_3 + G_2G_3H_2 + G_4 - G_2G_4H_1H_2} \qquad \textbf{Ans.}$$

Problem No. 4.26 Solve problem 4.8 by signal flow graph using Masons gain formula.

Solution

Converting the block diagram given in Fig. 4.47 into signal flow graph, the network obtained is shown in Fig. 4.93.

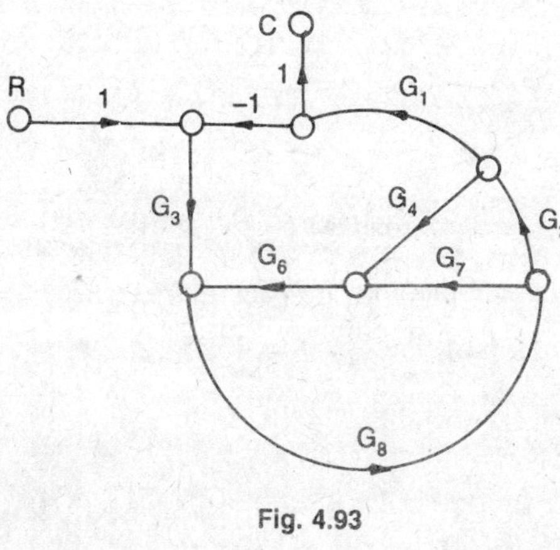

Fig. 4.93

Forward Path

$$P_1 = G_3G_8G_2G_5G_1$$

Loops

$$L_1 = -G_3G_8G_2G_5G_1$$

$$L_2 = -G_6G_8G_7$$

$$L_3 = -G_6G_8G_2G_5G_4$$

$$\therefore \quad \Delta = 1 + G_1G_2G_3G_5G_8 + G_6G_7G_8 + G_2G_4G_5G_6G_8$$

$$\Delta_1 = 1$$

$$T = \frac{P_1\Delta_1}{\Delta}$$

$$\therefore \quad T = \frac{G_1G_2G_3G_5G_8}{1 + G_1G_2G_3G_5G_8 + G_6G_7G_8 + G_2G_4G_5G_6G_8}$$ **Ans.**

Problem No. 4.27 Solve problem No. 4.25 and find out C/N assuming $R = 0$.

Solution

The signal flow graph is shown in Fig. 4.94.

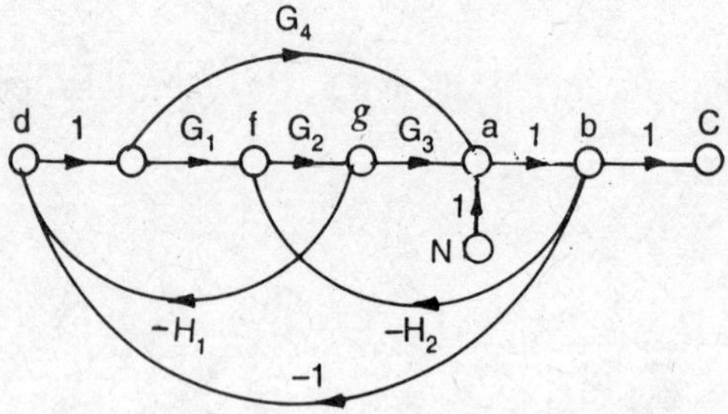

Fig. 4.94

Forward Path Their is only one forward path

$$P_1 = 1 \times 1 \times 1 = 1 = N - a - b - c$$

Loops

$$L_1 = a - b - d - e - f - g - a = -G_1G_2G_3$$

$$L_2 = d - e - f - g - d = -G_1G_2H_1$$

$$L_3 = a - b - f - g - a = -G_2G_3H_2$$

$$L_4 = a - b - d - e - a = -G_4$$

$$\Delta_1 = 1 - (L_2) = 1 + G_1G_2H_2$$

$$\frac{C}{N}(s) = \frac{1 + G_1G_2H_1}{1 + G_1G_2G_3 + G_1G_2H_1 + G_2G_3H_2 + G_4}$$

For $\frac{C}{N}(s) = 0$, the condition is

$$1 + G_1 G_2 H_1 = 0$$ **Ans.**

Problem No. 4.28 Solve problem 4.10 by signal flow technique using Masons Gain formula.

Solution

The block diagram of Fig. 4.56 is converted into signal flow as shown in Fig. 4.95.

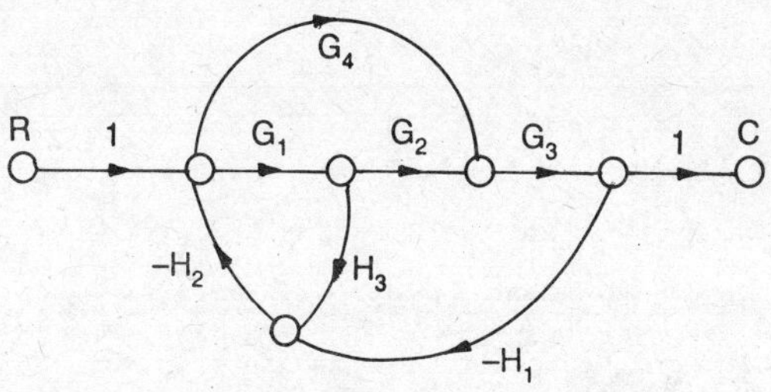

Fig. 4.95

Forward Paths

$$P_1 = G_1 G_2 G_3$$

$$P_2 = G_3 G_4$$

Loops

$$L_1 = G_3 G_4 H_1 H_2$$

$$L_2 = G_1 G_2 G_3 H_1 H_2$$

$$L_3 = -G_1 H_2 H_3$$

$$\Delta = 1 + G_1 H_2 H_3 - G_3 H_4 H_1 H_2 - G_1 G_2 G_3 H_1 H_2$$

$$\Delta_1 = \Delta_2 = 1$$

$$T = \frac{P_1 \Delta_1 + P_2 \Delta_2}{\Delta}$$

$$= \frac{G_1 G_2 G_3 + G_3 G_4}{1 + G_1 H_2 (H_3 - G_2 G_3 H_1) - G_3 G_4 H_1 H_2}$$ **Ans.**

Problem No. 4.29 Find *C/R* for the following system using Mason's gain rule.

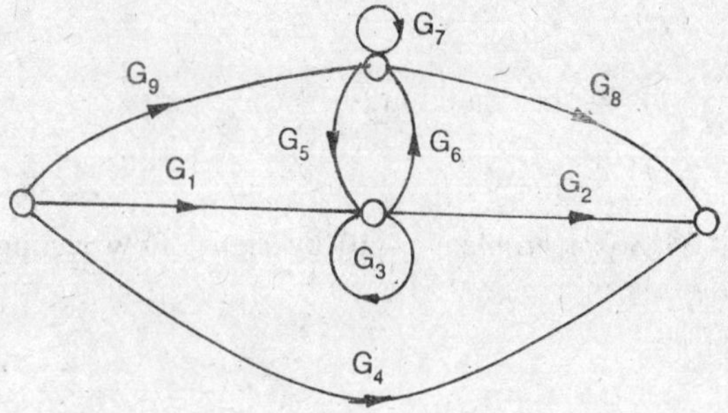

Fig. 4.96

Solution

Forward Paths

$$P_1 = G_1G_2$$
$$P_2 = G_4$$
$$P_3 = G_7G_8$$
$$P_4 = G_1G_5G_8$$
$$P_5 = G_7G_6G_2$$

Loops

$$L_1 = G_9$$
$$L_2 = G_3$$
$$L_3 = G_5G_6$$
$$\Delta = 1-(G_3+G_9+G_5G_6)+G_9G_3$$
$$\Delta_1 = 1-G_9$$
$$\Delta_2 = 1-(G_9+G_3+G_5G_6)+G_9G_3$$
$$\Delta_3 = 1-G_3$$
$$\Delta_4 = 1$$
$$\Delta_5 = 1$$

$$T \simeq \frac{P_1\Delta_1 + P_2\Delta_2 + P_3\Delta_3 + P_4\Delta_4 + P_5\Delta_5}{\Delta}$$

$$= \frac{G_1G_2(1-G_9)+G_4(1-G_9-G_3-G_5G_6+G_9G_3) + G_7G_8(1-G_3)+G_1G_5G_8+G_7G_6G_2}{1-G_3-G_9-G_5G_6+G_9G_3}$$ **Ans.**

Problem No. 4.30 Find C/R for the system shown in Fig. 4.97

(AMIE 1985)

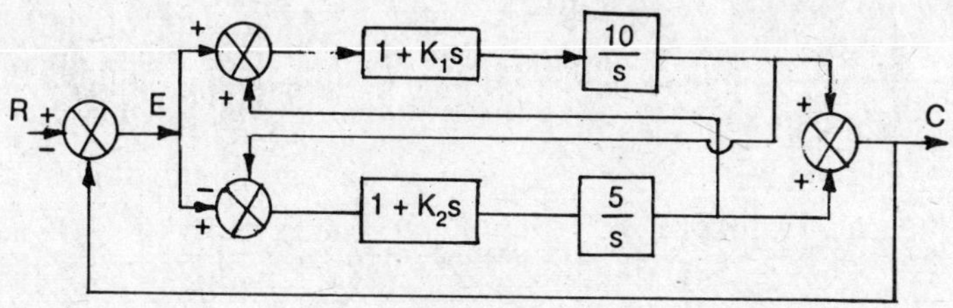

Fig. 4.97

Solution

Evaluation of C/R

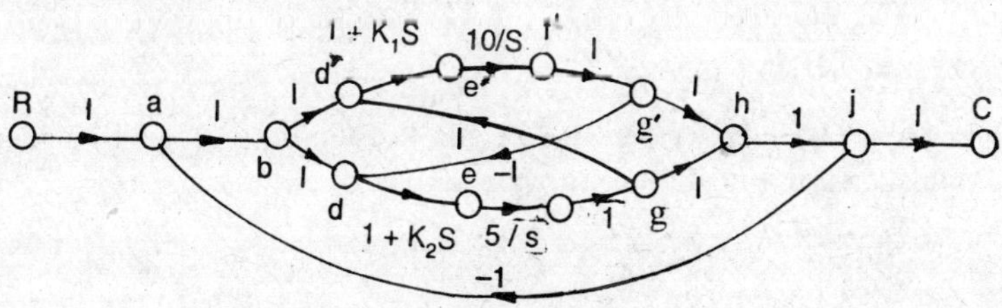

Fig. 4.98

Forward Paths

(1) $Rabdefghjc = \frac{5(1+K_2s)}{s}$

(2) $Rabdefgd'e'f'g'hjc = \frac{50(1+K_2s)(1+K_1s)}{s^2}$

(3) $Rabd'e'f'g'hjc = \frac{10(!+K_1s)}{s}$

(4) $Rabd'e'f'g'defgh = \frac{-50(1+K_1s)(1+K_2s)}{s^2}$

Loops

$$(1)\ abdefghja = \frac{5(1+K_2 s)}{s}$$

$$(2)\ abd'e'f'g'hja = \frac{10(1+K_1 s)}{s}$$

$$(3)\ abd'e'f'g'defghja = \frac{50(1+K_1 s)(1+K_2 s)}{s^2}$$

$$(4)\ abdefgd'e'f'g'hja = -\frac{50(1+K_1 s)(1+K_2 s)}{s^2}$$

$$\frac{C}{R} = \frac{\frac{5(1+K_2 s)}{s} + \frac{10(1+K_1 s)}{s} + \frac{50(1+K_2 s)(1+K_1 s)}{s^2} - \frac{50(1+K_1 s)(1+K_2 s)}{s^2}}{1 + \frac{5(1+K_2 s)}{s} + \frac{10(1+K_1 s)}{s} + \frac{50(1+K_2 s)(1+K_1 s)}{s^2} - \frac{50(1+K_1 s)(1+K_2 s)}{s^2}}$$

$$\frac{C}{R} = \frac{5(1+K_2 s) + 10(1+K_1 s)}{s + 5(1+K_2 s) + 10(1+K_1 s)}$$

$$= \frac{5s(K_2 + 2K_1) + 15}{s(1 + 5K_2 + 10K_1) + 15} \quad \textbf{Ans.}$$

Problem No. 4.31 Find $C(s) / R(s)$ for a control system represented by signal flow graph shown in Fig. 4.99 $C(s)$ is the output.

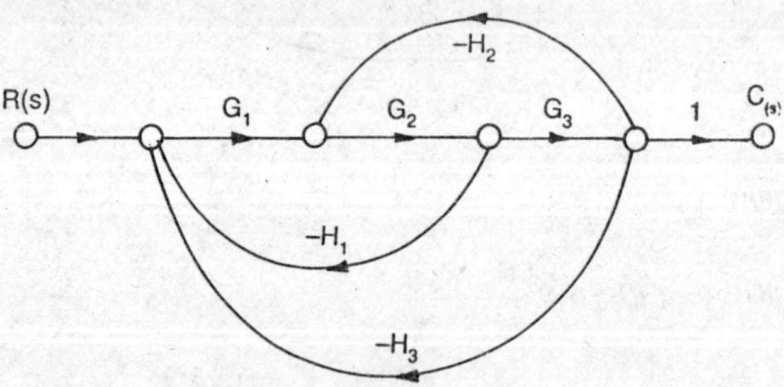

Fig. 4.99

Solution

Forward Paths

Only one and the gain is $G_1 G_2 G_3$

Loops

(1) $-G_1G_2H_1$

(2) $-G_1G_2G_3H_3$

(3) $-G_2G_3H_2$

Non touching Loops NIL

$$T = \frac{G_1G_2G_3}{1+G_1G_2H_1+G_1G_2G_3H_3+G_2G_3H_2}$$ **Ans.**

Problem No. 4.32 Use Masons gain formula for determining the overall transfer function of the system shown in Fig. 4.100. **(AMIE)**

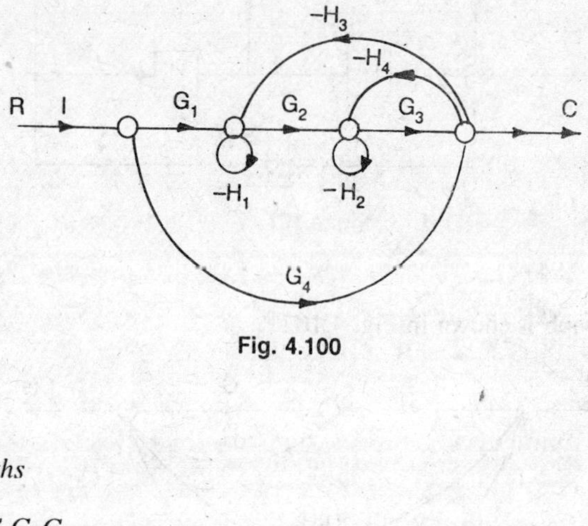

Fig. 4.100

Solution

Forward Paths

(1) $G_1G_2G_3$

(2) G_4

Loops

(1) $-H_1$

(2) $-H_2$

(3) $-G_3H_4$

(4) $-G_2G_3H_4$

Non Touching Loops

(1) $-H_1$ and $-H_2$

(2) $-H_1$ and $-G_3H_4$

$$T = \frac{G_1G_2G_3 + G_4(1 + H_1 + H_2 + H_1H_2)}{1 + H_1 + H_2 + G_3G_4 + G_2G_3H_3 + H_1H_2 + G_1H_1H_4}$$ **Ans.**

Problem No. 4.33 Draw the signal flow graph of the system shown in Fig. 4.101 and find the ratio $\frac{C(s)}{R(s)}$ **(AMIE)**

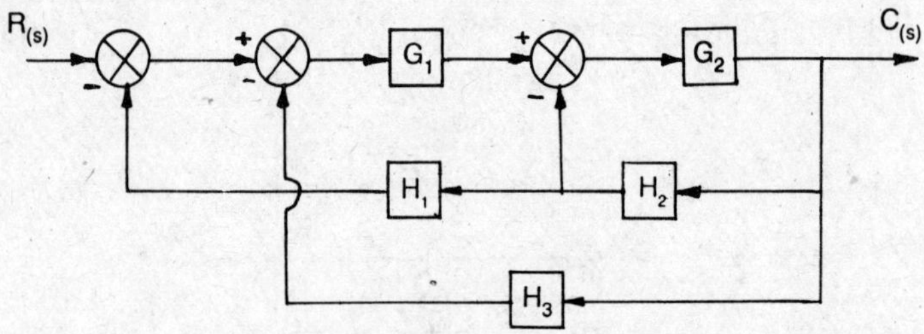

Fig. 4.101

Solution

Signal flow graph is shown in Fig. 4.102

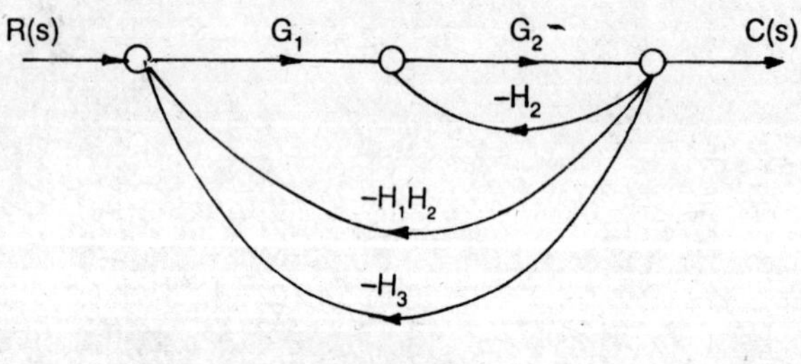

Fig. 4.102

By inspection the ratio

$$\frac{C(s)}{R(s)} = \frac{G_1G_2}{1 + G_1G_2H_3 + G_1G_2H_1H_2 + G_2H_2}$$ **Ans.**

Problem No. 4.34 Solve problem No. 4.12 by Signal flow graph. **(AMIE)**

Solution

The signal flow graph is shown in Fig. 4.103.

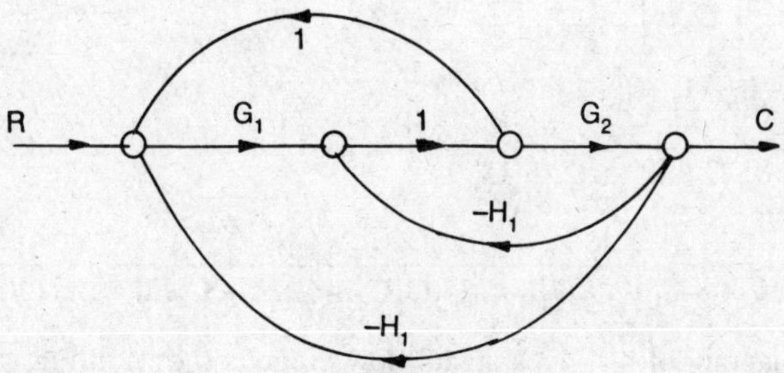

Fig. 4.103

By inspection

$$\frac{C}{R} = \frac{G_1 G_2}{1 + G_1 G_2 H_1 + G_2 H_1 - G_1}$$

Problem No. 4.35 Draw block diagram of the following signal flow graph shown in Fig. 4.104 below. Find the overall transfer function by block diagram reduction. Verify the result by Mason's gain formula. **(AMIE)**

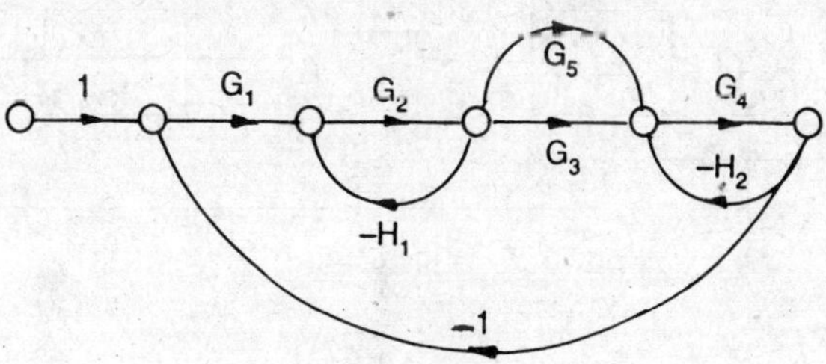

Fig. 4.104

Solution *Mason's gain formula*

Forward paths

(1) $G_1 G_2 G_3 G_4$

(2) $G_1 G_2 G_5 G_4$

Loops

(1) $- G_1 G_2 G_3 G_4$

(2) $-G_1G_2G_5G_4$

(3) $-G_2H_1$

(4) $-G_4H_2$

Non Touching Loops

(1) $-G_2H_1$ and $-G_4H_2$

$$\frac{C}{R} = \frac{G_1G_2G_3G_4 + G_1G_2G_4G_5}{1 + G_1G_2G_3G_4 + G_1G_2G_4G_5 + G_2H_1 + G_4H_2 + G_2G_4H_1H_2}$$

Block Diagram of the given signal flow graph is shown in Fig. 4.105.

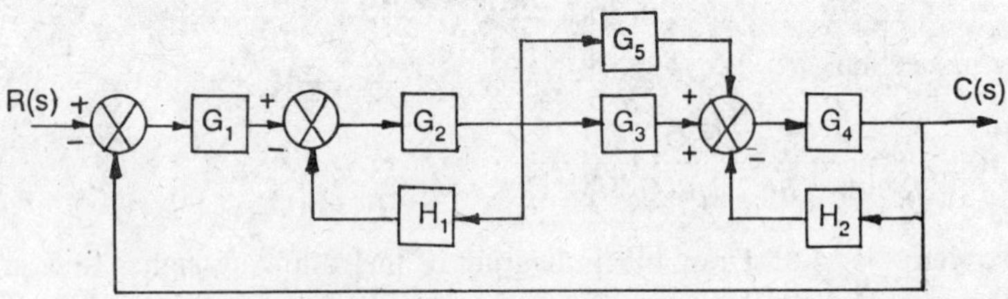

Fig. 4.105

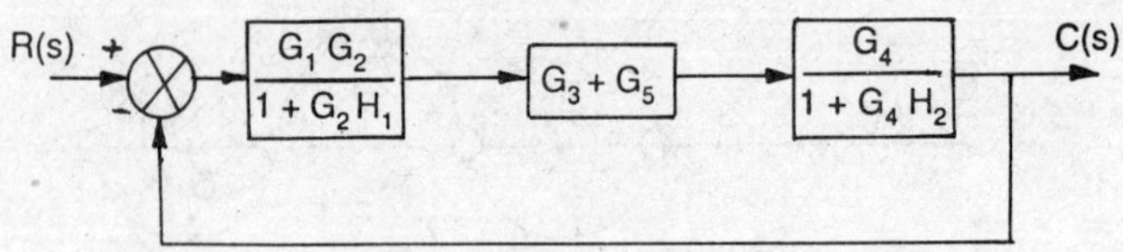

Fig. 4.106

$$\frac{C(s)}{R(s)} = \frac{G}{1+GH} = \frac{\dfrac{G_1G_2(G_3+G_5)G_4}{(1+G_2H_1)(1+G_4H_2)}}{1+\dfrac{G_1G_2(G_3+G_5)G_4}{(1+G_2H_1)(1+G_4H_2)}}$$

$$= \frac{G_1G_2G_3G_4 + G_1G_2G_4G_5}{1 + G_1G_2G_3G_4 + G_1G_2G_4G_5 + G_2H_1 + G_4H_2 + G_2G_4H_1H_2}$$

Problem No. 4.36 Using signal flow graph and Mason's gain formula, obtain the overall gain of the system depicted in Fig. 4.107.

(AMIE)

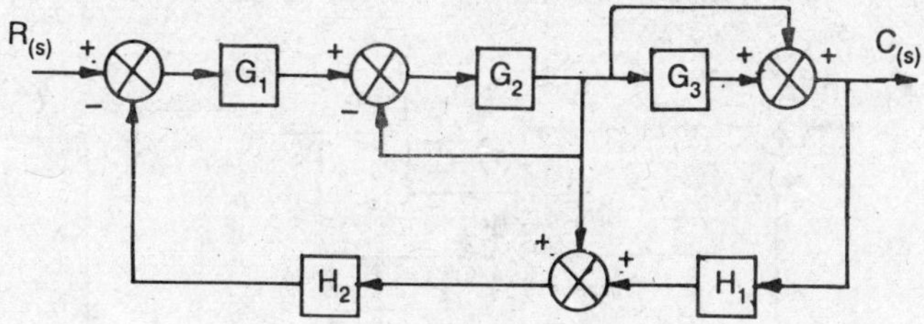

Fig. 4.107

Solution

Signal flow graph of the given block diagram is shown in Fig. 4.108.

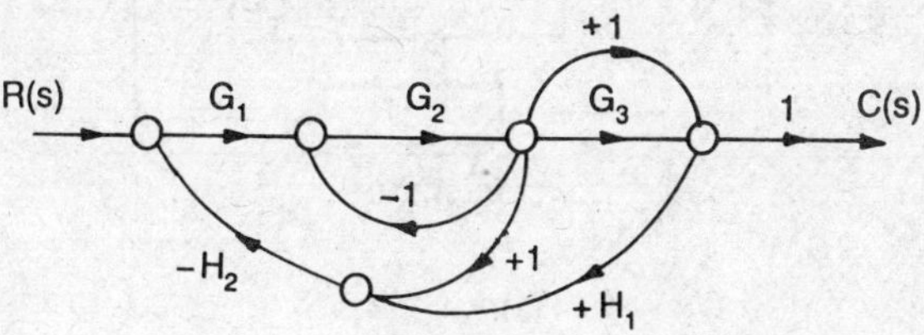

Fig. 4.108

Forward Paths

(1) $G_1G_2G_3$ and

(2) G_1G_2

Loops

(1) $-G_1G_2G_3H_1H$

(2) $-G_1G_2H_2$

(3) $-G_2$ and

(4) $-G_1G_2H_1H_2$

Overall gain

$$= \frac{G_1G_2 + G_1G_2G_3}{1 + G_2 + G_1G_2H_2 + G_1G_2H_1H_2 + G_1G_2G_3H_1H_2} \qquad \textbf{Ans.}$$

Block Diagram Reduction Technique

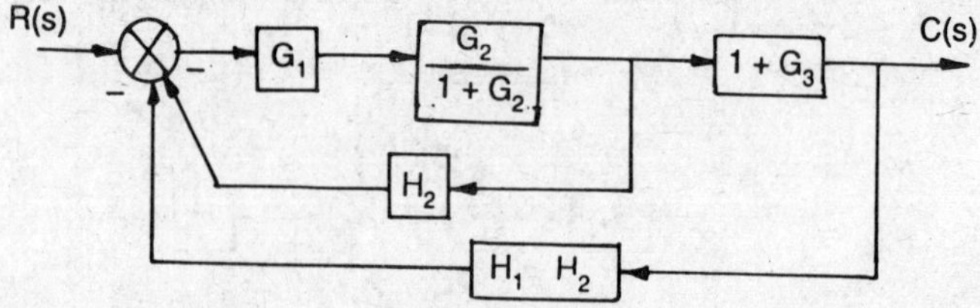

Fig. 4.109

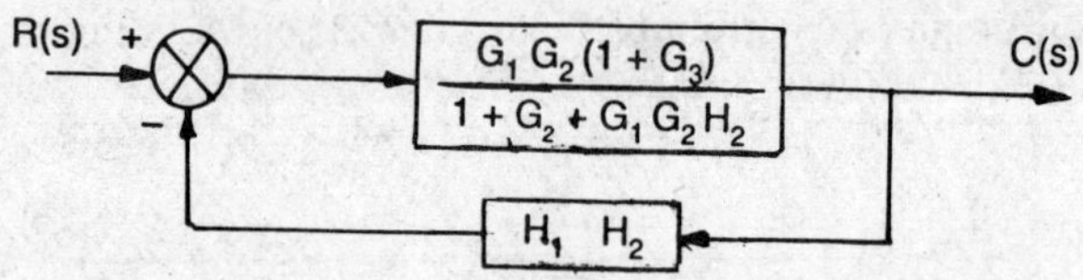

Fig. 4.110

$$\frac{C(s)}{R(s)} = \frac{\dfrac{G_1G_2(1+G_3)}{1+G_2+G_1G_2H_2}}{1+\dfrac{G_1G_2(1+G_3)}{1+G_2+G_1G_2H_2}H_1H_2}$$

$$= \frac{G_1G_2+G_1G_2G_3}{1+G_2+G_1G_2H_2+G_1G_2H_1H_2+G_1G_2G_3H_1H_2}$$ **Ans.**

Problem No. 4.37 Using the Masons gain formula obtain the overall transfer function whose signal flow graph is given in Fig. 4.111. State the number of forward paths, loops and the pair of non-touching loops. **(Pune University)**

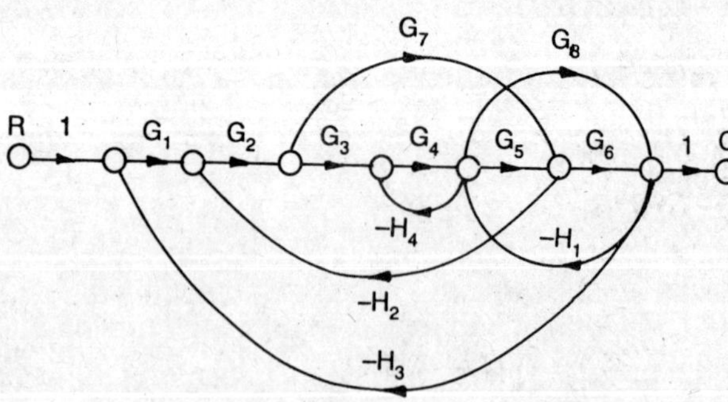

Fig. 4.111

Solution

Forward Paths

(1) $G_1G_2G_3G_4G_5G_6$

(2) $G_1G_2G_7G_6$

(3) $G_1G_2G_3G_4G_8$

Loops

(1) $-G_1G_2G_3G_4G_5G_6H_3$

(2) $-G_2G_3G_4G_5H_2$

(3) $-G_2G_7H_2$

(4) $-G_4H_4$

(5) $-G_8H_1$

(6) $-G_5G_6H_1$

Pair of Nontouching Loops

(1) $-G_4H_4$ and $-G_2G_7H_2$

(2) $-G_2G_7H_2$ and $-G_8H_1$

Overall gain

$$= \frac{G_1G_2G_3G_4G_5G_6 + G_1G_2G_6G_7(1+G_4H_4) + G_1G_2G_3G_4G_8}{\begin{array}{l}1+G_1G_2G_3G_4G_5G_6H_3 + G_2G_3G_4G_5H_2 + G_2G_7H_2 \\ + G_4H_4 + G_8H_1 + G_5G_6H_1 + G_4H_4G_2G_7H_2 \\ + G_2G_7H_2G_8H_1\end{array}} \qquad \textbf{Ans.}$$

5

Network Theorems

Problem No. 5.1 In the network shown in Fig. 5.1

$Z_1 = (8 + j8)\Omega$

$Z_2 = (8 - j8)\Omega$

$Z_3 = (2 + j20)\Omega$

$Z_L = (\ \ j10)\Omega$

$V = 10\angle 0°$

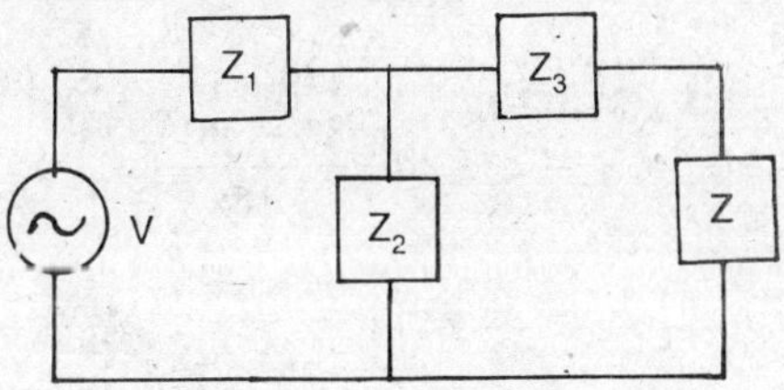

Fig. 5.1

Find the current through the load Z_L using Thevenin's theorem

Solution Remove Z_L and short circuit the voltage source V as shown in Fig. 5.2

$$V_{OC} - I_1 \times Z_2 \qquad ...(1)$$

$$I_1 = \frac{V}{Z_1 + Z_2}$$

$$= \frac{10\angle 0}{8 + j8 + 8 - j8}$$

$$= \frac{10\angle 0}{16}$$

$$= 0.625\angle 0$$

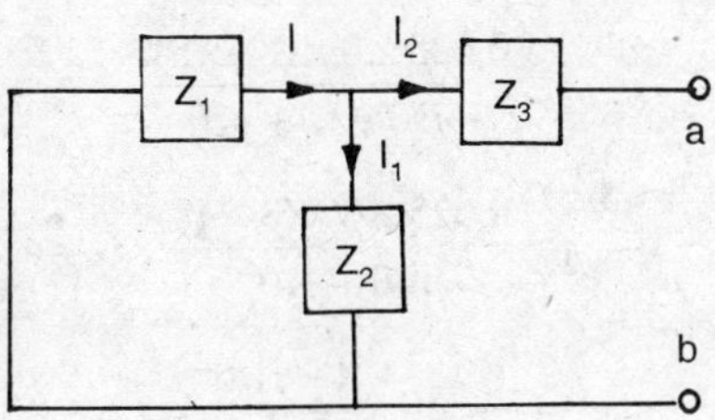

Fig. 5.2

Therefore

$$V_{OC} = 0.625\angle 0 \times (8 - j8) \qquad ...(2)$$

Calculation of Z_{th} (Fig. 5.3)

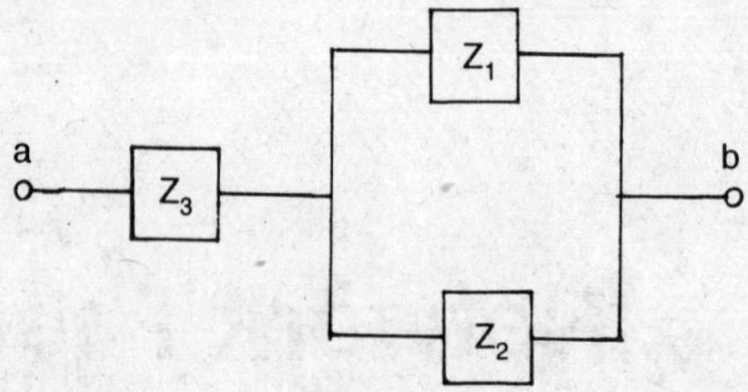

Fig. 5.3

$$Z_{th} = Z_3 + \frac{Z_1 \times Z_2}{Z_1 + Z_2}$$

$$= 2 + j20 + \frac{(8 + j8)(8 - j8)}{8 + j8 + 8 - j8}$$

$$= 2 + j20 + \frac{8^2 - (j8)^2}{16}$$

$$= 2 + j20 + \frac{64 + 64}{16}$$

$$= 2 + j20 + \frac{128}{16}$$

$$= 10 + j20$$

Current I_2 flowing in Z_L

$$I_2 = \frac{V_{OC}}{Z_L + Z_{TH}}$$

$$= \frac{0.625\angle 0 \times (8 - j8)}{-j10 + 10 + j20}$$

$$= \frac{0.625\angle 0 \times (8 - j8)}{10 + j10}$$

$$= \frac{5}{10}\angle 0 \times \frac{1 - j1}{1 + j1}$$

$$= \frac{1}{2}\angle 0 \times \left(\frac{(1-j1)(1-j1)}{1^2-(j)^2} \right)$$

$$= \frac{1}{2}\angle 0 \times \left(\frac{1-j-j-1}{2} \right)$$

$$= \frac{-2j}{4}$$

$$= \frac{-j}{2} \text{ amps.}$$

Problem No.5.2 In the network of Fig. 5.4, the switch K is closed at $t = 0$, a steady-state having previously existed. Find the current in the resistor R using Therenin's theorem.

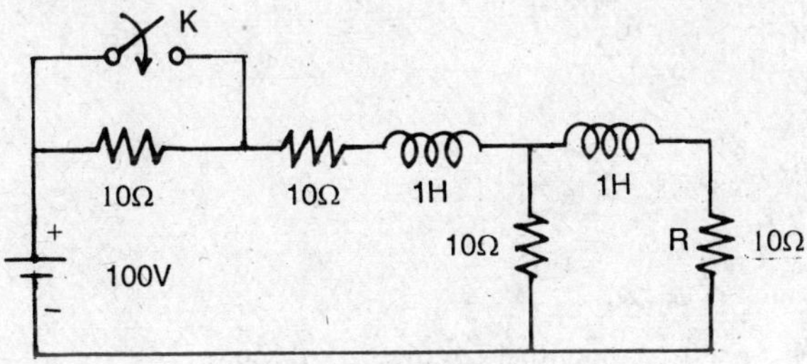

Fig. 5.4

Solution

(a) **When K is opcn**

Both the inductors act as short-circuit and the network is shown in Fig. 5.5.

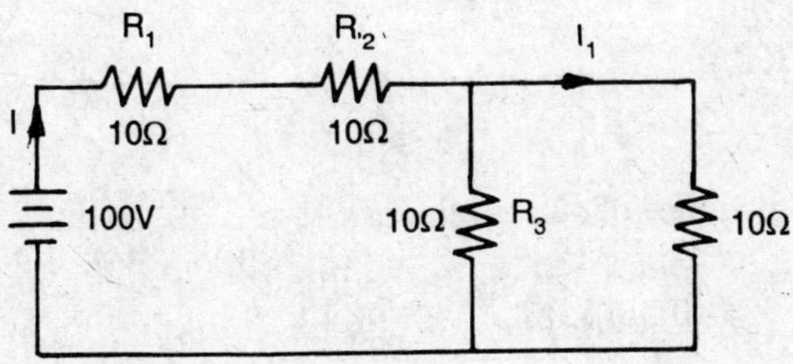

Fig. 5.5

Current I

$$= \frac{V}{R_1 + R_2 + \dfrac{R \times R_3}{R + R_3}}$$

$$= \frac{100}{10 + 10 + \dfrac{10 \times 10}{10 + 10}}$$

$= 4A$

Current I_1

$$= I \times \frac{R_3}{R + R_3}$$

$$= 4 \times \frac{10}{10 + 10}$$

$$= \frac{4 \times 10}{20}$$

$= 2A.$

When Switch K is closed at $t = 0$. It will short-circuit the resistance connected across switch K and the circuit in s-domain will be as shown in Fig. 5.6.

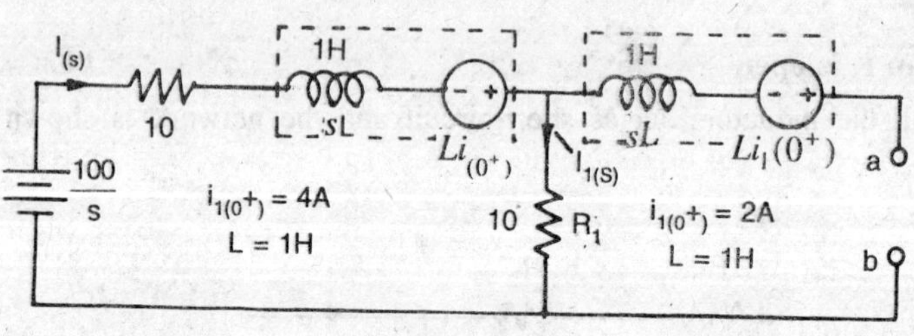

Fig. 5.6

Loop equations give the following relations

$$\frac{100}{s} = 10I(s) + sLI(s) - Li_{(0^+)} + 10I(s)$$

or
$$\frac{100}{s} = 20I(s) + sI(s) - 4$$

or $$I(s)(s+20) = \frac{4s+100}{s}$$

or $$I(s) = \frac{4(s+25)}{s(s+20)}$$

Open circuit Voltage V_{0C} across terminals *a-b*

$$V_{0C} = I_1(s) \times R - Li_1(0^+)$$

$$= 10 \times \frac{4(s+25)}{s(s+20)} - (-2)$$

$$= \frac{40(s+25)}{s(s+20)} + 2$$

$$= \frac{40(s+25) + 2s(s+20)}{s(s+20)}$$

$$= \frac{40s + 1000 + 2s^2 + 40s}{s(s+20)}$$

$$= \frac{2s^2 + 80s + 1000}{s(s+20)}$$

Thevenin's Impedence Z_{TH} (Fig. 5.7)

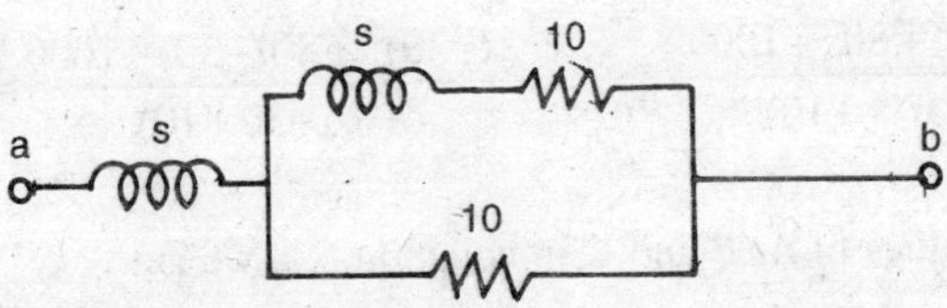

Fig. 5.7

$$Z_{TH} = s + \frac{(s+10) \times 10}{(s+10) + 10}$$

$$= s + \frac{10s + 100}{s + 20}$$

$$= \frac{s^2 + 30s + 100}{s + 20}$$

Current through '*R*'

$$I_R = \frac{V_{OC}}{Z_{TH} + R}$$

$$= \frac{\dfrac{2s^2 + 80s + 1000}{s(s+20)}}{\dfrac{s^2 + 30s + 100}{s+20} + 10}$$

$$= \frac{2s^2 + 80s + 1000}{s(s^2 + 30s + 100 + 10s + 200)}$$

$$= \frac{2s^2 + 80s + 1000}{s(s^2 + 40s + 300)} = \frac{2s^2 + 80s + 1000}{s(s+10)(s+30)}$$

$$= \frac{A}{s} + \frac{B}{s+10} + \frac{C}{s+30} \qquad \text{...(1)}$$

$$A = \left.\frac{2s^2 + 80s + 1000}{(s+10)(s+30)}\right|_{s=0} = \frac{1000}{300} = \frac{10}{3}$$

$$B = \left.\frac{2s^2 + 80s + 1000}{s(s+30)}\right|_{s=-10} = \frac{2(-10)^2 + 80(-10) + 1000}{-10(-10+30)} = -2$$

$$C = \left.\frac{2s^2 + 80s + 1000}{s(s+10)}\right|_{s=-30} = \frac{2(-30)^2 + 80(-30) + 1000}{-30(-30+10)} = \frac{2}{3}$$

Substituting values of *A*, *B* and *C* in equation (1), we get

$$I_R = \frac{10/3}{s} + \frac{-2}{s+10} + \frac{2/3}{s+30}$$

or
$$i_{R(t)} = \frac{10}{3} - 2e^{-10t} + \frac{2}{3}e^{-30t}$$

Problem No. 5.3 Obtain the Thevenin's equivalent for the bridge circuit shown in Fig. 5.8, at terminal *xy*.

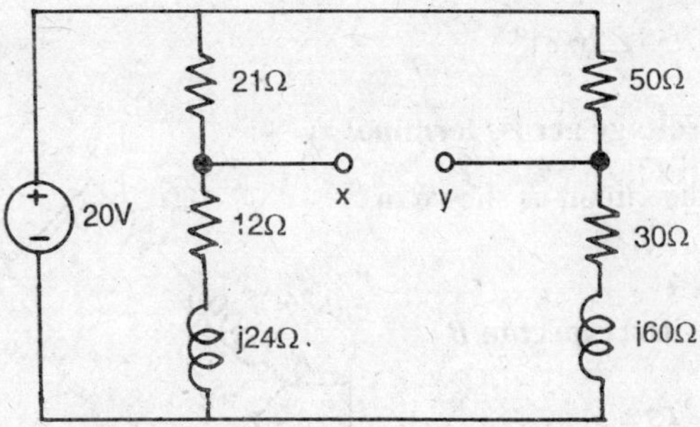

Fig. 5.8

Solution

Thevenin's Impedence (Z_{th}) (Fig. 5.9)

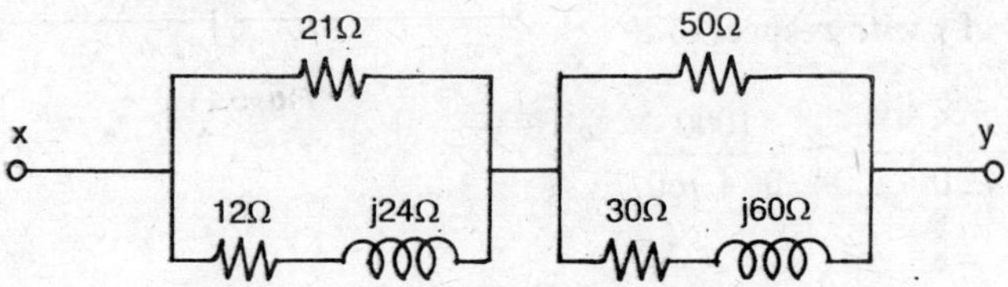

Fig. 5.9

$$Z_{TH} = \frac{21(12+j24)}{21+12+j24} + \frac{50(30+j60)}{50+30+j60}$$

$$= \frac{252+j504}{33+j24} + \frac{1500+j3000}{80+j60}$$

$$= \frac{84(1+j2)}{11+j8} + \frac{75(1+j2)}{4+j3}$$

$$= \frac{(1+j2)(1161+j852)}{20+j65}$$

$$= \frac{-543+j3174}{20+j65}$$

$$= \frac{3220\angle 180°-80.29°}{68\angle 72.89°}$$

$$= 47.35\angle 180°-80.29°-72.89°$$

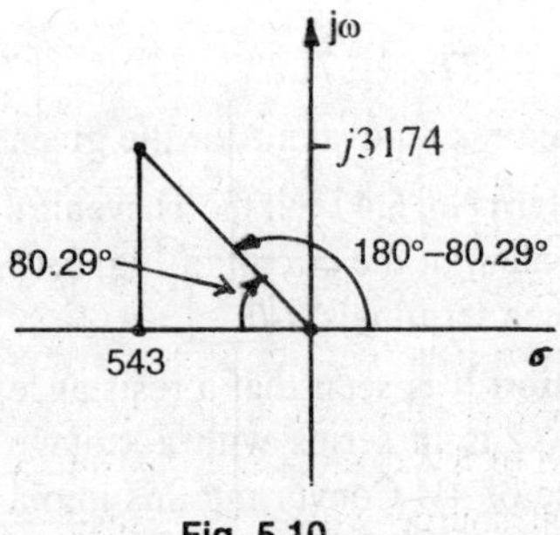

Fig. 5.10

$$= 47.35\angle 26.81°$$

Open Circuit Voltage across terminal *xy*

Let us redraw the circuit as shown in Fig. 5.11

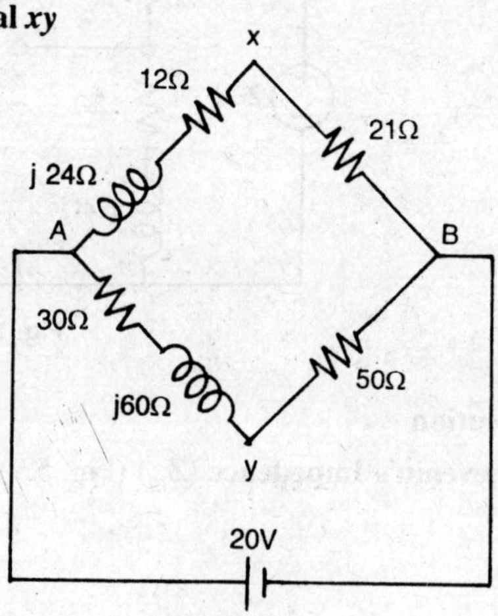

Fig. 5.11

Potential of *x* with respect to *B*

$$V_{xB} = \frac{20 \times 21}{12 + j24 + 21}$$

$$= \frac{420}{33 + j24}$$

$$= 8.324 - j6.05$$

Potential of *y* with respect to *B*

$$V_{yB} = \frac{20 \times 50}{50 + 30 + j60} = \frac{1000}{80 + j60}$$

$$= 8 - j6$$

$$V_{xy} = V_{xB} - V_{yB}$$

$$= 8.324 - j6.05 - 8 + j6$$

$$= 0.324 - j0.05$$

$$= 0.327\angle 180° - \tan^{-1}\frac{0.05}{0.324}$$

$$= 0.327\angle 180° - 8.773°$$

$$= 0.327\angle +171.23°$$

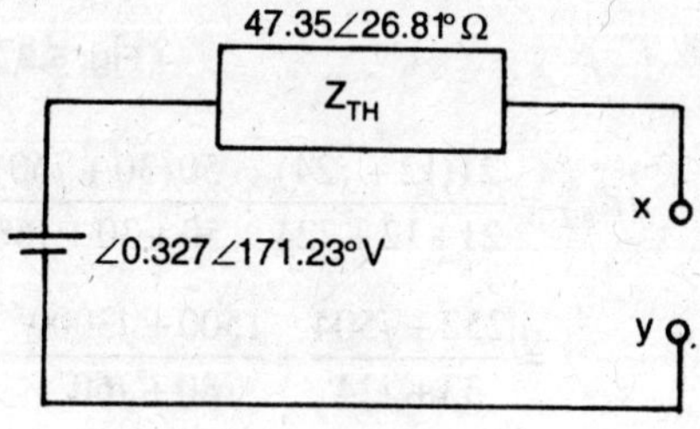

Fig. 5.12

Thevenin's equivalent for the given bridge circuit is shown is Fig. 5.12

Problem No. 5.4 Find the Thevenin's equivalent of the circuit of Fig. 5.13 between terminals *a-b*.

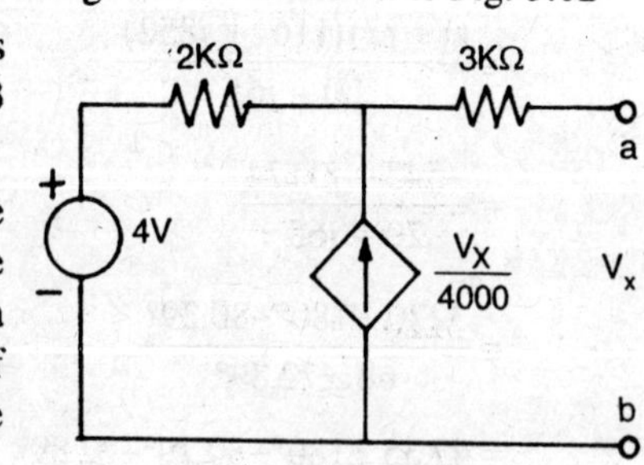

Fig. 5.13

Solution It is seen that a resistance of $2K\Omega$ is in series with a voltage source of $4V$. Converting this into a current source with a resistance of $2K\Omega$ across the current source. The current source will have value

$$= \frac{4}{2000}$$

The network changes to as shown in Fig. 5.14.

V_{ab} = Voltage across 2KΩ resistor

$$= I \times 2000$$

$$= \left(\frac{4}{2000} + \frac{V_x}{4000}\right) 2000$$

$$= \frac{V_x + 8}{2} V$$

$$Z_{ab} = 3 + 2$$

$$= 5 K\Omega$$

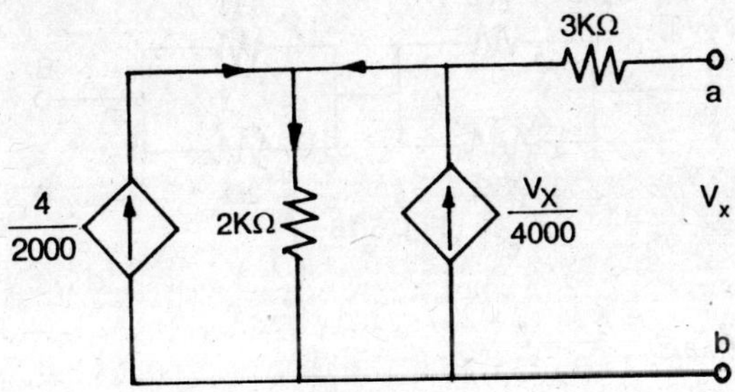

Fig. 5.14

Thevenin's equivalent is shown in 5.16

Fig. 5.15

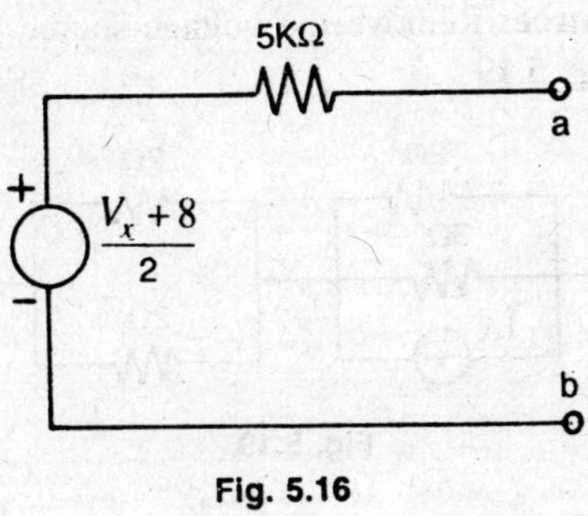

Fig. 5.16

Problem No. 5.5 Determine the current in the 1.0 ohm resistor across *A*-*B* of the network in Fig. 5.17.

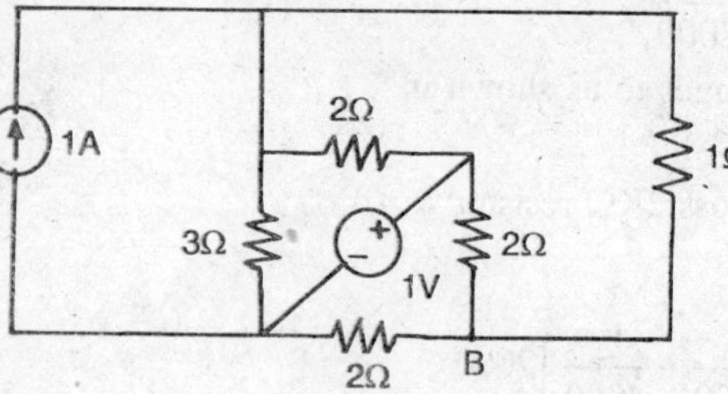

Fig. 5.17

Solution

Short circuiting the voltage source and open-circuiting the current source and removing the 1Ω resistor between A-B. We get a network shown in Fig. 5.18

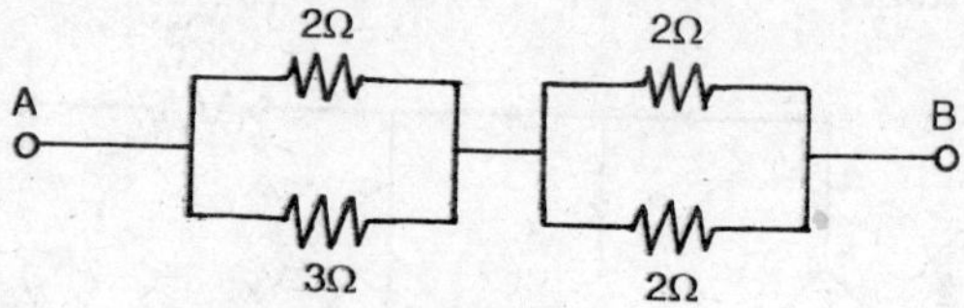

Fig. 5.18

$$Z_{TH} = Z_{AB} = \frac{2\times3}{3+2} + \frac{2\times2}{2+2}$$

$$= \frac{6}{5} + 1$$

$$= \frac{11}{5}$$

$$= 2.2\ \Omega$$

Let us now use superposition theorem to find out open-circuit voltage.

Due to Current Source: Remove the voltage-source and redraw the network as shown in Fig. 5.19

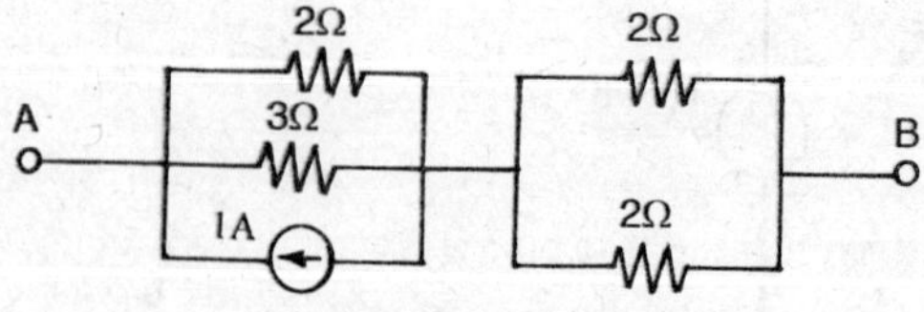

Fig. 5.19

Combining the parallel resistances, we get a simplified network shown in Fig. 5.20.

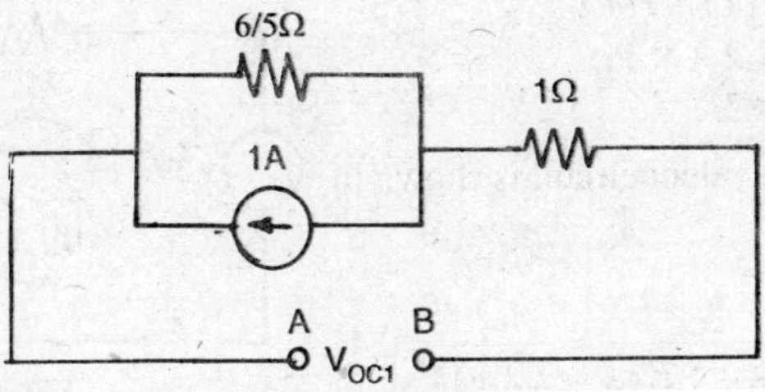

Fig. 5.20

$$V_{OC1} = IR = 1 \times \frac{6}{5} = 1.2V$$

Due to Voltage Source (Fig. 5.21)

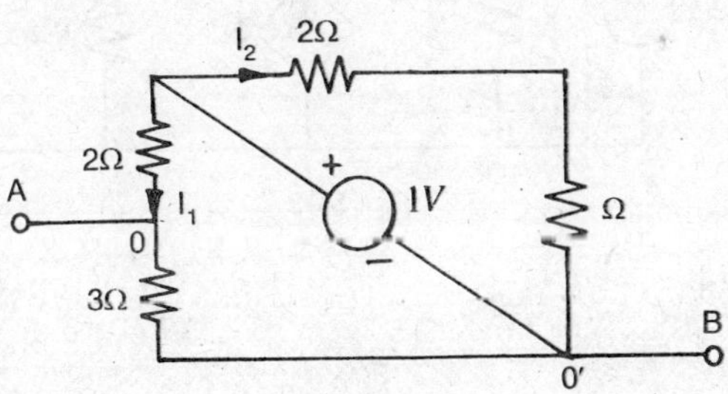

Fig. 5.21

$$I_1 = \frac{V}{2+3} = \frac{1}{5}A = 0.2A$$

$$I_2 = \frac{V}{2+2} = \frac{1}{4}A = 0.25A$$

$$V_{AO} = V - I_1R$$
$$= 1 - 0.2 \times 2$$
$$= 0.6V$$
$$V_{BO'} = V - I_2R$$
$$= 1 - 0.25 \times 2$$
$$= 0.5V$$
$$V_{OC2} = V_{AB} = V_{AO} - V_{BO'}$$

$= 0.6 - 0.5 = 0.1V$

Therefore open circuit voltage is

$$V_{OC} = V_{OC1} + V_{OC2}$$
$$= 1.2 + 0.1$$
$$= 1.3$$

Thevenin's equivalent circuit is shown in Fig. 5.22

$$I_{AB} = \frac{V_{OC}}{Z_{TH} + R_{AB}} = \frac{1.3}{2.2+1} = 0.41A$$

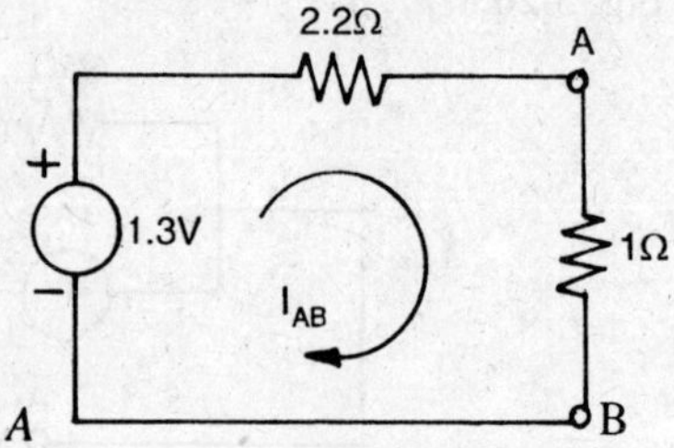

Fig. 5.22

Problem No. 5.6 Find the equivalent Thevenin's network for the network shown in Fig. 5.23, for the pair

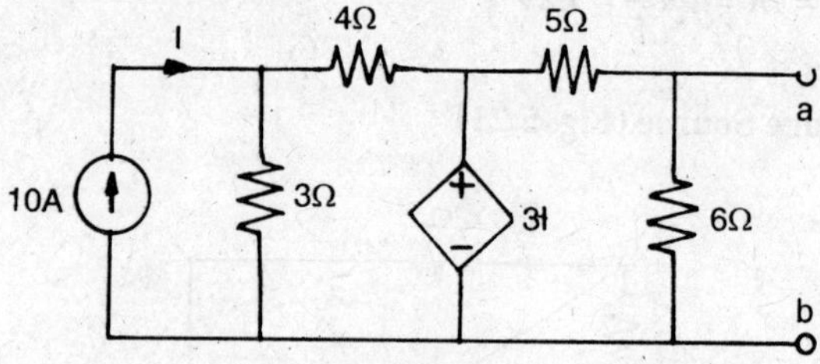

Fig. 5.23

of terminals a-b.

Solution

Thevenin's Impedance (Z_{th}). Short-circuiting the voltage source and open-circuiting the current-source, the network reduces to Fig. 5.24

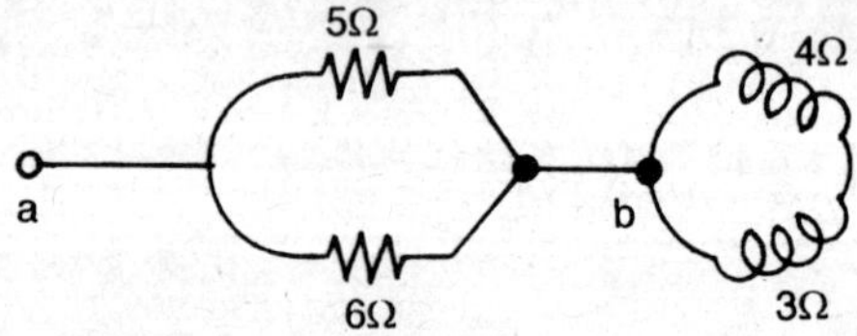

Fig. 5.24

Therefore

$$Z_{TH} = Z_{ab} = \frac{5 \times 6}{5+6} = \frac{30}{11}\Omega$$

Open-circuit Voltage Since these are two sources, we will use principle of superposition to determine open-circuit voltage.

(a) **Voltage Source (*V_{OC1}*)** network is shown in Fig. 5.25

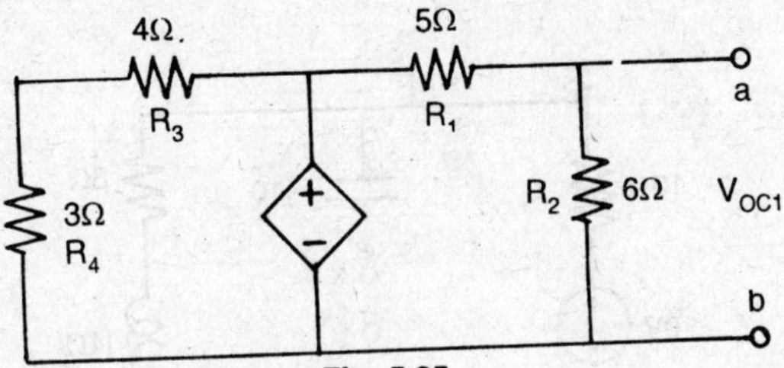

Fig. 5.25

V_{OC1} will be zero since it is going to be contributed by voltage source $3I$ and $I = 0$. Therefore, $3I = 0$ and hence, $V_{OC1} = 0$.

(b) **Current-Source (*V_{OC2}*)** Short-circuiting the voltage source as shown in Fig. 5.26, we see that current

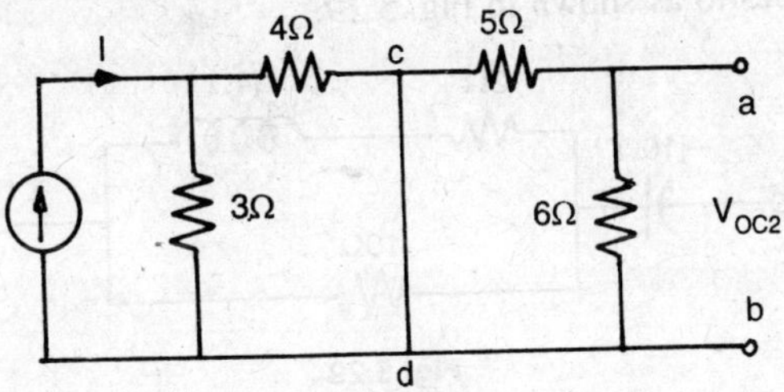

Fig. 5.26

supplied by current source will flow through the short circuit c-d and hence

$$V_{OC2} = 0$$

Therefore

$$\begin{aligned} V_{OC} &= V_{OC1} + V_{OC2} \\ &= 0 + 0 \\ &= 0 \end{aligned}$$

Thevenin's equivalent is shown in Fig. 5.27

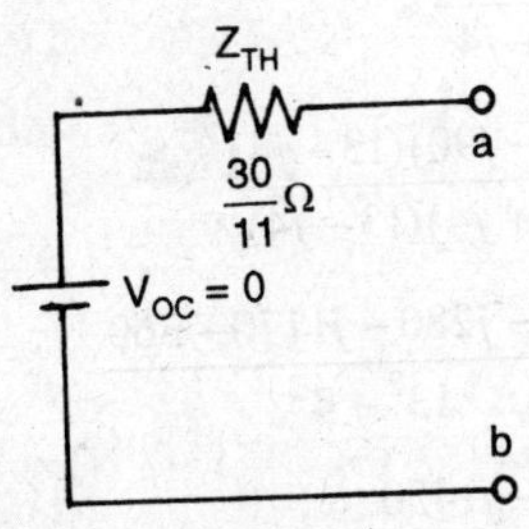

Fig. 5.27

Problem No. 5.7 Obtain the Thevenin equivalent of network shown in Fig. 5.28.

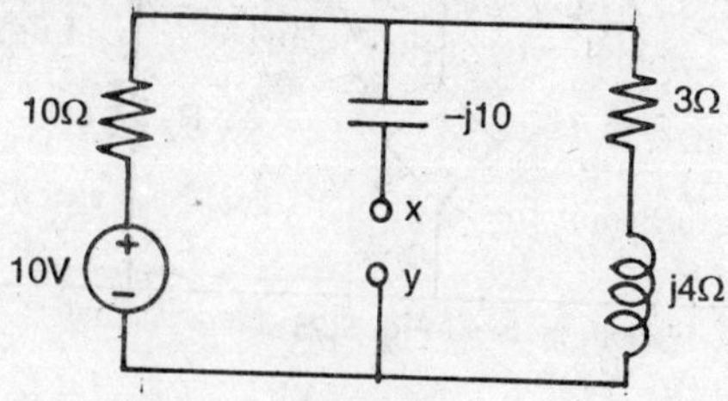

Fig. 5.28

Solution

Thevenin's Impedance (Z_{th}) Short-circuiting the voltage-source the given network reduces to as shown in Fig. 5.29.

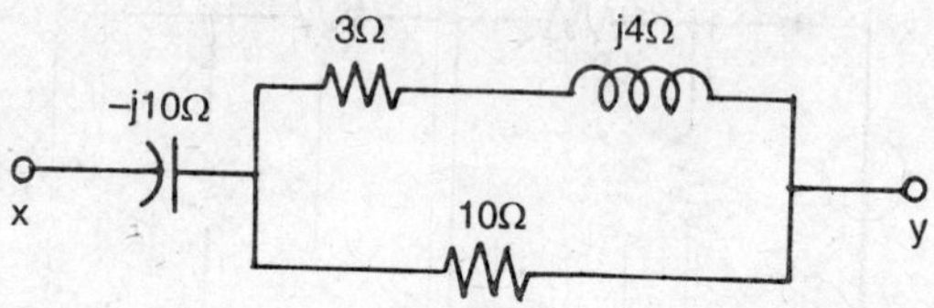

Fig. 5.29

$$Z_{TH} = Z_{xy} = \frac{(3+j4)\times 10}{3+j4+10} - \frac{j10}{1}$$

$$= \frac{30+j40}{13+j4} - j10$$

$$= \frac{30+j40-j130+40}{13+j4}$$

$$= \frac{70-j90}{13+j4}$$

$$= \frac{(70-j90)(13-j4)}{(13+j4)(13-j4)}$$

$$= \frac{910-j280-j1170-360}{13^2+4^2}$$

$$= \frac{550-j1450}{185}$$

$$= 2.973 - j7.839$$

$$= \sqrt{2.973^2 + 7.839^2} \angle \tan^{-1} - \frac{7.839}{2.973}$$

$$= 8.38\angle -69.23°\,\Omega$$

Open-Circuit Voltage

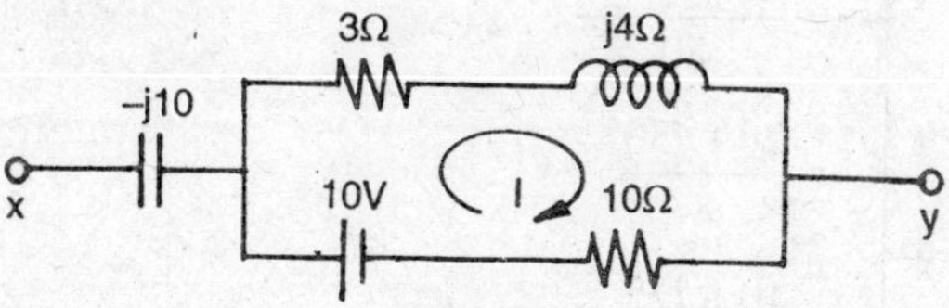

Fig. 5.30

$$I = \frac{10}{3 + j4 + 10} = \frac{10}{13 + j4} = \frac{130 - j4}{185} = 0.703 - j0.216$$

$$V_{OC} = V_{xy} = (0.703 - j0.216)(3 + j4)$$

$$= 2.973 + j2.172$$

$$= \sqrt{2.973^2 + 2.172^2} \angle \tan^{-1} \frac{2.172}{2.973}$$

$$= 3.68\angle 36.15°$$

Problem No. 5.8 Obtain Thevenin's equivalent of the network shown in Fig. 5.31.

Solution

Thevenin's Impedance

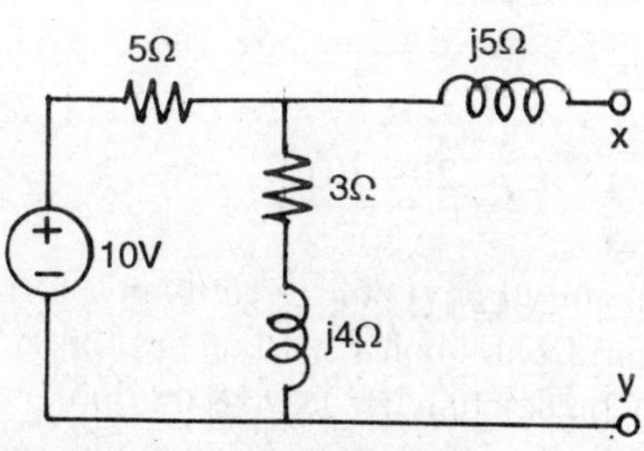

Fig. 5.31

$$Z_{TH} = Z_{xy} = j5 + \frac{(3 + j4) \times 5}{3 + j4 + 5}$$

$$= j5 + \frac{15 + j20}{8 + j4}$$

$$= \frac{j5(8 + j4) + 15 + j20}{8 + j4}$$

$$= \frac{-20 + j40 + 15 + j20}{8 + j4}$$

$$= \frac{-5 + j60}{8 + j4}$$

$$= \frac{(-5 + j60)(8 - j4)}{64 + 16}$$

$$= \frac{-40 + j20 + j480 + 240}{80}$$

$$= \frac{200 + j500}{80}$$

$$= \frac{20 + j50}{8}$$

$$= \frac{10 + j25}{4}$$

$$= 2.5 + j6.25$$

$$= \sqrt{2.5^2 + 6.25^2} \angle \tan^{-1} \frac{16.25}{2.5}$$

$$= 6.73 \angle 68.19° \Omega$$

Open-Circuit Voltage 'V_{oc}'

$$I = \frac{10}{5 + 3 + j4}$$

$$= \frac{10}{8 + j4}$$

$$= \frac{5}{4 + j2}$$

$$= \frac{20 - j10}{20}$$

$$= 1 - j0.5$$

$$V_{0C} = V_{xy} = (3 + j4)(1 - j0.5)$$

$$= 3 - j1.5 + j4 + 2$$

$$= 5 - j2.5$$

$$= 5.59 \angle \tan^{-1} \frac{2.5}{5} = 5.59 \angle 26.56°$$

Problem No. 5.9 Find Thevenin's equivalent at terminals A and B of the network given in Fig. 5.32. Then determine the current through the resistor R_L.

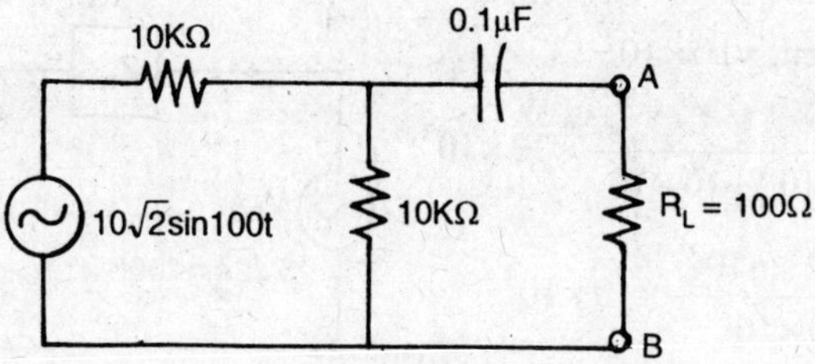

Fig. 5.32

Solution

Thevenin's Impedance Short-circuiting the voltage source and redrawing the network as shown in Fig. 5.33, we get

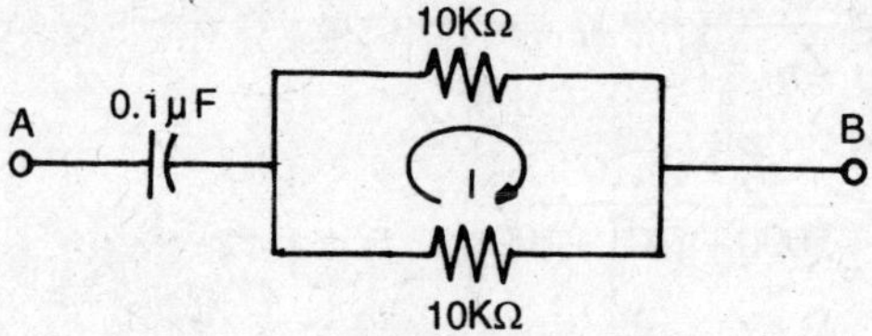

Fig. 5.33

$$Z_{TH} = Z_{AB} = \frac{-j}{0.1\times10^{-6}} + \frac{10\times10^3\times10\times10^3}{10\times10^3+10\times10^3}$$

$$= -10^6\left[\frac{j}{0.1} - \frac{100}{20\times10^3}\right]$$

$$= -10^6\left[j10 - \frac{1}{200}\right]$$

$$= -10^6\left[\frac{j2000-1}{200}\right]$$

$$= -0.5\times10^4\,[j2000-1]$$

$$= -5000[j2000-1]$$

$$= 5000 - j10^7$$

Open-Circuit Voltage 'V_{oc}'

$V_{OC} = V_{AB}$ = Voltage across 10KΩ resistor.

$$= \text{Current} \times 10 \times 10^3$$

$$= \frac{V}{10\times10^3 + 10\times10^3} \times 10\times10^3$$

$$= \frac{10\sqrt{2}\sin 100t}{20\times10^3} \times 10\times10^3$$

$$= 5\sqrt{2}\sin 100t$$

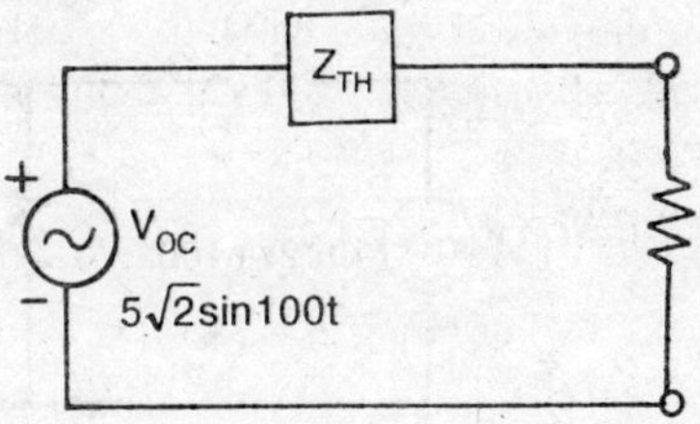

Fig. 5.34

Current through resitor 'R_L'

$$= \frac{V_{OC}}{Z_{TH} + R_L}$$

$$= \frac{5\sqrt{2}\sin 100t}{5000 - j10^7 + 100}$$

$$= \frac{5\sqrt{2}\sin 100t}{5100 - j10^7} A$$

Problem No. 5.10 Obtain the Thevenin's equivalent at terminals x-y of the network shown in Fig. 5.35.

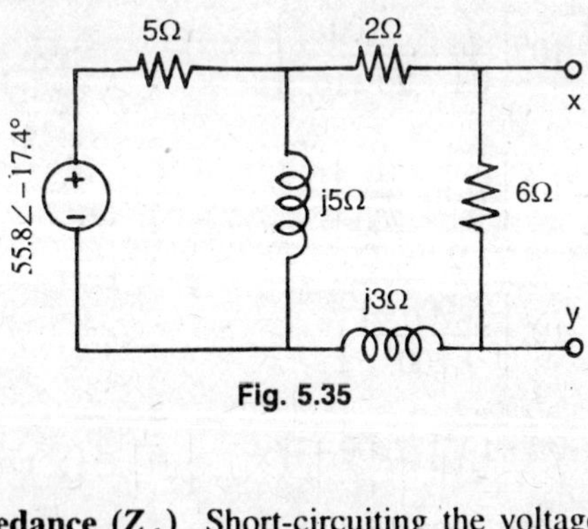

Fig. 5.35

Solution

Thevenin's Impedance (Z_{th}) Short-circuiting the voltage source the network reduces to as shown in Fig. 5.36.

$$Z_{ABCD} = 2 + \frac{5\times j5}{5 + j5} + j3$$

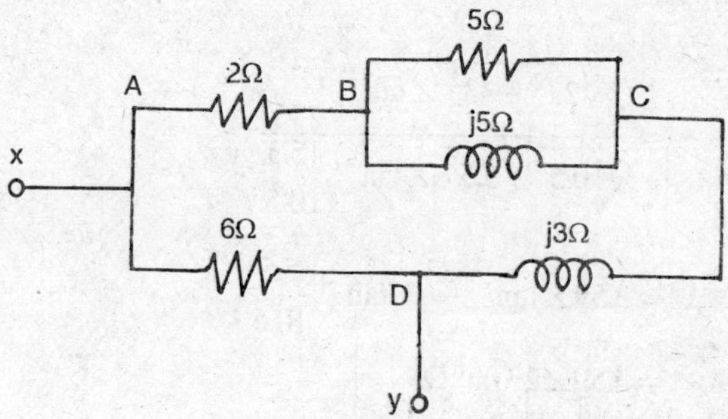

Fig. 5.36

$$= 2 + j3 + \frac{j25}{5 + j5}$$

$$= \frac{(2 + j3)(5 + j5) + j25}{5 + j5}$$

$$= \frac{10 + j10 + j15 - 15 + j25}{5 + j5}$$

$$= \frac{-5 + j50}{5 + j5}$$

$$= \frac{-1 + j10}{1 + j}$$

$$= \frac{(-1 + j10)(1 - j)}{1^2 - j^2}$$

$$= \frac{-1 + j + j10 + 10}{1 + 1}$$

$$= \frac{9 + j11}{2}$$

$$= 4.5 + j5.5$$

$$Z_{TH} = Z_{xy} = \frac{(4.5 + j5.5)6}{4.5 + j5.5 + 6}$$

$$= \frac{27 + j33}{10.5 + j5.5}$$

$$= \frac{\sqrt{27^2+33^2}\ \angle \tan^{-1}\frac{33}{27}}{\sqrt{10.5^2+5.5^2}\ \angle \tan^{-1}\frac{15.5}{10.5}}$$

$$= 3.59 \angle \tan^{-1}\frac{33}{27} - \tan^{-1}\frac{5.5}{10.5}$$

$$= 3.59 \angle 23.06° \Omega$$

$$= 3.3 + j1.41 \Omega$$

Open-Circuit Voltage 'V_{oc}'

Circuit of Fig. 5.35 is redrawn as shown in Fig. 5.37 and circulating currents marked

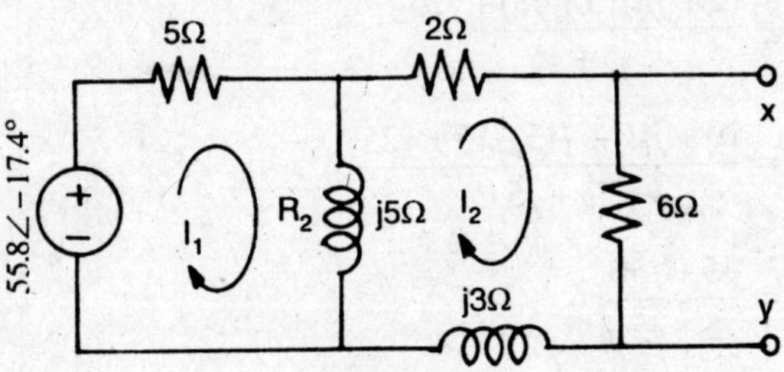

Fig. 5.37

$$5I_1 + j5(I_1 - I_2) = 55.8\angle -17.4°$$

or $\quad I_1(5+j5) - j5I_2 = 53.2 - j16.69$ ...(1)

and $\quad 2I_2 + 6I_2 + j(3+5)I_2 - j5I_1 = 0$

or $\quad I_2(2+6+j8) - j5I_1 = 0$

or $\quad (8+j8)I_2 - j5I_1 = 0$

or $\quad I_1 = \frac{8+j8}{j5}$ ...(2)

Substituting in equation (1), value of I_1

$$\frac{(5+j5)(8+j8)I_2}{j5} - j5I_2 = 53.2 - j16.69$$

or $\quad (5+j5)(8+j8)I_2 - j5 \times j5I_2 = (53.2 - j16.69)j5$

or $\quad \{(7.07\angle 45°) \times (11.31\angle 45°) + 25\} I_2 = (55.8\angle 17.4°) \times 5\angle 90°$

or $\{(7.07 \times 11.31 \angle 45°+45°)+25\} I_2 = 55.8 \times 5 \angle -17.4°+90°$

or $\{(79.96 \angle 90°)+25\} I_2 = 279 \angle 72.6°$

or $(25 + j79.96) I_2 = 279 \angle 72.6°$

or $I_2 = \dfrac{279 \angle 72.6°}{25 + j79.96}$

$= \dfrac{279 \angle 72.6°}{83.43 + \angle 72.6°}$

$= 3.34 \angle 0°$

$V_{OC} = V_{xy} = I_2 \times R_2$

$= 3.34 \angle 0° \times 6$

$= 20 \angle 0°$

Thevenin's equivalent circuit is shown in Fig. 5.38

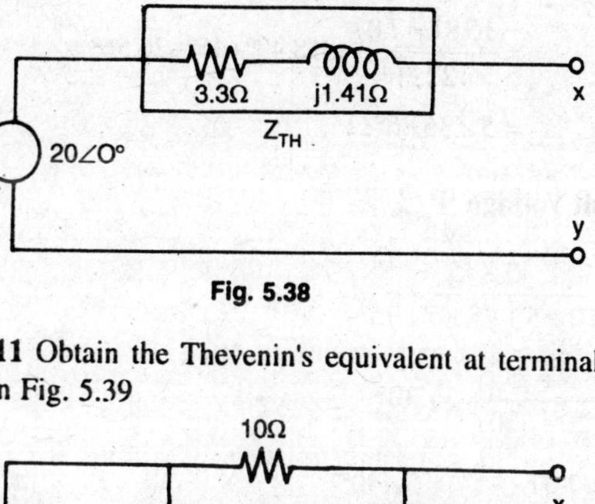

Fig. 5.38

Problem No. 5.11 Obtain the Thevenin's equivalent at terminal xy of the network shown in Fig. 5.39

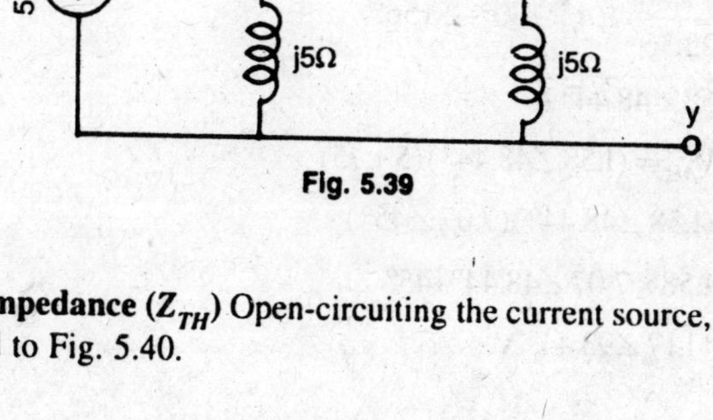

Fig. 5.39

Solution

Thevenin Impedance (Z_{TH}) Open-circuiting the current source, the network is simplified to Fig. 5.40.

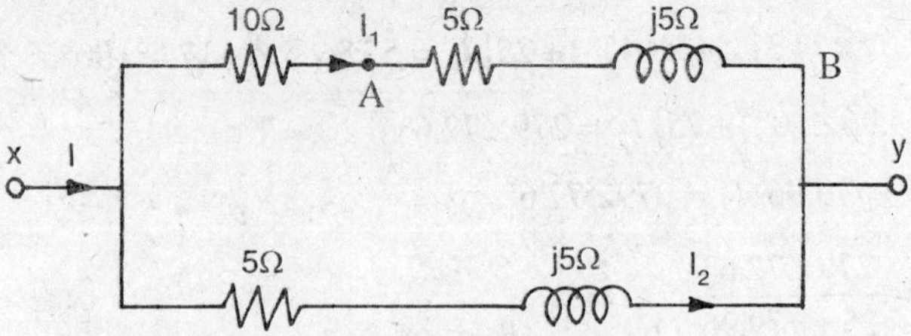

Fig. 5.40

$$Z_{TH} = Z_{xy} = \frac{(5+10+j5)(5+j5)}{5+10+j5+5+j5}$$

$$= \frac{(15+j5)(5+j5)}{20+j10}$$

$$= \frac{15.81\angle 18.43° \times 7.07\angle 45°}{22.36\angle 26.56°}$$

$$= \frac{15.81 \times 7.07}{22.36} \angle 18.43°+45°-26.56°$$

$$= 5\angle 36.86°\,\Omega$$

Open-Circuit Voltage 'V_{oc}'

$$I_1 = \frac{5+j5}{10+5+j5+5+j5} I$$

$$= \frac{5+j5}{20+j10} \times 5\angle 30°$$

$$= \frac{(7.07\angle 45°) \times 5\angle 30°}{22.36\angle 26.56°}$$

$$= \frac{7.07 \times 5}{22.36} \angle 45°+30°-26.56°$$

$$= 1.58\angle 48.44°\ A$$

$$V_{OC} = V_{AB} = (1.58\angle 48.44°)(5+j5)$$

$$= (1.58\angle 48.44°)(7.07\angle 45°)$$

$$= 1.58 \times 7.07 \angle 48.44°+45°$$

$$= 11.17\angle 93.44°\ \text{V}$$

Thevenin's equivalent circuit is shown in Fig. 5.41

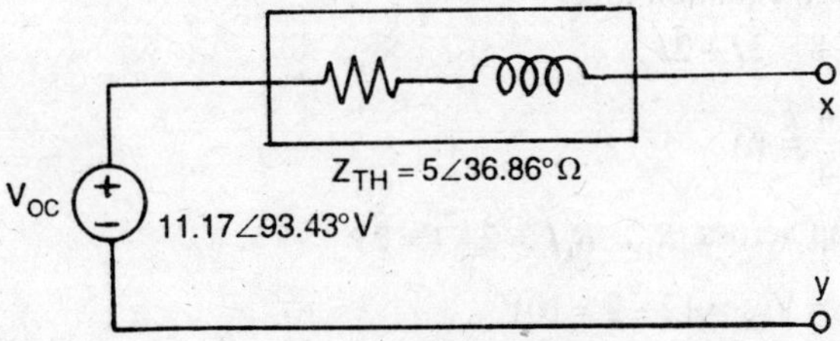

Fig. 5.41

Problem No. 5.12 Find Thevenin's equivalent of the network shown in Fig. 5.42 at terminals A–B. Determine the current through the load resistor of 4Ω connected across terminals A–B.

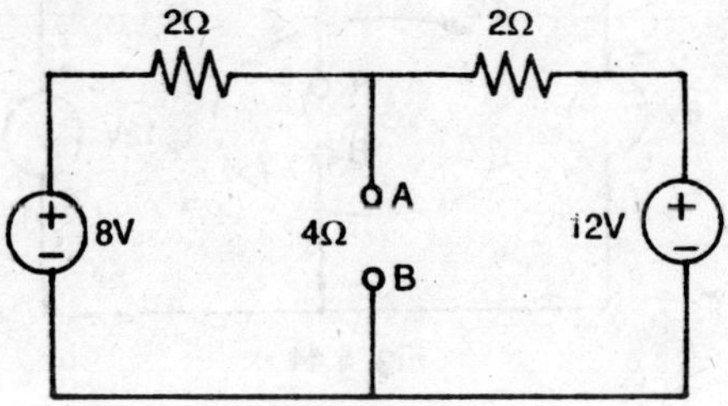

Fig. 5.42

Solution

Thevenin's Impedance (Z_{th})

Removing and short-circuiting the voltage source, the given network can be redrawn as shown in Fig. 5.43.

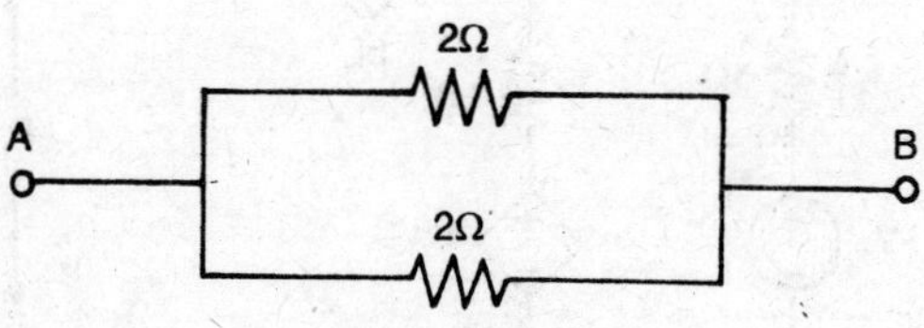

Fig. 5.43

$$Z_{TH} = Z_{AB} = \frac{2 \times 2}{2+2} = 1\Omega$$

Open-Circuit Voltage 'V_{oc}'

Writing mesh equation gives

$$12-8=2I+2I$$

or $$I=\frac{4}{4}=1A$$

Voltage drop across $R_1 = R_1 I = 2\times 1 = 2V$

$$V_{TH}=V_{AB}=12-2=10V$$

Therefore $$I_{AB}=\frac{V_{OC}}{Z_{TH}+R_{AB}}=\frac{10}{1+4}=2A$$

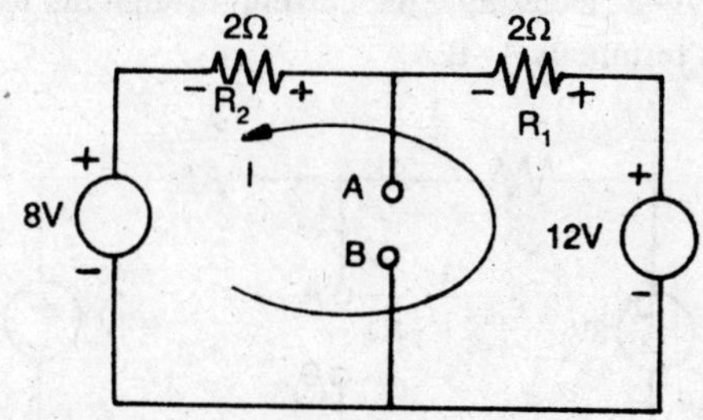

Fig. 5.44

Problem No. 5.13 For the network of Fig. 5.45, show that the equivalent Thevenin's circuit is represented by

$$V_\theta=\frac{V_1}{2}(1+a+b-ab)$$

and $$Z_\theta=\frac{3-b}{2}$$

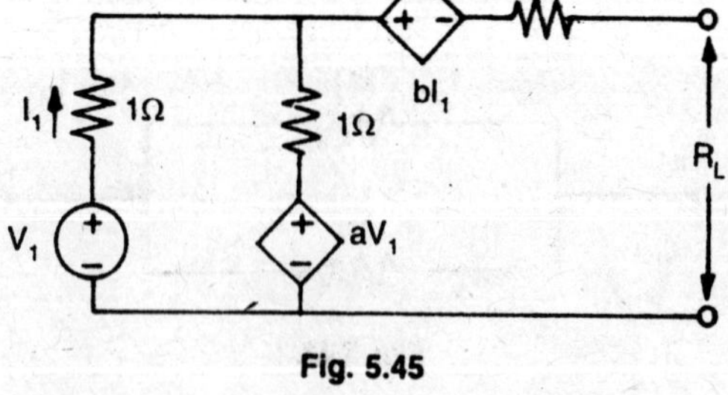

Fig. 5.45

Solution

Let the current distribution be as shown in Fig. 5.46.

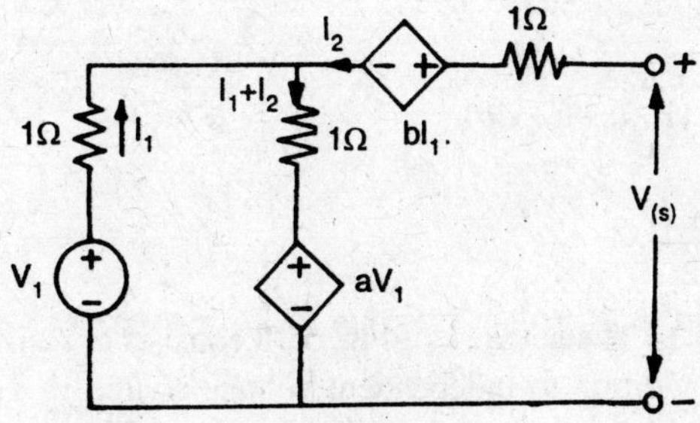

Fig. 5.46

$$V(s) = I_2 \times 1 + bI_1 + (I_1 + I_2) \times 1 + aV_1$$

or $$V(s) = 2I_2 + I_1(1+b) + aV_1 \qquad \text{...(1)}$$

and $$V_1 = I_1 + I_1 + I_2 + aV_1$$

or $$V_1 = 2I_1 + I_2 + aV_1 \qquad \text{...(2)}$$

Multiplying equations (2) by $\frac{1+b}{2}$ and substracting from equation (1), we get

$$V(s) - \frac{V_1(1+b)}{2} = 2I_2 - \frac{I_2(1+b)}{2} + aV_1 - \frac{aV_1(1+b)}{2}$$

or $$2V(s) - V_1(1+b) = 4I_2 - I_2(1+b) + 2aV_1 - aV_1(1+b)$$

or $$2V(s) - V_1(1+b) = I_2(3-b) + 2aV_1 - aV_1 - abV_1$$

or $$2V(s) = V_1(1+a+b-ab) + I_2(3-b)$$

or $$V(s) = \frac{V_1}{2}(1+a+b-ab) + I_2\frac{(3-b)}{2} \qquad \text{...(3)}$$

$$V(s) = V_\theta + I_2 Z_\theta \qquad \text{...(4)}$$

Equation (4) can be represented by a network shown in Fig. 5.47

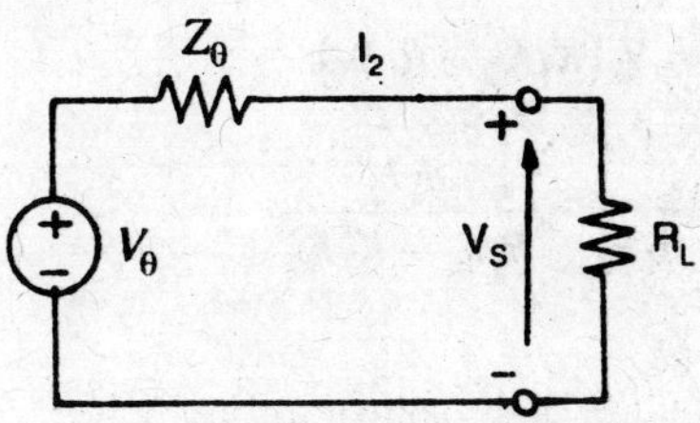

Fig. 5.47

where

$$V_\theta = \frac{V_1}{2}(1+a+b-ab)$$

and $$Z_\theta = \frac{3-b}{2}$$

Problem No. 5.14 The network of Fig. 5.18 consists of resistors and controlled sources in addition to independent voltage source V_s. For this network, find the Thevenin's equivalent network by determining an expression for the voltage V_θ and the Thevenin's equivalent resistance.

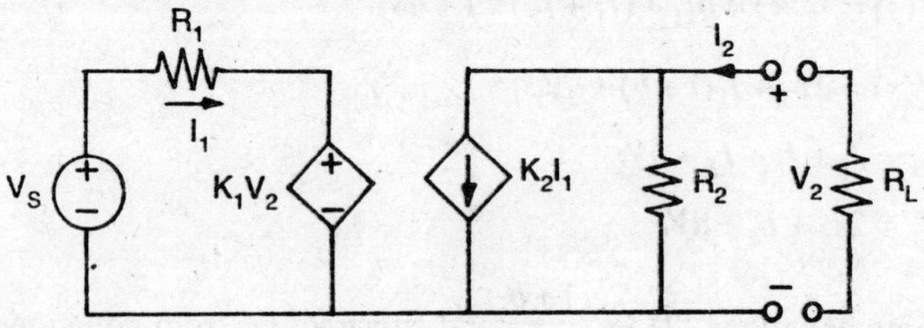

Fig. 5.48

Solution Mesh equations are

$$V_s = I_1R_1 + K_1V_2 \qquad ...(1)$$

$$I_2 = K_2I_1 + R_2V_2 \qquad ...(2)$$

Multiplying equation (1) by $\frac{K_2}{R_1}$ we get

$$\frac{V_sK_2}{R_1} = K_2I_1 + \frac{K_1K_2}{R_1}V_2 \qquad ...(3)$$

Subtracting equation (2) from equation (3), we get

$$\frac{V_sK_2}{R_1} - I_2 = V_2\left(\frac{K_1K_2}{R_1} - R_2\right)$$

or $$V_sK_2 - R_1I_2 = V_2(K_1K_2 - R_1R_2)$$

or $$V_2 = \frac{V_sK_2}{K_1K_2 - R_1R_2} - \frac{R_1I_2}{R_1R_2 - K_1K_2}$$

or $$V_2 = V_\theta + I_Z Z_\theta \qquad ...(4)$$

Equation (4) can be represented by a network shown in Fig. 5.49

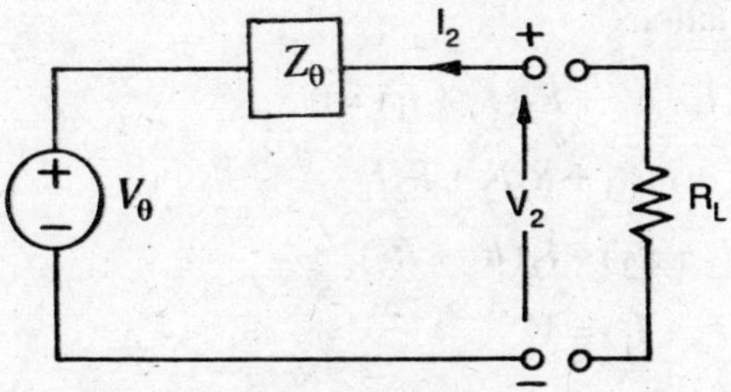

Fig. 5.49

where

$$V_\theta = \frac{V_s K_2}{K_1 K_2 - R_1 R_2}$$

and $$Z_\theta = \frac{R_1}{R_1 R_2 - K_1 K_2}$$

Problem No. 5.15 The network shown in Fig. 5.50 is a simple representation of a transistor. For this network, determine the Thevenin's equivalent network for the Load R_L.

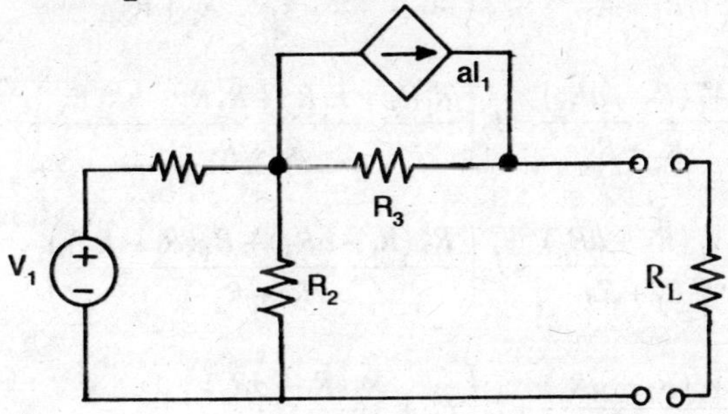

Fig. 5.50

Solution

Converting the current source into voltage source, we get a network shown in Fig. 5.51.

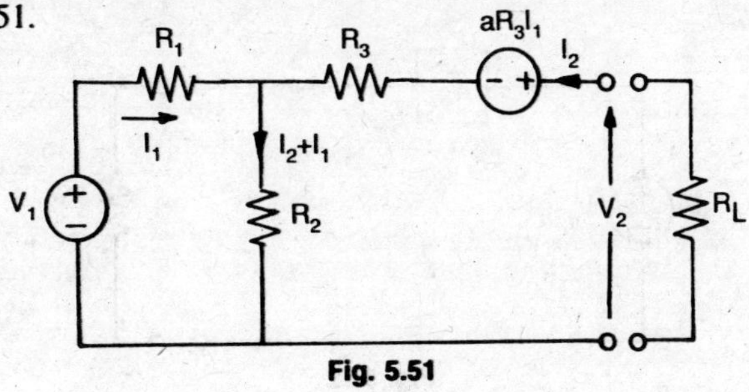

Fig. 5.51

Writing mesh equations

$$I_2R_3 + aR_3I_1 - V_2 + R_2(I_2 + I_1) = 0$$

or $$V_2 = aR_3I_1 + I_2R_3 + R_2I_2 + R_2I_1$$

or $$V_2 = I_1(aR_3 + R_2) + I_2(R_2 + R_3) \quad ...(1)$$

and $$I_1R_1 + R_2(I_2 + I_1) = V_1$$

or $$V_1 = I_1R_1 + I_1R_2 + R_2I_2$$

or $$V_1 = I_1(R_1 + R_2) + R_2I_2 \quad ...(2)$$

or $$I_1 = \frac{V_1 - I_2R_2}{R_1 + R_2} \quad ...(3)$$

Substituting the value of I_1 from equation (3) in (1), we get

$$V_2 = V_1\left(\frac{R_2 + aR_3}{R_1 + R_2}\right) + I_2\left(R_2 + R_3 - \frac{aR_2R_3}{R_1 + R_2} - \frac{R_2^2}{R_1 + R_2}\right)$$

or $$V_2 = \frac{V_1(R_2 + aR_3)}{R_1 + R_2} + I_2\left(\frac{R_1R_2 + R_1R_3 + R_2^2 + R_2R_3 - aR_2R_3 - R_2^2}{R_1 + R_2}\right)$$

or $$V_2 = \frac{V_1(R_2 + aR_3)}{R_1 + R_2} + I_2\left(\frac{R_1R_2 + R_1R_3 + R_2R_3 - aR_2R_3}{R_1 + R_2}\right)$$

or $$V_2 = \frac{V_1(R_2 + aR_3)}{R_1 + R_2} + I_2\left(\frac{R_2(R_1 - aR_3) + R_3(R_1 + R_2)}{R_1 + R_2}\right)$$

or $$V_2 = \frac{V_1(R_2 + aR_3)}{R_1 + R_2} + I_2\left(R_3 + \frac{R_2(R_1 - aR_3)}{R_1 + R_2}\right)$$

or $$V_2 = V_\theta + I_2Z_\theta \quad ...(4)$$

Equation (4) can be represented by a network shown in Fig. 5.52.

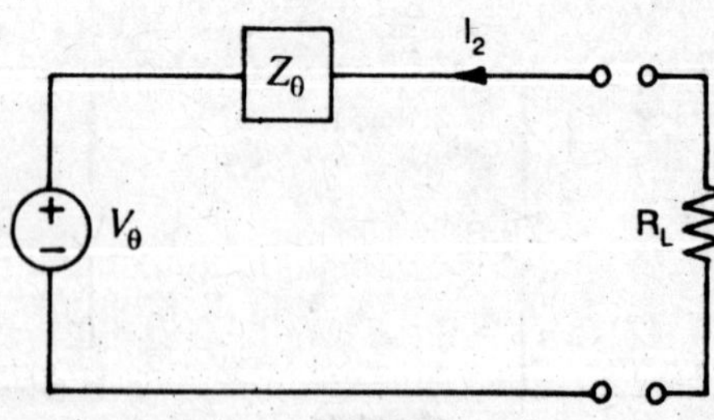

Fig. 5.52

where

$$Z_\theta = R_3 + \frac{R_2(R_1 - aR_3)}{R_1 + R_2}$$

and $$V_\theta = \frac{V_1(R_2 + aR_3)}{R_1 + R_2}$$

Problem No. 5.16 A vaccum tube amplifier circuit with input $e_{(t)}$ is shown in Fig. 5.53. Find the Thevenin's equivalent at cathode resistor R_K and determine the current through R_K.

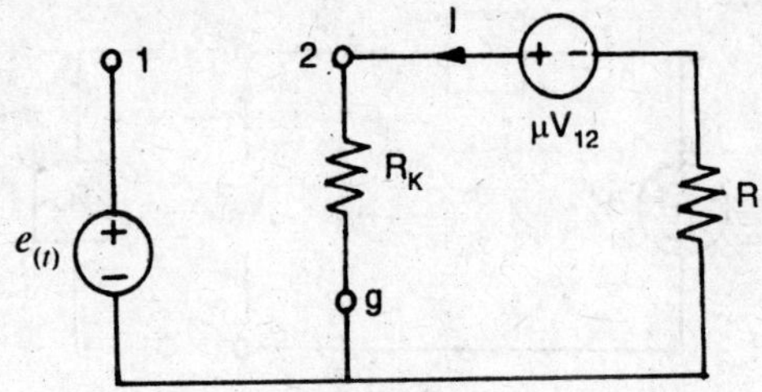

Fig. 5.53

Solution

Let us remove R_K as shown in Fig. 5.54

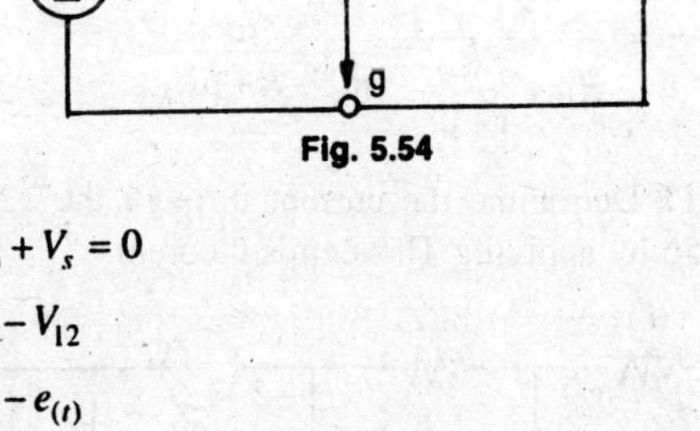

Fig. 5.54

$$V_{12} - e_{(t)} + V_s = 0$$

or $$V_s = e_{(t)} - V_{12}$$

or $$V_{12} = V_s - e_{(t)} \qquad ...(1)$$

Also

$$V_s = IR - \mu V_{12} \qquad ...(2)$$

Substituting value of V_{12} from equation (1) in (2), we get

$$V_s = \mu V_{12} + IR$$

$$V_s = \mu(e_{(t)} - V_s) + IR)$$

or $$V_s(1+\mu) = \mu e_{(t)} + IR$$

or $$V_s = \frac{\mu}{1+\mu} e_{(t)} + I\frac{R}{1+\mu}$$

or $$V_s = V_\theta + IZ_\theta \qquad ...(3)$$

Equation (3) is represented by a network as shown in Fig. 5.55.

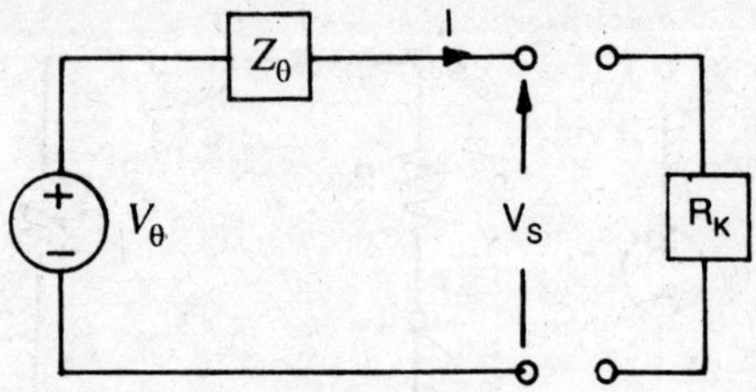

Fig. 5.55

where

$$V_\theta = \frac{\mu e_{(t)}}{1+\mu}$$

and $$Z_\theta = \frac{R}{1+\mu}$$

$$I = \frac{V_\theta}{R_K + Z_\theta} = \frac{\frac{\mu e_{(t)}}{1+\mu}}{R_K + \frac{R}{1+\mu}} = \frac{\mu e_{(t)}}{R_K + R + \mu R_K}$$

Problem No. 5.17 Detremine the current through the 12V source in the circuit of Fig. 5.56 by applying Thevenin's theorem.

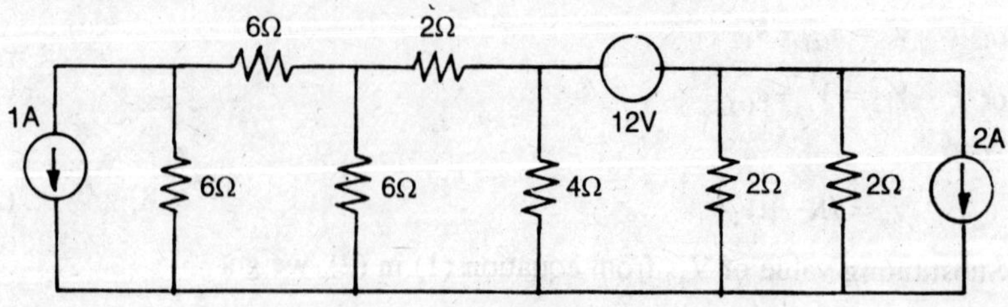

Fig 5.54

Solution

Converting the current sources

(a) 1 A having 6Ω resistance in parallel; and

(b) 2 A having 2Ω resistance in parallel, into voltage sources with resistances in series as shown in Fig. 5.57 and 5.58.

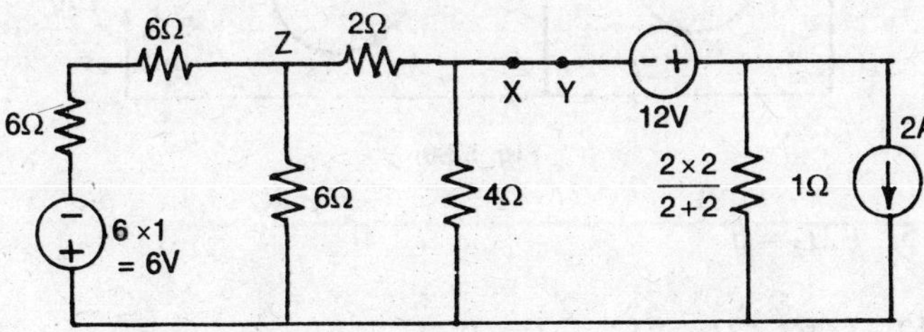

Fig. 5.57

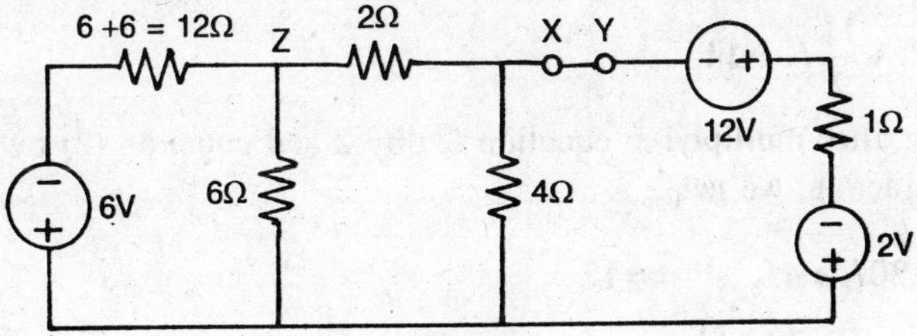

Fig. 5.58

Select two terminals xy as shown in Fig. 5.58.

Open Circuit Voltage (V_{oc})

Redrawing the network after converting the π into star as shown in Fig. 5.59 and 5.60 loop currents I_1 and I_2 are marked and calculated as given below

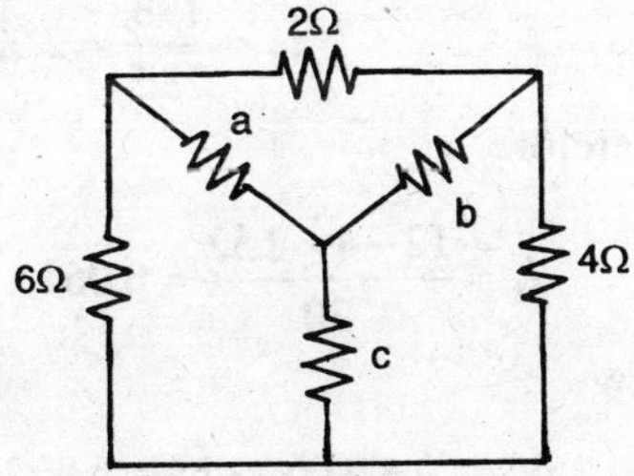

Fig. 5.59

$$a = \frac{6 \times 2}{6+2+4} = \frac{12}{12} = 1\Omega$$

$$b = \frac{4 \times 2}{6+2+4} = \frac{8}{12} = \frac{2}{3}\Omega$$

$$c = \frac{6 \times 4}{6+2+4} = \frac{24}{12} = 2\Omega$$

Mesh equation are

$$6 - 2I_1 - 2I_2 - I_1 - 12I_1 = 0$$

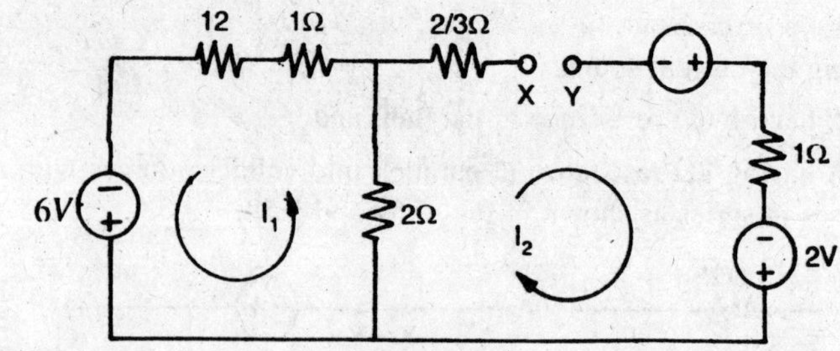

Fig. 5.60

or $$15I_1 + 2I_2 = 6 \quad \text{...(1)}$$

and $$12 - I_2 + 2 - 2I_1 - 2I_2 - \frac{2}{3}I_2 = 0$$

or $$2I_1 + \frac{11}{2}I_2 = 14 \quad \text{...(2)}$$

Rewriting after multiplying equation (1) by 2 and equation (2) by 15, and then subtracting, we get

$$30I_1 + 4I_2 = 12$$

$$30I_1 + \frac{15 \times 11}{2}I_2 = 15 \times 14$$

$$- \quad - \quad -$$

$$78.5I_2 = 198$$

or $$I_2 = \frac{198}{78.5} = 2.52A$$

Therefore,

$$I_1 = \frac{12 - 4 \times 2.52}{30} = 0.064A$$

Now

$$V_A = 6 + 0.064 \times 13 = 6.832V$$

$$V_x = 6.832 - 2.52 \times \frac{2}{3} = 5.15V$$

$$V_y = 6 - 2 - 12 = -8V$$

$$V_{OC} = V_{xy} = 5.15 - (-8) = 13.15V$$

Thevenin's Impedance (Z_{TH}). Short-circuiting the voltage sources of the

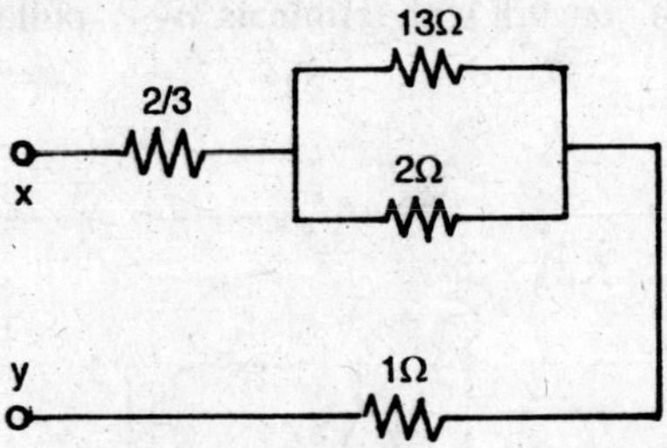

Fig. 5.61

network reduces to Fig. 5.61

$$Z_{TH} = Z_{xy} = \frac{13\times 2}{13+2}+1+\frac{2}{3}$$

$$= \frac{26}{15}+1+\frac{2}{3}$$

$$= \frac{26}{15}+1+\frac{2}{3}$$

$$= \frac{26+15}{15}+\frac{2}{3}$$

$$= 2.73 + 0.66$$

$$= 3.4\Omega$$

Therefore,

$$I_{xy} = \frac{V_{OC}}{Z_{TH}} = \frac{13.15}{3.4} = 3.87A$$

Problem No. 5.18 Network given in Fig. 5.62 contains a controlled source for the element values given with $v_1(t) = u(t)$ and for zero initial conditions, determine the equivalent Thevenin's network at $a - a'$

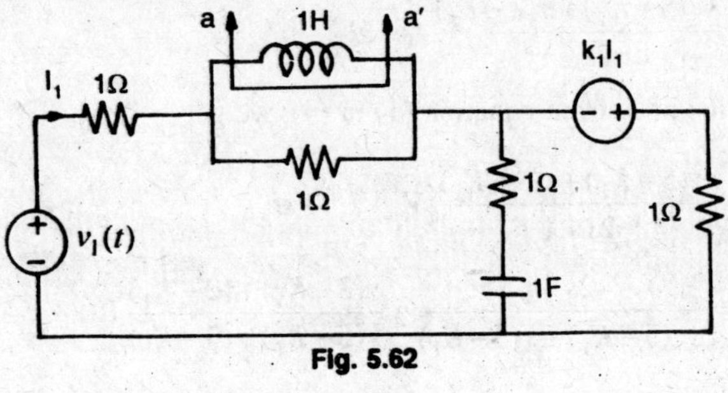

Fig. 5.62

Solution

Re-drawing the given network with terminals $a-a'$ pulled out and element 1 H removed,

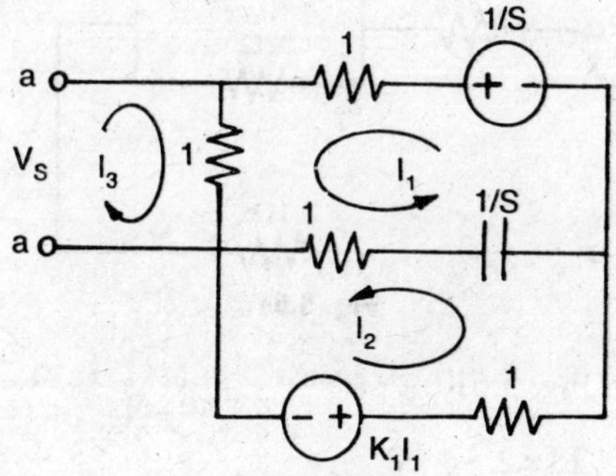

Fig. 5.63

Writing mesh equations for Fig. 5.63, we get

$$V_s = I_1 + I_3 \qquad \text{...(1)}$$

$$\frac{1}{s} = I_1\left(3+\frac{1}{s}\right) - I_2\left(1+\frac{1}{s}\right) + I_3$$

or $$1 = (3s+1)I_1 - (s+1)I_2 + sI_3 \qquad \text{...(2)}$$

and $$K_1 I_1 = I_2\left(2+\frac{1}{s}\right) - I_1\left(1+\frac{1}{s}\right)$$

or $$I_2 = \frac{sK_1 + s + 1}{2s+1} I_1 \qquad \text{...(3)}$$

Putting the value of I_2 from equation (3) in (2), we get

$$1 = (3s+1)I_1 - \frac{(s+1)(sK_1+s+1)}{2s+1} I_1 + sI_3$$

$$1 = \frac{s^2(5-K_1)+s(3-K_1)}{2s+1} I_1 + sI_3 \qquad \text{...(4)}$$

Putting value of I_1 from equation (1) in (4), we get

$$1 = \frac{s^2(5-K_1)+s(3-K_1)}{2s+1}\left[V_s - I_3\right] + sI_3$$

or $$V_s = \frac{2s+1}{s^2(5-K_1)+s(3-K_1)} + \frac{s(3-K_1)+(2-K_1)}{s(5-K_1)+(3-K_1)} I_3 \qquad \text{...(5)}$$

or $V_s = V_\theta + Z_\theta I_3$ ...(6)

Equation (6) can be represented by a network shown in Fig. 5.64

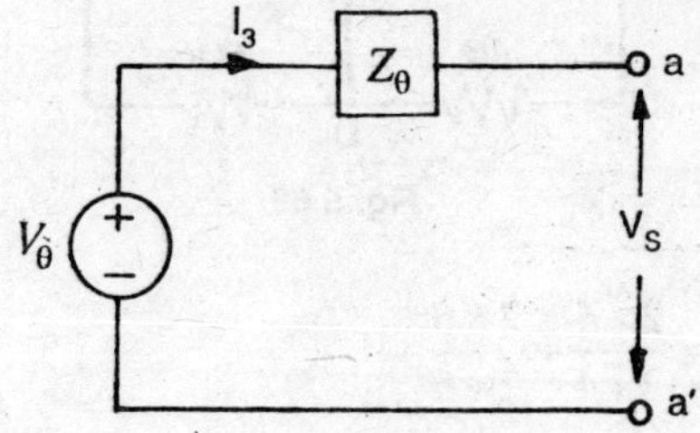

Fig. 5.64

where

$$V_\theta = V_{TH} = \frac{2s+1}{s^2(5-K_1)+s(3-K_1)}$$

and

$$Z_\theta = Z_{TH} = \frac{s(3-K_1)+(2-K_1)}{s(5-K_1)+(3-K_1)}$$

Problem No. 5.19 Determine the current through the ammeter of resistance 2Ω, connected in the wheatstone bridge at Fig. 5.65.

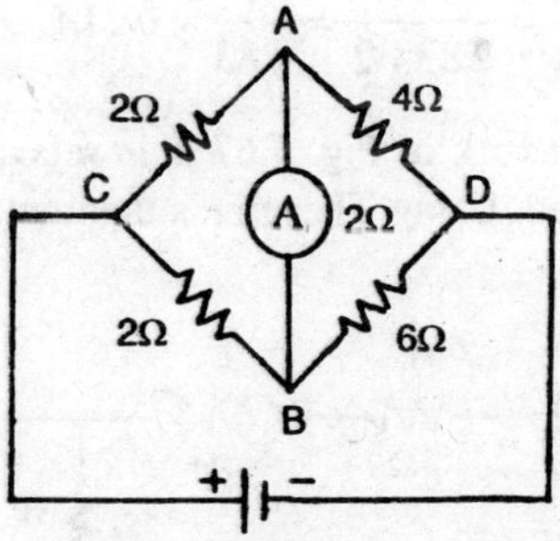

Fig. 5.65

Solution Remove ammeter between A and B and short circuit the voltage source 12V as shown in Fig. 5.66.

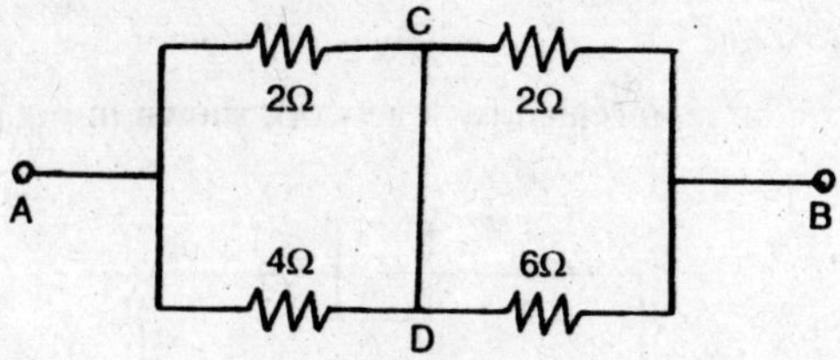

Fig. 5.66

$$Z_{TH} = Z_{AB} = \frac{2\times4}{2+4} + \frac{2\times6}{2+6}$$

$$= \frac{8}{6} + \frac{12}{8}$$

$$= 2.83\Omega$$

Open-Circuit Voltage 'V_{OC}' Refer Fig. 5.66

Potential of A w.r.t $D = 12\times\frac{4}{6} = 8V$

Potential of B w.r.t $D = \frac{12\times6}{8} = 9V$

Therefore,

$$V_{TH} = V_{OC} = 9 - 8 = 1V$$

$$I_2 = \frac{V_{OC}}{Z_{TH} + R_{AB}} = \frac{1}{2.83+2} = \frac{1}{4.83} = 0.21A$$

Problem No. 5.20 Network of Fig. 5.67 is in relaxed state until time $t = 0$ when switch K is closed. Using Thevenin's theorem, find the current in the resistor R_3.

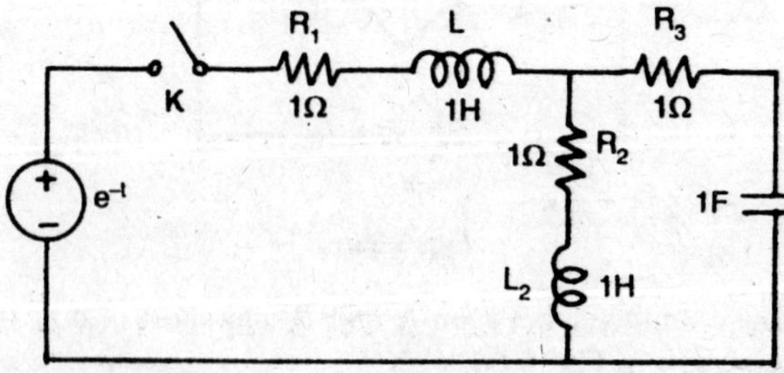

Fig. 5.67

Solution

Fig. 5.68 shows the transformed network of Fig. 5.67

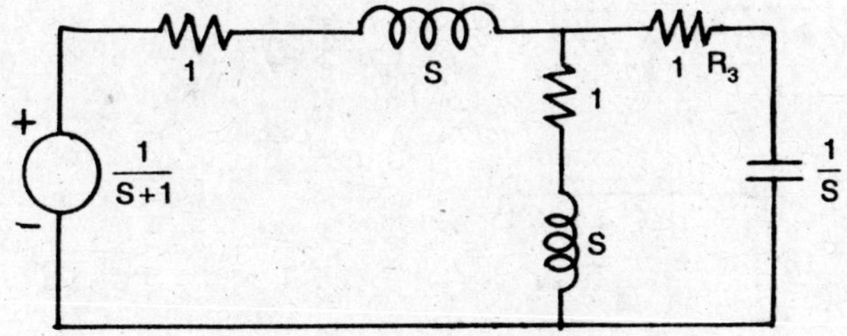

Fig. 5.68

Fig. 5.68 is further simplified to as shown in Fig. 5.69

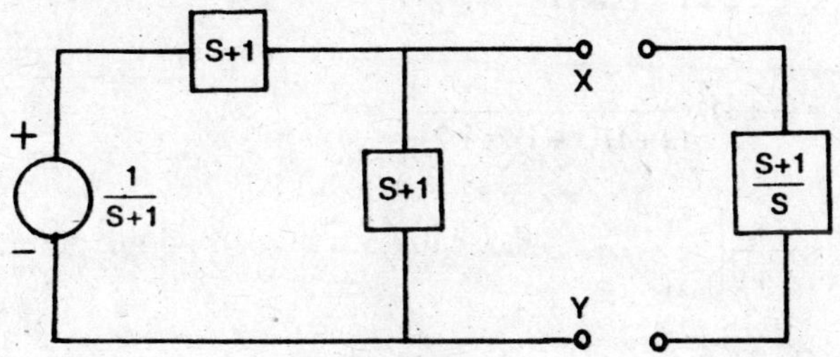

Fig. 5.69

Thevenin's impedance 'Z_{TH}'

$$Z_{TH} = Z_{xy} = \frac{(s+1)(s+1)}{(s+1)+(s+1)} = \frac{(s+1)(s+1)}{2(s+1)} = \frac{(s+1)}{2}$$

Open-Circuit Voltage 'V_{oc}'

$$V_{OC} = \frac{\frac{1}{s+1}\times(s+1)}{(s+1)+(s+1)} = \frac{1}{2(s+1)}$$

Current through R_3 is

$$I_{R3}(s) = \frac{V_{OC}}{Z_{TH}+Z_L}$$

$$= \frac{\frac{1}{2(s+1)}}{\frac{s+1}{2}+\frac{s+1}{s}}$$

$$= \frac{\frac{1}{2(s+1)}}{\frac{s^2+s+2s+2}{2s}}$$

$$= \frac{s}{(s+1)(s^2+3s+2)}$$

$$= \frac{s}{(s+1)(s+1)(s+2)}$$

$$= \frac{A}{s+1} + \frac{B}{(s+1)^2} + \frac{C}{s+2}$$

$$B = (s+1)^2 \left.\frac{s}{(s+1)(s+1)(s+2)}\right|_{s=-1}$$

$$= \left.\frac{s}{s+2}\right|_{s=-1}$$

$$= \frac{-1}{-1+2} = -1$$

$$C = (s+2)\left.\frac{s}{(s+1)^2(s+2)}\right|_{s=-2}$$

$$= \frac{-2}{(-2+1)^2} = -2$$

$$A = \left.\left\{\frac{d}{ds}\left[(s+1)^2 \frac{s}{(s+1)^2(s+2)}\right]\right\}\right|_{s=-1}$$

$$= \left.\left[\frac{d}{ds}\left(\frac{s}{s+2}\right)\right]\right|_{s=-1}$$

$$= \left.\frac{(s+2)\times 1 - s\times 1}{(s+2)^2}\right|_{s=-1}$$

$$= \left.\frac{s+2-s}{(s+2)^2}\right|_{s=-1}$$

$$= \left.\frac{2}{(s+2)^2}\right|_{s=-1}$$

$$= \frac{2}{(-1+2)^2} = 2$$

Therefore,

$$I_{R3}(s) = \frac{2}{s+1} + \frac{-1}{(s+1)^2} + \frac{-2}{s+2}$$

or $$i_{R3}(t) = \left(2e^{-t} - te^{-t} - 2e^{-2t}\right)u(t)$$

Problem No. 5.21 In the network of Fig. 5.70, capacitor C_1 is charged to 8 volts in the polarity shown. Capacitor C_2 is initially uncharged. At time $t = 0$, switch K is closed. Using Thevenin's theorem, find the transform current in resistor R_2.

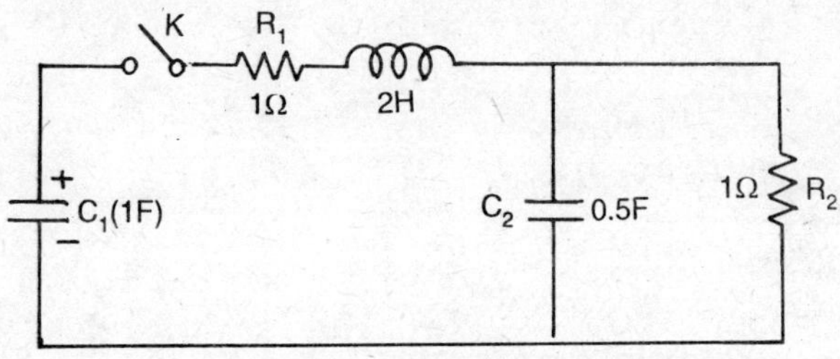

Fig. 5.70

Solution

Removing resistor R_2, the transform network of Fig. 5.70 is shown in Fig. 5.71

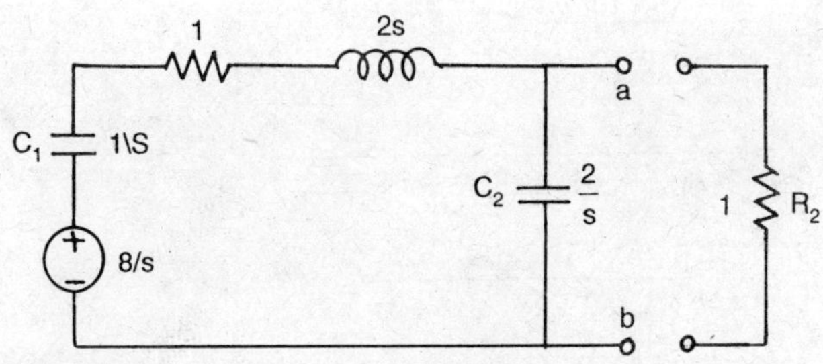

Fig. 5.71

Thevenin's Impedance (Z_{TH}) With voltage source short-circuited

$$Z_{TH} = Z_{ab} = \frac{\left(1+2s+\frac{1}{s}\right)\times\frac{2}{s}}{1+2s+\frac{1}{s}+\frac{2}{s}}$$

$$= \frac{2(2s^2+s+1)}{s(2s^2+s+3)}$$

Open-Circuit Voltage $V_{oc}(s)$

$$V_{OC}(s) = \frac{\frac{8}{s}\times\frac{2}{s}}{\frac{2}{s}+\frac{2s}{1}+\frac{1}{s}+1} = \frac{16/s^2}{\frac{2+2s^2+1+s}{s}}$$

$$= \frac{16/s^2}{2s^2+s+3/s}$$

$$= \frac{16}{s(2s^2+s+3)}$$

Current in Resistor 'R_2'

$$I_{R2}(s) = \frac{V_{OC(s)}}{Z_{TH}+R_2}$$

$$= \frac{\frac{16}{s(2s^2+s+3)}}{\frac{2(2s^2+s+1)}{s(2s^2+s+3)}+1}$$

$$= \frac{16}{2(2s^2+s+1)+s(2s^2+s+3)}$$

$$= \frac{16}{4s^2+2s+2+2s^3+s^2+3s}$$

$$= \frac{16}{2s^3+6s^2+5s+2}$$

Problem No. 5.23 In the network of Fig. 5.72, there is no initial voltage on the capacitor and no initial current across the inductor. Switch K is closed at $t = 0$. Using Thevenin's theorem find the current in the inductor L_2.

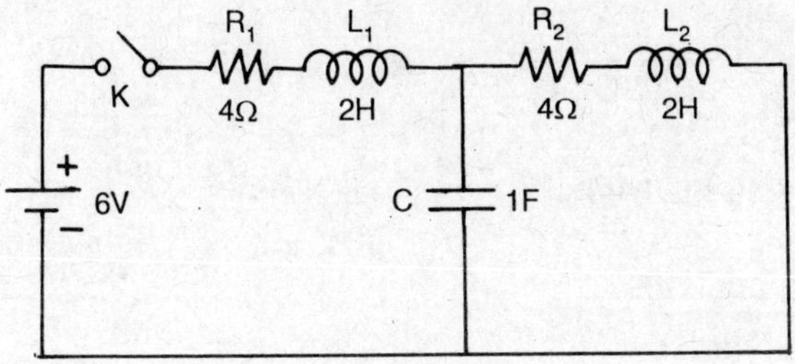

Fig. 5.72

Solution The transform network is shown in Fig. 5.73

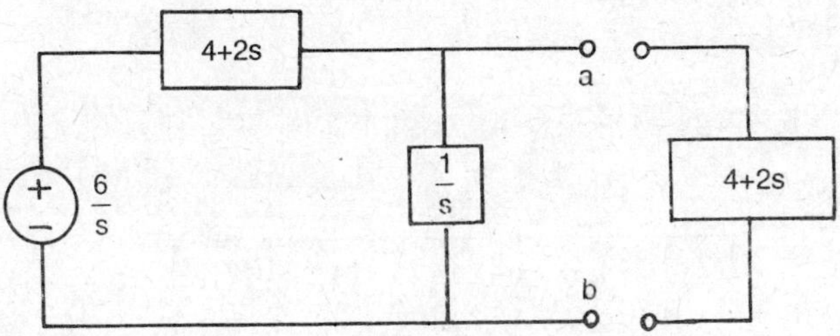

Fig. 5.73

Thevenin's Impedance (Z_{TH}) With voltage source short-circuited

$$Z_{TH} = Z_{ab} = \frac{(4+2s)\frac{1}{s}}{4+2s+\frac{1}{s}}$$

$$= \frac{2s+4}{2s^2+4s+1}$$

Open-Circuit Voltage '$V_{OC}(s)$'

$$V_{OC}(s) = \frac{\frac{6}{s}\times\frac{1}{s}}{4+2s+\frac{1}{s}}$$

$$= \frac{6/s^2}{\dfrac{4s+2s^2+1}{s}}$$

$$= \frac{6}{s(2s^2+4s+1)}$$

Current through Inductor

$$I_L(s) = \frac{V_{OC}(s)}{Z_{TH}+(2s+4)}$$

$$= \frac{\dfrac{6}{s(2s^2+4s+1)}}{\dfrac{2s+4}{2s^2+4s+1}+(2s+4)}$$

$$= \frac{6}{s\left\{(2s+4)+(2s+4)(2s^2+4s+1)\right\}}$$

$$= \frac{6}{s\left\{(2s+4)+4s^3+8s^2+2s+8s^2+16s+4\right\}}$$

$$= \frac{6}{s\left\{4s^3+16s^2+20s+8\right\}}$$

$$= \frac{6}{4s(s^3+4s^2+5s+2)}$$

$$= \frac{3/2}{s(s+2)(s+1)^2}$$

$$= \frac{3}{2}\left[\frac{A}{s}+\frac{B}{s+2}+\frac{C}{s+1}+\frac{D}{(s+1)^2}\right]$$

$$A = \left.\frac{1}{(s+2)(s+1)^2}\right|_{s=0} = \frac{1}{2}$$

$$B = \left.\frac{1}{s(s+1)^2}\right|_{s=-2} = -\frac{1}{2}$$

$$D = \left.\frac{1}{s(s+2)}\right|_{s=-1} = -1$$

$$C = \left\{\frac{d}{ds}\left[\frac{(s+1)^2}{(s+1)^2 \times s(s+2)}\right]\right\}\Bigg|_{s=-1}$$

$$= \left[\frac{d}{ds}\left\{\frac{1}{s(s+2)}\right\}\right]_{s=-1}$$

$$= \left[\frac{d}{ds}\left\{\frac{1}{s^2+2s)}\right\}\right]_{s=-1}$$

$$= \left.\frac{(s^2+2s)\times 0 - 1\times(2s+2)}{(s^2+2s)^2}\right|_{s=-1}$$

$$= \left.\frac{-(2s+2)}{(s^2+2s)^2}\right|_{s=-1} = 0$$

Therefore,

$$I_L(s) = \frac{3}{2}\left[\frac{1/2}{s} + \frac{-1/2}{s+2} + \frac{-1}{(s+1)^2}\right]$$

or $$i_L(t) = \frac{3}{2}\left[\frac{1}{2} - \frac{1}{2}e^{-2t} \quad te^{-t}\right]u(t)$$

$$= \frac{3}{4}\left[1 - e^{-2t} - 2te^{-t}\right]u(t)$$

Problem No. 5.22 Determine the current expression $i_2(t)$ in the network of Fig. 5.74 when switch is closed at $t = 0$. The inductor initially is deenergised.

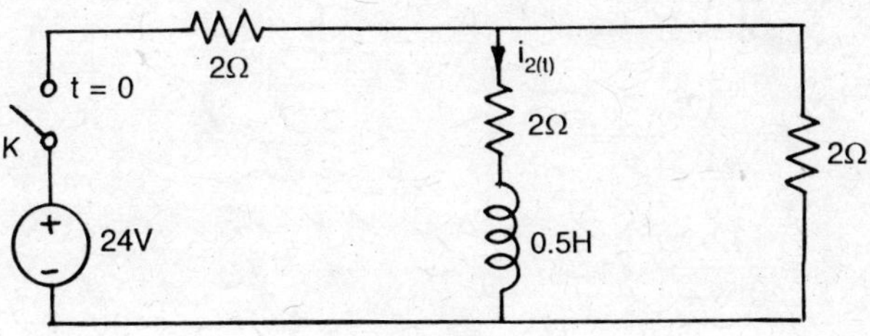

Fig. 5.74

Solution Removing the branch elements in which $i_2(t)$ is flowing, the transform network is shown in Fig. 5.75.

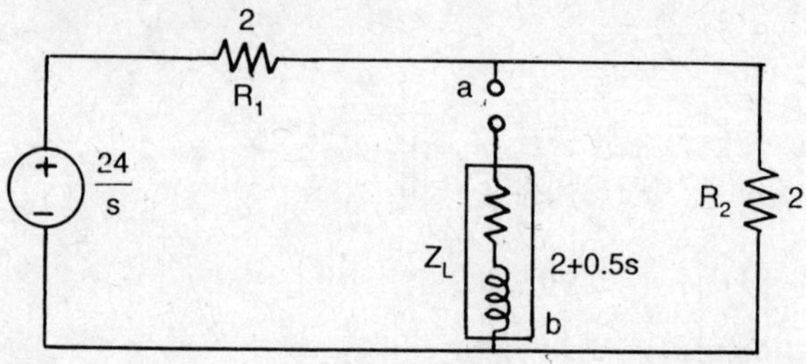

Fig. 5.75

Thevenin's Impedance 'Z_{TH}'

With voltage source short-circuited

$$Z_{TH} = Z_{ab} = \frac{R_1 R_2}{R_1 + R_2} = \frac{2 \times 2}{2+2} = \frac{4}{4} = 1\Omega$$

Open-Circuit Voltage ($V_{oc}(s)$)

$$V_{OC}(s) = V_{ab} = \frac{24}{s} - \left[\frac{24/s}{2+2}\right] \times R_1$$

$$= \frac{24}{s} - \left[\frac{24/s}{4}\right] \times 2$$

$$= \frac{24}{s} - \frac{12}{s}$$

$$= \frac{12}{s}$$

$$I_2(s) = \frac{V_{OC}(s)}{Z_L + Z_{TH}}$$

$$= \frac{12/s}{2 + 0.5s + 1}$$

$$= \frac{12/s}{0.5s + 3}$$

$$= \frac{24}{s(s+6)}$$

$$= 24\left(\frac{A}{s} + \frac{B}{s+6}\right)$$

$$A = \left.\frac{1}{s+6}\right|_{s=0} = \frac{1}{6}$$

$$B = \left.\frac{1}{s}\right|_{s=-6} = -\frac{1}{6}$$

Therefore,

$$I_2(s) = 24\left(\frac{1/6}{s} + \frac{-1/6}{s+6}\right)$$

$$= 4\left(\frac{1}{s} - \frac{1}{s+6}\right)$$

or $\quad i_2(t) = 4(1 - e^{-6t})$

Problem No. 5.23 In the network of Fig. 5.76, switch K is moved from position 1 to 2 at t = 0, after steady state is reached. With the help of Thevenin's theorem, determine current through inductor L.

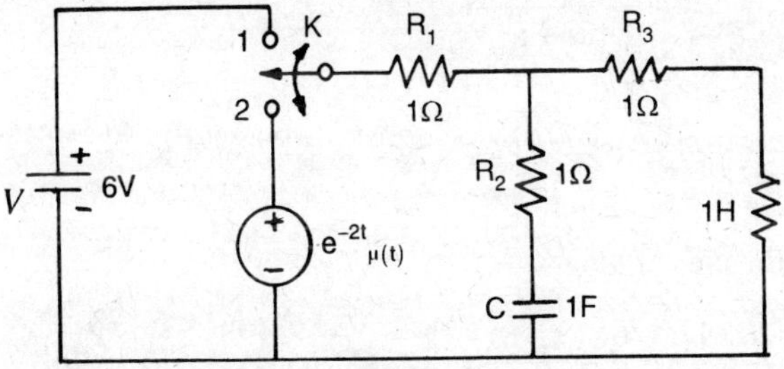

Fig. 5.76

Solution

Before t = 0 The switch K is in position 1 and steady state is reached. Capacitor C gets changed fully to 6V and no current flows through it. Current through inductor L

$$i_{L(0-)} = \frac{V}{R_1 + R_3} = \frac{6}{1+1} = 3A$$

At t = 0 Transform of the network is shown in Fig. 5.77.

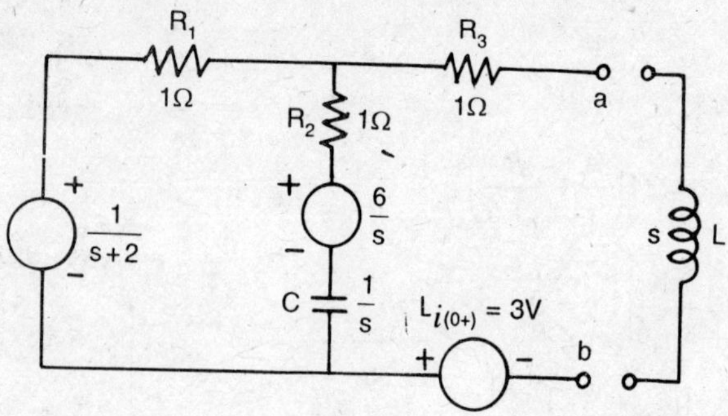

Fig. 5.77

Thevenin's Impedance 'Z_{TH}'

With L removed and voltage sources short-circuited

$$Z_{TH} = Z_{ab} = 1 + \frac{1 \times \left(1 + \frac{1}{s}\right)}{1 + 1 + \frac{1}{s}}$$

$$= 1 + \frac{s+1}{2s+1}$$

$$= \frac{2s+1+s+1}{2s+1}$$

$$= \frac{3s+2}{2s+1}$$

$$Z_s = s$$

Voltage in the closed loop

$$V(s) = \frac{1}{s+2} - \frac{6}{s} = \frac{s-6s-12}{s(s+2)}$$

$$= \frac{-5s-12}{s(s+2)}$$

$$= -\frac{5s+12}{s(s+2)}$$

Current in closed loop

$$I(s) = \frac{V(s)}{Z(s)}$$

Impedance of closed loop responsible for current $I(s)$

$$Z(s) = 1 + 1 + \frac{1}{s}$$

$$= \frac{2s+1}{s}$$

Therefore,

$$I(s) = \frac{-\dfrac{5s+12}{s(s+2)}}{\dfrac{2s+1}{s}}$$

$$= \frac{-(5s+12)}{(s+2)(2s+1)}$$

Open-Circuit Voltage 'V_{OC}'

$$V_{OC}(s) = V_{ab} = 3 + \frac{1}{s+2} - R_1 V(s)$$

$$= 3 + \frac{1}{s+2} - 1\left\{\frac{-(5s+12)}{s(s+2)}\right\}$$

$$= \frac{3s+6+1}{s+2} + \frac{5s+12}{s(s+2)}$$

$$= \frac{3s+7}{s+2} + \frac{5s+12}{s(s+2)}$$

$$= \frac{3s^2+7s+5s+12}{s(s+2)}$$

$$= \frac{3s^2+12s+12}{s(s+2)}$$

Therefore,

$$I_L(s) = \frac{V_{OC}(s)}{Z_{TH}+Z_L}$$

$$= \frac{\dfrac{3s^2+12s+12}{s(s+2)}}{\dfrac{3s+2}{2s+1}+s}$$

$$= \frac{\dfrac{3s^2+12s+12}{s(s+2)}}{\dfrac{3s+2+12s+6}{2s+1}}$$

$$= \frac{(3s^2+12s+12)(15s+8)}{s(s+2)(2s+1)}$$

Problem No. 5.24 The Capacitor C in the network of Fig. 5.75 is intially charged to 2Volts in the polarity shown. Switch K is closed at time $t = 0$. Using Thevenin's theorem find the current $i_{(t)}$ in the resistor R_3.

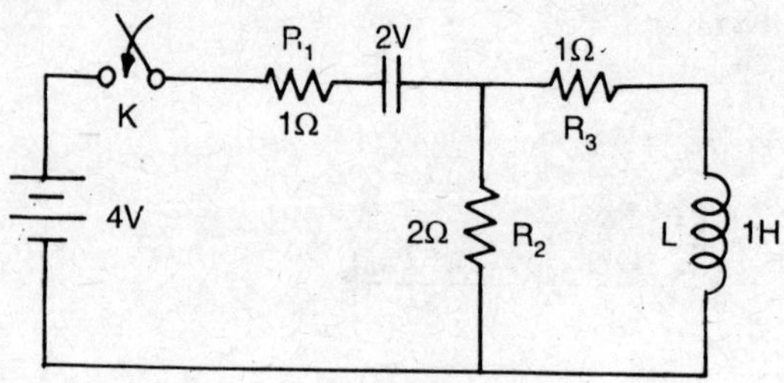

Fig. 5.78

Solution

(a) **Before $t = 0$** Current through the inductor

$$i_{L(0^-)} = \frac{V_C}{R_3} = \frac{2}{1} = 2A$$

At $t = 0$ The transform network is shown in Fig. 5.79.

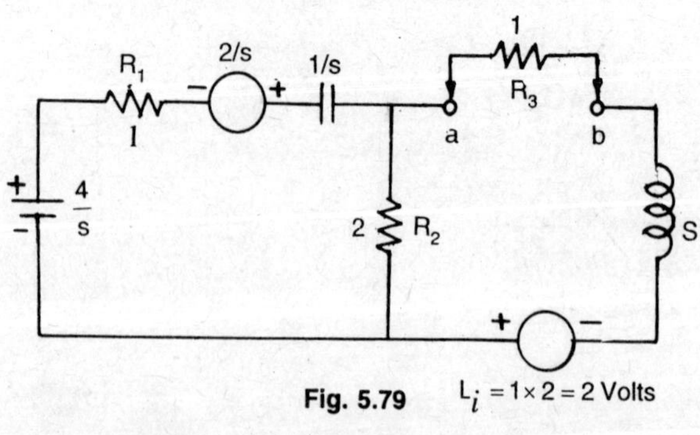

Fig. 5.79

Thevenin's Impedance (Z_{TH}) With voltage source short-circuited and R_3 removed

$$Z_{TH} = Z_{ab} = s + \frac{\left(1+\frac{1}{s}\right)\times 2}{1+\frac{1}{s}+2} = \frac{2(s+1)}{3s+1} + s = \frac{3s^2+3s+2}{3s+1}$$

Closed-loop driving Voltage ($V(s)$)

$$V(s) = \frac{4}{s} + \frac{2}{s} = \frac{6}{s}$$

Closed-loop Impedance ($Z(s)$)

$$Z(s) = 1 + \frac{1}{s} + 2 = \frac{3s+1}{s}$$

Closed-loop current ($I(s)$)

$$I(s) = \frac{V(s)}{Z(s)} = \frac{6/s}{\frac{3s+1}{s}} = \frac{6}{3s+1}$$

Open-Circuit Voltage (V_{oc})

$$V_{OC} = V_{ab} = R_3 I(s)$$

$$= \frac{2\times 6}{3s+1} = \frac{12}{3s+1}$$

Therefore,

$$I_{R3} = \frac{V_{OC}}{Z_{TH} + R_3}$$

$$= \frac{\frac{12}{3s+1}}{\frac{3s^2+3s+2}{3s+1} + 1}$$

$$= \frac{12}{3s^2+6s+3}$$

$$= \frac{4}{s^2 + 2s + 1}$$

$$= \frac{4}{(s+1)^2}$$

Therefore,

$$i_{R3}(t) = 4te^{-t}u(t) \qquad \text{...(1)}$$

Let us check the result by finding out current in the inductor.

$$Z_{TH} = 1 + \frac{\left(1+\frac{1}{s}\right)^2}{1+\frac{1}{s}+2}$$

$$= 1 + \frac{2(s+1)}{3s+1}$$

$$= \frac{3s+1+2s+2}{3s+1} = \frac{5s+3}{3s+1}$$

Therefore,

$$I_L = \frac{V_{OC}}{Z_{TH} + X_L} = \frac{\frac{6}{s} \times 2}{\frac{5s+3}{3s+1} + s}$$

$$= \frac{12}{5s+3+3s^2+s}$$

$$= \frac{12}{3s^2+6s+3}$$

$$= \frac{4}{(s+1)^2}$$

or $\quad i_L(t) = 4te^{-t}$ ...(2)

which is same as equation (1) because resistor R_3 and inductor L are in series connected in the same branch of the network.

Problem No. 5.25 Obtain Thevenin's equivalent of the circuit shown in Fig. 5.80, at terminals AB.

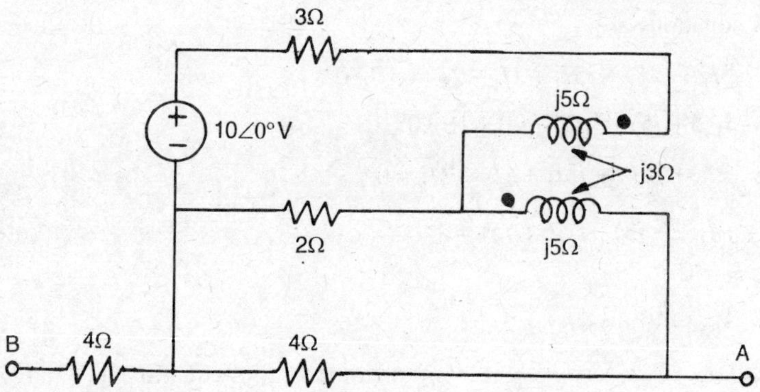

Fig. 5.80

Solution

Open Circuit Voltage 'V_{oc}'

V_{OC} or V_{ab} is obtained by writing mesh equations. Equivalent circuit of the network is shown in Fig. 5.81.

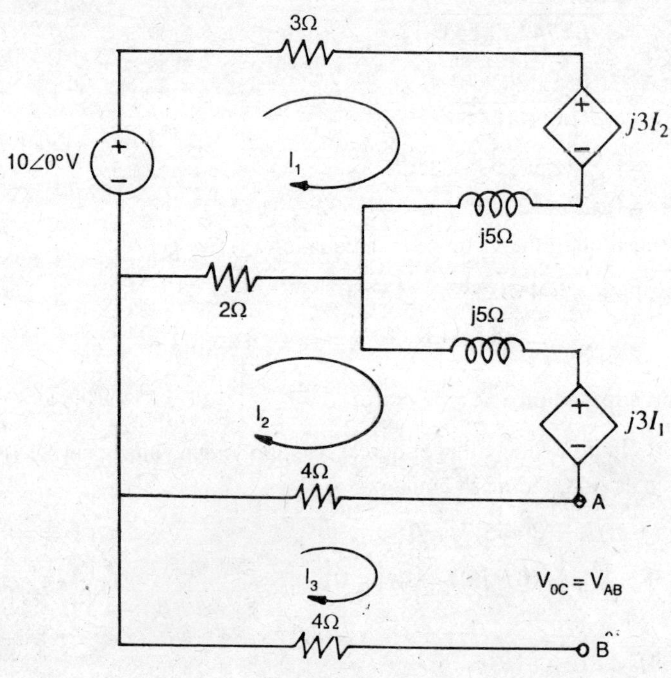

Fig. 5.81

Mesh equations are

$$3I_1 + j3I_2 + j5I_1 + 2I_1 - 2I_2 = 10\angle 0°$$

or $$I_1(5 + j5) - I_2(2 - j3) = 10\angle 0° \qquad ...(1)$$

and, $$2I_2 + j5I_2 + j3I_1 + 4I_2 - 2I_1 = 0$$

or $$-I_1(2 - j3) + I_2(6 + j5) = 0 \qquad ...(2)$$

Since $I_3 = 0$

$$V_{OC} = V_{AB} = 4I_2 \qquad ...(3)$$

Let us now solve for I_2 to get 'V_{OC}' with the help of equations (1) and (2).

$$I_2 = \begin{bmatrix} \dfrac{\begin{matrix} 5+j5 & 10 \\ -2+j3 & 0 \end{matrix}}{\begin{matrix} 5+j5 & -2+j3 \\ -2+j3 & 6+j5 \end{matrix}} \end{bmatrix} = \frac{20 - j30}{10 + j67}$$

$$= \frac{36.055\angle 360° - 56.31°}{67.742\angle 81.51°}$$

$$= \frac{36.055\angle 303.69°}{67.742\angle 81.51°}$$

$$= 0.532\angle 303.69° - 81.51°$$

$$= 0.532\angle 222.18° \quad \text{or} \quad = 0.532\angle -137.80°$$

Hence, substituting the value of I_2 in equation (3), we get

$$V_{OC} = V_{AB} = 4 \times 0.532\angle -137.80°$$

$$= 21.28\angle -137.80° \, V$$

Thevenin's Impedance 'Z_{TH}'

Short-circuiting the voltage sources and applying a voltage say $1\angle 0°$ at the terminals A-B, the mesh equations are

$$(5 + j5)I_1 - (2 - j3)I_2 = 0 \qquad ...(4)$$

$$(-2 + j3)I_1 + (6 - j5)I_2 - 4I_3 = 0 \qquad ...(5)$$

and

$$-4I_2 + 8I_3 = -1\angle 0° \qquad ...(6)$$

Substituting value of I_3 from equation (6) which is

$$I_3 = \frac{-1}{8}\angle 0° + \frac{1}{2} I_2; \qquad ...(7)$$

in equation (2), we get

$$(-2+j3)I_1 + (6+j5)I_2 - 4\left(-\frac{1}{8}\angle 0° + \frac{1}{2}I_2\right) = 0$$

or $$(-2+j3)I_1 + (6+j5)I_2 - 2I_2 = -\frac{1}{2}\angle 0°$$

or $$(-2+j3)I_1 + (4+j5)I_2 = -\frac{1}{2}\angle 0° \qquad ...(8)$$

Rewriting equations (4) and (8) again

$$(5+j5)I_1 + (-2+j3)I_2 = 0$$

and $$(-2+j3)I_1 + (4+j5)I_2 = -0.5\angle 0°$$

Solving for I_2

$$I_2 = \left[\frac{\begin{matrix} 5+j5 & 0 \\ -2+j3 & -0.5 \end{matrix}}{\begin{matrix} 5+j5 & -2+j3 \\ -2+j3 & 4+j5 \end{matrix}}\right] = \frac{0.5(5+j5)}{(5+j5)(4+j5)-(-2+j3)^2}$$

$$= \frac{-2.5-j2.5}{20+j25+j20-25-\left((j3)^2+2^2-2\times j3\times 2\right)}$$

$$= \frac{-2.5-j2.5}{-5+j45+9-4+j12}$$

$$= \frac{-2.5-j2.5}{+j57}$$

$$= \frac{(-2.5-j2.5)\times j}{(j57)\times j}$$

$$= \frac{2.5-j2.5}{-57}$$

$$= -0.044 + j0.044$$

$$= 0.062\angle 135°$$

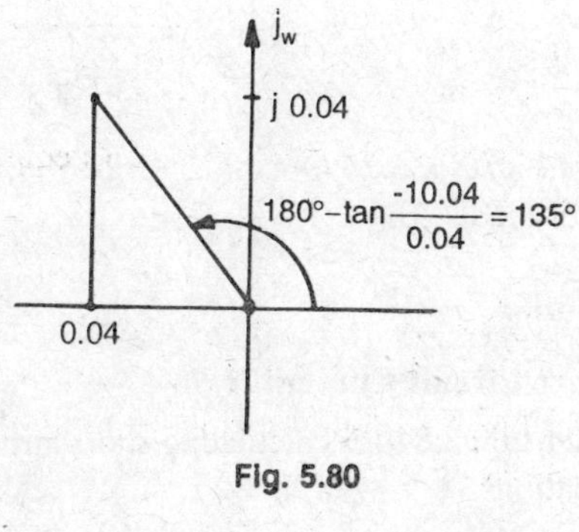

Fig. 5.80

From equation (7), we have

$$I_3 = -\frac{1}{8}\angle 0° + \frac{1}{2} I_2$$

$$= -\frac{1}{8}\angle 0° + \frac{1}{2}[-0.044 + j0.044]$$

$$= -0.125 - 0.022 + j0.022$$

$$= -0.147 + j0.022$$

$$Z_{TH} = \frac{-1\angle 0°}{I_3} = \frac{-1 + j0}{-0.147 + j0.022}$$

$$= \frac{1}{0.147 - j0.022}$$

$$= \frac{0.147 + j0.022}{0.147^2 + 0.022^2}$$

$$= \frac{0.147 + j0.022}{0.022}$$

$$= 6.654 + j$$

$$= 6.73\angle 8.55°\,\Omega$$

Thevenin's equivalent network is shown in Fig. 5.83.

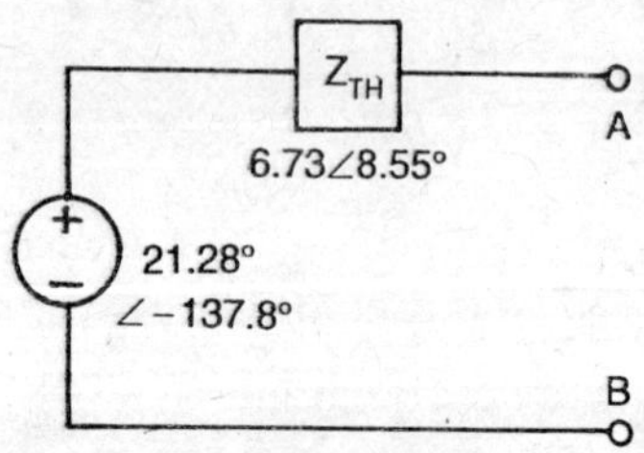

Fig. 5.83

Problem No. 5.26 Obtain Norton's equivalent circuit for Problem No. 5.16, at terminals AB.

Solution

Short Circuit Current 'I_{sc}'

Refer Fig. 5.84 for calculating short-circuit current I_{sc} with the help of mesh equations.

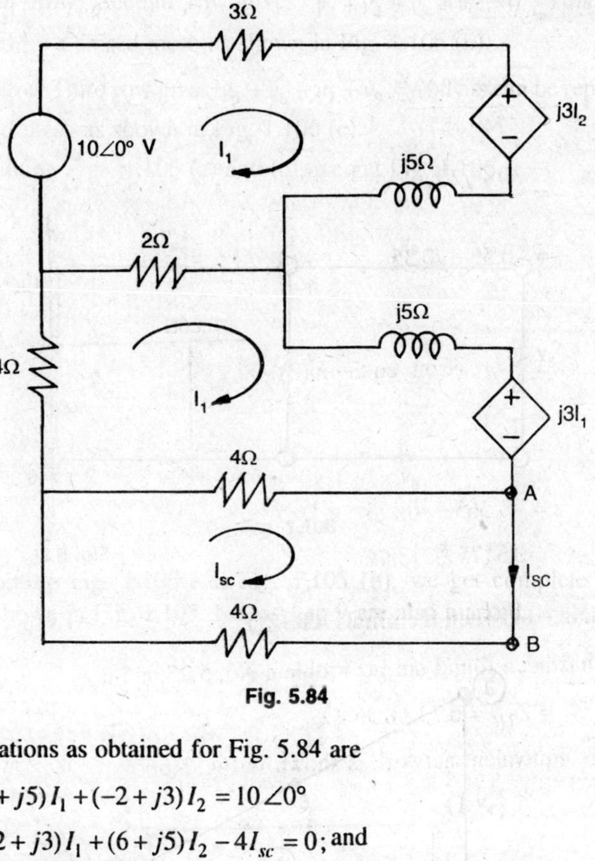

Fig. 5.84

Mesh equations as obtained for Fig. 5.84 are

$$(5+j5)I_1+(-2+j3)I_2=10\angle 0° \quad ...(1)$$

$$(-2+j3)I_1+(6+j5)I_2-4I_{sc}=0; \text{ and} \quad ...(2)$$

$$-4I_2+8I_{sc}=0 \quad ...(3)$$

From equation (3)

$$4I_{sc}=2I_2 \quad ...(4)$$

Substituting value of $4I_3$ from equations (4) in (2), we get

$$(-2+j3)I_1+(4+j5)I_2=0 \quad ...(5)$$

From equations (1) and (5)

$$I_2=\frac{\begin{vmatrix}5+j5 & 10\\ -2+j3 & 0\end{vmatrix}}{\begin{vmatrix}5+j5 & -2+j3\\ -2+j3 & 4+j5\end{vmatrix}}$$

or $$I_2 = \frac{20 - j30}{j57}$$

$$= \frac{(20 - j30)\,j}{j \times (j57)}$$

$$= \frac{30 + j20}{-57}$$

$$= -0.53 - j0.35$$

Now

$$I_{sc} = \frac{4}{8} I_2 \quad \text{(from equation(3))}$$

$$= \frac{1}{2}[-0.53 - j0.35]$$

$$= -0.265 - j0.175$$

$$= 0.3175\angle 213.44°$$

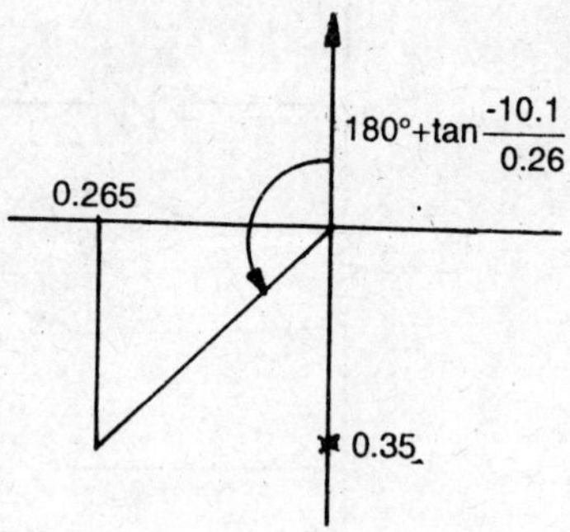

Fig. 5.85

Impedance between terminals AB (Z_T)

It has been found out in Problem No. 5.25 and is

$$Z_T = Z_{TH} = 6.73\angle 8.55°\,\Omega$$

Norton's equivalent network is shown in Fig. 5.86

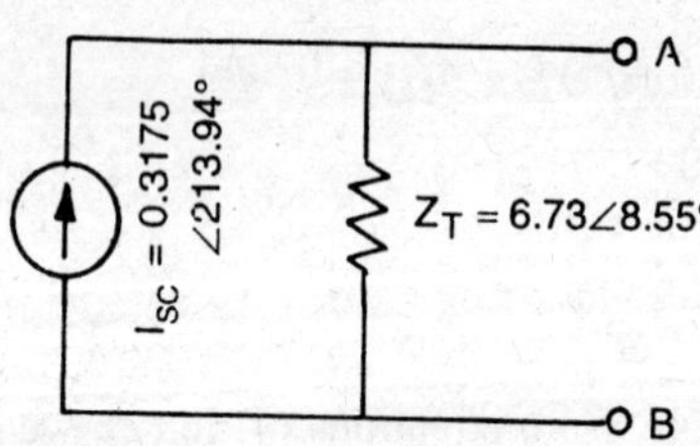

Fig. 5.86

Problem No. 5.27 Obtain Norton's equivalent for Problem No. 5.7.

Solution Fig. 5.87 is used for calculating short-circuit current I_1

Now

$$I = \frac{10}{10 + \dfrac{-j10(3 + j4)}{-j10 + 3 + j4}}$$

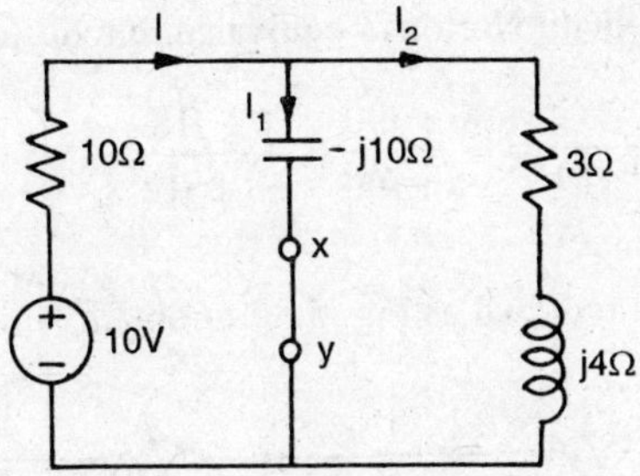

Fig. 5.87

$$= \frac{30 - j60}{70 - j90}$$

Therefore, short-circuit current

$$I_1 = I \times \frac{3 + j4}{3 + j4 - j10}$$

$$= \left(\frac{30 - j60}{70 - j90}\right) \times \left(\frac{3 + j4}{3 - j6}\right)$$

$$= \frac{(67.1 \angle -63.43^\circ) \times (5 \angle -306.87^\circ)}{(114 \angle -52.12^\circ) \times (6.71 \angle -63.43^\circ)}$$

$$= \frac{10}{114} \angle -254.75^\circ$$

$$= 0.44 \angle -254.75^\circ \quad \text{or} \quad 0.44 \angle 360^\circ - 254.75^\circ$$

$$\text{or} \quad 0.44 \angle 105.25^\circ$$

$Z_{xy} = Z_T = 8.38 \angle -69.23^\circ$ (This has been calculated in Problem No. 5.7)

Norton's equivalent network is shown in Fig. 5.88.

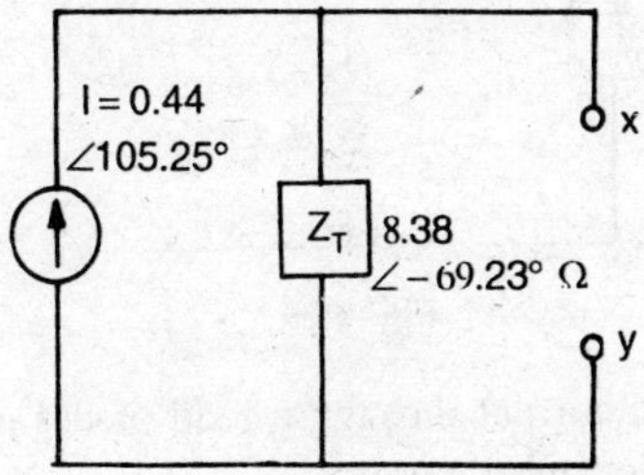

Fig. 5.88

Problem No. 5.28 Obtain Nortons's equivalent circuit for Problem No. 5.8.

$$I = \frac{10}{5 + \dfrac{j5(3+j4)}{j5+3+j4}} = \frac{30+j90}{-5+j60} = \frac{6+j18}{-1+j12}$$

Solution Fig. 5.29 is redrawn as Fig. 5.89 to obtain Norton's equivalent.

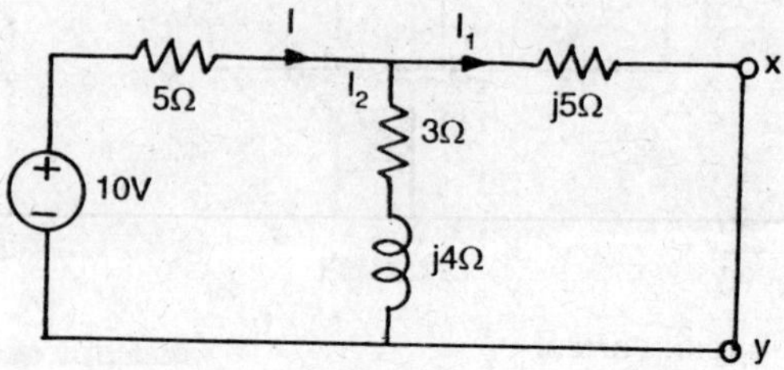

Fig. 5.89

Therefore,

$$I_1 = I_{xy} = \frac{6+j18}{-1+j12} \times \frac{3+j4}{3+j4+j5}$$

$$= \frac{(6+j18)(3+j4)}{(-1+j12)(3+j9)} = \frac{(18.97\angle 71.56^\circ)(5\angle 53.13^\circ)}{(12.04\angle 94.76^\circ)(9.5\angle 71.56^\circ)}$$

$$= 0.829\angle -41.69^\circ$$

Impedance 'Z_{xy}' It has been calculated in Problem No. 5.8 as

$$Z_{xy} = Z_{TH} = 2.5 + j6.25\,\Omega$$

Norton's equivalent circuit is shown in Fig. 5.90

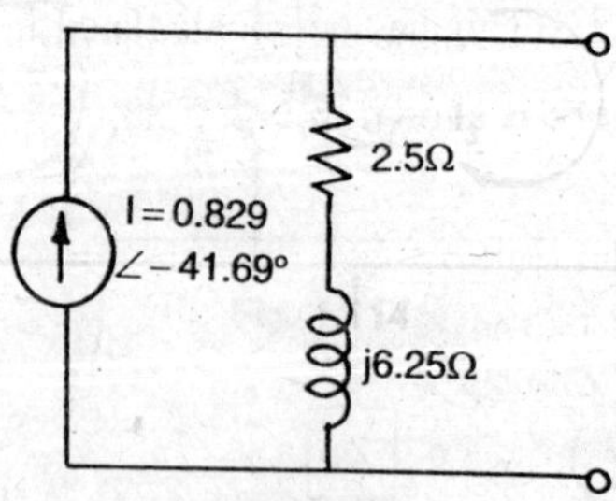

Fig. 5.90

Problem No. 5.31 Find current through a load of 8Ω in the circuit of Fig. 5.91 with the help of Norton's theorem.

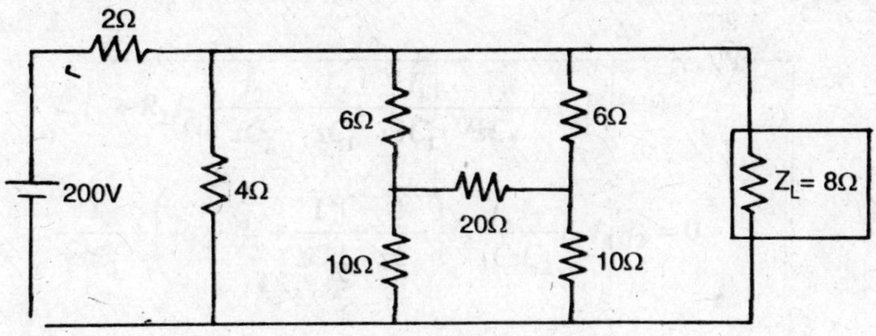

Fig. 5.91

Solution

Impedance through Load Terminals

Short-circuiting the voltage source and removing $Z_L = 8\Omega$, impedance through load terminals is calculated. Network is further simplified by delta formed by 10Ω, 20Ω and 10Ω resistors into star as shown in Fig. 5.92.

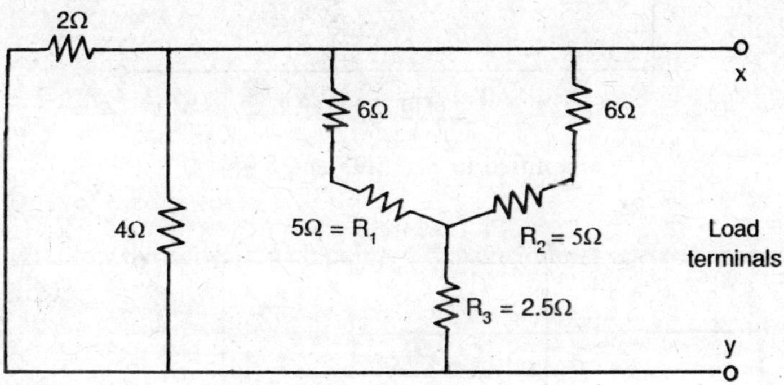

Fig. 5.92

$$R_3 = \frac{10 \times 10}{10 + 10 + 20} = 2.5\Omega$$

$$R_1 = \frac{10 \times 20}{10 + 10 + 20} = 5\Omega$$

$$R_2 = \frac{10 \times 20}{10 + 10 + 20} = 5\Omega$$

Fig. 5.92 is further simplified to Fig. 5.93 and 5.94.

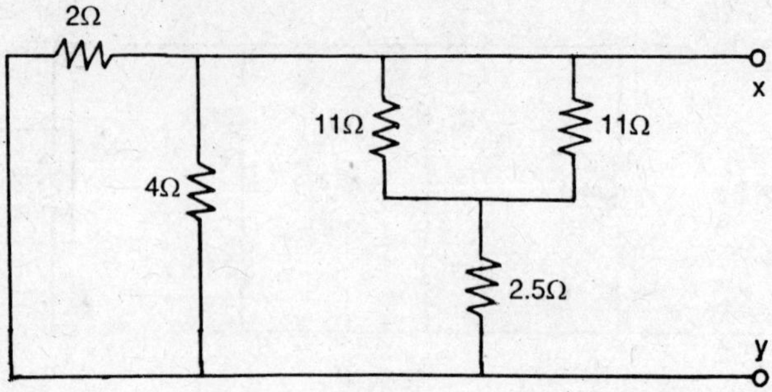

Fig. 5.93

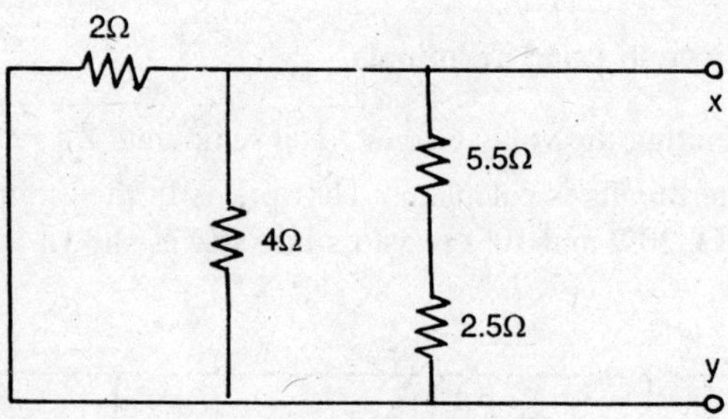

Fig. 5.94

Fig. 5.94 is further simplified to Fig. 5.95 and 5.96

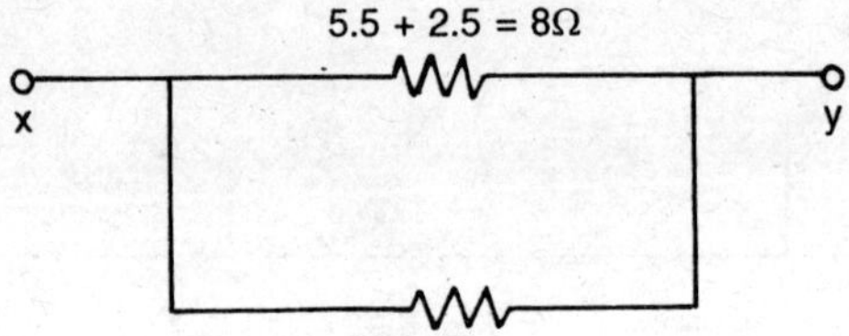

$\frac{4 \times 2}{2+4} = 1.33\Omega$

Fig. 5.95

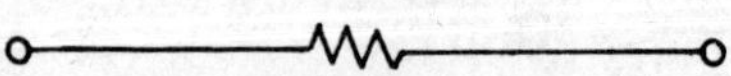

$\frac{8 \times 1.33}{8+1.33} = 1.14\Omega$

Fig. 5.96

Short-Circuit Current Transforming voltage source into current sources of Fig. 5.91 and short-circuiting the load terminals, we get a network as shown in Fig. 5.97

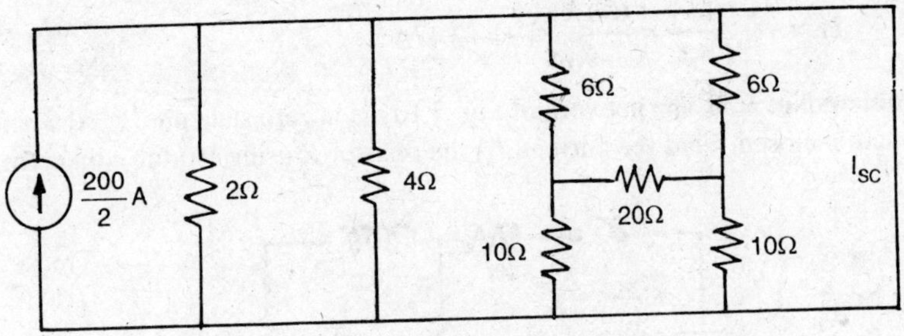

Fig. 5.97

Converting delta into star and further simplifying as done before Fig. 5.97 is simplified to Fig. 5.98 and Fig. 5.99

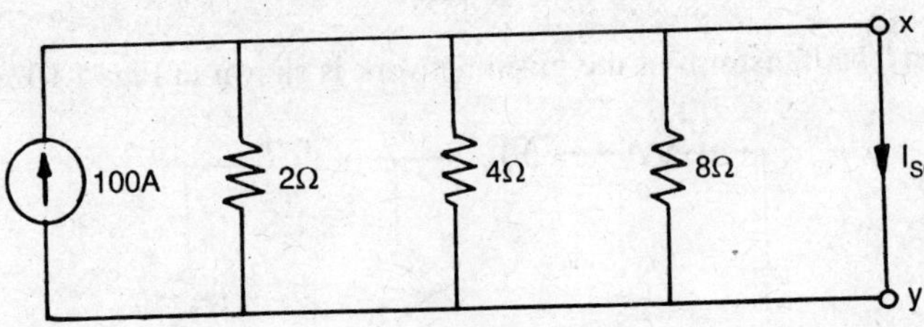

Fig. 5.98

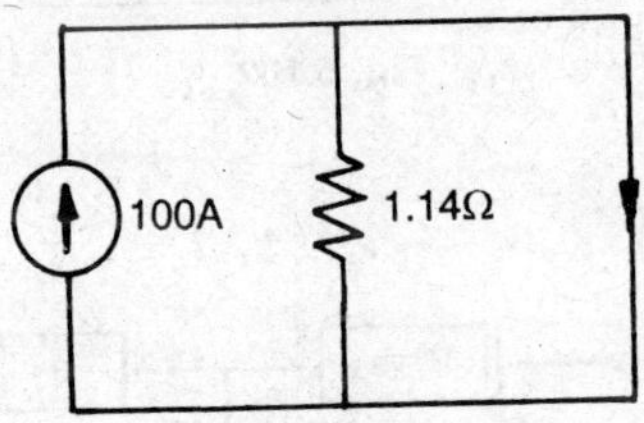

Fig. 5.99

Therefore,

$$I_{sc} = 100A$$

and Norton's equivalent circuit is shown in Fig. 5.100

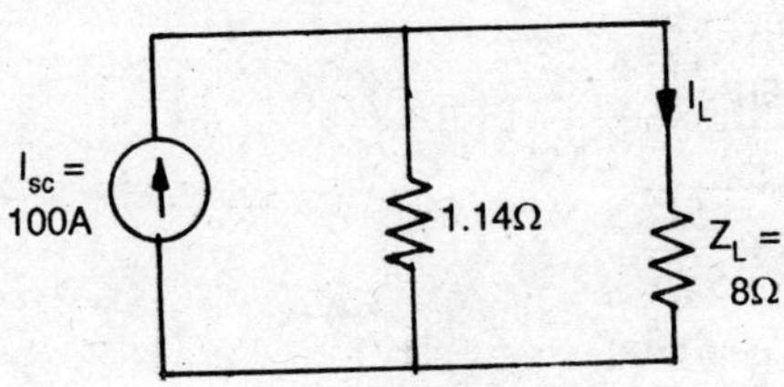

Fig. 5.100

$$I_L = \frac{I_{sc} \times 1.14}{1.14+8} = \frac{100 \times 1.14}{9.14} = 12.47A$$

Problem No. 5.30 The network of Fig. 5.101 is in zero state until $t = 0$ when switch is closed. Find the current $i(t)$ the resistor R using Norton's theorem.

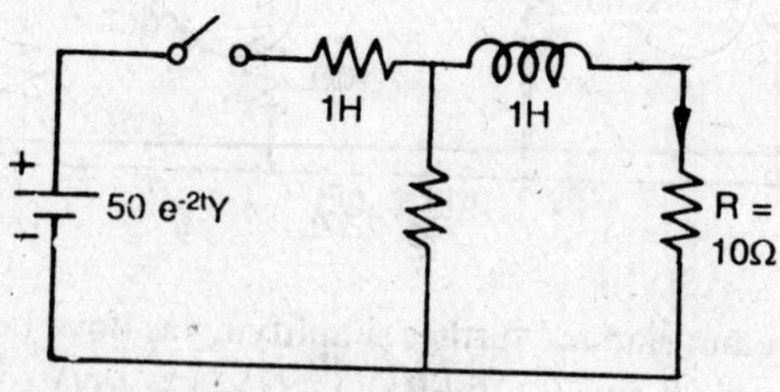

Fig. 5.101

Solution The transform of the given network is shown in Fig. 5.102

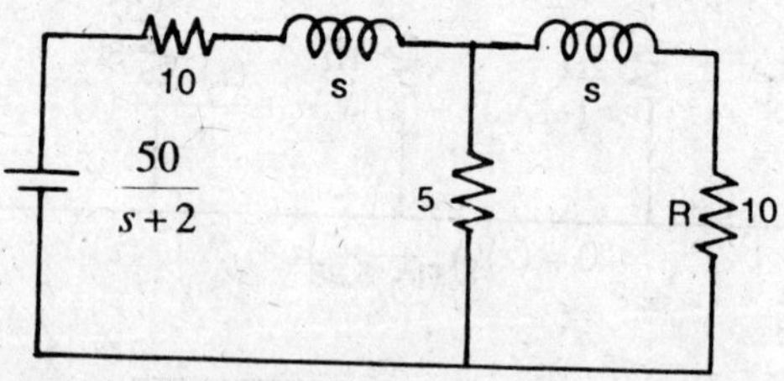

Fig. 5.102

Short-Circuit Current

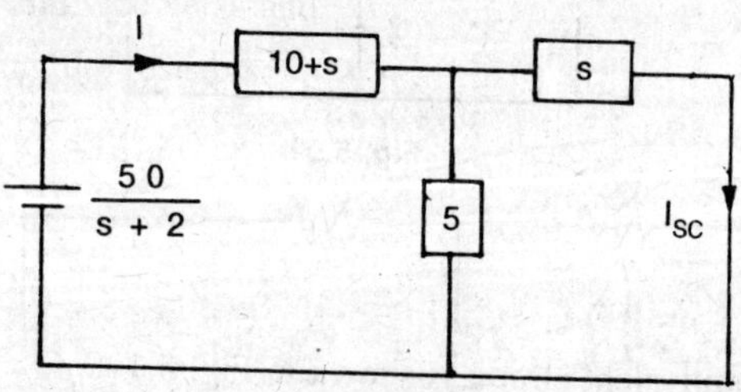

Fig. 5.103

$$I = \frac{\dfrac{50}{s+2}}{10+s+\dfrac{5s}{5+s}}$$

$$= \frac{\dfrac{50}{s+2}}{\dfrac{50+10s+5s+s^2+5s}{s+5}}$$

$$= \frac{50(s+5)}{(s+2)(s^2+20s+50)}$$

Therefore,

$$I_{sc} = \frac{I\times 5}{s+5} = \frac{50(s+5)}{(s+2)(s^2+20s+50)} \times \frac{5}{(s+5)} = \frac{250}{(s+2)(s^2+20s+50)}$$

Impedance through terminals of '*R*'

It is obtained from Fig. 5.104

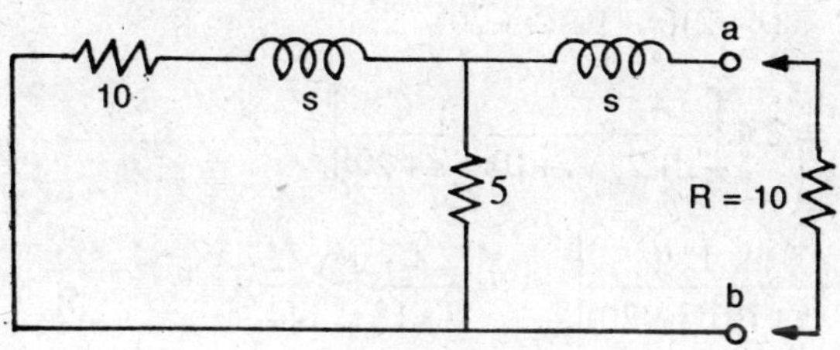

Fig. 5.104

$$Z_{ab} = s + \frac{5(s+10)}{5+s+10}$$

$$= \frac{5s+s^2+10s+5s+50}{s+15}$$

$$= \frac{s^2+20s+50}{s+15}$$

Norton's equivalent circuit is shown in Fig. 5.105

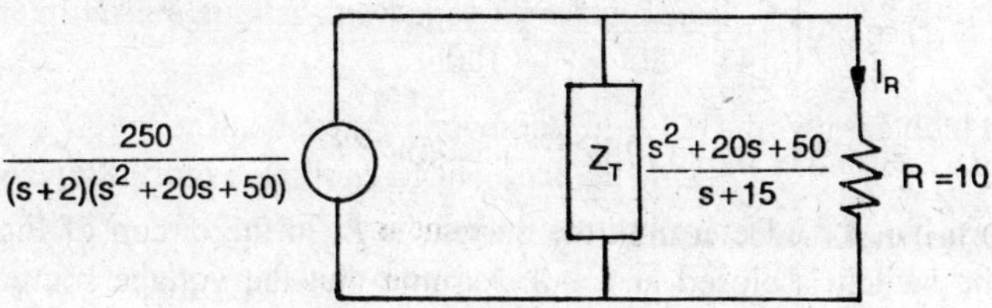

Fig. 5.105

$$I_R(s) = \frac{\dfrac{250}{(s+2)(s^2+20s+50)} \times \dfrac{(s^2+20s+50)}{s+15}}{10 + \dfrac{s^2+20s+50}{s+15}}$$

$$= \frac{250/s+2}{10s+150+s^2+20s+50}$$

$$= \frac{250/s+2}{s^2+30s+200}$$

$$= \frac{250}{(s+2)(s^2+30s+200)}$$

$$= \frac{250}{(s+2)(s+10)(s+20)}$$

$$= 250\left[\frac{A}{s+2} + \frac{B}{s+10} + \frac{C}{s+20}\right]$$

$$A = \left.\frac{1}{(s+10)(s+20)}\right|_{s=-2} = \frac{1}{8\times 18} = \frac{1}{144}$$

$$B = \left.\frac{1}{(s+2)(s+20)}\right|_{s=-10} = \frac{1}{-8\times 10} = -\frac{1}{80}$$

$$C = \left.\frac{1}{(s+2)(s+10)}\right|_{s=-20} = \frac{1}{-18\times -10} = -\frac{1}{180}$$

Therefore,

$$I_R(s) = 250\left[\frac{1/144}{s+2} + \frac{-1/80}{s+10} + \frac{1/180}{s+20}\right]$$

or $$i_R(t) = 250\left(\frac{e^{-2t}}{144} - \frac{1}{80}e^{-10t} + \frac{1}{180}e^{-20t}\right)$$

$$= 0.354e^{-2t} - 3.125e^{-10t} + 1.389e^{-20t}$$

Problem No. 5.31 Determine the current is L_2 in the circuit of Fig. 5.106 after the switch is closed at $t = 0$. Assume that the voltage source $v(t)$ is applied at $t = -\infty$.

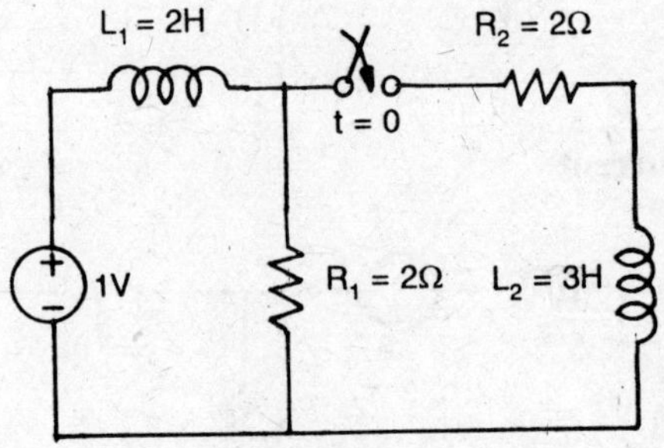

Fig. 5.106

Solution

Before t = 0 With the switch open and $v(t)$ applied at $t = -\infty$, steady state current flows through the loop consisting of $L_1 = 2H$ and $R_1 = 2\Omega$ with voltage source $v(t)$

Value of steady state current is

$$i_{(0^-)} = \frac{V_0}{R_1} = \frac{1}{2} = 0.5A$$

At t = 0 Transform network is shown in Fig. 5.107

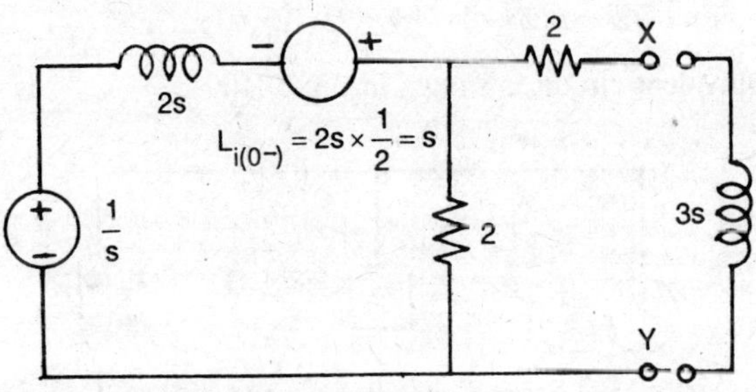

Fig. 5.107

Impedance through terminals *xy* (Z_T)

With voltage sources short-circuited

$$Z_T = \frac{2s \times 2}{2s + 2} + 2 = \frac{4s + 4s + 4}{2s + 2}$$

$$= \frac{8s + 4}{2s + 2}$$

$$= \frac{4s+2}{s+1} = \frac{2(2s+1)}{s+1}$$

Short-Circuit Current

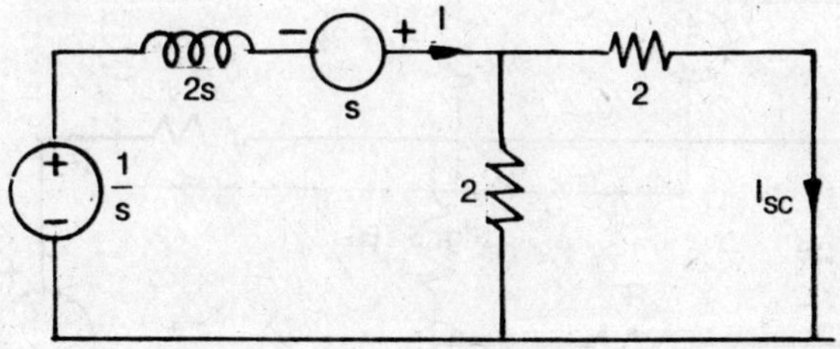

Fig. 5.108

$$I = \frac{\frac{1}{s} - s}{2s + \frac{2 \times 2}{2+2}} = \frac{1-s^2}{s(2s+1)}$$

Therefore,

$$I_{sc} = I \times \frac{2}{2+2} = \frac{(1-s)^2}{s(2s+1)} \times \frac{2}{4} = \frac{1-s^2}{2s(2s+1)}$$

Norton's equivalent circuit is shown in Fig. 5.109

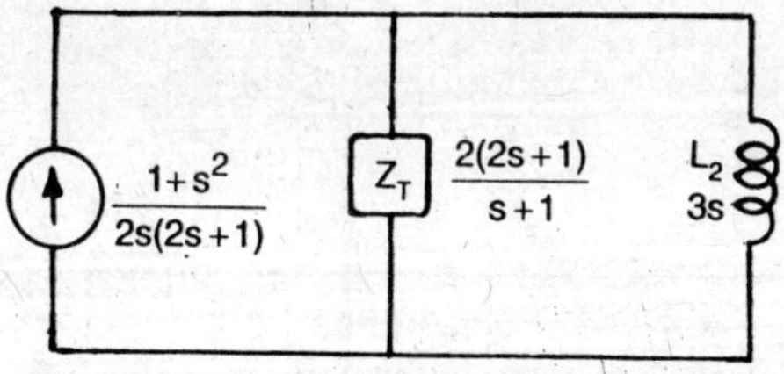

Fig. 5.109

$$I_{L2}(s) = \frac{1-s^2}{2s(2s+1)} \times \frac{\frac{2(2s+1)}{s+1}}{\frac{2(2s+1)}{s+1} + 3s}$$

$$= \frac{1-s^2}{2s(2s+1)} \times \frac{2(2s+1)}{2(2s+1)+3s^2+3s}$$

$$= \frac{1-s^2}{s(4s+2+3s^2+3s)}$$

$$= \frac{1-s^2}{s(3s^2+7s+2)}$$

$$= \frac{(1-s)(1+s)}{s(s+2)\left(s+\frac{1}{3}\right)}$$

Verification by Mesh Analysis (Fig. 5.110)

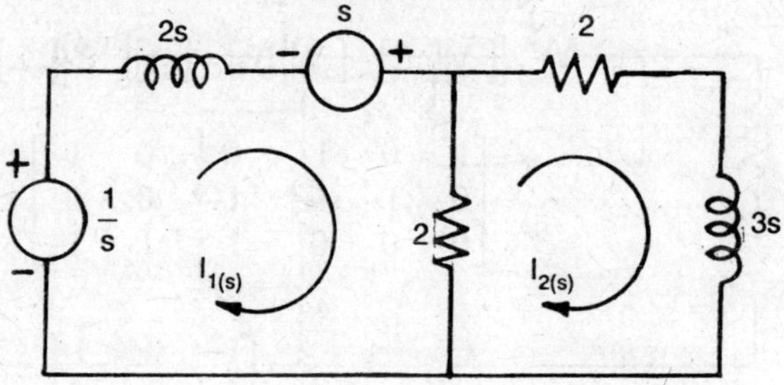

Fig. 5.110

Mesh equations are

$$\frac{1}{s} - s = (2s+2)I_1(s) - 2I_2(s) \qquad \text{...(1)}$$

and

$$0 = (4+3s)I_2(s) - 2I_1(s) \qquad \text{...(2)}$$

or $$I_1(s) = \frac{4+3s}{2} I_2(s) \qquad \text{...(3)}$$

Putting the value of $I_1(s)$ in equation (1), we get

$$\frac{1-s^2}{s} = \frac{(2s+2)(4+3s)}{2} I_2(s) - 2I_2(s)$$

$$\frac{2(1-s^2)}{s} = (6s^2+14s+4)I_2(s)$$

$$I_2(s) = \frac{2(1-s^2)}{s(6s^2+14s+4)}$$

$$= \frac{2(1+s)(1-s)}{s(3s^2+7s+2)}$$

$$= \frac{2(1-s)(1+s)}{s(s+2)\left(s+\frac{1}{3}\right)}$$

which is same as found out earlier.

Problem No. 5.32 Find the Thevenin's equivalent circuit of Fig. 5.111.

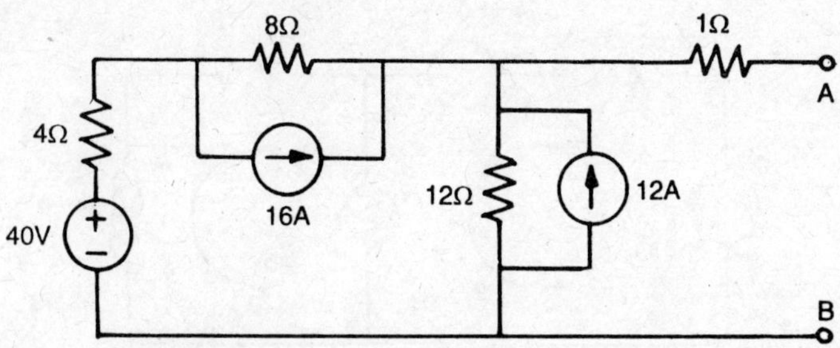

Fig. 5.111

Solution Converting the current sources into voltage sources, the network gets simplified to Fig. 5.112

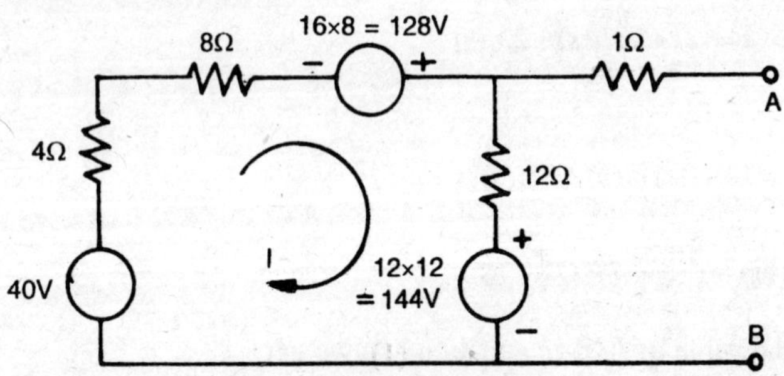

Fig. 5.112

Thevenin's Impedance (Z_{TH}) Short-circuiting the voltage sources,

$$Z_{TH} = Z_{AB} = \frac{12\times 12}{12+12}+1$$

$$= \frac{144}{24} + 1$$

$$= 6 + 1 = 7\Omega$$

Open Circuit-Voltage (V_{oc})

Mesh equation for the loop is

$$40 + 128 - 144 = (4 + 8 + 12)I$$

or $24 = 24I$

or $I = \frac{24}{24} = 1A$

Therefore,

$$V_{OC} = V_{AB} = 1 \times 6 = 6 \text{ Volts}$$

Thevenin's equivalent circuit is shown in Fig. 5.113

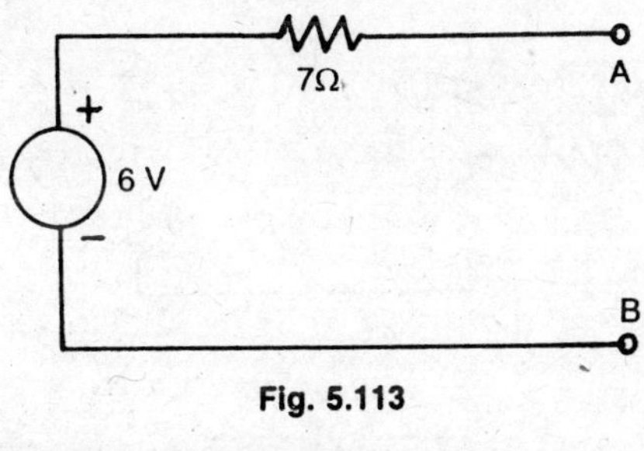

Fig. 5.113

Problem No. 5.33 Find the current in the 10Ω resistor in the network of Fig. 5.114 by Thevenin's theorem. Resistances are in ohms.

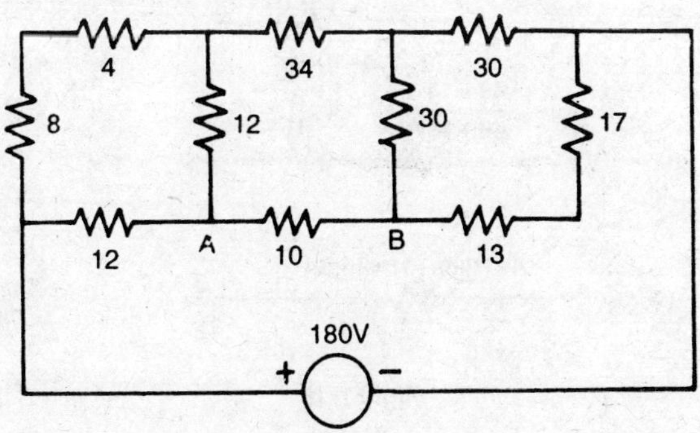

Fig. 5.114

Solution Replacing the two delta's on the left and right with star, the network simplifies to Fig. 5.115.

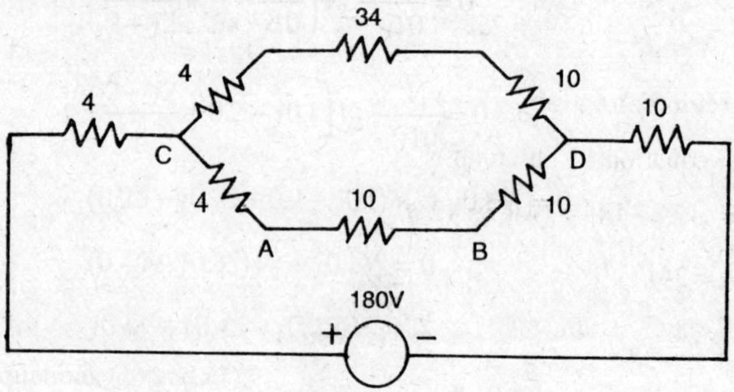

Fig. 5.115

Thevenin's Impedance (Z_{TH}) Removing 10Ω resistance and short-circuiting voltage source, we get simplified network shown in Fig. 5.116

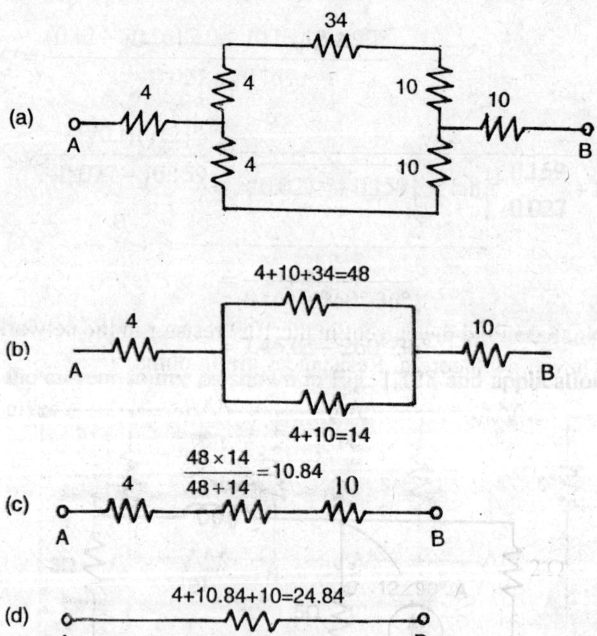

Fig. 5.116

From Fig. 5.115

$$V_{OC} = V_{AB} = V_{CD} = 180 \times \frac{4+34+10}{4+34+10+4+10}$$

$$= \frac{4320}{31} = 139.35V$$

Therefore,

Current through 10Ω resistor is

$$= \frac{V_{OC}}{Z_{TH} + R_{10}} = \frac{139.35}{24.84+10}$$

$$= 3.99A$$

Problem No. 5.34 Using Thevenin's theorem find current in a 2Ω resistor connected between A and B in the network of Fig. 5.117.

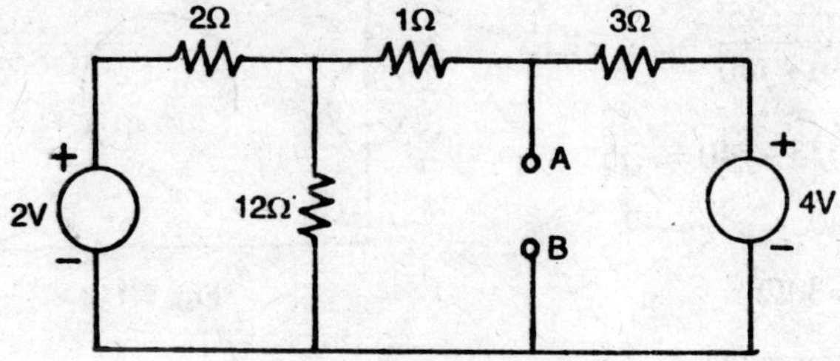

Fig. 5.117

Solution If I_1 and I_2 are mesh currents, then mesh equations are

$$14I_1 + 12I_2 = 2 \quad ...(1)$$

and $$12I_1 + 16I_2 = 4 \quad ...(2)$$

solving for I_1 and I_2 gives

$$I_2 = 0.4A$$

Therefore,

$$V_{OC} = V_{AB} = 4 - 0.4 \times 3 = 2.8V$$

Short circuiting the voltage sources gives

$$Z_{TH} = R_{AB} = \left(1 + \frac{2 \times 12}{14}\right)$$ in parallel with 3Ω resistor.

$$= \frac{\frac{19}{7} \times 3}{\frac{19}{7} + 3}$$

$$= \frac{57}{40}\Omega$$

$$= 1.425\Omega$$

$$I_{R2} = \frac{V_{OC}}{Z_{TH} + R_2} = \frac{2.8}{1.425 + 2} = 0.82A$$

Problem No. 5.35 Calculate the change in the current in the circuit of Fig. 5.118 by using the compensation theorem when the resistance has changed to $j35\Omega$.

Solution

$$\delta I = \frac{I\delta Z}{Z + \delta Z + Z_{TH}} \quad ...(1)$$

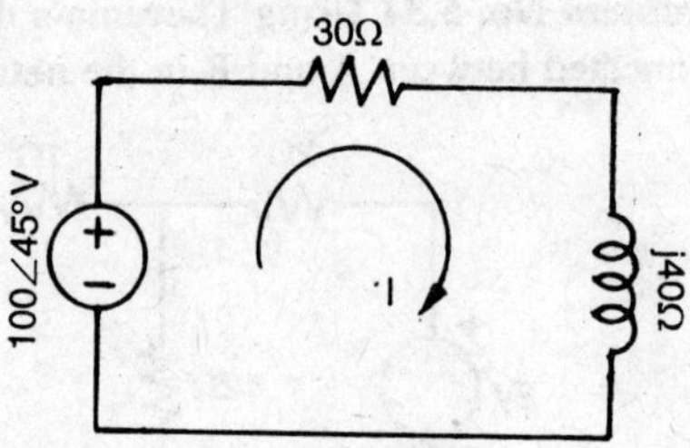

Fig. 5.118

$$I = \frac{100\angle 45°}{30 + j40} = 2\angle -8.13°$$

$$\delta Z = j35 - j40 = -j5$$

$$Z = j40$$

$$Z_{TH} = 30\Omega$$

Substituting the values in equation (1), we get

$$\delta I = \frac{(2\angle -8.13°)\times(-j5)}{j40 - j5 + 30}$$

$$= \frac{10\angle -8.13° + 90°}{30 + j35}$$

$$= \frac{10\angle -8.13^0 + 90°}{46.1\angle 49.4°} = 0.217\angle -8.13° + 90° + 49.4°$$

$$= 0.217\angle 32.47°$$

Problem No. 5.36 The 5Ω resistor has been changed to 8Ω resistor in the circuit of Fig. 5.119. Determine the compensation source V_c and calculate the current through the 3Ω resistor.

$$I = \frac{10\angle 0°}{5 + \frac{(3 + j4)j5}{3 + j4 + j5}} = \frac{10\angle 0°}{5.83 + j.25}$$

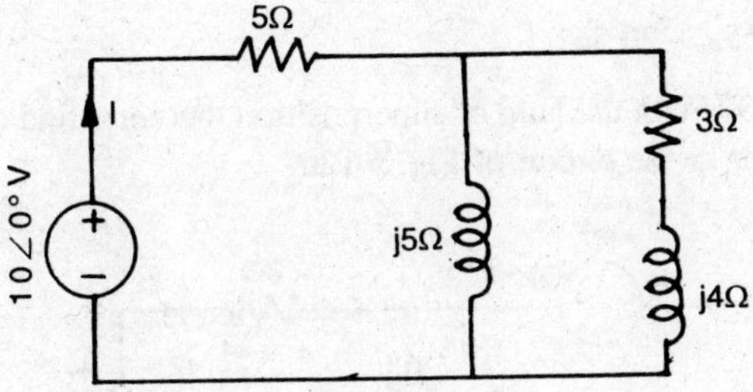

Fig. 5.119

$$= \frac{10\angle 0^\circ}{6.34\angle 23.21^\circ}$$

$$= 1.576\angle -23.21^\circ$$

Compensation source $V_C = (1.576\angle -23.21^\circ)\times \delta Z$

$$= (1.576\angle -23.21^\circ)\times (8-5)$$

$$= 4.728\angle -23.21^\circ\, V$$

$$\delta I = \frac{I\delta Z}{Z+\delta Z+Z_{TH}}$$

$$= \frac{1.576\angle -23.21^\circ \times 3}{5+3+0.83+j2.5}$$

$$= \frac{4.728\angle -23.21^\circ}{8.83+j2.5}$$

$$= \frac{4.728}{9.18}\angle -23.21^\circ - 15.81^\circ$$

$$= 0.515\angle -39^\circ A$$

In the 3Ω resistor

$$\delta I = \frac{(0.515\angle -39^\circ)(j5)}{3+j4+j5}$$

$$= \frac{2.575\angle 51^\circ}{3+j9}$$

$$= \frac{2.575\angle 51^\circ}{9.49\angle 71.56^\circ}$$

$= 0.271\angle 51° - 71.56°$

$= 0.271\angle -20.56°$

Problem No. 5.37 With the help of superposition theorem, find the current I in the 3Ω resistor in the circuit of Fig. 5.120.

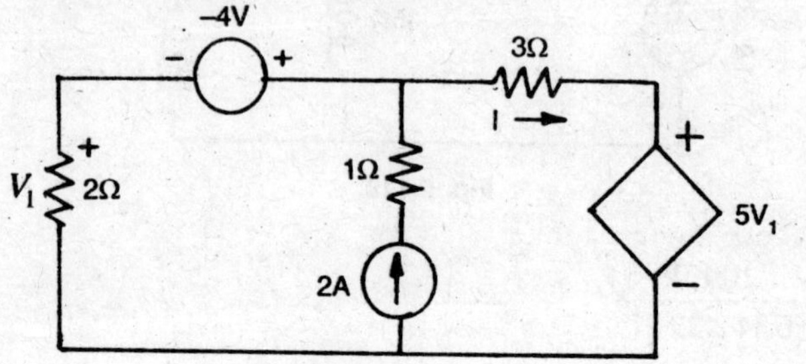

Fig. 5.120

Solution

(a) **Voltage Source '–4V' in the circuit.** Other voltage sources are removed, keeping their internal resistances and open-circuiting the current source as shown in Fig. 5.121, mesh equations are

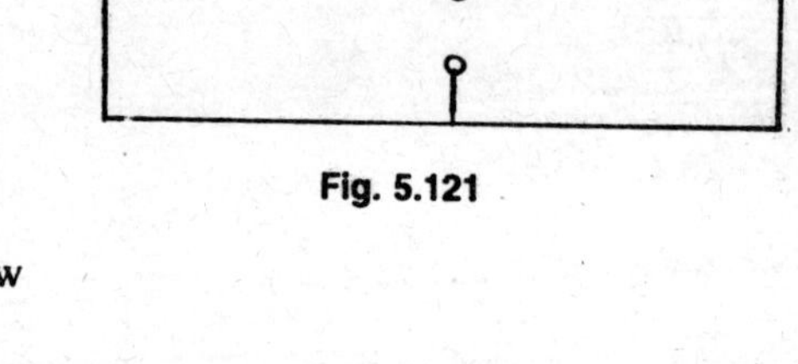

Fig. 5.121

written as below

$$I_1 = \frac{4}{3+2} = 0.8A \qquad ...(1)$$

(b) **Dependent Voltage Source 'V_1' in the circuit.** Removing the $-4V$ source and open-circuiting the current source, as shown in Fig. 5.122, mesh equations

are written as given below

$$5V_1 = 3I_2 + 2I_2$$

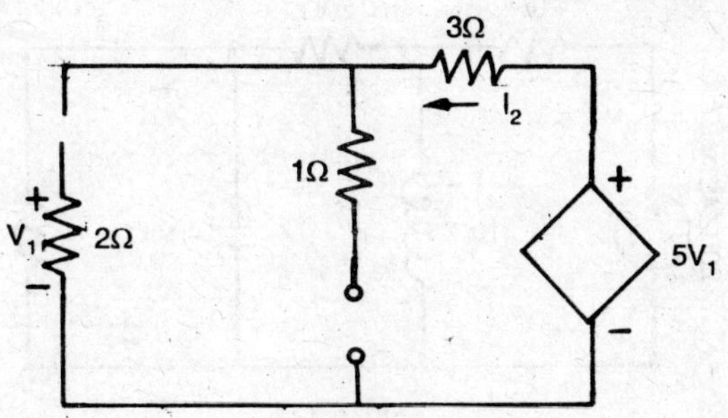

Fig. 5.122

or $\quad 5V_1 = 5I_2$ ...(2)

Also,

$$V_1 = 2I_2$$

Therefore, the dependent voltage source which is equal to $5V_1$ is

$$= 5 \times 2I_2$$

$$= 10I_2 \quad ...(3)$$

Substituting the value of $5V_1 = 10I_2$ form equation (3) in (2), we get

$$10I_2 = (3+2)I_2 \quad ...(4)$$

If we see the R.H.S. of equation (4), we see that if $I_2 = 0$ then, R.H.S. equates with L.H.S. Therefore,

$$I_2 = 2A$$

(c) **Current source 2A in the Circuit.** Short-circuiting other voltage sources as shown in Fig. 5.123, the mesh equation is

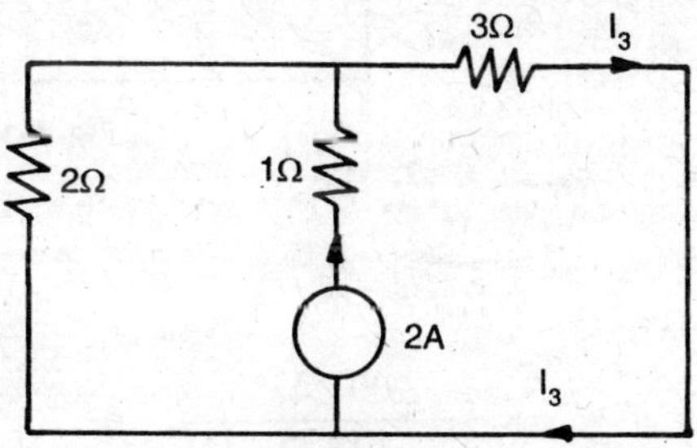

Fig. 5.123

$$I_3 = \frac{2 \times 2}{2 \times 3} = 0.67A$$

Therefore

$$I = I_1 + I_2 + I_3$$

$$= 0.8 - 2 + 0.67$$

$$= -0.53A$$

Problem No. 5.38 Verify reciprocity theorem for the network shown in Fig. 5.124.

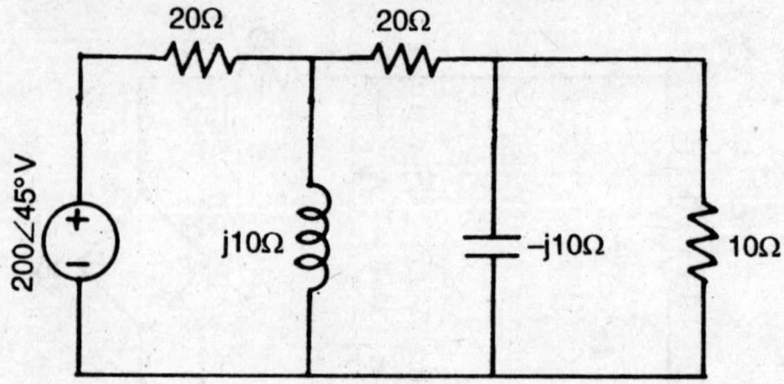

Fig. 5.124

Solution It is seen that 10Ω resistance and −j10Ω capacitive reactance −j10Ω in paralllel and in series with 20Ω resistance.

$$Z = 20 + \frac{10 \times -j10}{10 + (-j10)} = 25 - j5$$

Network of Fig. 5.124 is redrawn as Fig. 5.125.

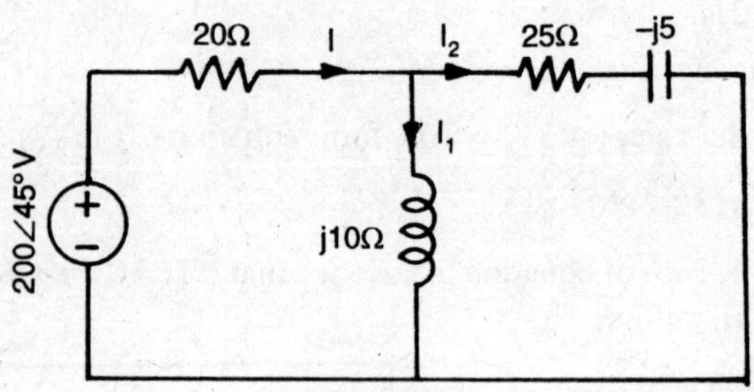

Fig. 5.125

$$I = \frac{V}{Z_{net}}$$

$$= \frac{200\angle 45^\circ}{20 + \dfrac{j10(25 - j5)}{j10 + 25 - j5}}$$

$$= \frac{200\angle 45^\circ}{20 + \dfrac{50 + j250}{25 + j5}}$$

$$= \frac{200\angle 45^\circ}{550 + j350} \times 25 + j5$$

$$= \frac{(200\angle 45^\circ)(25.5\angle 11.31^\circ)}{652\angle 32.5^\circ}$$

$$= 7.82\angle 45^\circ + 11.31^\circ - 32.5^\circ$$

$$= 7.82\angle 23.81^\circ$$

$$I_2 = \frac{I \times j10}{j10 + 25 - j5} = \frac{7.82\angle 23.81^\circ \times j10}{25 + j5}$$

$$= \frac{78.2\angle 23.81^\circ + 90^\circ}{25.5\angle 11.31^\circ}$$

$$= 3.07\angle 23.81^\circ + 90^\circ - 11.31^\circ$$

$$= 3.07\angle 102.5^\circ\ A \qquad \ldots(1)$$

Verification Generator $200\angle 45^\circ$ is replaced for current I_2 as shown in Fig. 5.126

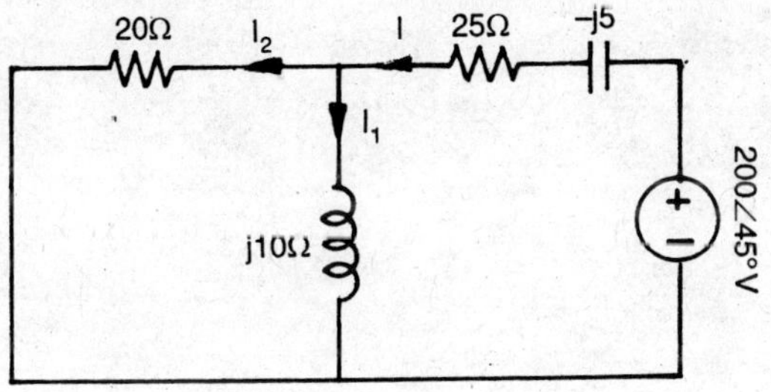

Fig. 5.126

$$I - \frac{200\angle 45^\circ}{(25 - j5) + \dfrac{j10 \times 20}{20 + j10}}$$

$$= \frac{200\angle 45^\circ \times (20 + j10)}{500 - j100 + j250 + 50 + j200}$$

$$= \frac{(200\angle 45^\circ) \times 22.36\angle 26.56^\circ}{550 + j350}$$

$$= \frac{4472.14\angle 71.56^\circ}{651.9\angle 32.47^\circ} = 6.86\angle 39.09^\circ$$

$$I_2 = \frac{6.86\angle 39.09^\circ}{1} \times \frac{j10}{20 + j10}$$

$$= \frac{(6.86\angle 39.09^\circ) \times (10\angle 90^\circ)}{22.36\angle 26.56^\circ}$$

$$= \frac{6.86\angle 129.09^\circ}{22.36\angle 26.56^\circ}$$

$$= 3.067\angle 129.09^\circ - 26.56^\circ$$

$$= 3.067\angle 102.53^\circ\, A$$

which is same as found out in equation (1) and hence reciprocity theorem is verified.

6

Network Functions

Problem No. 6.1 Derive the voltage transfer function of the network shown in Fig. 6.1.

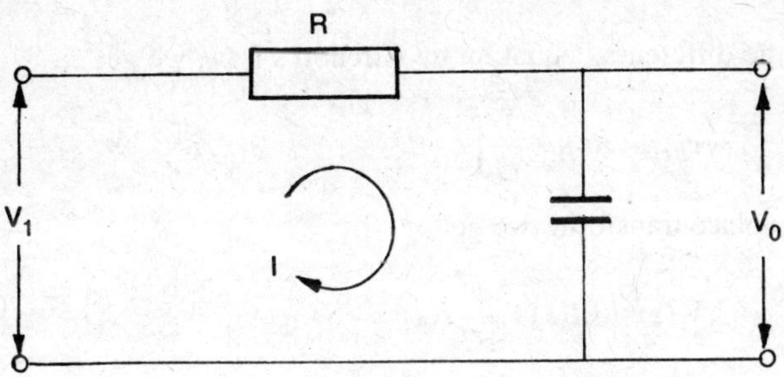

Fig. 6.1

Solution

Writing the differential equations with the help of Kirchoff's law

$$v_{i(t)} = Ri_{(t)} + v_{0(t)};$$

$$v_{o(t)} = \frac{1}{C}\int i_{(t)}\, dt$$

Taking Laplace transform, we get

$$V_i(s) = RI(s) + V_o(s);$$

and $$V_o(s) = \frac{1}{sC} I(s)$$

or $$I(s) = sCV_o(s)$$

or $$V_i(s) = sRCV_o(s) + V_o(s)$$

$$\frac{V_o(s)}{V_i(s)} = \frac{1}{1+s\,RC} = \frac{1}{1+sT}$$

where $\quad T = RC$ **Ans.**

Problem No. 6.2 Derive the voltage transfer function of the network shown in Fig. 6.2.

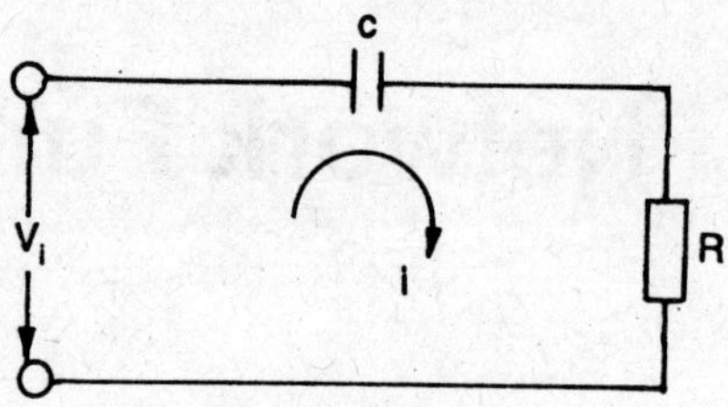

Fig. 6.2

Solution

Writing the differential equation by Kirchoff's Law, we get

$$v_{i(t)} = Ri_{(t)} + \frac{1}{C}\int i_{(t)}\,dt$$

Taking Laplace transform, we get

$$V_i(s) = RI(s) + \frac{1}{sC}I(s)$$

$$\frac{I(s)}{V_i(s)} = \frac{1}{R+\frac{1}{sC}} = \frac{sC}{1+sCR}$$ **Ans.**

Problem No. 6.3 Derive the voltage transfer function of the network shown in Fig. 6.3.

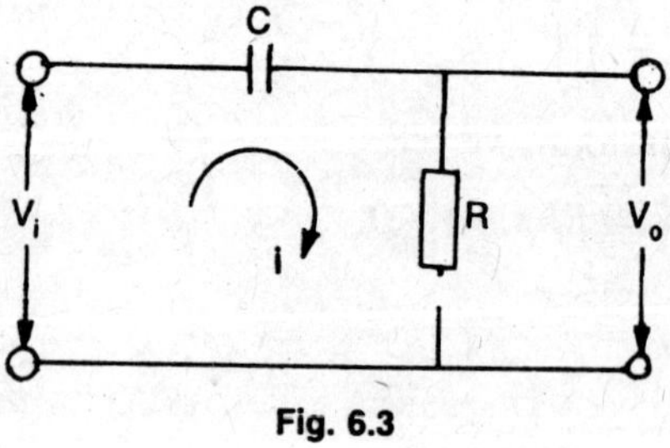

Fig. 6.3

Solution

Writing the differential equations by Kirchoff's Law, we get

$$V_{i(t)} = Ri_{(t)} + \frac{1}{C}\int i_{(t)}\, dt \qquad ...(1)$$

$$V_{o(t)} = Ri_{(t)} \qquad ...(2)$$

Taking Laplace transform, we get

$$V_i(s) = RI(s) + \frac{I(s)}{sC} \qquad ...(3)$$

and

$$V_o(s) = RI(s)$$

$$I(s) = \frac{V_o(s)}{R} \qquad ...(4)$$

Substituting the value of $I_1(s)$ from (4) in (3) we get

$$V_i(s) = I(s)\left[R + \frac{1}{sC}\right]$$

or

$$V_i(s) = \frac{V_o(s)}{R}\left[R + \frac{1}{sC}\right]$$

$$V_i(s) = V_o(s)\left[1 + \frac{1}{sCR}\right]$$

$$\frac{V_o(s)}{V_i(s)} = \frac{1}{1 + \frac{1}{sRC}}$$

$$= \frac{sRC}{1 + sRC}$$

$$= \frac{sT}{1 + sT} \qquad \textbf{Ans.}$$

where $T = RC$

Problem No. 6.4 Determine voltage transfer function of the network shown in Fig. 6.4 in phase lead form.

Solution

If Z_1 is the equivalent impedance of the parallel combination of R_1 and C then

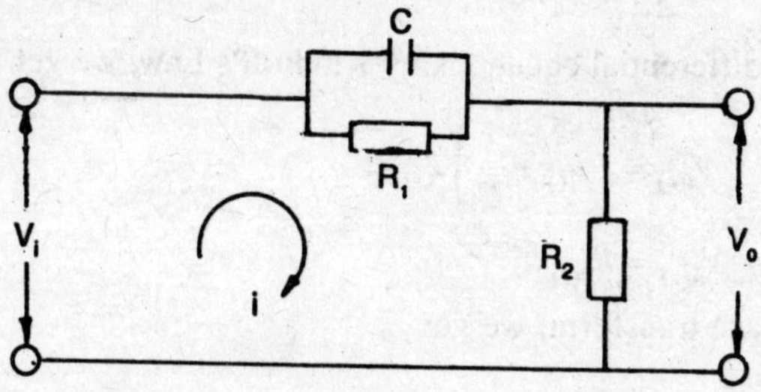

Fig. 6.4

$$\frac{1}{Z} = \frac{1}{R_1} + \frac{j\omega C}{1}$$

$$\therefore \qquad Z_1 = \frac{R_1}{1 + j\omega CR_1} = \frac{R_1}{1 + sCR_1}, \ (j\omega = s).$$

Writing differential equations by Kirchoff's laws, we get

and $\qquad v_{i(t)} = Z_1 i_{(t)} + R_2 i_{(t)}$

$$v_{0(t)} = R_2 i_{(t)}$$

Taking Laplace transform of the above equations, we get

$$V_i(s) = Z_1 I(s) + R_2 I(s) \quad \text{and}$$

$$V_0(s) = R_2 I(s) \quad \therefore\ I(s) = \frac{V_0(s)}{R_2}$$

or $\qquad V_i(s) = Z_1 \frac{V_0(s)}{R_2} + V_0(s)$

or $\qquad V_i(s) = V_0(s)\left(\frac{Z_1}{R_2} + 1\right)$

Substituting the value of Z_1, we get

$$V_i(s) = V_0(s)\left(\frac{R_1}{1 + sCR_1} \times \frac{1}{R_2} + 1\right)$$

$$\frac{V_0(s)}{V_i(s)} = \frac{R_2(1 + sCR_1)}{R_1 + R_2 + sCR_1 R_2}$$

$$= \frac{R_2}{R_1 + R_2}\left[\frac{1 + sCR_1}{1 + \frac{sCR_1R_2}{R_1 + R_2}}\right]$$

put $\frac{R_2}{R_1 + R_2} = a$ and $CR_1 = T$, then we get

$$\frac{V_0(s)}{V_i(s)} = \frac{a(1 + sT)}{1 + saT}$$ **Ans.**

Problem No. 6.5 Find the voltage transfer function of the electrical network shown in Fig. 6.5.

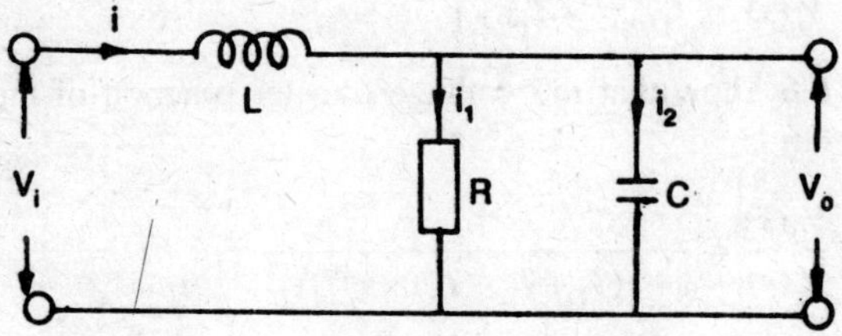

Fig. 6.5

Solution

Assuming no external load

$\therefore \quad i = i_1 + i_2$

Applying Kirchoff's laws to the electrical network, we get

$$v_{i(t)} = \frac{Ldi_{(t)}}{dt} + Ri_{1(t)}$$

and $\quad Ri_{1(t)} = v_{0(t)}$

$$v_{0(t)} = \frac{1}{C}\int i_{2(t)}dt = Ri_{1(t)}$$

Taking Laplace transform, we get

$$V_i(s) = sLI(s) + V_0(s) \qquad \text{and}$$

$$V_0(s) = \frac{I_2(s)}{sC} = RI_1(s)$$

or $\quad V_i(s) = V_0(s) + sL(I_1(s) + I_2(s))$

or
$$V_i(s) = V_0(s) + sL\left(\frac{V_0(s)}{R} + sCV_0(s)\right)$$

or
$$V_i(s) = V_0(s)\left[1 + \frac{sL}{R} + s^2LC\right]$$

or
$$\frac{V_0(s)}{V_i(s)} = \frac{1}{s^2LC + \frac{sL}{R} + 1}$$

putting $\frac{L}{R} = T_1$ and $CR = T_2$, we get

$$\frac{V_0(s)}{V_i(s)} = \frac{1}{T_1T_2s^2 + T_1s + 1}$$ **Ans.**

Problem No. 6.6 Show that the votltage transfer function of the network shown in Fig. 6.6 is

$$\frac{V_0(s)}{V_i(s)} = \frac{1}{1 + s(T_1 + T_2 + T_3) + s^2T_1T_2}$$

where

$T_1 = R_1C_1$;

$T_2 = R_2C_2$; and $T_3 = R_1C_2$

Solution

Let $R_1 = Z_1$

$\frac{1}{sC_1} = Z_2$

$R_2 = Z_3$

$\frac{1}{sC_2} = Z_4$

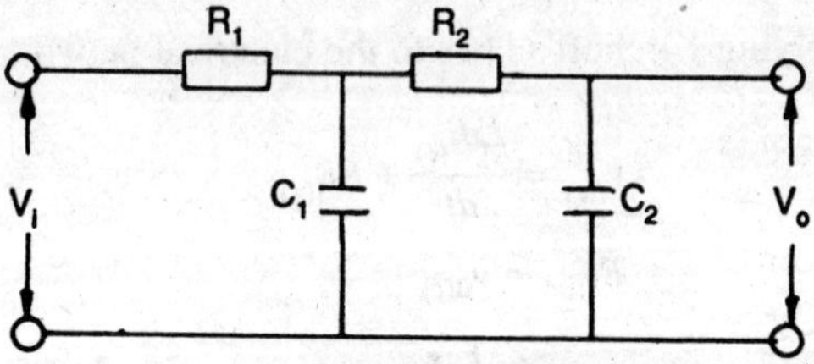

Fig. 6.6

Redrawing the Fig. 6.6 after substituting the values, we get (Fig. 6.7)

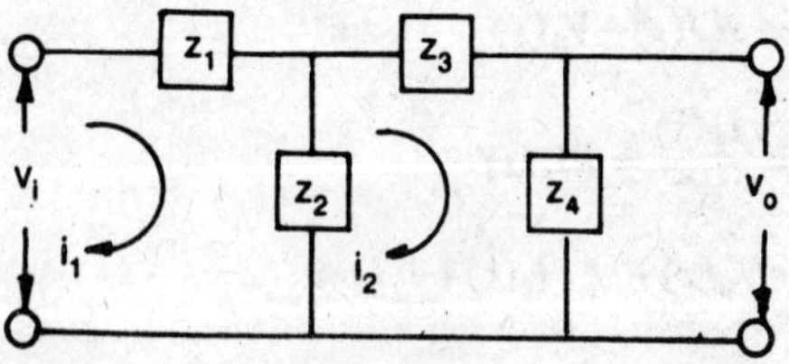

Fig. 6.7

Writing the differential equations with the help of Kirchoff's laws, and after taking Laplace transform, we get

$$V_i(s) = (Z_1 + Z_2)I_1(s) - Z_2 I_2(s) \quad ...(1)$$

$$0 = Z_3 I_2(s) + Z_4 I_2(s) + Z_2 I_2(s) - Z_2 I_1(s) \quad ...(2)$$

$$V_0(s) = Z_4 I_2(s) \quad ...(3)$$

From equation (2)

$$I_1(s) = \frac{(Z_2 + Z_3 + Z_4)I_2(s)}{Z_2} \quad ...(4)$$

Substituting the value of $I_1(s)$ in equation (1), we get

$$V_i(s) = \frac{(Z_1 + Z_2)(Z_2 + Z_3 + Z_4)I_2(s) - Z_2 I_2(s)}{Z_2}$$

$$\frac{V_i(s)}{I_2(s)} = \frac{(Z_1 + Z_2)(Z_2 + Z_3 + Z_4) - Z_2{}^2}{Z_2}$$

But from equation (3)

$$I_2(s) = \frac{V_0(s)}{Z_4},$$

Therefore,

$$\frac{V_i(s)}{V_o(s)/Z_4} = \frac{(Z_1 + Z_2)(Z_2 + Z_3 + Z_4) - Z_2^2}{Z_2}$$

or

$$\frac{V_o(s)}{V_i(s)} = \frac{Z_2 Z_4}{Z_1 Z_2 + Z_1 Z_3 + Z_1 Z_4 + Z_2 Z_3 + Z_2 Z_4}$$

Substituting the values of Z_1, Z_2, Z_3 and Z_4, we get

$$\frac{V_0(s)}{V_i(s)} = \frac{\frac{1}{sC_1}\frac{1}{sC_2}}{\frac{R_1}{sC_1} + \frac{R_1 R_2}{1} + \frac{R_1}{sC_2} + \frac{R_2}{sC_1} + \frac{1}{s^2 C_1 C_2}}$$

$$= \frac{1}{sR_1C_2 + s^2 R_1 R_2 C_1 C_2 + sR_1C_1 + sR_2C_2 + 1}$$

$$= \frac{1}{1 + s(R_1C_1 + R_2C_2 + R_1C_2) + s^2 R_1R_2C_1C_2}$$

$$= \frac{1}{1 + s(T_1 + T_2 + T_3) + s^2 T_1T_2}$$ **Ans.**

Problem No. 6.7 Derive the transfer function of the circuit shown in Fig. 6.8. If $v_2 = 8 \sin 10\,t$ volts, $R_1 = 50$ k ohms, $R_2 = 5$ k ohms and $C = 1\mu$F. Calculate the output voltage in magnitude and in phase relative to input voltage.

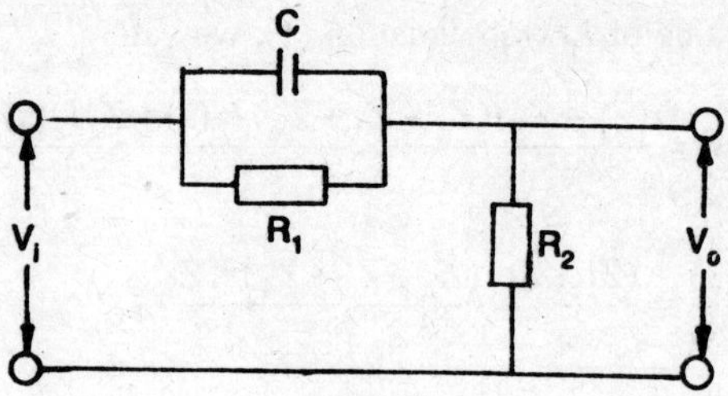

Fig. 6.8

Solution

The transfer function of the circuit is derived in problem No. 6.4 and is

$$G(s) = \frac{V_0(s)}{V_i(s)} = \frac{R_2(1 + sCR_1)}{R_1 + R_2 + sCR_1R_2}$$

$$\therefore \quad G(s) = \frac{5000(1 + s \times 1 \times 10^{-6} \times 50 \times 1000)}{50 \times 1000 + 5000 + s \times 1 \times 10^{-6} \times 50 \times 1000 \times 5000}$$

$$= \frac{0.1(1 + 0.05s)}{1 + 0.0045s}$$ **Ans.**

put $\quad s = j\omega$,

Therefore,

$$G_{j(\omega)} = \frac{0.1(1 + 0.05j\omega)}{1 + 0.0045j\omega}$$

$$\angle G_{j(\omega)} = \tan^{-1} 0.05\omega - \tan^{-1} 0.0045\omega$$

since $\quad v_i = 8 \sin 10t$, therefore, $\omega = 10$

$$\therefore \quad \angle G_{j(\omega)} = \tan^{-1} 0.05 \times 10 - \tan^{-1} 0.0045 \times 10$$

$$= 26.265° - 2.576°$$

$$= 23.989° \qquad \textbf{Ans.}$$

Also,

$$M = \angle G(j\omega) = \frac{0.1\sqrt{1+(0.05\omega)^2}}{\sqrt{1+(0.0045\omega)^2}}$$

putting

$\omega = 10$, we get

$$M = \frac{0.1\sqrt{1+0.5^2}}{\sqrt{1+0.045^2}}$$

$$= 0.1019$$

Therefore,

$$v_0 = 8 \times 0.1019$$

$$= 0.8156 \text{ volts} \qquad \textbf{Ans.}$$

Problem No. 6.8 If C = 1 μF in the circuit shown if Fig. 6.9. What values of R_1 and R_2 will give T = 0.6 sec. and **a** = 0.1.

Solution

The transfer function of the network shown in Fig. 6.9 has been found in Problem No. 6.4 and is of the form

$$\frac{V_0(s)}{V_i(s)} = \frac{a(1+sT)}{1+asT}$$

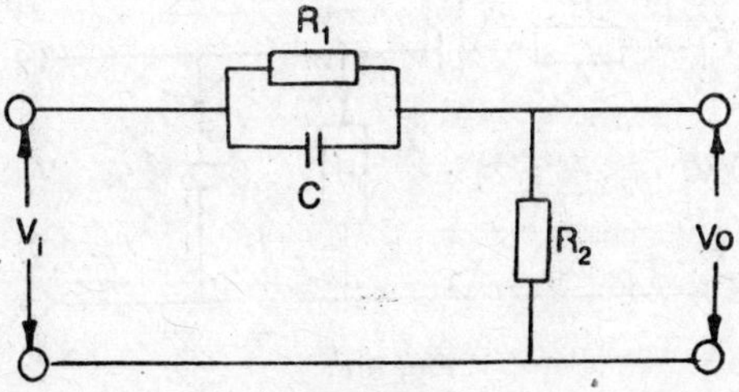

Fig. 6.9

where $\quad T = CR_1$ and $a = \dfrac{R_2}{R_1 + R_2}$

when $T = 0.6$ sec.

$$T = CR_1$$

$$0.6 = 1 \times 10^{-6} \times R_1$$

$$\therefore \quad R_1 = \frac{0.6}{1 \times 10^{-6}} = 0.6\ \text{M}\Omega \qquad \textbf{Ans.}$$

Also $$a = \frac{R_2}{R_1 + R_2}$$

or $$0.1 = \frac{R_2}{0.6 + R_2}$$

or $$0.06 + 0.1R_2 = R_2$$

$$\therefore \quad R_2 = \frac{0.06}{0.9} = 0.066\ \text{M}\Omega \qquad \textbf{Ans.}$$

Problem 6.9 Find the transfer function of the network shown in Fig. 6.10. Plot its Poles and Zeros for R = C = 1.

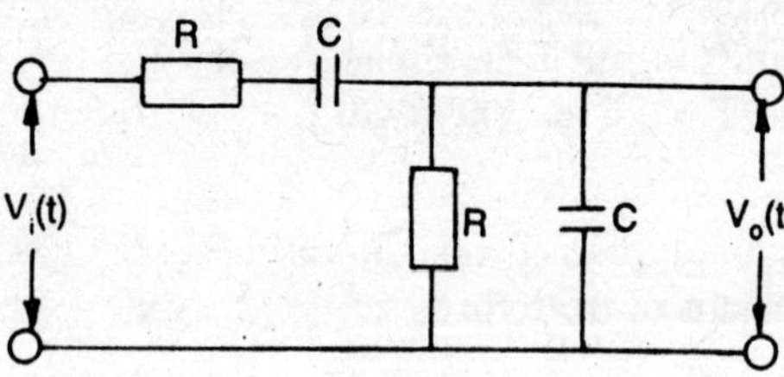

Fig. 6.10

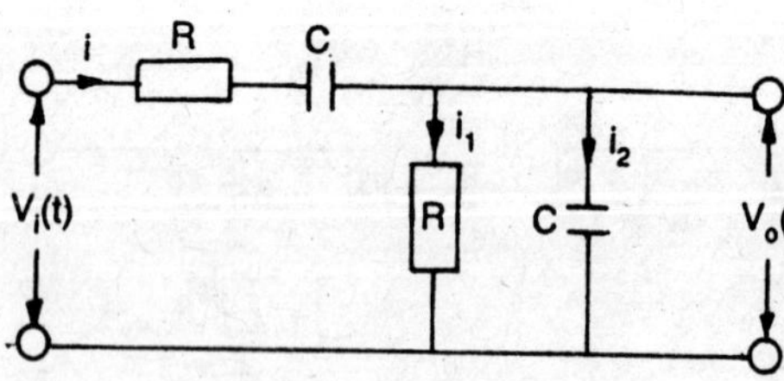

Fig. 6.11

Solution

Assume current distribution as shown in Fig. 6.11

Assuming no external load

$$i = i_1 + i_2$$

Applying Kirchoff's laws to the electrical network, we get

$$v_{i(t)} = Ri_{(t)} + \frac{1}{C}\int i_{(t)}dt + Ri_{1(t)}$$

and $$v_{0(t)} = \frac{1}{C}\int i_{2(t)}dt = Ri_{1(t)}$$

Taking Laplace transform, we get

$$V_i(s) = RI(s) + \frac{I(s)}{sC} + V_0(s)$$

and $$V_0(s) = \frac{I_2(s)}{sC} = RI_1(s)$$

or $$V_i(s) = (I_1(s) + I_2(s))\left[R + \frac{1}{sC}\right] + V_0(s)$$

or $$V_i(s) = V_0(s)\left[\frac{1}{R} + sC\right]\left[R + \frac{1}{sC}\right] + V_0(s)$$

or $$V_i(s) = V_0(s)\left[1 + \frac{1}{sRC} + sRC + 1 + 1\right]$$

or $$V_i(s) = V_0(s)\left[3 + \frac{1 + s^2R^2C^2}{sRC}\right]$$

or $$\frac{V_0(s)}{V_i(s)} = \frac{sRC}{s^2R^2C^2 + 3sRC + 1}$$

But, $R = C = 1$

$$\frac{V_0(s)}{V_i(s)} = \frac{s}{s^2 + 3s + 1}$$ **Ans.**

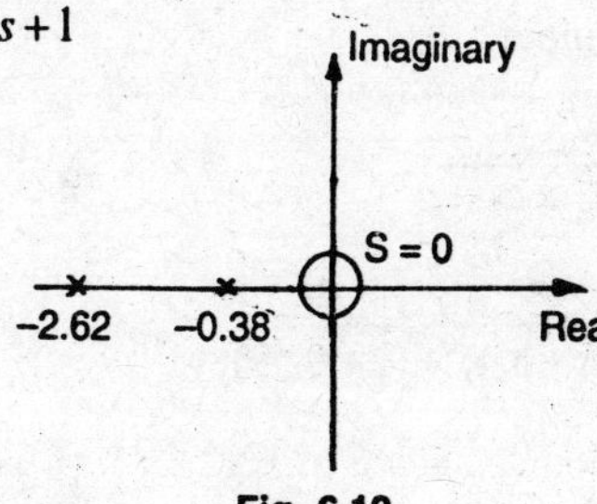

Fig. 6.12

There is one zero at origin and poles at –2.62 and –0.38. The plot is shown in Fig. 6.12

Problem 6.10 Determine the transfer function relation $V_0(s)$ to $V_i(s)$ for the network shown in Fig. 6.10.

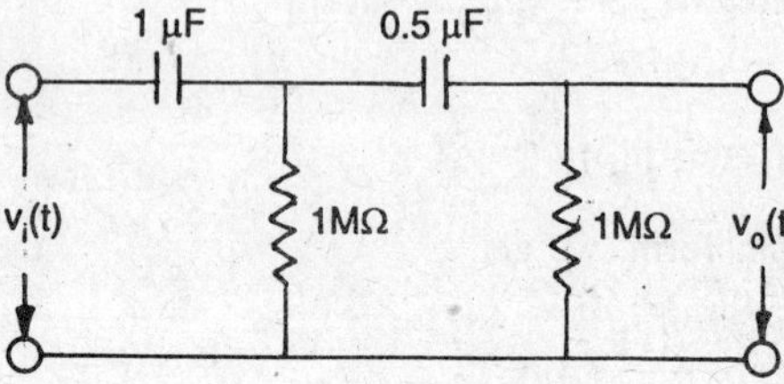

Fig. 6.13

Calculate output voltage, $t \geq 0$, for a unit a step voltage input at $t = 0$.

Solution

This is similar to problem No. 6.6. The transfer function is

$$\frac{V_0}{V_i}(s) = \frac{Z_2 Z_4}{Z_1 Z_2 + Z_1 Z_3 + Z_1 Z_4 + Z_2 Z_3 + Z_2 Z_4}$$

where $$Z_1 = \frac{1}{s}$$

$$Z_2 = 1$$

$$Z_3 = \frac{1}{0.5s}$$

$$Z_4 = 1$$

Substituting the values we get

$$\frac{V_0}{V_i}(s) = \frac{s^2}{s^2 + 4s + 2}$$ **Ans.**

For a unit step voltage input

$$V_0(s) = \frac{s}{s^2 + 4s + 2}$$

$$= \frac{A}{(s + 3.4)} + \frac{B}{(s + 0.59)}$$

where $A = 1.2$

and $B = -0.21$

$$V_0(s) = \frac{1.2}{(s+3.4)} + \frac{-0.21}{(s+0.59)}$$

or $$v_{0(t)} = 1.2e^{-3.4t} - 0.21e^{-0.59t}$$ **Ans.**

Problem 6.11 Determine the transfer function relating $V_0(s)$ to $V_i(s)$ for network shown in Fig. 6.14. Calculate output voltage, for a unit step voltage input at $t \geq 0$ when C_1 = 1μF, R_1 = 1MΩ, C_2 = 0.5μF and R_2 = 1MΩ.

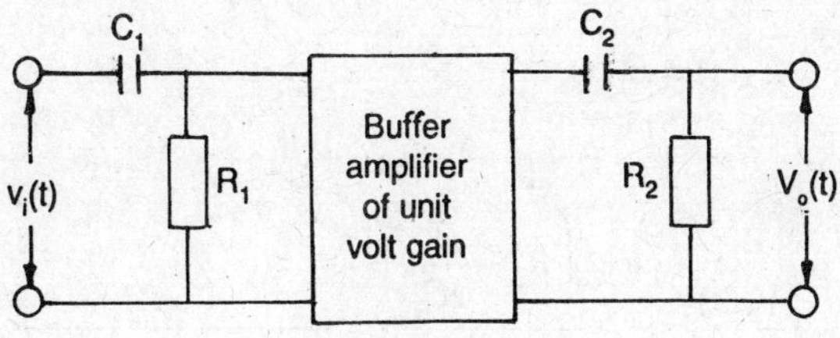

Fig. 6.14

Solution

Redrawing the circuit diagram as shown in Fig. 6.15 and applying Kirchoff's laws, we get

$$V_i(s) = I_1(s)R_1 + \frac{I_1(s)}{sC_1}$$

$$E_2(s) = I_2(s)R_2 + \frac{I_2(s)}{sC_2};$$

and $$E_1(s) = I_1(s)\,R_1$$

$$V_0(s) = I_2(s)\,R_2$$

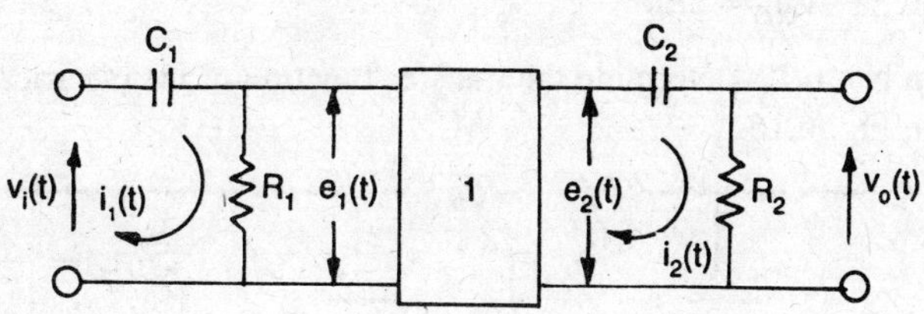

Fig. 6.15

Transfer function is

$$\frac{V_0(s)}{V_i(s)} = \frac{I_2(s)R_2}{I_1(s)\left(R_1 + \frac{1}{sC_1}\right)}$$

and
$$\frac{I_2(s)}{I_1(s)} = \frac{E_2(s)\Big/\left(R_2 + \frac{1}{sC_2}\right)}{E_1(s)/R_1}$$

But $\quad E_2(s) = E_1(s)$

Therefore

$$\frac{V_0(s)}{V_i(s)} = \frac{R_2\Big/\left(R_2 + \frac{1}{sC_2}\right)}{\left(R_1 + \frac{1}{sC_1}\right)\Big/R_1} = \frac{R_2}{\left(R_2 + \frac{1}{sC_2}\right)\left(1 + \frac{1}{sR_1C_1}\right)}$$

or
$$\frac{V_0(s)}{V_i(s)} = \frac{1}{\left(1 + \frac{1}{sR_2C_2}\right)\left(1 + \frac{1}{sR_1C_1}\right)}$$

Substituting the values of R_1, R_2, C_1 and C_2, we get

$$\frac{V_0(s)}{V_i(s)} = \frac{s^2}{s^2 + 3s + 2}$$ **Ans.**

For unit step voltage input

$$V_0(s) = \frac{s}{s^2 + 3s + 2} = \frac{2}{(s+2)} - \frac{1}{(s+1)}$$

Therefore, $\quad v_{0(t)} = 2e^{-2t} - e^{-t}$ **Ans.**

Problem No. 6.12 Determine the transfer function of the electrical network shown in Fig. 6.16.

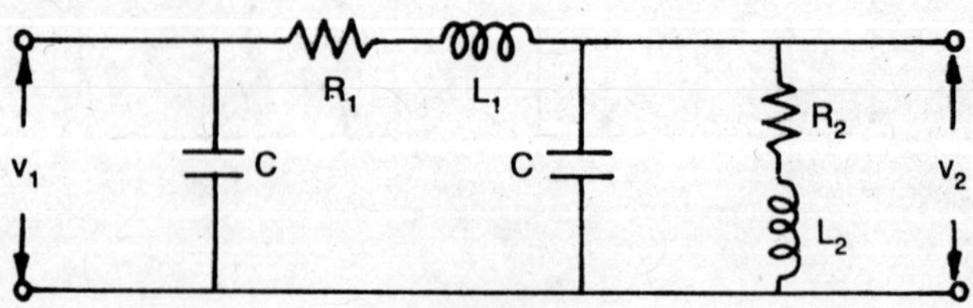

Fig. 6.16

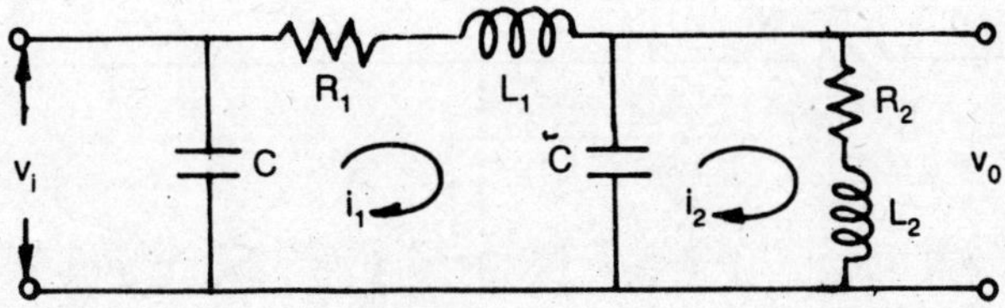

Fig. 6.17

Solution

Assuming current distribution as shown in Fig. 6.17 the differential equations can be written as

$$v_{i(t)} = R_1 i_{1(t)} + L_1 \frac{di_{1(t)}}{dt} + \frac{1}{C}\int (i_{1(t)} - i_{2(t)})\,dt,$$

$$0 = \frac{1}{C}\int (i_{1(t)} - i_{2(t)})\,dt + R_2 i_{2(t)} + L_2 \frac{di_{2(t)}}{dt}$$

and $$v_{0(t)} = R_2 i_{2(t)} + L_2 \frac{di_{2(t)}}{dt}$$

Assuming initial conditions to be zero, the Laplace transform of the above equations can be written as

$$V_i(s) = RI_1(s) + sL_1 I_1(s) + \frac{1}{sC}(I_1(s) - I_2(s)),$$

$$0 = \frac{1}{sC}(I_2(s) - I_1(s)) + R_2 I_2(s) + sL_2 I_2(s)$$

and $$V_0(s) = R_2 I_2(s) + sL_2 I_2(s)$$

$$\frac{V_0(s)}{V_i(s)} = \frac{R_2 + sL_2}{s^3 CL_1L_2 + s^2 C(R_1L_2 + L_1R_2) + s(L_1 + L_2 + CR_1R_2) + (R_1 + R_2)}$$

Problem No. 6.13 Show that the voltage transfer function of the network shown in Fig. 6.18 can be written as

$$\frac{V_2(s)}{V_1(s)} = \frac{1}{R_1C_2}\,\frac{s}{\left(s^2 + \frac{R_1C_1 + R_2(C_1 + C_2)}{R_1R_2C_1C_2}s + \frac{1}{R_1R_2C_1C_2}\right)}$$

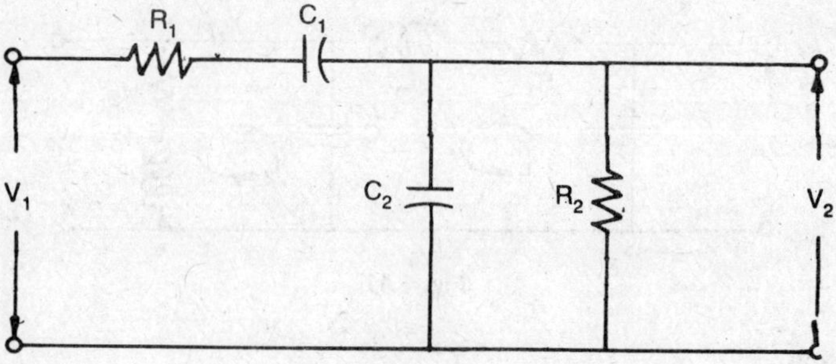

Fig. 6.18

Solution

Assume no external load and current distribution as shown in Fig. 6.19.

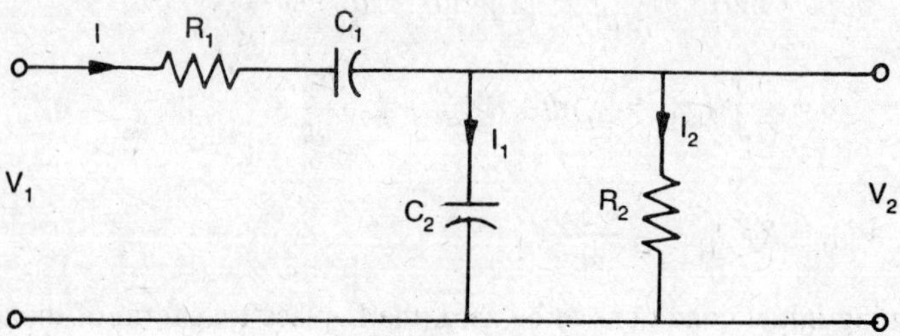

Fig. 6.19

$$I(s) = I_1(s) + I_2(s); \quad \text{...(1)}$$

$$V_1(s) = R_1 I(s) + \frac{I(s)}{sC_1} + \frac{I_1(s)}{sC_2} \quad \text{...(2)}$$

$$V_2(s) = R_2 I_2(s) + \frac{I_1(s)}{sC_2} \quad \text{...(3)}$$

Substituting the value of $\frac{I_1(s)}{sC_2}$ from equation (3) in (1), we get

$$V_1(s) = R_1 I(s) + \frac{I(s)}{sC_1} + V_2(s) \quad \text{...(4)}$$

Putting the value of $I(s)$ from equation (2) in (4), we get

$$V_1(s) = I_1(s) + I_2(s)\left[R_1 + \frac{1}{sC_1}\right] + V_2(s) \quad \text{...(5)}$$

Putting the values of $I_1(s)$ and $I_2(s)$ from equation (3) in (5), we get

$$V_1(s) = V_2(s)\left(sC_2 + \frac{1}{R_2}\right)\left(R_1 + \frac{1}{sC_1}\right) + V_2(s)$$

or $$V_1(s) = V_2(s)\left(sR_1C_2 + \frac{C_2}{C_1} + \frac{R_1}{R_2} + \frac{1}{sR_2C_1}\right) + V_2(s)$$

or $$V_1(s) = V_2(s)\left(\frac{s^2R_1R_2C_1C_2 + sR_2C_2 + sR_1C_1 + 1}{sR_2C_1}\right) + V_2(s)$$

or $$V_1(s) = V_2(s)\left(\frac{s^2R_1R_2C_1C_2 + sR_2C_2 + sR_1C_1 + sR_2C_1 + 1}{sR_2C_1}\right)$$

or $$\frac{V_2(s)}{V_1(s)} = \frac{sR_2C_1}{\left(s^2R_1R_2C_1C_2 + sR_2C_2 + sR_1C_1 + sR_2C_1 + 1\right)}$$

or $$\frac{V_2(s)}{V_1(s)} = \frac{sR_2C_1}{R_1R_2C_1C_2\left(s^2 + \frac{sR_2C_2 + sR_1C_1 + sR_2C_1}{R_1R_2C_1C_2} + \frac{1}{R_1R_2C_1C_2}\right)}$$

or $$\frac{V_2(s)}{V_1(s)} = \frac{\dfrac{sR_2C_1}{R_1R_2C_1C_2}}{s^2 + \dfrac{R_1C_1 + R_2(C_1 + C_2)}{R_1R_2C_1C_2}s + \dfrac{1}{R_1R_1C_1C_2}}$$

or $$\frac{V_2(s)}{V_1(s)} = \frac{1}{R_1C_2}\,\frac{s}{s^2 + \dfrac{R_1C_1 + R_2(C_1 + C_2)}{R_1R_2C_1C_2}s + \dfrac{1}{R_1R_1C_1C_2}}$$

Problem No. 6.14 Write nodal equations for the *RC* ladder network shown in Fig. 6.20 and hence determine the function $\dfrac{V_0}{V_1}$.

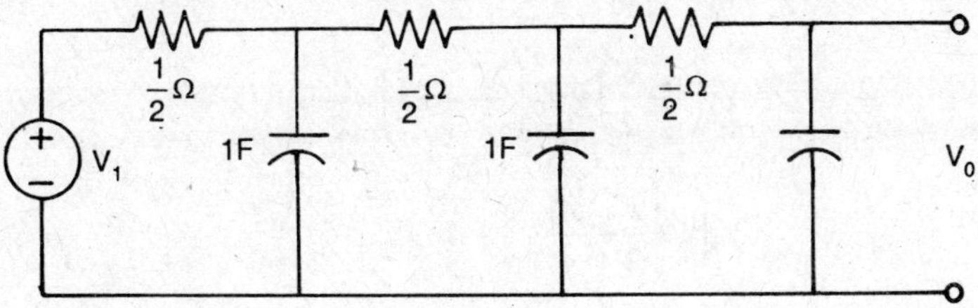

Fig. 6.20

Solution

Laplace transformed network is shown in Fig. 6.21

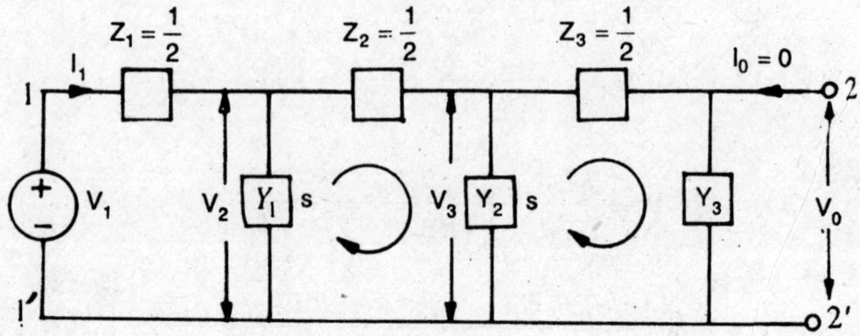

Fig. 6.21

Starting from port $2-2'$ we have,

$$I_3 = Y_3 V_0 = sV_0 \qquad ...(1)$$

$$V_3 = V_0 + I_3 Z_3$$

$$= V_0 + (sV_0) \times \frac{1}{2}$$

$$= V_0\left(1+\frac{s}{2}\right) \qquad ...(2)$$

$$I_2 = I_3 + V_3 Y_2$$

$$= sV_0 + V_0\left(1+\frac{s}{2}\right)s$$

$$= V_0\left(s+s+\frac{s^2}{2}\right)$$

$$= V_0\left(\frac{s^2}{2}+2s\right) \qquad ...(3)$$

$$V_2 = V_3 + I_2 Z_2$$

$$= V_0\left(1+\frac{s}{2}\right) + V_0\left(\frac{s^2}{2}+2s\right) \times \frac{1}{2}$$

$$= V_0\left(1+\frac{s}{2}+\frac{s^2}{4}+s\right)$$

$$= V_0\left(\frac{4+2s+s^2+4s}{4}\right)$$

$$= V_0\left(\frac{s^2+6s+4}{4}\right) \qquad \text{...(4)}$$

$$I_1 = I_2 + V_2Y_1$$

$$= V_0\left(\frac{s^2}{2}+2s\right) + V_0\left(\frac{s^2+6s+4}{4}\right)s$$

$$= V_0\left(\frac{s^2}{2}+2s+\frac{s^3+6s^2+4s}{4}\right)$$

$$= V_0\left(\frac{2s^2+8s+s^3+6s^2+4s}{4}\right)$$

$$= V_0\left(\frac{s^3+8s^2+12s}{4}\right) \qquad \text{...(5)}$$

$$V_1 = V_2 + I_1Z_1$$

$$= V_0\left(\frac{6s^2+6s+4}{4}\right) + V_0\left(\frac{s^3+8s^2+12s}{4}\right) \times \frac{1}{2}$$

$$= V_0\left(\frac{12s^2+12s+8+s^3+8s^2+12s}{8}\right)$$

$$= V_0\left(\frac{s^3+20s^2+24s+8}{8}\right) \qquad \text{...(6)}$$

Therefore,

$$\frac{V_0}{V_1} = \frac{V_0}{V_0\left(\dfrac{s^3+20s^2+24s+8}{8}\right)}$$

$$= \frac{8}{s^3+20s^2+24s+8}.$$

Problem No. 6.15 Express $\frac{V_0}{V_1}$ for the *LC* ladder network of Fig. 6.22 in the form $K\left[\dfrac{(s^2+a)(s^2+b)}{(s^2+c)(s^2+d)}\right]$ and find the values of k, a, b, c and d.

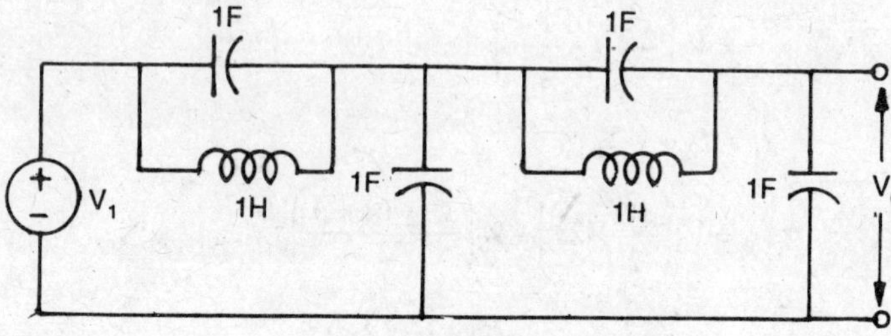

Fig. 6.22

Solution

Laplace transformed network is shown in Fig. 6.23.

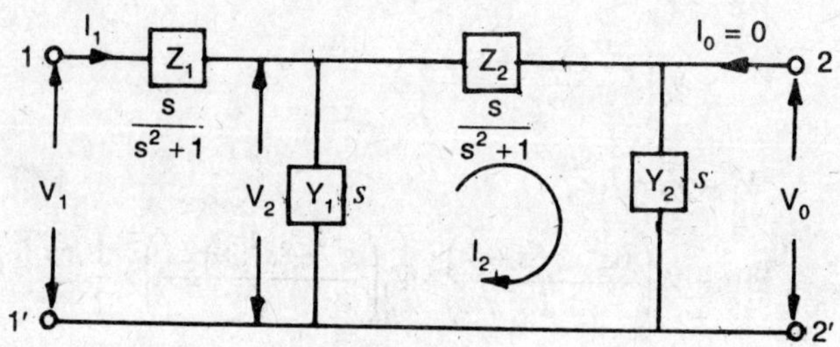

Fig. 6.23

Starting from port $2-2'$ we get,

$$I_2 = V_0 Y_2$$

$$= V_0 s = sV_0 \qquad \text{...(1)}$$

$$V_2 = V_0 + I_2 Z_2$$

$$= V_0 + (sV_0)\times\frac{s}{s^2+1}$$

$$= V_0\left(1+\frac{s^2}{s^2+1}\right)$$

$$= V_0\left(\frac{2s^2+1}{s^2+1}\right) \qquad \text{...(2)}$$

$$I_1 = I_2 + V_2 Y_1$$

$$= sV_0 + V_0\left(\frac{2s^2+1}{s^2+1}\right)s$$

$$= V_0\left[\frac{s(s^2+1)+s(2s^2+1)}{s^2+1}\right]$$

$$= V_0\left[\frac{s^3+s+2s^3+s}{s^2+1}\right]$$

$$= V_0\left[\frac{4s^3+2s}{s^2+1}\right] \qquad \text{...(3)}$$

$$V_1 = V_2 + I_1 Z_1$$

$$= V_0\left(\frac{2s^2+1}{s^2+1}\right)+V_0\left(\frac{4s^3+2s}{s^2+1}\right)\left(\frac{s}{s^2+1}\right)$$

$$= V_0\left[\frac{2s^2+1}{s^2+1}+\frac{4s^4+2s^2}{(s^2+1)^2}\right]$$

$$= V_0\left[\frac{(2s^2+1)(s^2+1)+4s^4+2s^2}{(s^2+1)^2}\right]$$

$$= V_0\left[\frac{2s^4+2s^2+s^2+1+4s^4+2s^2}{(s^2+1)^2}\right]$$

$$= V_0\left[\frac{6s^4+5s^2+1}{(s^2+1)^2}\right]$$

Therefore,

$$\frac{V_0}{V_1} = \frac{V_0}{V_0\left[\dfrac{6s^4+5s^2+1}{(s^2+1)^2}\right]}$$

$$= \frac{(s^2+1)^2}{6s^4+5s^2+1}$$

$$= \frac{(3s^2+1)(2s^2+1)}{(s^2+1)^2}$$

$$= \frac{6\left(s^2+\frac{1}{3}\right)\left(s^2+\frac{1}{2}\right)}{(s^2+1)(s^2+1)} \qquad ...(4)$$

This can be compared with

$$\frac{V_0}{V_1} = K\left[\frac{(s^2+a)(s^2+b)}{(s^2+c)(s^2+d)}\right] \qquad ...(5)$$

Comparing equations (4) and (5) we see that

$$K = 6\ ;\ a = \frac{1}{3};\ \ b = \frac{1}{2};\ c = d = 1$$

Problem No. 6.16 Find $\frac{V_0}{V_1'}$ for the circuit shown in Fig. 6.24. If $C_1 = C_2 = 1F$, find a set of values for R_1, R_2 and R_3 to realise the function

$$\frac{V_0}{V_1} = \frac{-2}{s^2+16.625s+8}$$

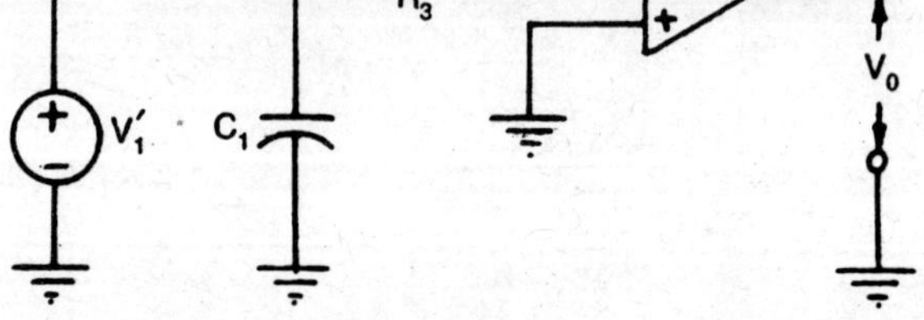

Fig. 6.24

Solution Let

V_1 = voltage at node 1; and

V_2 = voltage at node 2.

Node equations By making using of KCL, node equations are

node 1:

$$V_1(Y_1+Y_2+Y_3+Y_4) - V_0Y_4 - V_1'Y_1 - V_2Y_1 = 0 \qquad ...(1)$$

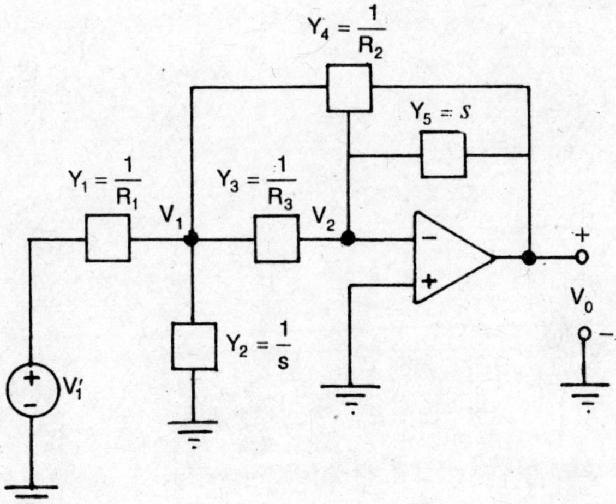

Fig. 6.25

Since $V_2 = 0$, as node 2 is grounded, we have

$$V_1(Y_1 + Y_2 + Y_3 + Y_4) - V_0Y_4 - V_1'Y_1 = 0 \quad ...(2)$$

node 2:

$$V_2(Y_3 + Y_5) - V_1Y_3 - V_0Y_5 = 0 \quad ...(3)$$

Since $V_2 = 0$

$$-V_1Y_3 - V_0Y_5 = 0 \quad ...(4)$$

or

$$V_1 = -\frac{V_0Y_5}{Y_3} \quad ...(5)$$

Equation (2) is

$$V_1(Y_1 + Y_2 + Y_3 + Y_4) - V_0Y_4 - V_1'Y_1 = 0 \quad ...(6)$$

Substituting value of V_1 from equation (5) in (6), we get

$$-\frac{V_0Y_5}{Y_3}(Y_1 + Y_2 + Y_3 + Y_4) - V_0Y_4 = V_1'Y_1$$

or $$\frac{V_0}{V_1'} = \frac{-Y_1Y_3}{Y_5(Y_1 + Y_2 + Y_3 + Y_4) + Y_3Y_4}$$

Putting $$Y_1 = \frac{1}{R_1};$$

$$Y_4 = \frac{1}{R_2};$$

$$Y_3 = \frac{1}{R_3};$$

$$Y_5 = Y_2 = s$$

$$\frac{V_0}{V_1'} = \frac{-\dfrac{1}{R_1 R_3}}{s\left(\dfrac{1}{R_1} + s + \dfrac{1}{R_3} + \dfrac{1}{R_2}\right) + \dfrac{1}{R_2 R_3}}$$

$$= \frac{-\dfrac{1}{R_1 R_3}}{s^2 + \dfrac{s}{R_1} + \dfrac{s}{R_2} + \dfrac{s}{R_3} + \dfrac{1}{R_2 R_3}}$$

$$= \frac{-\dfrac{1}{R_1 R_3}}{\dfrac{R_1 R_2 R_3 s^2 + R_2 R_3 s + R_1 R_3 s + R_1 R_2 s + R_1}{R_1 R_2 R_3}}$$

or
$$\frac{V_0}{V_1'} = \frac{-R_2}{R_1 R_2 R_3 s^2 + s(R_1 R_2 + R_2 R_3 + R_3 R_1) + R_1} \qquad ...(7)$$

Comparing relation obtained in equation (7) with the required relation given in equation (8) below,

$$\frac{V_0}{V_1'} = \frac{-2}{s^2 + 16.625s + 8} \qquad ...(8)$$

we have

$$\left.\begin{aligned} R_2 &= 2\Omega; \text{ and} \\ R_1 &= 8\Omega; \text{ and} \\ R_1 R_2 R_3 &= 1 \\ \text{or } 8 \times 2 \times R_3 &= 1 \\ \text{or } \quad R_3 &= \frac{1}{16}\Omega \end{aligned}\right\} \qquad ...(9)$$

Check

$$R_1R_2 + R_2R_3 + R_3R_1 = 16.625$$

$$8\times 2 + 2\times\frac{1}{16} + \frac{1}{16}\times 8 = 16.625$$

$$16 + \frac{1}{8} + \frac{1}{2} = 16.625$$

$$\frac{128+1+4}{8} = 16.625$$

$$\frac{133}{8} = 16.625$$

$$16.625 = 16.625$$

Since L.H.S. = R.H.S.

values of R_1, R_2 and R_3 found out are correct

or

If we substitute the values of R_1, R_2 and R_3 obtained in equation (9), we get

$$\frac{V_0}{V_1'} = \frac{-2}{\left(8\times 2\times\frac{1}{16}\right)s^2 + \left(8\times 2 + 2\times\frac{1}{16} + \frac{1}{16}\times 8\right)s + 8}$$

$$= \frac{-2}{s^2 + 16.625s + 8}$$

which is the function required to be realised.

Problem No. 6.17 Obtain the transfer function in the form $\dfrac{s}{K(s^2+as+b)}$ of the network shown in Fig. 6.26.

Solution

KCL equations at nodes '*a*' and '*b*' are

Node 'a'

$$V_a(Y_1 + Y_2 + Y_3) - V_bY_2 - V_0Y_3 - V_1Y_1 = 0$$

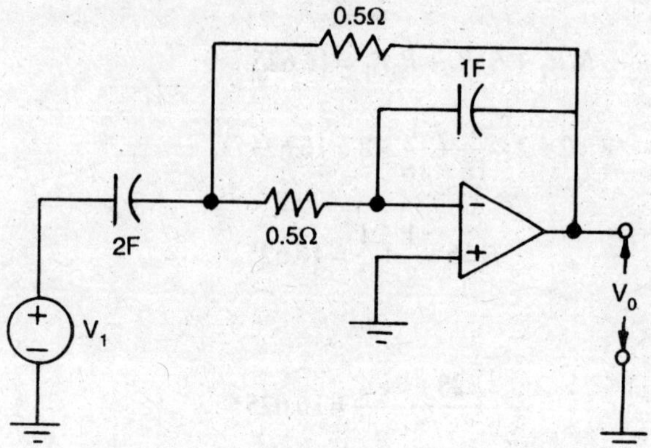

Fig. 6.26

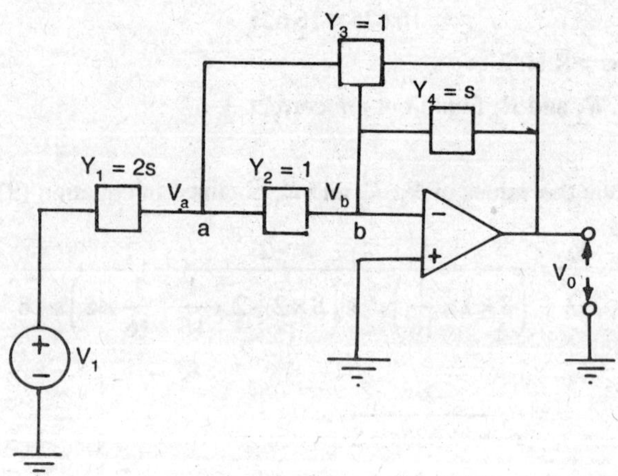

Fig. 6.27

Since $V_b = 0$ being grounded

$$V_a(Y_1 + Y_2 + Y_3) - V_0 Y_3 - V_1 Y_1 = 0 \qquad ...(1)$$

Node 'b'

$$V_b(Y_2 + Y_4) - V_0 Y_4 - V_a Y_2 = 0$$

Since $V_b = 0$, we have

$$-V_0 Y_4 - V_a Y_2 = 0$$

or

$$V_a = \frac{-V_0 Y_4}{Y_2} \qquad ...(2)$$

Substituting value of V_a from equation (2) in equation (1), we get

$$\frac{-V_0 Y_4}{Y_2}(Y_1 + Y_2 + Y_3) - V_0 Y_3 - V_1 Y_1 = 0$$

or
$$\frac{-V_0 Y_4}{Y_2}(Y_1 + Y_2 + Y_3) - V_0 Y_3 = V_1 Y_1$$

or
$$-V_0[Y_4(Y_1 + Y_2 + Y_3)] - V_0 Y_3 Y_2 = V_1 Y_1 Y_2$$

or
$$\frac{V_0}{V_1} = \frac{-Y_1 Y_2}{Y_4(Y_1 + Y_2 + Y_3) + Y_3 Y_2} \qquad \text{...(3)}$$

Now

$$Y_1 = 2s;$$
$$Y_2 = 1;$$
$$Y_3 = 1;$$
$$Y_4 = s$$

Substituting above given values in equations (3), we get

$$\frac{V_0}{V_1} = \frac{-2s}{s(2s+1+1)+1}$$

$$= \frac{-2s}{2s^2 + 2s + 1}$$

$$= \frac{-s}{s^2 + s + \frac{1}{2}}$$

This can be written as

$$\frac{V_0}{V_1} = \frac{s}{K(s^2 + as + b)}$$

where

$$K = -1;$$
$$a = 1; \quad \text{and}$$
$$b = \frac{1}{2}$$

Problem No. 6.18 Show that the voltage transfer function of the network shown in Fig. 6.28 can be written as $\dfrac{Z_1 Z_2 + Z_3(Z_1 + Z_2 + Z_4)}{Z_1 Z_3 + (Z_1 + Z_3)(Z_2 + Z_4)}$

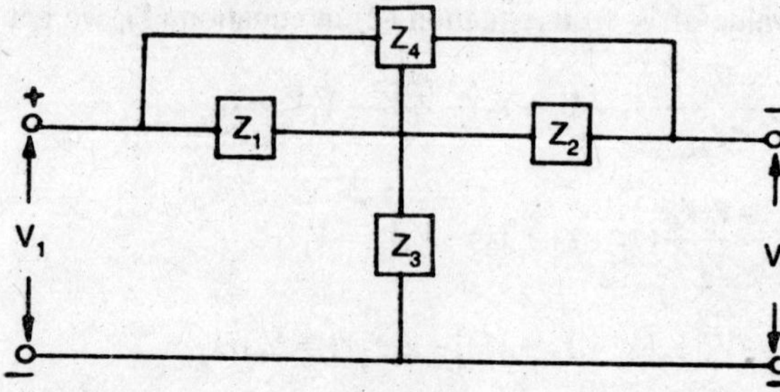

Fig. 6.28

The given network can be taken as a parallel combination of the two networks as shown in Fig. 6.29(a) and 6.29(b)

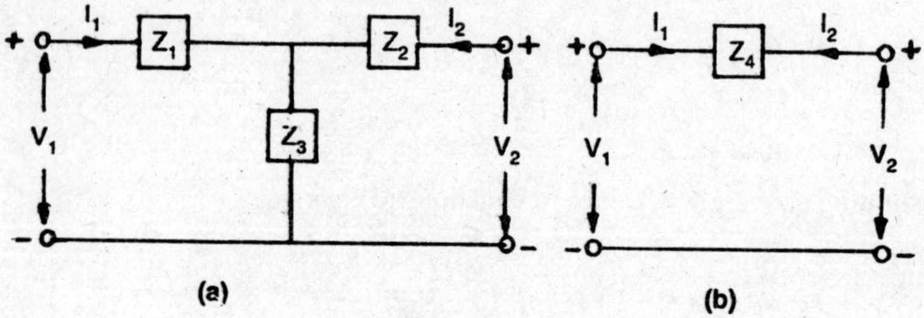

Fig. 6.29

Fig. 6.29(a)

$$V_1 = Z_1 I_1 + Z_3(I_1 + I_2) = (Z_1 + Z_3) I_1 + Z_3 I_2 \quad ...(1)$$

and $$V_2 = Z_3 I_1 + (Z_2 + Z_3) I_2 \quad ...(2)$$

Therefore,

$$Z_{11} = (Z_1 + Z_3);$$

$$Z_{12} = Z_3 = Z_{21};$$

$$Z_{22} = Z_2 + Z_3$$

$$\Delta Z = Z_{11} Z_{22} - Z_{21} Z_{12}$$

$$= (Z_1 + Z_3)(Z_2 + Z_3) - Z_3^2$$

$$Y_{21} = \frac{Z_{21}}{\Delta Z} = \frac{-Z_3}{\Delta Z}$$

$$Y_{22} = \frac{-Z_{11}}{\Delta Z} = \frac{Z_1 + Z_3}{\Delta Z} \quad \text{...(5)}$$

Fig. 6.29(b)

Y-parameter of the network are

$$Y_{11} = Y_{22} = \frac{1}{Z_4} \quad \text{...(6)}$$

$$Y_{12} = Y_{21} = -\frac{1}{Z_4} \quad \text{...(7)}$$

Y_{21} for the network is

$$Y_{21} = \frac{-Z_3}{\Delta Z} - \frac{1}{Z_4} \quad \text{...(8)}$$

$$Y_{22} = \frac{Z_1 + Z_3}{\Delta Z} + \frac{1}{Z_4}$$

Voltage Ratio

$$\frac{V_2}{V_1} = \frac{-Y_{21}}{Y_{22}}$$

$$= -\frac{\dfrac{-Z_3}{\Delta Z} - \dfrac{1}{Z_4}}{\dfrac{Z_1 + Z_3}{\Delta Z} + \dfrac{1}{Z_4}}$$

$$= -\frac{-Z_3 Z_4 - \Delta Z}{(Z_1 + Z_3) Z_4 + \Delta Z}$$

$$= -\frac{-Z_3 Z_4 + Z_3^2 - (Z_1 + Z_3)(Z_2 + Z_3)}{Z_1 Z_4 + Z_3 Z_4 + (Z_1 + Z_3)(Z_2 + Z_3) - Z_3^2}$$

$$= -\frac{-Z_3 Z_4 - Z_1 Z_2 - Z_1 Z_3 - Z_2 Z_3}{Z_1 Z_4 + Z_3 Z_4 + Z_1 Z_2 + Z_1 Z_3 + Z_2 Z_3}$$

$$= \frac{Z_1 Z_2 + Z_3 (Z_1 + Z_2 + Z_4)}{Z_1 Z_3 + Z_1 (Z_2 + Z_4) + Z_3 (Z_2 + Z_4)}$$

$$= \frac{Z_1 Z_2 + Z_3(Z_1 + Z_2 + Z_4)}{Z_1 Z_3 + (Z_1 + Z_3)(Z_2 + Z_4)}$$

Problem No. 6.19 Show that the driving point impedance of the network shown in Fig. 6.30 by continued fraction becomes

$$\frac{s^4 + 3s^2 + 1}{s^3 + 2s}$$

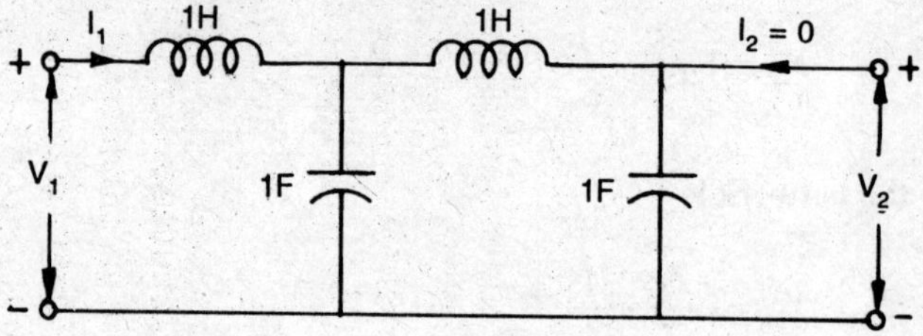

Fig. 6.30

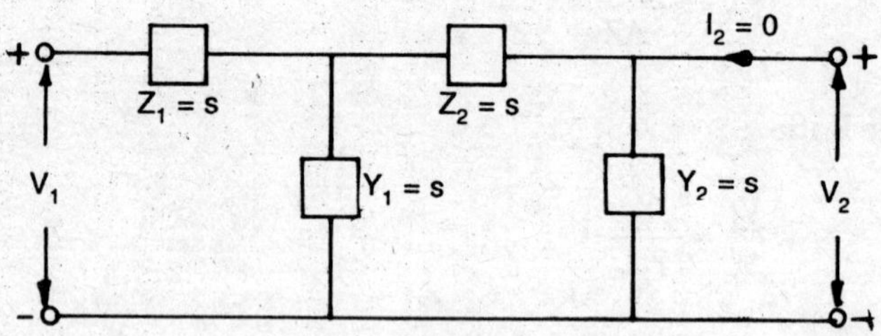

Fig. 6.31

Solution

$$Z = Z_1 + \cfrac{1}{Y_1 + \cfrac{1}{Z_2 + \cfrac{1}{Y_2}}}$$

$$= s + \cfrac{1}{s + \cfrac{1}{s + \cfrac{1}{s}}}$$

$$= s + \cfrac{1}{s + \cfrac{s}{s^2 + 1}}$$

$$= s + \frac{s^2 + 1}{s^3 + s + s}$$

$$= \frac{s^4 + 2s^2 + s^2 + 1}{s^3 + 2s}$$

$$= \frac{s^4 + 3s^2 + 1}{s^3 + 2s}$$

Problem No. 6.20 Find the open-circuit driving point impedance at terminals $1-1'$ of the ladder network shown in Fig. 6.32.

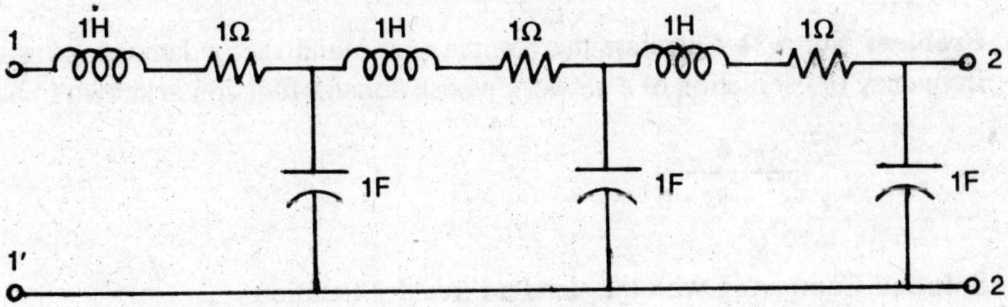

Fig. 6.32

Solution

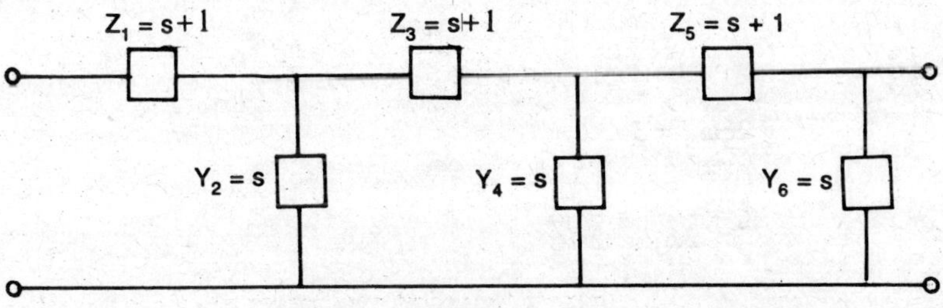

Fig. 6.33

$$Z = Z_1 + \cfrac{1}{Y_2 + \cfrac{1}{Z_3 + \cfrac{1}{Y_4 + \cfrac{1}{Z_5 + \cfrac{1}{Y_6}}}}}$$

$$= s+1+\cfrac{1}{s+\cfrac{1}{s+1+\cfrac{1}{s+\cfrac{1}{(s+1)+\cfrac{1}{s}}}}}$$

On simplification, it gives

$$Z = \frac{s^6+6s^5+11s^4+13s^3+10s^2+5s+1}{s^5+5s^4+5s^3+3s^2+3s}$$

Problem No. 6.21 Calculate the damping ratio and the undamped natural frequency of oscillation of a network whose transfer function is known to be

$$\frac{6}{s^2+3s+6}$$

Solution Comparing with the standard transfer function $\dfrac{\omega_n^{\,2}}{s^2+2\xi\omega_n s+6}$

We have

$$\omega_n = \text{undamped natural frequency}$$
$$= \sqrt{6}$$

and

$$2\xi\omega_n = 3$$

or

$$\xi = \frac{3}{2\omega_n} = \frac{3}{2\sqrt{6}}$$

Problem No. 6.22 The network shown in Fig. 6.34 is known as a shunt peaking network. Show that impedance has the form

$$Z(s) = \frac{K(s-z_1)}{(s-p_1)(s-p_2)}$$ and determine

z_1, p_1 and p_2 in terms of *R*, *L* and *C*. If the poles and zeros have the locations shown in Fig. 6.35 with $Z(j0) = 1$, find the values of *R*, *L* and *C*.

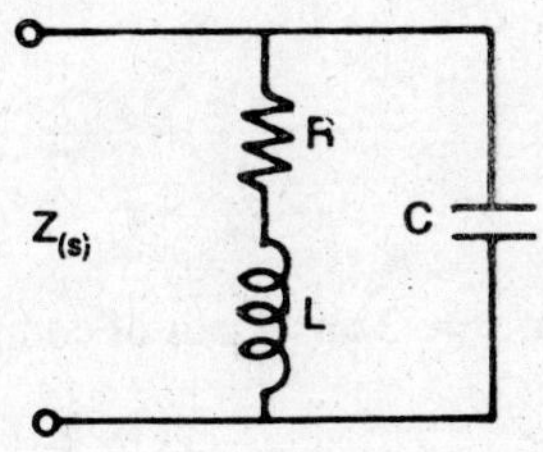

Fig. 6.34

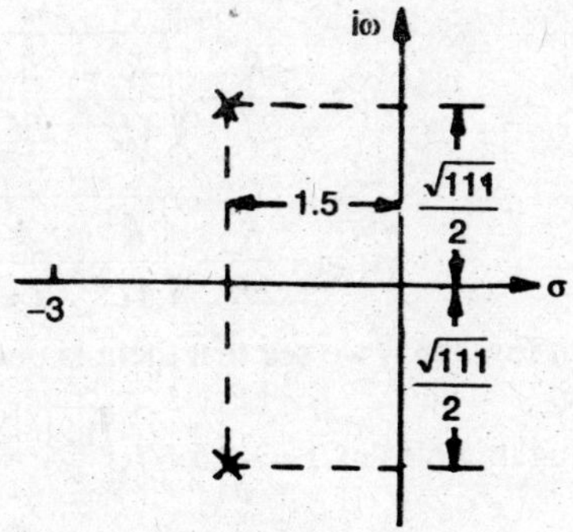

Fig. 6.35

Solution

In the transform form

$$Z(s) = \frac{(R+sL)\times\frac{1}{sC}}{R+sL+\frac{1}{sC}}$$

$$= \frac{sL+R}{s^2LC+sRC+1}$$

$$= \frac{L(s+R/L)}{s^2LC+sRC+1} = \frac{s+R/L}{s^2C+\frac{sRC}{L}+\frac{1}{L}} \quad ...(1)$$

$$= \frac{(s+R/L)}{\left(s+\frac{R}{2L}+\sqrt{\frac{R^2}{4L^2}-\frac{1}{LC}}\right)\left(s+\frac{R}{2L}-\sqrt{\frac{R^2}{4L^2}-\frac{1}{LC}}\right)} \quad ...(2)$$

Comparing with

$$Z(s) = \frac{K(s-z_1)}{(s-p_1)(s-p_2)}$$

we get

$$z_1 = -\frac{R}{L};$$

$$p_1 = -\frac{R}{2L} - \sqrt{\frac{R^2}{4L^2} - \frac{1}{LC}} \text{ ; and}$$

$$p_2 = -\frac{R}{2L} + \sqrt{\frac{R^2}{4L^2} - \frac{1}{LC}}$$

From Fig. 6.35, we see that there is one zero at $s = -3$ ans a pair of complex conjugate poles at $s = -1.5 \pm j\sqrt{\frac{111}{2}}$

Therefore,

$$Z(s) = \frac{K(s+3)}{\left(s + 1.5 + j\frac{\sqrt{111}}{2}\right)\left(s + 1.5 - j\frac{\sqrt{111}}{2}\right)} \qquad ...(3)$$

$$Z(s) = \frac{K(s+3)}{(s+1.5)^2 + \left(\sqrt{111}/2\right)^2} = \frac{K(s+3)}{s^2 + 3s + 30} \qquad ...(4)$$

It is given that

$Z(j0) = 1$, Therefore,

$$Z(j0) = 1 = \frac{3K}{30}$$

or $\quad K = 10 \qquad ...(5)$

Comparing equations (1) and (4), we get

(a) $\quad C = \frac{1}{K} = \frac{1}{10}; \qquad ...(9)$

(b) $\quad \frac{1}{L} = \frac{30}{K}$

or $\quad \frac{1}{L} = \frac{30}{10}$ (because K = 10 found earlier)

or $\quad L = \frac{1}{3} \qquad ...(10)$

(c) $\quad \frac{RC}{L} = \frac{3}{K} \qquad ...(11)$

putting $K = 10$, $L = \dfrac{1}{3}$ and $C = \dfrac{1}{10}$, we get

$$\frac{R \times 1/10}{1/3} = \frac{3}{10}$$

or $$3\frac{R}{10} = \frac{3}{10}$$

or $R = 1$.

Problem No. 6.23 For the RC two-port network shown in Fig. 6.36 show that

$$G_{12} = \frac{1/R_1R_2C_1C_2}{\left(s^2 + (R_1C_1 + R_1C_2 + R_2C_2)s/R_1R_2C_1C_2 + 1/R_1R_2C_1C\right)}$$

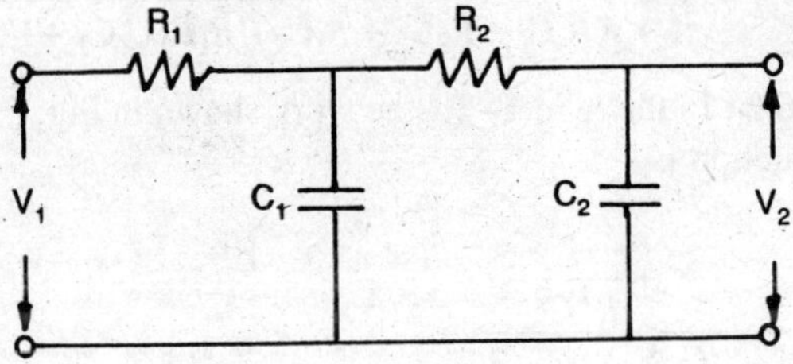

Fig. 6.36

Solution Transformed network is shown in Fig. 6.37

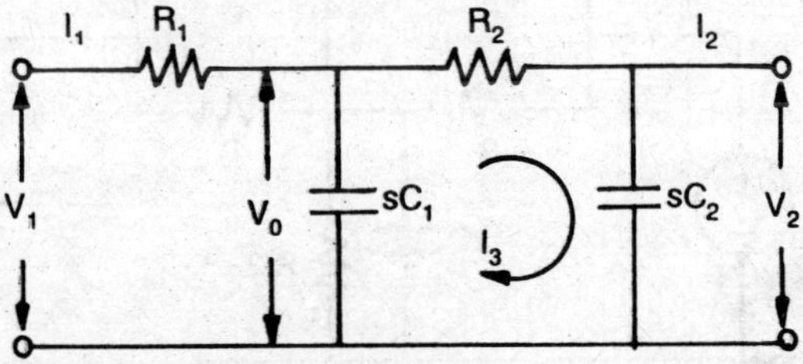

Fig. 6.37

$$I_3 = V_2sC_2$$

$$\begin{aligned} V_0 &= V_2 + I_3R_2 \\ &= V_2 + V_2s\,R_2C_2 \\ &= V_2(1 + sR_2C_2) \end{aligned}$$

$$I_1 = I_3 + V_0 sC_1$$
$$= V_2 sC_2 + V_2(1+sR_2C_2)sC_1$$
$$= V_2\left(sC_2 + sC_1 + s^2R_2C_1C_2\right)$$
$$V_1 = V_0 + I_1R_1$$
$$= V_2(1+sR_2C_2) + V_2(sC_2 + sC_1 + s^2R_2C_1C_2)R_1$$
$$= V_2(1+sR_2C_2 + sR_1C_2 + sR_1C_1 + s^2R_1R_2C_1C_2)$$

Therefore,

$$\frac{V_2}{V_1} = \frac{1}{s^2R_1R_2C_1C_2 + s(R_1C_1 + R_2C_2 + R_1C_2) + 1}$$

or $$\frac{V_2}{V_1} = G_{12} = \frac{1/R_1R_2C_1C_2}{s^2 + s(R_1C_1 + R_2C_2 + R_1C_2)/R_1R_2C_1C_2 + 1/R_1R_2C_1C_2}$$

Problem No. 6.24 In the twin T– RC network shown in Fig. 6.38, show that the transfer function is

$$\frac{s^2 + \dfrac{1}{R^2C^2}}{s^2 + \left(\dfrac{4}{RC}\right)s + \dfrac{1}{R^2C^2}}$$

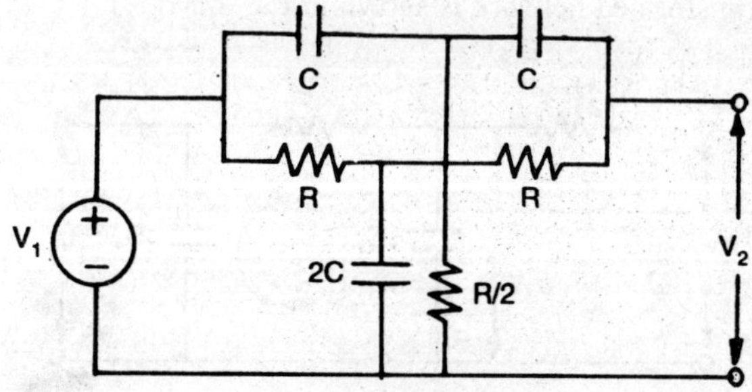

Fig. 6.38

Solution

$$Z_{1A} = \frac{\Sigma Z_1 Z_2}{Z_2} = \frac{\dfrac{1}{sC}\times\dfrac{1}{sC} + \dfrac{1}{sC}\times\dfrac{R}{2} + \dfrac{1}{sC}\times\dfrac{R}{2}}{\dfrac{1}{sC}}$$

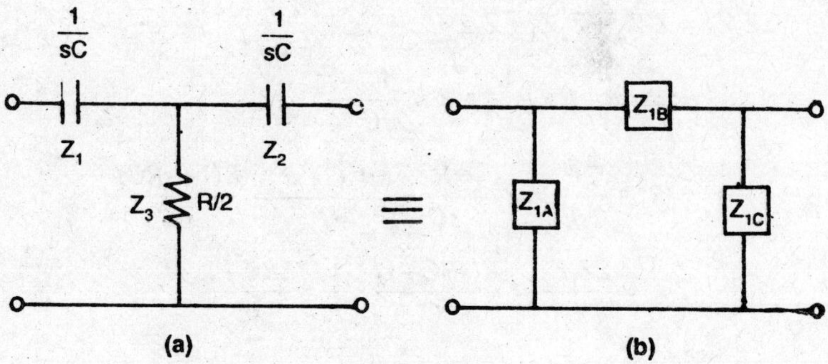

Fig. 6.39

$$= \frac{\dfrac{1}{s^2C^2} + \dfrac{R}{2sC} + \dfrac{R}{2sC}}{\dfrac{1}{sC}}$$

$$= \frac{\dfrac{2 + sRC + sRC}{2s^2C^2}}{\dfrac{1}{sC}} = \frac{2(1+sRC)}{2sC} = \frac{1+sRC}{sC}$$

$$Z_{1C} = \frac{\Sigma Z_1 Z_2}{Z_1} = Z_{1A} = \frac{1+sRC}{sC}$$

$$Z_{1B} = \frac{\Sigma Z_1 Z_2}{Z_3} = \frac{\dfrac{2(1+sRC)}{2s^2C^2}}{R/2} = \frac{2(1+sRC)}{s^2RC^2}$$

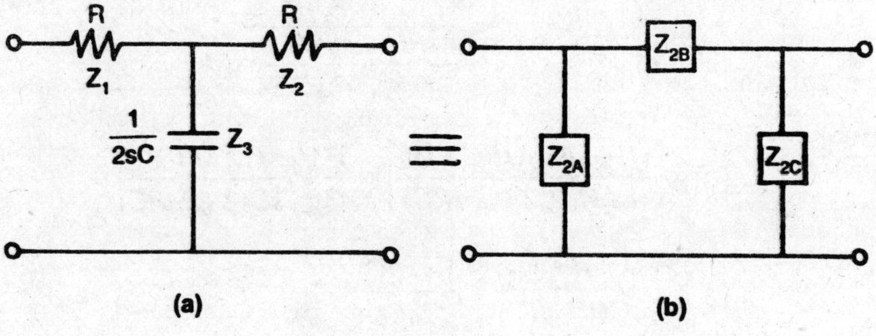

Fig. 6.40

$$\Sigma Z_1Z_2 = Z_1Z_2 + Z_2Z_3 + Z_3Z_2$$

$$= R \times R + R \times \frac{1}{2sC} + R \times \frac{1}{2sC}$$

$$= R^2 + \frac{2R}{2sC} = R^2 + \frac{R}{sC} = \frac{sR^2C + R}{sC} = \frac{R(1+sRC)}{sC}$$

$$Z_{2A} = \frac{\Sigma Z_1Z_2}{Z_2} = \frac{R(sRC+1)}{sC} \times \frac{1}{R} = \frac{sRC+1}{sC}$$

$$Z_{2C} = \frac{\Sigma Z_1Z_2}{Z_1} = Z_{2A} = \frac{sRC+1}{sC}$$

$$Z_{2B} = \frac{\Sigma Z_1Z_2}{Z_3} = \frac{\frac{R(1+sRC)}{sC}}{1/2sC} = 2R(1+sRC)$$

The two π-networks are placed in parallel to obtain equivalent π-network shown in Fig. 6.41.

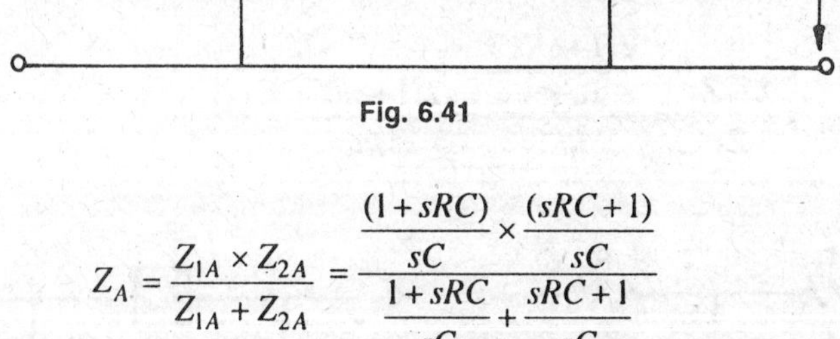

Fig. 6.41

$$Z_A = \frac{Z_{1A} \times Z_{2A}}{Z_{1A} + Z_{2A}} = \frac{\frac{(1+sRC)}{sC} \times \frac{(sRC+1)}{sC}}{\frac{1+sRC}{sC} + \frac{sRC+1}{sC}}$$

$$= \left[\frac{(1+sRC)(sRC+1)}{(1+sRC)+(1+sRC)}\right] \times \frac{1}{sC} = \frac{(1+sRC)^2}{sC(2+2sRC)}$$

$$= \frac{(1+sRC)^2}{2sC(1+sRC)} = \frac{1+sRC}{2sC}$$

$$Z_C = \frac{Z_{1C} \times Z_{2C}}{Z_{1C} + Z_{2C}} = Z_A = \frac{1+sRC}{2sC} \quad Z_B = \frac{Z_{1B} \times Z_{2B}}{Z_{1B} + Z_{2B}}$$

$$= \frac{\dfrac{2(1+sRC)}{s^2RC^2} \times \dfrac{2R(1+sRC)}{1}}{\dfrac{2(1+sRC)}{s^2RC^2} + \dfrac{2R(1+sRC)}{1}}$$

$$= \frac{4R(1+sRC)^2}{2(1+sRC) + 2s^2R^2C^2(1+sRC)}$$

$$= \frac{4R(1+sRC)^2}{2(1+sRC)\left[1+s^2R^2C^2\right]} = \frac{2R(1+sRC)}{1+s^2R^2C^2}$$

$$\frac{V_2}{V_1} = \frac{I_2 Z_B}{I_1 Z_A}, \quad \text{Also } I_2(Z_B + Z_C) = I_1 Z_A$$

or
$$\frac{I_2}{I_1} = \frac{Z_A}{Z_B + Z_C}$$

Therefore,
$$\frac{V_2}{V_1} = \frac{Z_C}{Z_B + Z_C} \qquad \text{...(1)}$$

$$= \frac{\dfrac{1+sRC}{2sC}}{\dfrac{2R(1+sRC)}{1+s^2R^2C^2} + \dfrac{1+sRC}{2sC}}$$

$$= \frac{(1+sRC)(1+s^2R^2C^2)}{4sRC(1+sRC) + (1+sRC)(1+s^2R^2C^2)}$$

$$= \frac{(1+sRC)(1+s^2R^2C^2)}{(1+sRC)(4sRC + 1 + s^2R^2C^2)}$$

$$= \frac{1+s^2R^2C^2}{1+4sRC+s^2R^2C^2}$$

$$= \frac{s^2 + 1/R^2C^2}{s^2 + \dfrac{4s}{RC} + \dfrac{1}{R^2C^2}} = \frac{s^2 + \dfrac{1}{R^2C^2}}{s^2 + \left(\dfrac{4}{RC}\right)s + \left(\dfrac{1}{R^2C^2}\right)}$$

Problem No. 6.25 Show that the voltage transfer function of the network shown in Fig. 6.42 can be written as

$$\frac{V_2}{V_1} = \frac{b_3 s^3 + b_2 s^2 + b_1 s + b_0}{a_3 s^3 + a_2 s^2 + a_1 s + a_0}$$

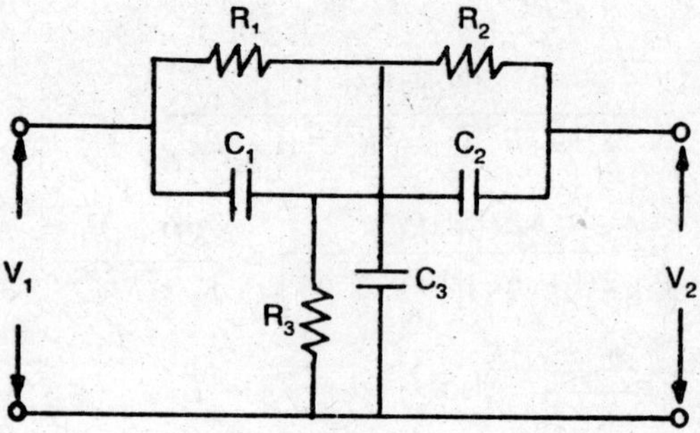

Fig. 6.42

where

$$a_3 = b_3 = R_1 R_2 R_3 C_1 C_2 C_3$$

$$a_2 = R_3[R_1 C_3 (C_1 + C_2) + (R_1 + R_2) C_1 C_2] + R_1 R_2 C_2 C_3$$

$$b_2 = R_3 (R_1 + R_2) C_1 C_2$$

$$a_1 = R_3 (C_1 + C_2) + R_2 C_2 + (C_2 + C_3) R_1$$

$$b_1 = R_3 (C_1 + C_2)$$

$$a_0 = b_0 = 1$$

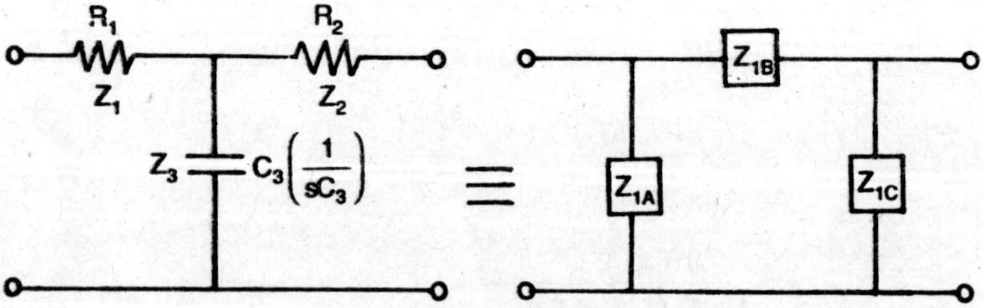

Fig. 6.43

$$\Sigma Z_1 Z_2 = Z_1 Z_2 + Z_2 Z_3 + Z_3 Z_1$$

$$= R_1 R_2 + \frac{R_2}{sC_3} + \frac{R_1}{sC_3} = \frac{sR_1 R_2 C_3 + R_1 + R_2}{sC_3}$$

$$Z_{1A} = \frac{\Sigma Z_1 Z_2}{Z_2} = \frac{\dfrac{sR_1R_2C_3 + R_1 + R_2}{sC_3}}{R_2}$$

$$= \frac{sR_1R_2C_3 + R_1 + R_2}{sR_2C_3}$$

$$Z_{1B} = \frac{\Sigma Z_1 Z_2}{Z_3} = \frac{\dfrac{sR_1R_2C_3 + R_1 + R_2}{sC_3}}{\dfrac{1}{sC_3}}$$

$$= R_1 + R_2 + sR_1R_2C_3$$

$$Z_{1C} = \frac{\Sigma Z_1 Z_2}{Z_1} = \frac{\dfrac{sR_1R_2C_3 + R_1 + R_2}{sC_3}}{R_1}$$

$$= \frac{R_1 + R_2 + sR_1R_2C_3}{sR_1C_3}$$

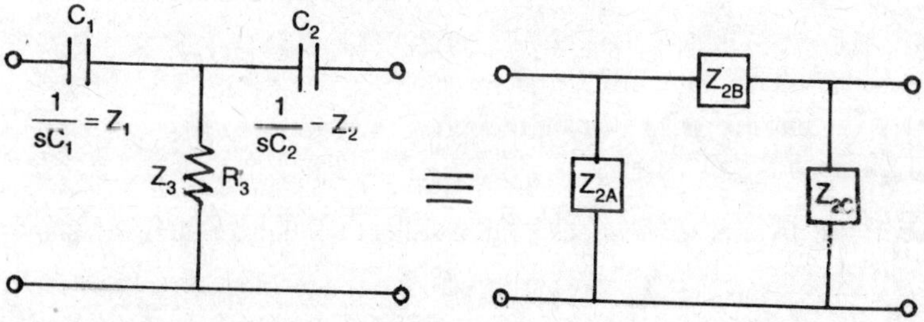

Fig. 6.44

$$\Sigma Z_1 Z_2 = Z_1Z_2 + Z_2Z_3 + Z_3Z_1$$

$$= \frac{1}{sC_1} \times \frac{1}{sC_2} + \frac{1}{sC_2} \times \frac{R_3}{1} + \frac{R_3}{1} \times \frac{1}{sC_1}$$

$$= \frac{1}{s^2C_1C_2} + \frac{R_3}{sC_2} + \frac{R_3}{sC_1}$$

$$= \frac{1 + sR_3C_1 + sR_3C_2}{s^2C_1C_2}$$

$$= \frac{1+s(C_1R_3+C_2R_3)}{s^2C_1C_2}$$

$$Z_{2A} = \frac{\Sigma Z_1Z_2}{Z_2} = \frac{\dfrac{1+s(C_1R_3+C_2R_3)}{s^2C_1C_2}}{\dfrac{1}{sC_2}}$$

$$= \frac{1+sR_3\,(C_1+C_2)}{sC_1}$$

$$Z_{2B} = \frac{\Sigma Z_1Z_2}{Z_3} = \frac{\dfrac{1+sR_3\,(C_1+C_2)}{s^2C_1C_2}}{R_3}$$

$$= \frac{1+sR_3\,(C_1+C_2)}{s^2R_3C_1C_2}$$

$$Z_{2C} = \frac{\Sigma Z_1Z_2}{Z_1} = \frac{\dfrac{1+sR_3\,(C_1+C_2)}{s^2C_1C_2}}{\dfrac{1}{sC_1}}$$

$$= \frac{1+sR_3\,(C_1+C_2)}{s^2C_2}$$

The two π-networks are now placed in parallel to obtain a π-network shown in Fig. 6.45.

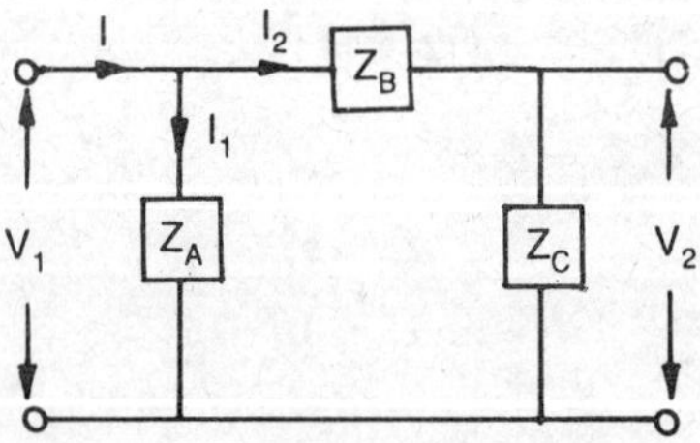

Fig. 6.45

$$Z_B = \frac{Z_{1B}\times Z_{2B}}{Z_{1B}+Z_{2B}}$$

$$= \frac{(R_1 + R_2 + sR_1R_2R_3)\dfrac{(1+sR_3(C_1+C_2))}{s^2R_3C_1C_2}}{(R_1 + R_2 + sR_1R_2C_3)\dfrac{1+sR_3(C_1+C_2)}{s^2R_3C_1C_2}}$$

$$= \frac{(R_1 + R_2 + sR_1R_2C_3)(1+sR_3C_1 + sR_3C_2)}{s^2R_1R_3C_1C_2 + s^2R_2R_3C_1C_2 + s^3R_1R_2R_3C_1C_2C_3 + 1 + sR_3C_1 + sR_3C_2}$$

$$= \frac{R_1 + sR_1R_3C_1 + sR_1R_3C_2 + R_2 + sR_2R_3C_1 + sR_2R_3C_2 + sR_1R_2C_3 + s^2R_1R_2R_3C_1C_3 + s^2R_1R_2R_3C_2C_3}{s^3R_1R_2R_3C_1C_2C_3 + s^2(R_1R_3C_1C_2 + R_2R_3C_1C_2) + s(R_3C_1 + R_3C_2) + 1}$$

$$= \frac{s^2R_1R_2R_3C_3(C_1+C_2) + sR_1R_3(C_1+C_2) + sR_2R_3(C_1+C_2) + sR_1R_2C_3 + R_1 + R_2}{s^3R_1R_2R_3C_1C_2C_3 + s^2R_3C_1C_2(R_1+R_2) + sR_3(C_1+C_2) + 1} = \frac{a}{b} \qquad \text{...(1)}$$

$$Z_C = \frac{Z_{1C} \times Z_{2C}}{Z_{1C} + Z_{2C}}$$

$$= \frac{\dfrac{R_1 + R_2 + sR_1R_2C_3}{sR_1C_3} \times \dfrac{1+sR_3(C_1+C_2)}{sC_2}}{\dfrac{R_1 + R_2 + sR_1R_2C_3}{sR_1C_3} + \dfrac{1+sR_3(C_1+C_2)}{sC_2}}$$

$$= \frac{(R_1 + R_2 + sR_1R_2C_3)(1+sR_3C_1 + sR_3C_2)}{sR_1C_2 + sR_2C_2 + s^2R_1R_2C_2C_3 + sR_1C_3 + s^2R_1R_3C_1C_3 + s^2R_1R_3C_2C_3}$$

$$= \frac{s^2R_1R_2R_3C_3(C_1+C_2) + sR_1R_3(C_1+C_2) + sR_2R_3(C_1+C_2) + sR_1R_2C_3 + R_1 + R_2}{s^2(R_1R_2C_2C_3 + R_1R_3C_1C_3 + R_1R_3C_2C_3) + s(R_1C_2 + R_2C_2 + R_1C_3)}$$

$$= \frac{a}{c} \qquad \text{...(2)}$$

From Fig. 6.45

$$\frac{V_2}{V_1} = \frac{I_2Z_B}{I_1Z_A}$$

Also $\quad I_2(Z_B + Z_C) = I_1Z_A$

or
$$\frac{I_2}{I_1} = \frac{Z_A}{Z_B + Z_C}$$

Therefore,
$$\frac{V_2}{V_1} = \frac{Z_C}{Z_B + Z_C} \qquad ...(3)$$

By putting the values of Z_B and Z_C from equations (1) and (2) in equation (3), we get

$$\frac{V_2}{V_1} = \frac{\frac{a}{c}}{\frac{a}{c} + \frac{a}{b}} = \frac{ab}{ab + ac} = \frac{b}{b + c}$$

Therefore,

$$\frac{V_2}{V_1} = \frac{s^3 R_1 R_2 R_3 C_1 C_2 C_3 + s^2 R_3 C_1 C_2 (R_1 + R_2) + sR_3 (C_1 + C_2) + 1}{s^3 R_1 R_2 R_3 C_1 C_2 C_3 + s^2 R_3 C_1 C_2 (R_1 + R_2) + sR_3 (C_1 + C_2) + 1 + s^2 (R_1 R_2 C_2 C_3 + R_1 R_3 C_1 C_3 + R_1 R_3 C_2 C_3) + s(R_1 C_2 + R_2 C_2 + R_1 C_3)}$$

Putting

$$a_3 = b_3 = R_1 R_2 R_3 C_1 C_2 C_3$$
$$b_2 = R_3 C_1 C_2 (R_1 + R_2)$$
$$b_1 = R_3 (C_1 + C_2)$$
$$a_0 = b_0 = 1$$

we get

$$\frac{V_2}{V_1} = \frac{b_3 s^3 + b_2 s^2 + b_1 s + b_0}{a_3 s^3 + s^2 (R_3 C_1 C_2 (R_1 + R_2) + R_3 (C_1 + C_2) + R_1 R_2 C_2 C_3 + R_1 R_3 C_1 C_3 + R_1 R_3 C_2 C_3) + s(R_3 (C_1 + C_2) + R_1 (C_2 + C_3) + R_2 C_2) + a_0}$$

or

$$\frac{V_2}{V_1} = \frac{b_3 s^3 + b_2 s^2 + b_1 s + b_0}{a_3 s^3 + s^2 (R_3 [R_1 C_3 (C_1 + C_2) + (R_1 + R_2) C_1 C_2] + R_1 R_2 C_2 C_3) + s(R_3 (C_1 + C_2) + R_2 C_2 (C_2 + C_3) R_1) + a_0}$$

putting

$$a_2 = R_3 [R_1 C_3 (C_1 + C_2) + (R_1 + R_2) C_1 C_2] + R_1 R_2 C_2 C_3$$

and

$$a_1 = R_3(C_1 + C_2) + R_2C_2 + (C_2 + C_3)R_1$$

we get

$$\frac{V_2}{V_1} = \frac{b_3s^3 + b_2s^2 + b_1s + b_0}{a_3s^3 + a_2s^2 + a_1s + a_0}$$

Problem No. 6.26 A two terminal network consists of a coil having an inductance L and resistance R shunted by a capacitor C.

The poles and zeros of the driving point impedance function Z of this network are:

poles at $-\frac{1}{2} \pm j\frac{\sqrt{3}}{2}$; zeros at -1.

If $Z(j0) = -1$, determine the value of R, L and C.

Solution

The given network in transformed form is shown in Fig. 6.46 and the pole-zero pattern is plotted in Fig. 6.47.

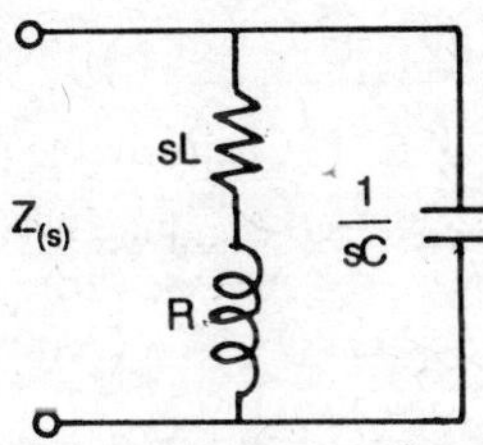

Fig. 6.46

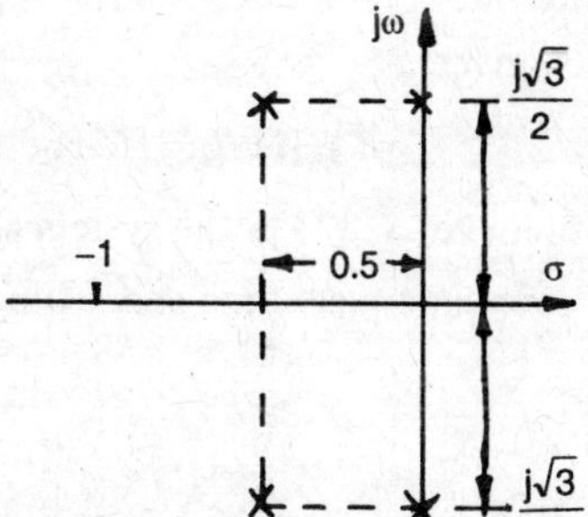

Fig. 6.47

$$Z(s) = \frac{s + R/L}{s^2C + s\frac{RC}{L} + \frac{1}{L}} \quad ...(1)$$

(Refer Problem No. 6.22, equation (1))

From Fig. 6.47, we conclude that

$$Z(s) = \frac{K(s+1)}{\left(s + 0.5 + j\frac{\sqrt{3}}{2}\right)\left(s + 0.5 - j\frac{\sqrt{3}}{2}\right)}$$

$$= \frac{K(s+1)}{(s+0.5)^2 + \left(\frac{\sqrt{3}}{2}\right)^2}$$

$$= \frac{K(s+1)}{s^2 + s + 1} \qquad \text{...(2)}$$

It is given that

$Z(j0) = 1$; Therefore

$$Z(j0) = 1 = \frac{K \times 1}{1} \qquad (\because \omega = 0)$$

or $\qquad K = 1$

Therefore,

$$Z(s) = \frac{s+1}{s^2 + s + 1} \qquad \text{...(3)}$$

Comparing equations (1) and (3), we get

$$\frac{R}{L} = 1 \; ; \; \frac{RC}{L} = 1 \text{ and } \frac{1}{L} = 1$$

which gives

$$R = 1\Omega,\ L = 1H \text{ and } C = 1F.$$

Problem No. 6.27 For the system shown in Fig. 6.48. Caclulate the percentage overshoot peak time and setting time, given $\xi = 0.1$ and $\omega_n = 1$.

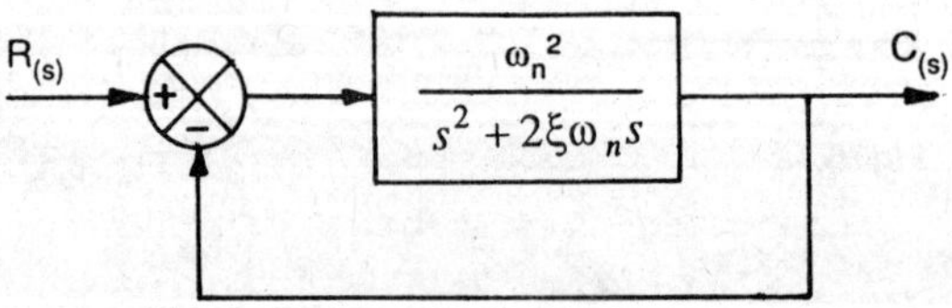

Fig. 6.48

Solution

$$G(s) = \frac{\omega_n^2}{s^2 + 2\xi\omega_n s} \text{ and } H(s) = 1$$

Substituting $\omega_n = 1$ and $\xi = 0.1$, we get

$$G(s) = \frac{1}{s^2 + 0.2s}$$

Therefore,

$$\frac{C(s)}{R(s)} = \frac{G(s)}{1 + G(s)H(s)}$$

$$= \frac{\frac{1}{s^2 + 0.2s}}{1 + \frac{1}{s^2 + 0.2s} \times 1}$$

$$= \frac{1}{s^2 + 0.2s + 1}$$

Peak Overshoot

$$M_p = e^{\left(-\xi/\sqrt{1-\xi^2}\right)\pi}$$

$$= e^{\left(-0.1/\sqrt{1-0.1^2}\right)\pi}$$

$$= 0.729 \text{ or } 72.9\%$$

Peak Time

$$t_p = \frac{\pi}{\omega_n\sqrt{1-\xi^2}} = \frac{\pi}{1\sqrt{1-0.1^2}} = 3.16 \text{ seconds}$$

Setting time

$$t_s \text{ (for 2\% tolerance band)} = \frac{3.91}{\xi\omega_n} = \frac{3.91}{0.1 \times 1} = 39.1 \text{ seconds}$$

Problem No. 6.28 Determine the driving point impedence of the network shown in Fig. 6.49.

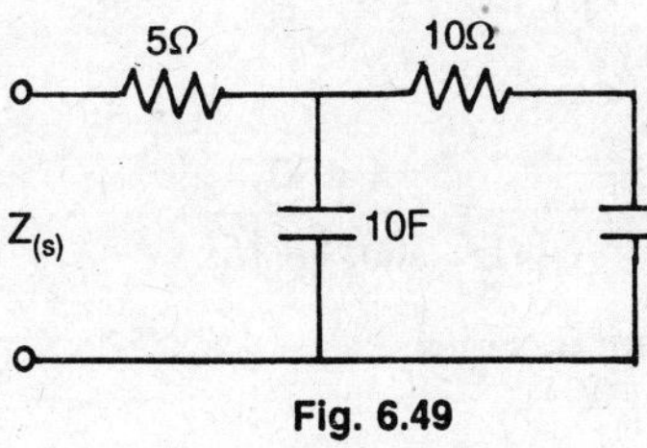

Fig. 6.49

Solution

Transform network corresponding to the given network is shown in Fig.6.50.

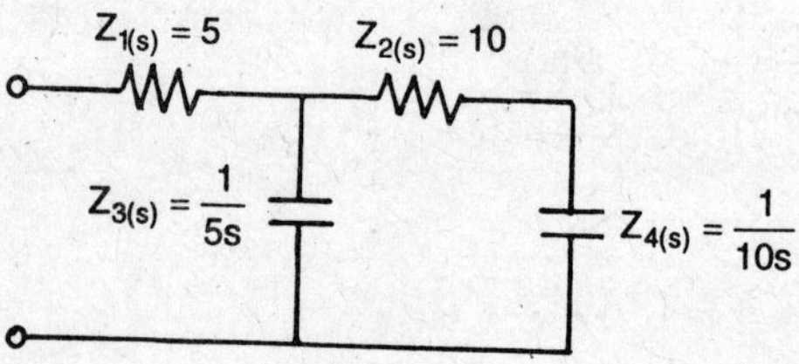

Fig. 6.50

$Z_2(s)$ and $Z_3(s)$ are in series

$$Z_{23}(s) = Z_2(s) + Z_3(s)$$

$$= 10 + \frac{1}{5s} = \frac{1+50s}{5s}$$

$$Y_{23}(s) = \frac{1}{Z_{23}(s)} = \frac{1}{\frac{1+50s}{5s}} = \frac{5s}{1+50s}$$

$$Y_4(s) = \frac{1}{Z_4(s)} = \frac{1}{1/10s} = 10s$$

Now $Z_{23}(s)$ and $Z_4(s)$ are in parallel. Therefore

$$Y_{234}(s) = Y_{23}(s) + Y_4(s)$$

$$= \frac{5s}{1+50s} + \frac{10s}{1}$$

$$= \frac{5s + 10s + 500s^2}{1+50s}$$

$$= \frac{500s^2 + 15s}{1+50s}$$

Therefore,

$$Z_{234}(s) = \frac{1}{Y_{234}(s)} = \frac{1+50s}{500s^2 + 15s}$$

Now, $Z_1(s)$ and $Z_{234}(s)$ are in series

Therefore,

$$Z(s) = Z_1(s) + Z_{234}(s)$$

$$= \frac{5}{1} + \frac{1+50s}{500s^2+15s}$$

$$= \frac{2500s^2+75s+1+50s}{500s^2+15s}$$

$$= \frac{2500s^2+125s+1}{500s^2+15s}$$

Problem No. 6.29 A series RLC circuit has its driving point admittance pole-zero diagram as shown in Fig. 6.51. Obtain time domain response.

Solution

The diagram depicts one zero at $s = 0$, and poles at $s = -5 \pm j15$. Also scale factor is equal to 10. Therefore

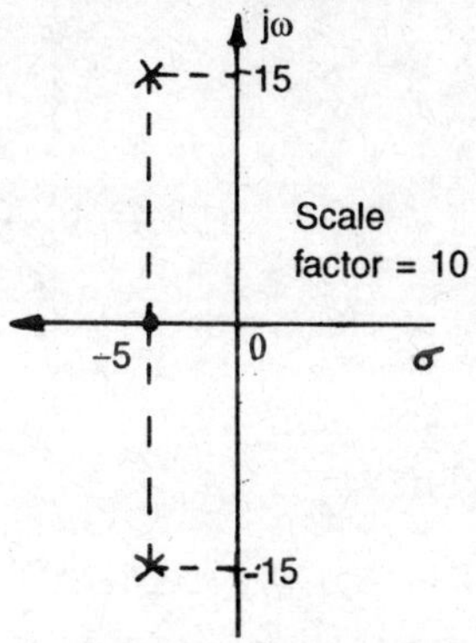

Fig. 6.51

$$Y(s) = \frac{10s}{(s+5+j15)(s+5-j15)}$$

$$y_{(t)} = K_1 e^{-(5+j15)t} + K_2 e^{-(5-j15)t}$$

Value of K_1 Refer Fig. 6.52

$$OA = \sqrt{5^2+15^2} = 15.8$$

$$\theta = 90° + \tan^{-1}\frac{AB}{AC}$$

$$= 90° + \tan^{-1}\frac{5}{15}$$

$$= 108.43°$$

$$K_1 = H\frac{M_{OA}\angle\theta}{M_{DA}\angle\theta}$$

$$= 10\frac{15.8\angle 108.43°}{30\angle 90°}$$

$$= 5.26\angle 18.43°$$

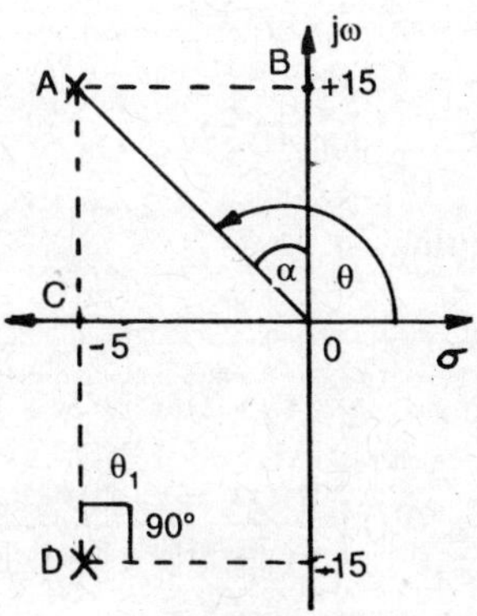

Fig. 6.52

Value of K_2 (Fig. 6.53)

$$K_2 = H\frac{M_{OA}\angle\theta}{M_{DA}\angle\theta_1}$$

$$= \frac{10 \times 15.8\angle -108.43°}{30\angle -90°}$$

$$= 5.26\angle -18.43°$$

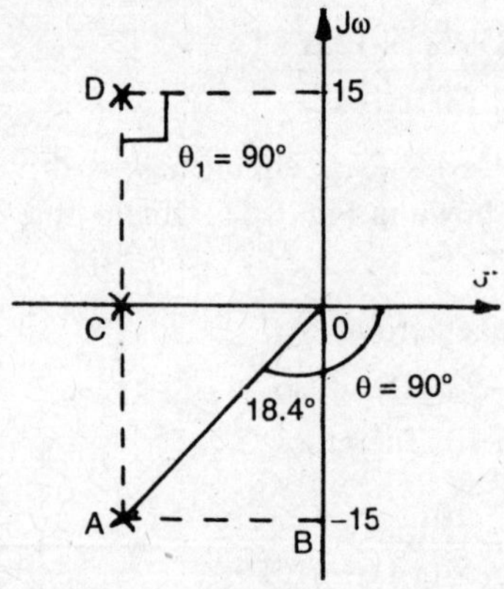

Fig. 6.53

Therefore,

$$y_{(t)} = 5.26e^{+8.43°}\, e^{-(5+j15)t} + 5.26e^{-18.43°}e^{-(5-j15)t}$$

Problem No. 6.30 The transform voltage $V(s)$ of a network is given by

$V(s) = \dfrac{3s}{(s+2)(s^2+2s+2)}$. Plot its pole-zero diagram and hence obtain $v(t)$

Solution

$$V(s) = \frac{3s}{(s+2)(s^2+2s+2)}$$

or $$V(s) = \frac{3s}{(s+2)(s+1-j)(s+1+j)}$$

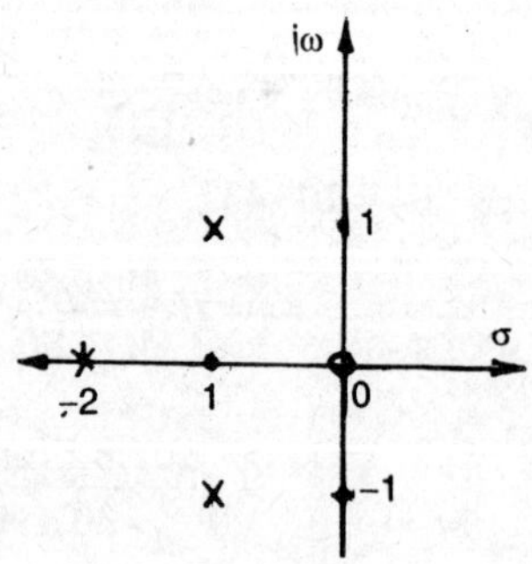

Fig. 6.54

Pole-zero diagram is shown in Fig. 6.54

$$v(t) = K_1e^{-2t} + K_2e^{-(1-j)t} + K_3e^{-(1+j)t}$$

Value of K_1 (Fig. 6.55)

$$\theta = \tan^{-1}\frac{1}{1} = 45°$$

Therefore $\theta = 90°+45°$

$$= 135°$$

$$BC = AC = \sqrt{1^2 + 1^2}$$

$$= \sqrt{2}$$

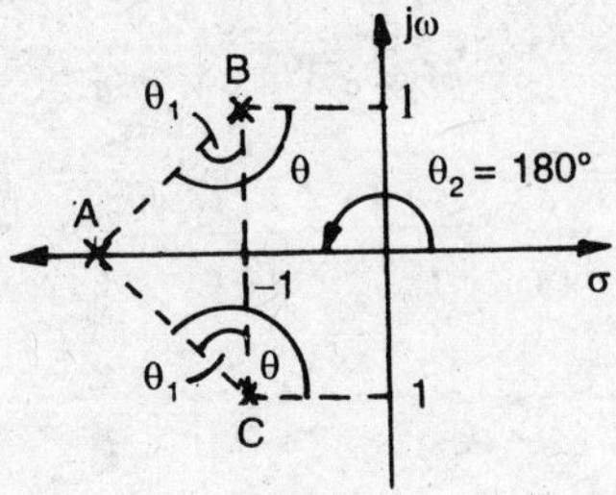

Fig. 6.55

$$K_1 = \frac{HM_{OA}e^{\theta_1}}{M_{BA}e^{-\theta}M_{CA}e^{\theta}}$$

$$= 3 \times \frac{2e^{180°}}{\sqrt{2}\,e^{-135°} \times \sqrt{2}\,e^{+135°}}$$

$$= \frac{3\times 2\,e^{180°}}{2\,e^{0}}$$

$$= 3e^{180°}$$

$$= 3(\cos 180° + j\sin 180°)$$

$$= -3$$

Value of K_2 (Fig. 6.56)

$$K_2 = \frac{H\,M_{OB}\,e^{135°}}{M_{AB}\,e^{45°}\,M_{CB}\,e^{90°}}$$

$$= \frac{3\times\sqrt{2}\,e^{135°}}{\sqrt{2}\,e^{45°} \times 2\,e^{90°}}$$

$$= \frac{3}{2}e^{0}$$

$$= \frac{3}{2}$$

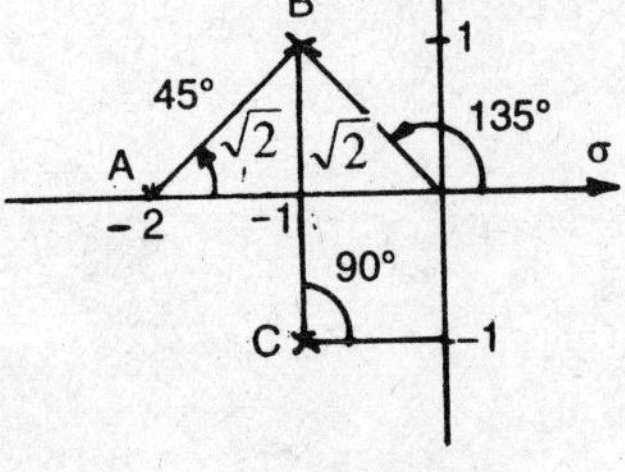

Fig. 6.56

Value of K_3 (Fig. 6.57)

$$k_3 = \frac{H \times M_{OC}\, e^{-135^\circ}}{M_{BC}\, e^{-90^\circ} \times M_{AC}\, e^{-45^\circ}}$$

$$= \frac{3 \times \sqrt{2}\; e^{-135^\circ}}{2\, e^{-90^\circ} \times \sqrt{2}\, e^{-45^\circ}}$$

$$= \frac{3}{2} e^{0} = \frac{3}{2}$$

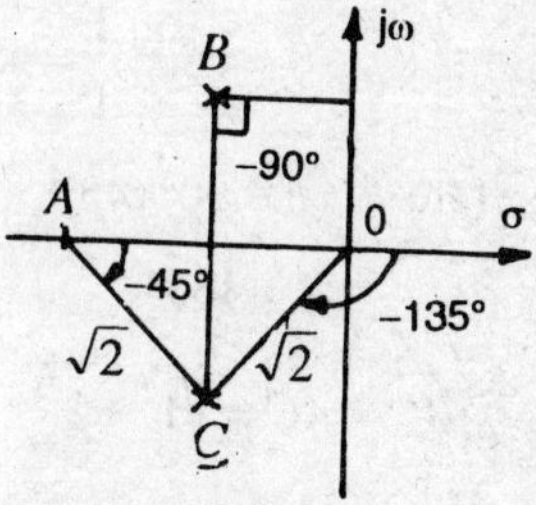

Fig. 6.57

Therefore,

$$v(t) = -3e^{-2t} + \frac{3}{2} e^{-(1-j)t} + \frac{3}{2} e^{-(1+j)t}$$

$$= -3e^{-2t} + 3\left[\left(e^{-t}\right)\left(\frac{e^{jt} + e^{-jt}}{2}\right)\right]$$

$$= -3e^{-2t} + 3e^{-t} \cos t$$

7

Network Synthesis

Problem No. 7.1 Test whether the following polynomials are Hurwitz

(a) $s^3 + 4s^2 + 5s + 2$

(b) $s^7 + 2s^6 + 2s^5 + s^4 + 4s^3 + 8s^2 + 8s + 4$

(c) $s^4 + 7s^3 + 6s^2 + 21s + 8$

(d) $s^4 + 5s^3 + 5s^2 + 4s + 10$

(e) $s^3 + s^2 + 2s + 2$

(f) $s^4 + s^2 + s + 1$

(g) $s^5 + 2s^3 + s$

(h) $s^7 + s^5 + s^3 + s$

Solution

(a) $s^3 + 4s^2 + 5s + 2$

$$P(s) = s^3 + 5s$$

$$Q(s) = 4s^2 + 2$$

$$4s^2 + 2 \overline{\big)\, s^3 + 5} \quad \frac{s}{4}$$

$$\underline{s^2 + \frac{s}{2}}$$

$$4.5s \overline{\big)\, 4s^2 + 2} \quad 0.88s$$

$$\underline{4s^2}$$

$$2 \overline{\big)\, 4.5s} \quad 225s$$

$$\underline{4.5s}$$

Hence, the continued fraction expansion is

$$= \frac{s}{4} + \frac{1}{0.88s + \dfrac{1}{2.25s}}$$

Since all the quotient terms $\frac{1}{4}$, 0.88, 2.25 are positive, the given polynomial is Hurwitz.

(b) $s^7 + 2s^6 + 2s^5 + s^4 + 4s^3 + 8s^2 + 8s + 4$

$$P(s) = s^7 + 2s^5 + 4s^3 + 8s$$

$$Q(s) = 2s^6 + s^4 + 8s^2 + 4$$

$$2s^6 + s^4 + 8s^2 + 4 \,\big)\, s^7 + 2s^5 + 4s^3 + 8s \,\big(\, \frac{s}{2}$$

$$s^7 + \frac{s^5}{2} + 4s^3 + 2s$$

$$\frac{3}{2}s^5 + 6s \,\big)\, 2s^6 + s^4 + 8s^2 + 8 \,\big(\, \frac{4s}{3}$$

$$2s^6 + 8s^2$$

$$s^4 + 4 \,\big)\, \frac{3}{2}s^5 + 6s \,\big(\, \frac{s}{2}$$

$$\frac{3}{2}s^5 + 6s$$

$$\text{X}$$

The division has terminated suddenly and (s^4 + 4) is a factor, therefore, the given polynomial $N(s)$ can be written as

$$N(s) = s^7 + 2s^6 + 2s^5 + s^4 + 4s^3 + 8s^2 + 8s + 4$$

$$= (s^2 + 4)(s^3 + 2s^2 + 2s + 1)$$

$$= (s^2 + 4)(s + 1)(s^2 - s + 1)$$

$$= (s^2 + 2s + 2)(s^2 - 2s + 2)(s + 1)(s^2 - s + 1)$$

It is seen that factors $(s^2 - 2s + 2)$ and $(s^2 - s + 1)$ are not Hurwitz as coefficients of s are negative. Hence, the given polynomial is not Hurwitz.

(c) $s^4 + 7s^3 + 6s^2 + 21s + 8$

$$P(s) = s^4 + 6s^2 + 8$$

$$Q(s) = 7s^3 + 21s$$

$$7s^3 + 21s\overline{\big)\, s^4 + 6s^2 + 8 \,\big(}\ \frac{s}{7}$$
$$\underline{s^4 + 3s^2}$$
$$3s^2 + 8\overline{\big)\, 7s^3 + 21s \,\big(}\ \frac{7}{3}s$$
$$\underline{7s^3 + \frac{56}{3}s}$$
$$\frac{7}{3}s\overline{\big)\, 3s^3 + 8 \,\big(}\ \frac{9}{7}s$$
$$\underline{3s^2}$$
$$8\overline{\big)\, \frac{7}{3}s \,\big(}\ \frac{7}{24}s$$
$$\underline{\frac{7}{3}s}$$
$$\text{X}$$

Hence $N(s) = \dfrac{P(s)}{Q(s)}$ can be written as

$$= \frac{s}{7} + \cfrac{1}{\frac{7}{3}s + \cfrac{1}{\frac{9}{7}s + \cfrac{1}{\frac{7}{24}s}}}$$

Since all the quotient terms $\frac{1}{7}, \frac{7}{3}, \frac{9}{7}$ and $\frac{7}{24}$ are not negative, the given polynomial is Hurwitz.

(d) $\boldsymbol{s^4 + 5s^3 + 5s^2 + 4s + 10}$

$$N(s) = \frac{P(s)}{Q(s)}$$

$$P(s) = s^4 + 5s^2 + 10$$

$$Q(s) = 5s^3 + 4s$$

$$5s^3+4s\,\overline{\big)\,s^4+5s^2+10\,\big(}\,\frac{s}{5}$$
$$\underline{s^4+\frac{4s^2}{5}}$$
$$\frac{21}{5}s^2+10\,\overline{\big)\,5s^3+4s\,\big(}\,\frac{25}{21}s$$
$$\underline{5s^3+\frac{250s}{21}}$$
$$-\frac{161}{21}s\,\overline{\big)\,\frac{21}{5}s^2+10\,\big(}\,-\frac{440}{830}s$$
$$\underline{\frac{21}{5}s^2}$$
$$10\,\overline{\big)\,-\frac{166}{21}s\,\big(}\,-\frac{166}{210}s$$
$$\underline{\frac{166}{21}s}$$
$$\text{X}$$

Therefore,

$$N(s)=\frac{P(s)}{Q(s)}$$

$$=\frac{s}{5}+\cfrac{1}{\frac{25}{21}s+\cfrac{1}{\frac{-440}{830}s+\cfrac{1}{\frac{-166}{210}s}}}$$

Since all the quotients are not positive, i.e., $\frac{-440}{830}$ and $\frac{-166}{210}$ are negative, the given polynomial is not Hurwitz.

(e) s^3+s^2+2s+2

$$P(s)=s^3+2s$$

$$Q(s)=s^2+2$$

$$s^2+2\overline{)\,s^3+2s\,}\big(\,s$$
$$\underline{s^3+2s}$$
$$\times$$

We see the division has been terminated abruptly by a common factor (s^3+2s). The polynomial can be written as

$$N(s)=(s^3+2s)\left(1+\frac{1}{s}\right)$$

We see that both the factors are Hurwitz and hence, given polynomial is Hurwitz

(f) s^4+s^2+s+1

$$s^2+1\overline{)\,s^4+s\,}\big(\,s$$
$$\underline{s^4+s^2}$$
$$-s^2+s\overline{)\,s^2+1\,}\big(-1$$
$$\underline{+s^2-s}$$
$$s+1\overline{)\,-s^2+s\,}\big(-s$$
$$\underline{-s^2-s}$$
$$2s\overline{)\,s+1\,}\big(\,1/2$$
$$\underline{s}$$
$$1\overline{)\,2s\,}\big(\,2s$$
$$\underline{2s}$$
$$\times$$

$$N(s)=s^2+\cfrac{1}{-1+\cfrac{1}{-s+\cfrac{1}{\cfrac{1}{2}+\cfrac{1}{2s}}}}$$

It is seen that one term related s^3 is missing. Moreover, continued fraction gives quotients terms negative. Hence, given polynomial is not Hurwitz.

(g) $s^5 + 2s^3 + s$

Since the polynomial

$N(s) = s^5 + 2s^3 + s$, contains odd function only, hence it is not possible to perform continued fraction expansion, therefore,

$$N(s) = \frac{P(s)}{P'(s)}$$

$$P'(s) = 5s^4 + 6s^2 + 1$$

$$
\begin{array}{l}
5s^4 + 6s^2 + 1 \,\big)\, s^5 + 2s^3 + s \,\big(\, \frac{s}{5} \\
\qquad\qquad\quad\; s^5 + \frac{6}{5}s^3 + \frac{s}{5} \\
\hline
\frac{4}{5}s^3 + \frac{4}{5}s \,\big)\, 5s^4 \; 6s^2 + 1 \,\big(\, -1 \\
\qquad\qquad\qquad 5s^4 + 5s^2 \\
\hline
\qquad\qquad\qquad s^2 + 1 \,\big)\, \frac{4}{5}s^3 + \frac{4}{5}s \,\big(\, \frac{4}{5}s \\
\qquad\qquad\qquad\qquad\qquad \frac{4}{5}s^3 + \frac{4}{5}s \\
\hline
\qquad\qquad\qquad\qquad\qquad\quad \times
\end{array}
$$

Since, division has terminated abruptly $N(s) = s^5 + 2s^3 + s = (s^2 + 1)$ $(s^2 + 1)s$. Since all the three factors are Hurwitx, because all the co-efficients are positive, the given polynomial is Hurwitz.

(h) $s^7 + s^5 + s^3 + s$

Since the polynomial $N(s) = s^7 + s^5 + s^3 + s$ contains, odd functions only, hence, continued fraction expansion is not possible, therefore

$$N(s) = \frac{P(s)}{P'(s)}$$

and $\qquad P'(s) = 7s^6 + 5s^4 + 3s^2 + 1$

$$7s^6+5s^4+3s^2+1\,\big)\,s^7+s^5+s^3+s\,\big(\,\frac{s}{7}$$

$$s^7+\frac{5}{7}s^5+\frac{3}{7}s^3+\frac{s}{7}$$

$$\frac{2}{7}s^5+\frac{4}{7}s^3+\frac{6}{7}s\,\big)\,7s^6+5s^4+3s^2+1\,\big(\,\frac{49}{2}s$$

$$7s^6+14s^4+21s^2$$

$$-9s^4-18s^2+1\,\big)\,\frac{2}{7}s^5+\frac{4}{7}s^3+\frac{6}{7}s\,\big(\,-\frac{2}{63}s$$

$$\frac{2}{7}s^5+\frac{4}{7}s^3-\frac{2}{63}s$$

$$\frac{56}{63}s\,\big)\,-9s^4-18s^2+1$$

There is no need to proceed with further division as one of the quotient term $\frac{-2}{63}$ is negative. Hence given polynomial is not Hurwitz.

Problem No. 7.2 Find out the condition for $F(s)=\dfrac{s+a}{s^2+bs+c}$ to be positive real.

Solution

(a) **Condition No. 1** "The poles and zeros of $F(s)$ to be in the left-half plane or on the jω axis." This implies that the coefficients a, b, c must be greater or equal to zero.

(b) **Condition No. 2** " $F(s)$ may have only simple poles on the $j\omega$ axis with real and positive residues. $F(s)$ will have poles on the $j\omega$ axis, if $b = 0$. Therefore $F(s)$ can be written as

$$F(s)=\frac{s}{s^2+c}+\frac{a}{s^2+c} \qquad \text{...(1)}$$

(c) **Condition No. 3** " $\operatorname{Re} F(j\omega)\geq 0$ "

$$M_1(j\omega)\,M_2(j\omega)-N_1(j\omega)\,N_2(j\omega)\geq 0$$

$M_1(j\omega)=a$;

$M_2(j\omega)=s^2+c$;

$N_1(j\omega)=s$;

$N_2(j\omega)=bs$ and

Therefore,

$$a(s^2+c)-s(bs)\geq 0$$

putting $s=j\omega$

$$a\left\{(j\omega)^2+c\right\}-j\omega\left\{b(j\omega)\right\}\geq 0$$

$$a(-\omega^2+c)+b\omega^2\geq 0$$

Let

$$A(\omega)=a(-\omega^2+c)+b\omega^2\geq 0$$

or $$A(\omega)=(b-a)\omega^2+ac\geq 0$$

In order to ensure that $A(\omega)$ does not have positive real roots of ω,

$$b\geq a$$

Therefore,

if $b=0$, then $a>0$

Therefore, the conditions which have to be fulfilled so that $F(s)$ is positive real are

(a) $a, b, c\geq 0$

(b) $b\geq a$

Problem No. 7.3 Determine the positive realness of the function

$$F_{(s)}=\frac{s^2+s+6}{s^2+s+1}$$

Solution

$$F_{(s)}=\frac{s^2+s+6}{s^2+s+1}$$

$$=\frac{s^2+s+1+5}{s^2+s+1}$$

$$=\frac{s^2+s+1}{s^2+s+1}+\frac{5}{s^2+s+1}=1+\frac{5}{s^2+s+1}$$

$$=1+\frac{A}{s+\frac{1}{2}+j\frac{\sqrt{3}}{2}}+\frac{B}{s+\frac{1}{2}-j\frac{\sqrt{3}}{2}}$$

$$A = \left.\frac{5}{s+\frac{1}{2}+j\frac{\sqrt{3}}{2}}\right|_{s=-\frac{1}{2}+j\frac{\sqrt{3}}{2}}$$

$$= \frac{5}{-\frac{1}{2}+j\frac{\sqrt{3}}{2}+\frac{1}{2}+j\frac{\sqrt{3}}{2}} = \frac{5}{2j\frac{\sqrt{3}}{2}} = \frac{-j5}{\sqrt{3}}$$

$$B = \left.\frac{5}{s+\frac{1}{2}-j\frac{\sqrt{3}}{2}}\right|_{s=-\frac{1}{2}-j\frac{\sqrt{3}}{2}}$$

$$= \frac{5}{-\frac{1}{2}-j\frac{\sqrt{3}}{2}+\frac{1}{2}-j\frac{\sqrt{3}}{2}}.$$

$$= -\frac{5}{2j\frac{\sqrt{3}}{2}} = \frac{j5}{\sqrt{3}}$$

Therefore,

$$F(s) = 1+\frac{-j5/\sqrt{3}}{s+\frac{1}{2}+j\frac{\sqrt{3}}{2}}+\frac{j5/\sqrt{3}}{s+\frac{1}{2}-j\frac{\sqrt{3}}{2}}$$

Thus it is seen that residue at pole $s = -\frac{1}{2}+j\frac{\sqrt{3}}{2}$ is $-j\frac{5}{\sqrt{3}}$ i.e., negative.

Hence, the given function $F(s)$ is not real positive.

Problem No 7.4 Check the real positiveness of function

$$F(s) = \frac{s^2+6s+5}{s^2+9s+14}$$

Solution

$$F(s) = \frac{s^2+6s+5}{s^2+9s+14}$$

$$= \frac{s^2+9s+14-3s-9}{s^2+9s+14} = \frac{s^2+9s+14}{s^2+9s+14} - \frac{3s+9}{s^2+9s+14}$$

$$= 1 - \frac{3s+9}{s^2+9s+14}$$

$$= 1 - \frac{3s+9}{(s+2)(s+7)}$$

$$= 1 - 3\left[\frac{A}{s+2} + \frac{B}{s+7}\right]$$

$$A = \left.\frac{s+3}{s+7}\right|_{s=-2} = \frac{-2+3}{-2+7} = \frac{1}{5}$$

$$B = \left.\frac{s+3}{s+2}\right|_{s=-7} = \frac{-7+3}{-7+2} = \frac{-4}{-5} = \frac{4}{5}$$

Therefore,

$$F(s) = 1 - 3\left[\frac{1/5}{s+2} + \frac{4/5}{s+7}\right]$$

$$= 1 + \frac{-3/5}{s+2} + \frac{-12/5}{s+7}$$

From the above, it is seen that residues at poles s = –2 and –7 are $-\frac{3}{5}$ and $-\frac{12}{5}$ respectively. Since, the residues are negative, given function is not real positive.

Problem No. 7.5 Check positive realness of function $F(s) = \frac{s^2+1}{s^2+4s}$

Solution

$$F_{(s)} = \frac{s^2+1}{s^2+4s} = \frac{s^2+4s-4s+1}{s^2+4s}$$

$$= \frac{s^2+4s}{s^2+4s} - \frac{4s+1}{s^2+4s}$$

$$= 1 - \frac{4s+1}{s^2+4s} = 1 - \frac{4s+1}{s(s+4)} - \frac{1}{s(s+4)}$$

$$= 1 - \frac{4}{s+4} - \left[\frac{A}{s} + \frac{B}{s+4}\right]$$

$$= 1 - \frac{4}{s+4} - \frac{-1/4}{s} - \frac{1/4}{s+4}$$

$$= 1 + \frac{-4}{s+4} + \frac{1/4}{s} + \frac{-1/4}{s+4}$$

$$= 1 + \frac{-17/4}{s+4} + \frac{1/4}{s}$$

It is seen that residue at pole $s = -4$ is $-17/4$ which is negative and hence given function is not positive real.

Probelm No. 7.6 Determine positive realness of the function

$$F(s) = \frac{(s+2)(s+4)}{(s+1)(s+3)}$$

Solution

$$F(s) = \frac{(s+2)(s+4)}{(s+1)(s+3)} = \frac{s^2+6s+8}{s^2+4s+3}$$

$$= \frac{s^2+4s+3+2s+5}{s^2+4s+3} = 1 + \frac{2s+5}{s^2+4s+3}$$

$$= 1 + \frac{2s+5}{(s+1)(s+3)}$$

$$= 1 + \left[\frac{A}{s+1} + \frac{B}{s+3}\right]$$

$$A = \left.\frac{2s+5}{s+3}\right|_{s=-1} = \frac{3}{2}$$

$$B = \left.\frac{2s+5}{s+3}\right|_{s=-3} = \frac{1}{2}$$

or

$$F(s) = 1 + \frac{3/2}{s+1} + \frac{1/2}{s+5}$$

The residues at poles $s = -1$ and -5 are 3/2 and 1/2 and are positive. Hence, the given function is positive real.

Problem No. 7.7 Determine positive realness of function

$$F(s) = \frac{s^2+1}{s^3+4s}$$

Solution

$$F(s) = \frac{s^2+1}{s^3+4s}$$

$$= \frac{s^2}{s^3+4s} + \frac{1}{s^3+4s}$$

$$= \frac{s}{s^2+4s} + \frac{1}{s(s^2+4)}$$

$$= \frac{A}{(s+j2)} + \frac{B}{(s-j2)} + \frac{A_1}{s} + \frac{B_1}{(s+j2)} + \frac{C_1}{s-j2}$$

$$= \frac{1/2}{s+j2} + \frac{1/2}{s-j2} + \frac{1/4}{s} + \frac{-1/8}{s+j2} + \frac{-1/8}{s-j2}$$

$$= \frac{3/8}{s+j2} + \frac{3/8}{s-j2} + \frac{1/4}{s}$$

Since the residue of poles are positive, the given function is positive real.

Problem No. 7.8 Determine whether the function $\frac{s+2}{s^2+3s+2}$ is positive real or not.

Solution

$$F(s) = \frac{s+2}{s^2+3s+2}$$

$F(s)$ has zero at −2 and poles at −1 and −2. Therefore, $F(s)$ has no poles on the right hand side of s-plane.

$$F(s) = \frac{s+2}{s^2+3s+2}$$

put $s = j\omega$

$$F(j\omega) = \frac{j\omega+2}{(j\omega)^2+3j\omega+2}$$

$$= \frac{j\omega + 2}{-\omega^2 + 3j\omega + 2}$$

$$= \frac{j\omega + 2}{(2-\omega)^2 + 3j\omega}$$

$$= \frac{(j\omega + 2)\left[(2-\omega^2) - 3j\omega\right]}{\left[(2-\omega^2)\right]^2 + 9\omega^2}$$

$$= \frac{2j\omega - j\omega^3 + 3\omega^2 + 4 - 2\omega^2 - 6j\omega}{4 + \omega^4 - 4\omega^2 + 9\omega^2}$$

$$= \frac{\omega^2 - 4j\omega - j\omega^3 + 4}{\omega^4 + 5\omega^2 + 4}$$

$$= \frac{\omega^2 + 4}{\omega^4 + 5\omega^2 + 4} - j\omega\frac{(\omega^2 + 4)}{\omega^4 + 5\omega^2 + 4}$$

$\text{Re}\, F(j\omega) = \dfrac{\omega^2 + 4}{\omega^4 + 5\omega^2 + 2}$ is > 0 for all values of ω and hence $F(s)$ is real positive.

Problem No. 7.9 Determine whether the function $\dfrac{2s^2 + s + 1}{s^2 + s + 2}$ is minimum positive real or not

Solution

(a) At $s = 0$, the value of function is 1/2 and at $s = \infty$, the value of function is 2. Thus, it is seen that at $s = 0$ and $s = \infty$ the function has real, positive and finite values.

(b) The given function does not have any pole or zero on the imaginary axis of s-plane.

(c) putting $s = j\omega$, we get

$$F(s) = \frac{2(j\omega)^2 + j\omega + 1}{(j\omega)^2 + j\omega + 2} = \frac{-2\omega^2 + j\omega + 1}{-\omega^2 + j\omega + 2} = \frac{(1-2\omega^2) + j\omega}{(2-\omega^2) + j\omega}$$

$$= \frac{\left[(1-2\omega^2) + j\omega\right]\left[(2-\omega^2) - j\omega\right]}{(2-\omega^2) + (j\omega)^2}$$

$$= \frac{2-\omega^2 - j\omega - 4\omega^2 + 2\omega^4 + j2\omega^3 + 2j\omega - j\omega^3 + \omega^2}{4+\omega^4 - 4\omega^2 - \omega^2}$$

$$= \frac{2-4\omega^2 + 2\omega^4 + j\omega + j\omega^3}{\omega^4 - 3\omega^2 + 4}$$

$$\text{Re } F(j\omega) = \frac{2\omega^4 - 4\omega^2 + 2}{\omega^4 - 3\omega^2 + 4}$$

$$= \frac{2(\omega^4 - 2\omega^2 + 1)}{\omega^4 - 3\omega^2 + 4} = \frac{2(\omega^2 - 1)^2}{\omega^4 - 3\omega^2 + 4}$$

It can be seen that at $\omega = 1$, Re $F(j\omega) = 0$, i.e., the real part is zero at least at one finite real frequency or value of ω. Therefore, the given function is minimum positive real function.

Problem No. 7.10 Synthesisze the RL driving-point impedance.

$$Z(s) = \frac{2(s+1)(s+3)}{(s+2)(s+6)}$$

Solution

$$Z(s) = \frac{2(s+1)(s+3)}{(s+2)(s+6)} = \frac{2(s^2+4s+3)}{s^2+8s+12} = 2\left[\frac{s^2+4s+3}{s^2+8s+12}\right]$$

$$= 2\left[\frac{s^2+8s+12-4s-9}{s^2+8s+12}\right] = 2\left[1 - \frac{4s+9}{s^2+8s+12}\right]$$

$$= 2\left[1 - \frac{4s+9}{(s+2)(s+6)}\right]$$

$$= 2\left[1 - \left[\frac{A}{s+2} + \frac{B}{s+6}\right]\right]$$

$$A = \left.\frac{4s+9}{s+6}\right|_{s=-2} = \frac{1}{4}$$

$$B = \left.\frac{4s+9}{s+2}\right|_{s=-6} = \frac{-15}{-4} = \frac{15}{4}$$

Therefore,

$$Z(s) = 2\left[1 - \frac{1/4}{s+2} - \frac{15/4}{s+6}\right]$$

Since it yields negative residues, therefore let us expand $\frac{Z(s)}{s}$ by partial fraction method

$$\frac{Z(s)}{s} = \frac{2(s+1)(s+3)}{s(s+2)(s+6)}$$

$$= \frac{A}{s} + \frac{B}{s+2} + \frac{C}{s+6}$$

$$A = \left.\frac{2(s+1)(s+3)}{(s+2)(s+6)}\right|_{s=0} = \frac{2\times1\times3}{2\times6} = \frac{1}{2}$$

$$B = \left.\frac{2(s+1)(s+3)}{s(s+6)}\right|_{s=-2} = \frac{2\times-1\times1}{-2\times4} = \frac{1}{4}$$

$$C = \left.\frac{2(s+1)(s+3)}{s(s+2)}\right|_{s=-6} = \frac{2\times-5\times-3}{-6\times-4} = \frac{5}{4}$$

Therefore,

$$\frac{Z(s)}{s} = \frac{1/2}{s} + \frac{1/4}{s+2} + \frac{5/4}{s+6}$$

or

$$Z(s) = \frac{1}{2} + \frac{\frac{s}{4}}{s+2} + \frac{\frac{5s}{4}}{s+6}$$

The network realised by Foster first form is shown in Fig. 7.1

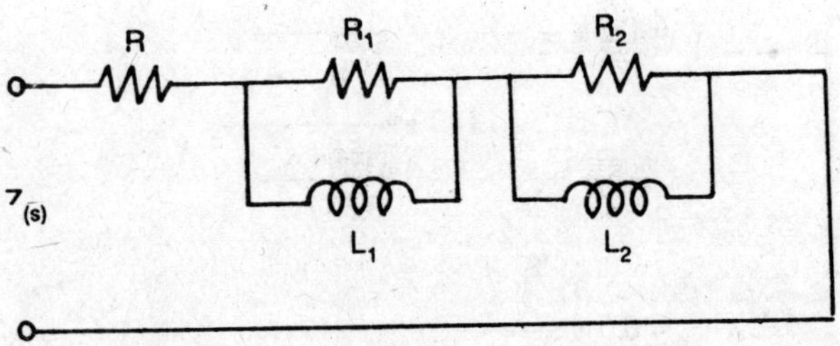

Fig. 7.1

where

$$R = \frac{1}{2}\,\Omega$$

$$R_1 = \frac{1}{4}\Omega;\ L_1 = \frac{1}{8}H$$

$$R_2 = \frac{5}{4}\Omega;\ L_2 = \frac{5}{24}H$$

Probelm No 7.11 Synthesize the function of Problem No. 7.10 by Cauer second form of network realization

Solution

$$Z(s) = \frac{2(s+1)(s+3)}{(s+2)(s+6)} = \frac{6+8s+2s^2}{12+8s+s^2}$$

$$= 0.5 + \frac{4s+1.5s^2}{12+8s+s^2} = 0.5 + \cfrac{1}{\cfrac{12+8s+s^2}{4s+1.5s^2}}$$

$$= 0.5 + \cfrac{1}{\cfrac{3}{s} + \cfrac{3.5s+s^2}{4s+1.5s^2}}$$

$$= 0.5 + \cfrac{1}{\cfrac{3}{s} + \cfrac{1}{1.143 + \cfrac{0.357s^2}{3.5s+s^2}}}$$

$$= 0.5 + \cfrac{1}{\cfrac{3}{s} + \cfrac{1}{1.143 + \cfrac{1}{\cfrac{3.5s+s^2}{0.357s^2}}}}$$

$$= 0.5 + \cfrac{1}{\cfrac{3}{s} + \cfrac{1}{1.143 + \cfrac{1}{\cfrac{10}{s} + \cfrac{s^2}{0.357s^2}}}}$$

$$= 0.5 + \cfrac{1}{\cfrac{3}{s} + \cfrac{1}{1.143 + \cfrac{1}{\cfrac{10}{s} + \cfrac{1}{0.35}}}}$$

$$= 0.5 + \cfrac{1}{\cfrac{1}{\cfrac{1}{0.33s} + \cfrac{1}{1.143 + \cfrac{1}{\cfrac{1}{0.1s} + \cfrac{1}{0.35}}}}}$$

The RL network realized with the elemental values is shown in Fig. 7.2.

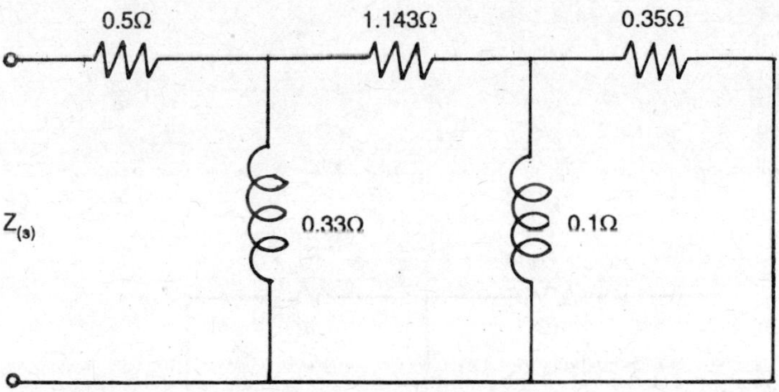

Fig. 7.2

Problem No. 7.12 Synthesize the RC driving point impedance

$Z(s) = \dfrac{(s+1)(s+3)}{s(s+2)}$ to get the Cauer first and second forms of network realisation.

Solution

$$Z(s) = \frac{(s+1)(s+3)}{s(s+2)} = \frac{s^2+4s+3}{s^2+2s}$$

First Cauer Form

$$
\begin{array}{l}
s^2+2s\,\overline{)\,s^2+4s+3\,}\,(\,1 \longrightarrow R_1 \\
\qquad\quad \underline{s^2+2s} \\
\qquad\quad 2s+3\,\overline{)\,s^2+2s\,}\,(\,\dfrac{s}{2} \longrightarrow C_1 s \\
\qquad\qquad\quad \underline{s^2+\dfrac{3s}{2}} \\
\qquad\qquad\qquad \dfrac{s}{2}\,\overline{)\,2s+3\,}\,(\,4 \longrightarrow R_2 \\
\qquad\qquad\qquad\quad \underline{2s} \\
\qquad\qquad\qquad\qquad 3\,\overline{)\,\dfrac{s}{2}\,}\,(\,\dfrac{s}{6} \longrightarrow C_2 s \\
\qquad\qquad\qquad\qquad\quad \underline{\dfrac{s}{2}} \\
\qquad\qquad\qquad\qquad\quad \text{X}
\end{array}
$$

The synthesised network is shown in Fig. 7.3

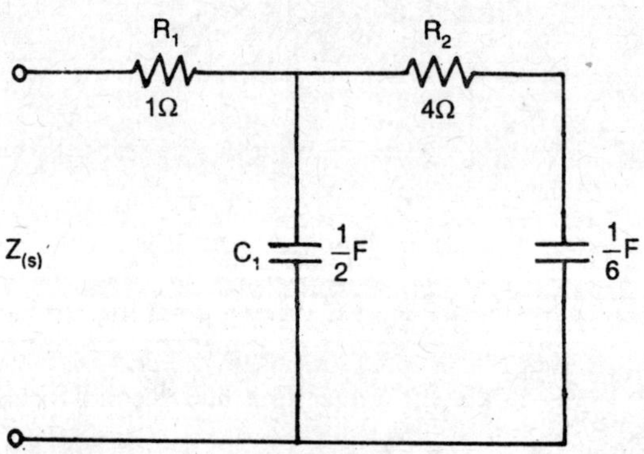

Fig. 7.3

Second Cauer Form

$$
Z(s) = \frac{3+4s+s^2}{2s+s^2}
$$

The synthesized network is shown in Fig. 7.4

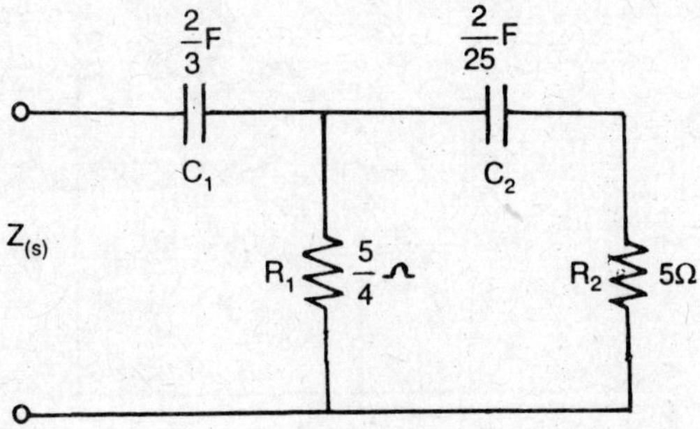

Fig. 7.4

Problem No. 7.13 Synthesize the LC driving point impedance function

$$Z(s) = \frac{s^4 + 10s^2 + 9}{s^3 + 4s}$$ to get the Cauer first and second form of network realisation.

Solution

First Cauer Form

$$s^3 + 4s \overline{\big)\, s^4 + 10s^2 + 9 \,\big(}\, s \longrightarrow L_1 = 1H$$
$$\underline{s^4 + 4s^2}$$
$$6s^3 + 9 \overline{\big)\, s^3 + 4s \,\big(}\, \frac{s}{6} \longrightarrow C_1 = \frac{1}{6}F$$
$$\underline{s^3 + \frac{3}{2}s}$$
$$\frac{5}{2}s \overline{\big)\, 6s^3 + 9 \,\big(}\, \frac{12s}{5} \longrightarrow L_2 = \frac{12}{5}H$$
$$\underline{6s^2}$$
$$9 \overline{\big)\, \frac{5}{2}s \,\big(}\, C_2 = \frac{5}{18}F$$
$$\underline{\frac{5}{2}s}$$
$$X$$

Synthesized network in shown in Fig. 7.5

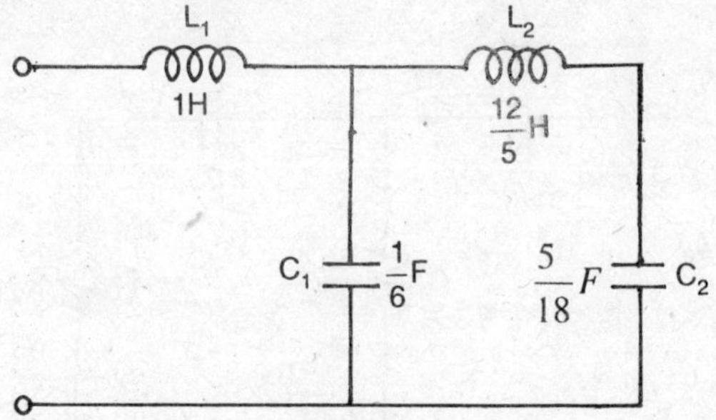

Fig. 7.5

Second Cauer Form

$$4s+s^3\overline{)9+10s^2+s^4}\left(\frac{9}{4s}\longrightarrow C_1=\frac{4}{9}F\right.$$

$$\underline{9+\frac{9}{4}s^2}$$

$$\frac{31}{4}s^2+s^4\overline{)4s+s^3}\left(\frac{16}{31s}\longrightarrow L_1=\frac{31}{16}H\right.$$

$$\underline{4s+\frac{16}{3}s^3}$$

$$\frac{15}{31}s^3\overline{)\frac{31}{4}s^2+s^4}\left(\frac{961}{60s}\longrightarrow C_2=\frac{60}{961}F\right.$$

$$\underline{\frac{31}{4}s^2}$$

$$s^4\overline{)\frac{15}{31}s^3}\left(\frac{16}{31}s\longrightarrow L_2=\frac{31}{15}H\right.$$

$$\underline{\frac{15}{31}s^3}$$

$$\text{X}$$

Synthesized network is shown in Fig. 7.6

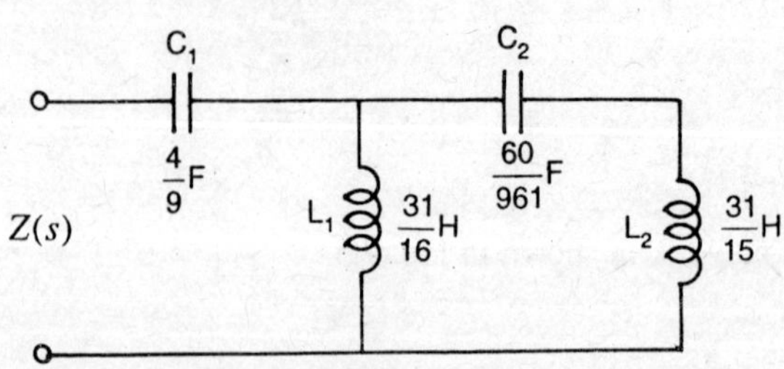

Fig. 7.6

Problem No. 7.14 Synthesize the Foster first and second forms of LC driving point impedance

$$Z(s) = \frac{(s^2+1)(s^2+9)}{s(s^2+4)}$$

Solution

$$Y(s) = \frac{s^2+3s}{s^2+2s+6}$$

Foster First form

$$Z(s) = \frac{(s^2+1)(s^2+9)}{s(s^2+4)}$$

$$\frac{Z(s)}{s} = \frac{(s^2+1)(s^2+9)}{s^2(s^2+4)}$$

Let $s^2 = x$

$$\frac{Z(s)}{s} = \frac{(x+1)(x+9)}{x(x+4)} = \frac{x^2+10x+9}{x^2+4x} = \frac{x^2+4x+6x+9}{x^2+4x}$$

$$= 1 + \frac{6x+9}{x^2+4x}$$

$$= 1 + \frac{3(2x+3)}{x(x+4)}$$

$$= 1 + 3\left[\frac{A}{x} + \frac{B}{x+4}\right]$$

$$A = \left.\frac{2x+3}{x+4}\right|_{x=0} = \frac{3}{4}$$

$$B = \left.\frac{2x+3}{x}\right|_{x=-4} = \frac{(2\times -4)+3}{-4} = \frac{5}{4}$$

Therefore,

$$\frac{Z_{(s)}}{s} = 1 + 3\left[\frac{3/4}{x} + \frac{5/4}{x+4}\right]$$

$$= 1 + \frac{9/4}{x} + \frac{15/4}{x+4}$$

Putting $x = s^2$, we have

$$\frac{Z(s)}{s} = 1 + \frac{9/4}{s^2} + \frac{15/4}{s^2+4}$$

$$Z(s) = s + \frac{9/4}{s} + \frac{15s/4}{s^2+4}$$

$$= s + \frac{1}{4s/9} + \frac{1}{\frac{4s}{15} + \frac{1}{\frac{15}{16}s}}$$

Synthesized network is shown in Fig. 7.7

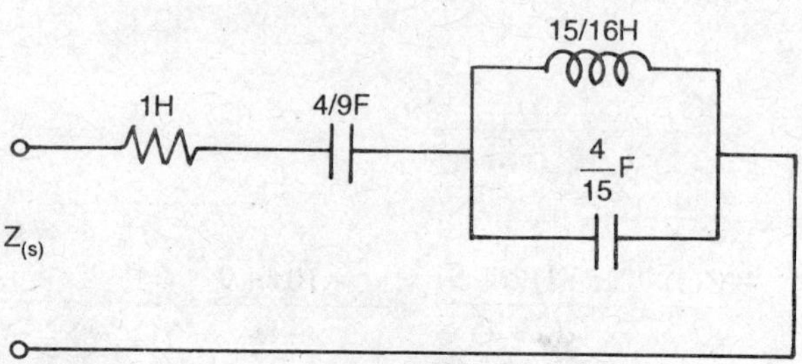

Fig. 7.7

Second Foster Form

$$Z(s) = \frac{(s^2+1)(s^2+9)}{s(s^2+4)}$$

$$\frac{Y(s)}{s} = \frac{s^2+4}{(s^2+1)(s^2+9)}$$

$s^2 = x$, Therefore,

$$\frac{Y(s)}{s} = \frac{x+4}{(x+1)(x+9)} = \frac{A}{x+1} + \frac{B}{x+9}$$

$$A = \left.\frac{x+4}{x+9}\right|_{x=-1} = \frac{-1+4}{-1+9} = \frac{3}{8}$$

$$B = \left.\frac{x+4}{x+1}\right|_{x=-9} = \frac{-9+4}{-9+1} = \frac{5}{8}$$

Therefore,

$$\frac{Y(s)}{s} = \frac{3/8}{x+1} + \frac{5/8}{x+9}$$

$$Y(s) = \frac{3s/8}{x+1} + \frac{5s/8}{x+9}$$

$$Y(s) = \frac{3s/8}{s^2+1} + \frac{5s/8}{s^2+9} = \frac{1}{\frac{8}{3}s + \frac{1}{3s/8}} + \frac{1}{\frac{8}{5}s + \frac{1}{5s/72}}$$

$$= \frac{1}{\frac{8}{3}s + \frac{1}{3s/8}} + \frac{1}{\frac{8}{5}s + \frac{1}{5s/72}}$$

Synthesized network is shown in Fig. 7.8

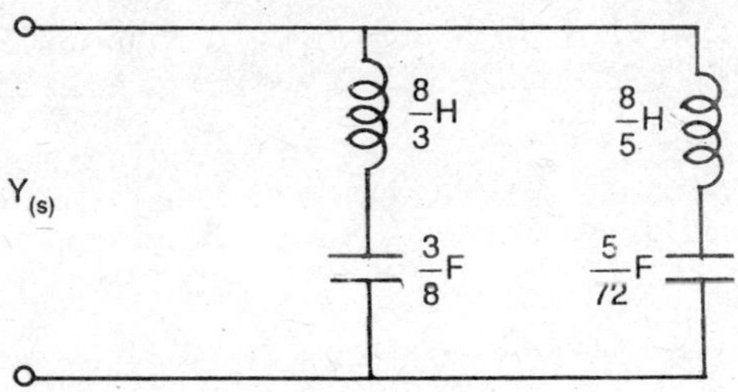

Fig. 7.8

Problem No. 7.15 Synthesize the network whose driving point impedance is

$$\frac{s^2+2s+6}{s(s+3)}$$

Solution

$$Z(s) = \frac{s^2+2s+6}{s(s+3)} = \frac{2}{s} + \frac{s}{s+3}$$

$$= \frac{1}{s/2} + \frac{1}{1+3/s}$$

$$= \frac{1}{s/2} + \frac{1}{1+\frac{1}{s/3}}$$

Synthesized network is shown in Fig. 7.9

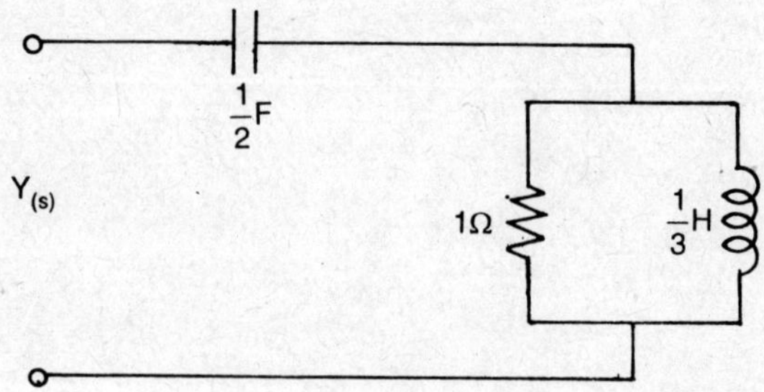

Fig. 7.9

8

Bode Plots

Sinusoidal transfer functions are commonly represented by Bode plot. It is a plot of magnitude and angle of transfer function against frequency. It is a log-arithmic plot and consists of two plots:

1. Plot of logarithm of magnitude against frequency in logarithmic scale.
2. Plot of phase angle against frequency in logarithmic scale.

Logarithmic magnitude is represented as 20 log $|G(j\omega)|$. The base of logarithm is 10 and the unit is 'decibel' usually abbreviated as 'db'.

The following are the advantages of Bode plot:

1. Plotting of Bode plot is relatively easier as compared to other methods as the loci of $(1+sT)$ and $\dfrac{K}{(1+sT)}$ can be represented by straight line asymptotes
2. Multiplication is converted into addition, so if

$$G(s) = \frac{K}{s(1+sT)}$$

and putting $s = j\omega$

then $20\log|G(j\omega)| = 20\log K - 20\log|j\omega| - 20\log|1 + j\omega T|$

3. Low and high frequencies characteristics can be represented on a single diagram.
4. Study of relative stability is easier as parameters of analysis of relative stability are gain and phase margins which are visibly seen on sketch.
5. If modification of an existing system is to be studied, it can be easily done on a Bode plot.

Basic terms that appear in a transfer function and method of plotting them is described below.

1. ***Constant term 'K'*** It gives a constant magnitude of 20 log K. It does not give any phase shift. It is represented by a line parallel to 0 db line (slope db/dec) and starts from a point having a magnitude 20 log K.

2. $(s)^{\pm 1}$ Its magnitude is 20 log ω. It is a straight line having a slope of ± 20 db/sec and passes through ω = 1 rad/sec where its magnitude is 0 db.

3. **(1 + sT)** $^{\pm n}$. Magnitude is given by $n \times 20 \log [1 + j\omega T]$ having a slope of $\pm n \times 20$ db/dec. Asymptotes are approximated by

(a) $\omega T << 1$, magnitude is $n \times 20 \log 1$ i.e. 0 db

(b) $\omega T >> 1$, magnitude is $\pm n$ 20 log ωT. It has a slope of $\pm n \times 20$ db/dec. Asymptotes meet at a point where

$$20 \log \omega T = 0 \qquad \text{i.e.} \qquad \omega T = 1$$

or
$$\omega = \frac{1}{T}$$

which is called the 'corner frequency'. Due to approximation, maximum error of 3db occurs at the corner frequency. Exact plot can be drawn by finding the magnitude one octave to the left and right of corner frequency.

4. $s^2 + 2\xi \omega_n s + \omega_n$ This is written as

(a) If $\zeta > 1$, it can be reduced to two first order factors

(b) If $\zeta < 1$, the magnitude is

$$= \pm 20 \log \left| 1 - \left(\frac{\omega}{\omega_n}\right)^2 + j2\zeta\frac{\omega}{\omega_n} \right|$$

$$= \pm 20 \log \left| \left\{1 - \left(\frac{\omega}{\omega_n}\right)^2\right\}^2 + 4\zeta^2\left(\frac{\omega}{\omega_n}\right)^2 \right|^{1/2}$$

This term is approximated by drawing two straight lines. Corner frequency is ω_n. Below ω_n, line follows 0 db line and beyond ω_n it has a slope of ± 40 db/dec.

Error due to approximation depends upon the value of ζ. If ζ is small the error is large.

5. ***Phase Plot*** If the transfer function is as given below

$$G(s) = \frac{K(1 + sT_1)}{s(1 + sT_2)(s^2 + 2\zeta\omega_n s + \omega_n^2)}$$

then $\underline{|G(j\omega)} = \phi = -90° + \tan^{-1}\omega T_1 - \tan^{-1}\omega T_2 - \tan^{-1}\dfrac{2\zeta\omega/\omega_n}{1-(\omega/\omega_n)^2}$

Problem 8.1 Sketch the Bode Plots and determine the gain cross-over and phase cross-over frequencies.

$$G(s) = \frac{10}{s(1+0.5s)(1+0.1s)}$$ **(Pune University)**

Solution

Corner frequencies The corner frequencies are 2 and 10.

Magnitude Plot

Ser. No.	*Factor*	*Corner frequency*	*Asymptotic log-magnitude Characteristic*
1	$\frac{1}{s}$	None	Straight line of constant slope (–20 db/dec) passing through at $\omega = 1$
2	$\frac{1}{(1+0.5s)}$	$\omega_1 = 2$	Straight line of constant slope (–20 db/dec) originating from $\omega_1 = 2$
3	$\frac{1}{(1+0.1s)}$	$\omega_2 = 10$	Straight line of constant slope (–20 db/dec) originating from $\omega_2 = 10$
4	10	None	Straight line of constant slope of 0 db/dec starting from 20 log 10 = 20 db point

Magnitude plots for individual factors are shown by dotted lines. Resultant line is shown by a firm line (Fig. 8.1).

Phase Plot $\phi = \;90°\;\tan^{-1} 0.5\omega - \tan^{-1} 0.1\omega$

Ser. No.	ω	ϕ
1	0	–90°
2	0.1	–93.43°
3	1	–122.3°
4	2	–146.31°
5	5	–184.76°
6	10	–213.7°
7	15	–228.7°

Magnitude and phase plots are shown in Fig. 8.1. From the plots

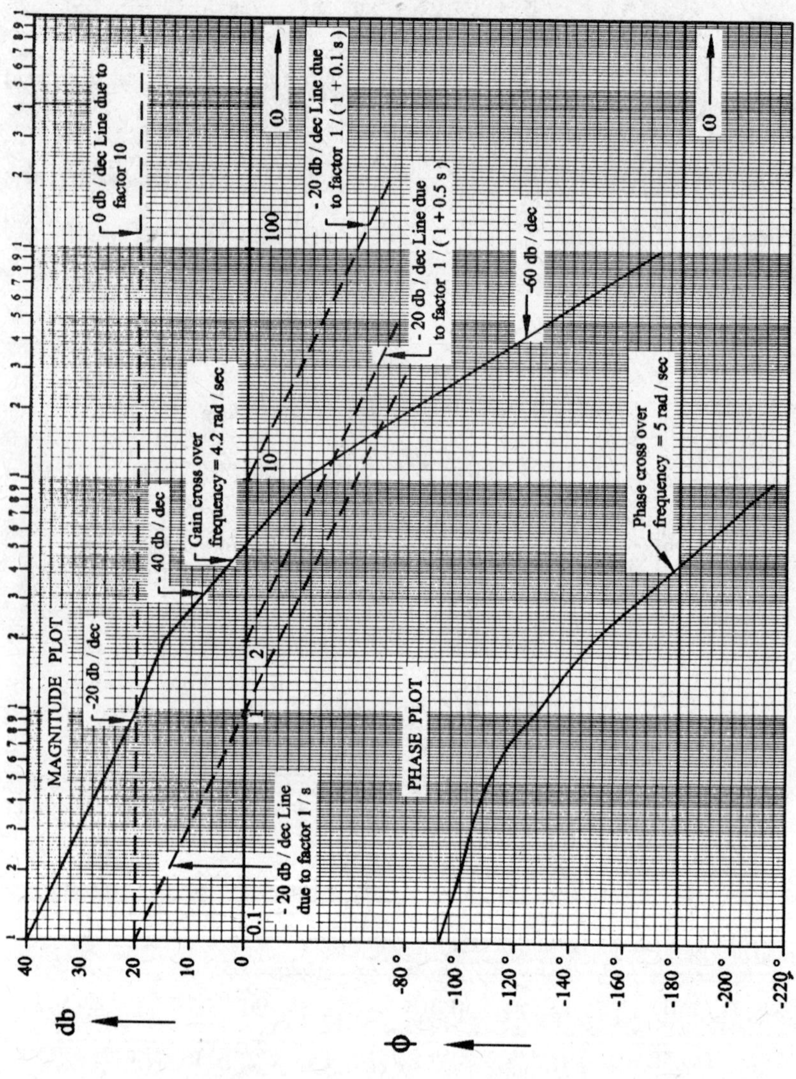

Fig. 8.1

1. Gain crossover frequency = 4.2 rad/sec **Ans.**
2. Phase crossover frequency = 4.5 rad/sec. **Ans.**

Problem 8.2 Sketch the Bode plot for the transfer function

$$G(s) = \frac{K s^2}{(1+0.2\,s)(1+0.02\,s)}$$

Determine the system gain K for the gain cross-over frequency to be 5 rad/sec.

Solution

Let, $K = 1$, then

$$G(s) = \frac{s^2}{(1+0.2\,s)(1+0.02\,s)}$$

Corner frequencies The corner frequencies are 5 and 50 rad/sec.

Magnitude Plot

Ser. No.	*Factor*	*Corner frequency*	*Asymptotic log-magnitude Characteristic*
1	s^2	None	Straight line of constant slope 40 db/dec passing through $\omega = 1$
2	$\frac{1}{(1+0.2\,s)}$	$\omega_1 = 5$	Straight line of constant slope –20 db/dec originating from $\omega = 5$
3	$\frac{1}{(1+0.02\,s)}$	$\omega_2 = 50$	Straight line of constant slope –20 db/dec originating from $\omega = 50$

The magnitude plot is shown in Fig. 8.2. If the gain cross-over frequency is required to be 5 rad/sec, then the magnitude plot must cross the 0 db line at 5 rad/sec. For this, the plot has to be brought down by 28 db. Hence

$$20 \log K = -28$$

$$K = 0.04$$

Ans.

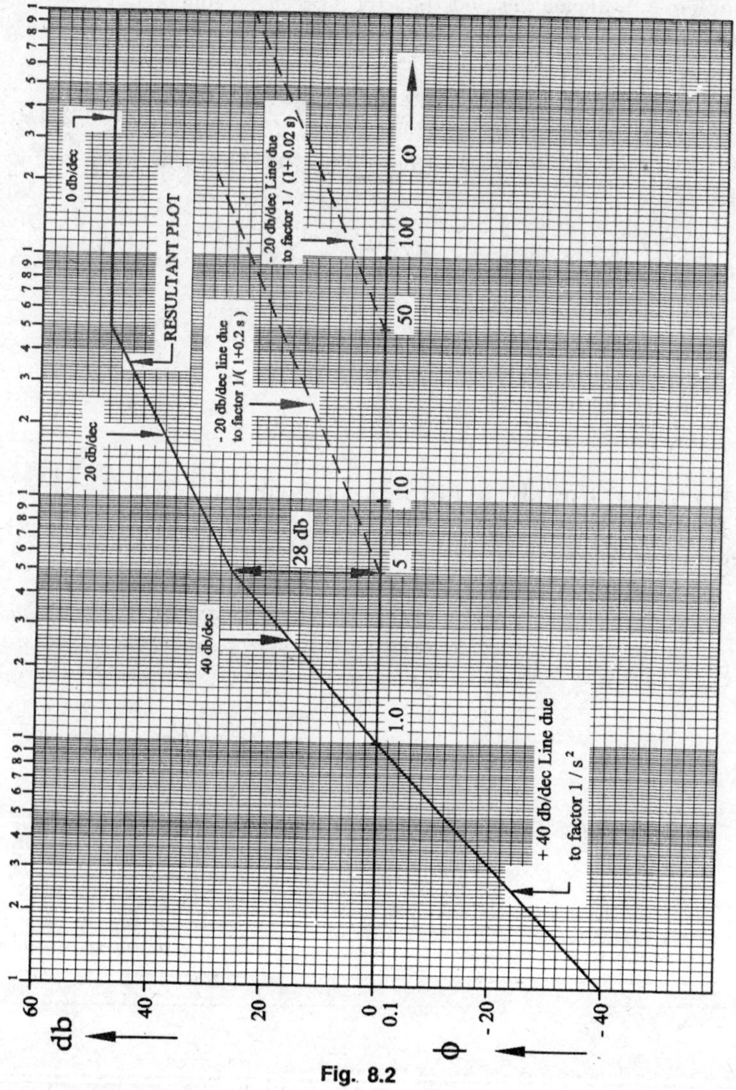

Fig. 8.2

Problem 8.3 Sketch the Bode plot for a unity feedback system characterised by the open-loop transfer function

$$G(s) = \frac{K(1+0.2\,s)(1+0.025s)}{s^3(1+0.001s)(1+0.005\,s)}$$

Show that the system is conditionally stable. Find the range of values of K for which the system is stable.

Solution Let $K = 1$

Magnitude Plot

Ser. No.	*Factor*	*Corner frequency*	*Asymptotic log-magnitude Characteristic*
1	$\frac{1}{s^3}$	None	Straight line of constant slope of –60 db/dec passing through $\omega = 1$
2	$(1 + 0.2\,s)$	5	Straight line of constant slope of 20 db/dec originating from $\omega = 5$
3	$(1 + 0.025\,s)$	40	Straight line of constant slope of 20 db/dec originating from $\omega = 40$
4	$\frac{1}{(1+0.005\,s)}$	200	Straight line of constant slope of –20 db/dec originating from $\omega = 200$
5	$\frac{1}{(1+0.001\,s)}$	1000	Straight line of constant slope of –20 db/dec originating from $\omega = 1000$

Phase Plot

$$\phi = -\,270° + \tan^{-1} 0.2\,\omega + \tan^{-1} 0.025\,\omega$$
$$-\tan^{-1} 0.005\,\omega - \tan^{-1} 0.001\,\omega.$$

Ser. No.	*ω rad/sec*	ϕ
1	0.1	–268°
2	0.8	–260°
3	1	–257°
4	3	–236°
5	10	–198°
6	15	–183°
7	30	–163°
8	60	–148°
9	100	–147°
10	300	–172°
11	400	–182°

Resultant Magnitude and phase plots are shown in Fig. 8.3.

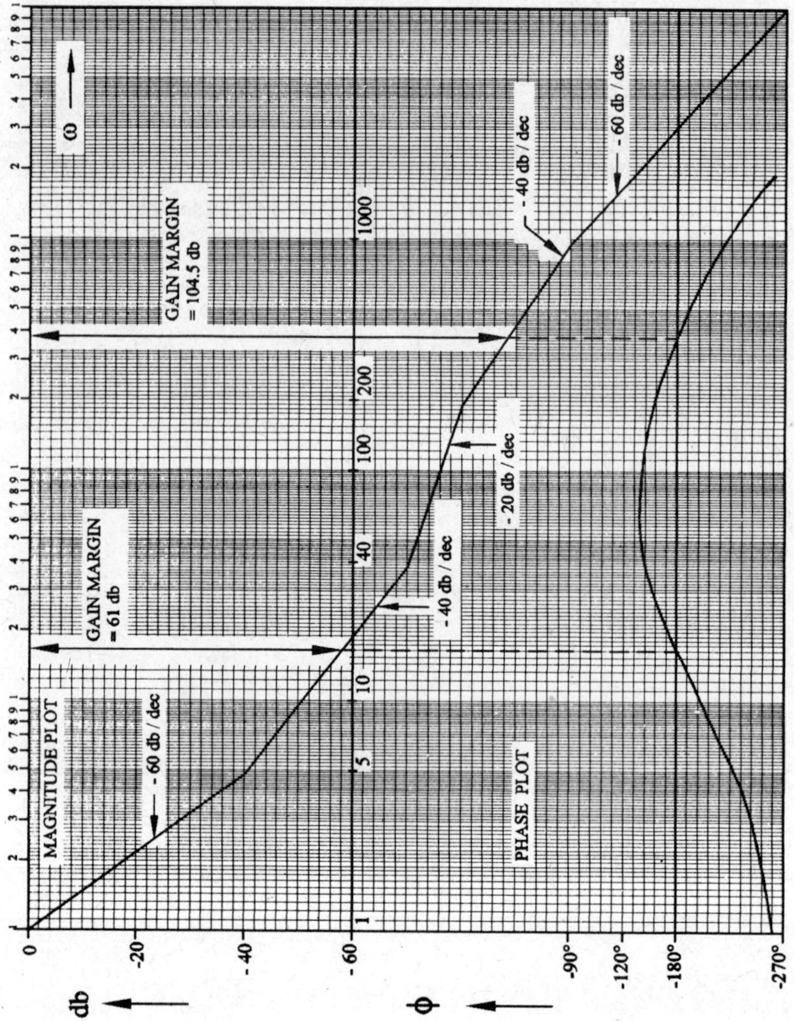

Fig. 8.3

The phase plot crosses the – 180° line twice indicating that the system under consideration is conditionally stable.

Gain Margins are 61 db and 104.5 db

Now $\quad 20 \log K = 61$

$\therefore \quad K = 1122$

Also $\quad 20 \log K = 104.5$

$$K = 167880$$

$\therefore$ Condition for stability is

$$1122 < K < 167880 \quad \textbf{Ans.}$$

Problem 8.4 Draw the Bode plot for a system having

$$G(s)\,H(s) = \frac{100}{s(s+1)(s+2)} \quad \text{Find}$$

(a) Gain Margin
(b) Phase margin
(c) Gain cross over frequency
(d) Phase cross over frequency **(Pune University)**

Solution

$$G(s)\,H(s) = \frac{50}{s(s+1)(1+0.5s)}$$

Magnitude Plot

Ser. No.	*Factor*	*Corner frequency rad/sec*	*Asymptotic log-magnitude Characteristic*
1	50	None	Straight line of slope 0 db/dec starting from point 20 log 50 = 34 db
2	$\frac{1}{s}$	None	Straight line of slope 20 db/dec passing through $\omega = 1$
3	$\frac{1}{(1+s)}$	1	Straight line of slope–20 db/dec originating from $\omega = 1$
4	$\frac{1}{(1+0.5s)}$	2	Straight line of slope–20 db/dec originating from $\omega = 2$

Phase Plot $\phi = -90° - \tan^{-1} \omega - \tan^{-1} 0.5\omega$

Ser. No.	*ω rad/sec*	ϕ
1	0	–90°
2	0.1	–98.6°
3	0.2	–107°
4	0.5	–130.6°
5	1	–161.6°
6	1.3	–175.5°
7	1.4	–179.5°
8	1.5	–183.2°
9	2	–198.4°
10	4.45	–233°

Magnitude and Phase plots as shown in Fig. 8.4

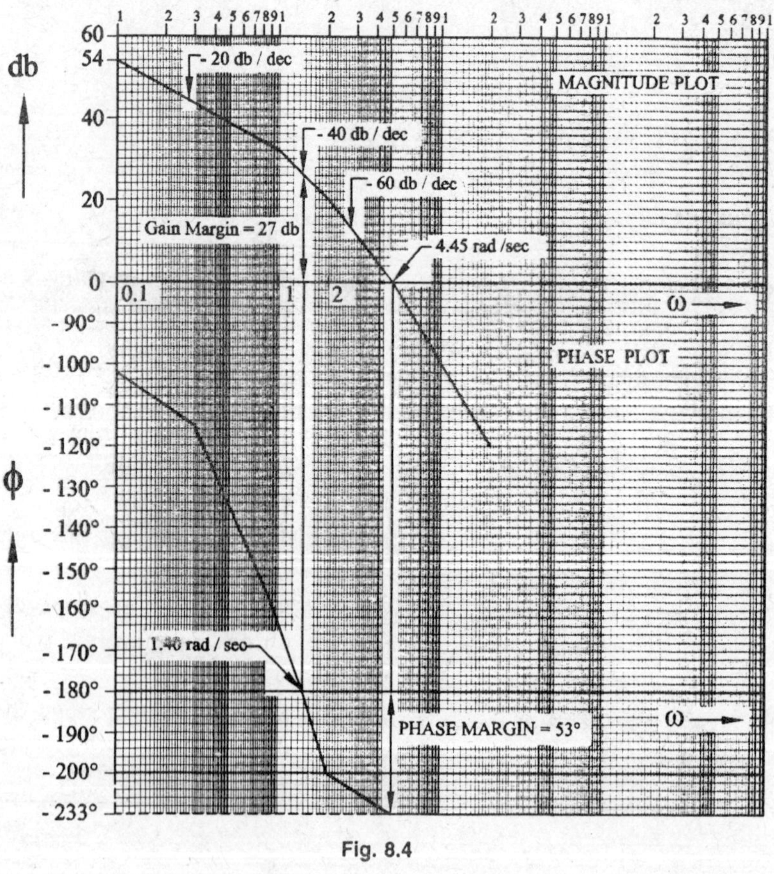

Fig. 8.4

Result

1. Gain Crossover frequency : 4.45 rad/sec
2. Phase Crossover frequency : 1.40 rad/sec
3. Gain Margin : 27 db
4. Phase Margin : 53°.

Problem 8.4 The Open-loop transfer function of a certain unity feedback system is

$$G(s) = \frac{K}{s(s+2)(s+10)}$$

Construct Bode plots and determine.

(a) Limiting value of K for system to be stable
(b) Value of K for gain margin to be 10 db
(c) Value of K for phase margin to be 50° **(Pune University)**

Solution $$G(s) = \frac{0.025\,K}{s(1+0.5s)(1+0.05s)}$$

Let = 0.025 K = 1, then $G(s) = \dfrac{1}{s(1+0.5s)(1+0.05s)}$

Magnitude Plot

Ser. No.	*Factor*	*Corner frequency rad/sec*	*Asymptotic log-magnitude Characteristic*
1	$\frac{1}{s}$	None	Straight line of 0 db/dec passing through ω = 1
2	$\frac{1}{(1+0.5s)}$	2	Straight line of –20 db/dec originating from ω = 2
3	$\frac{1}{(1+0.05s)}$	20	Straight line of –20 db/dec originating from ω = 4

Phase Plot $\phi = -90° - \tan^{-1} 0.5\,\omega - \tan^{-1} 0.05\,\omega$

Ser. No.	*ω rad/sec*	ϕ
1	0	–90°
2	1	–119°
3	2	–141°
4	2.5	–148.5°
5	3	–155°
6	4	–165°
7	4.5	–168.7°
8	5	–172°
9	6	–178.3°
10	6.5	–181°
11	10	–195°

Bode plots are shown in Fig. 8.5

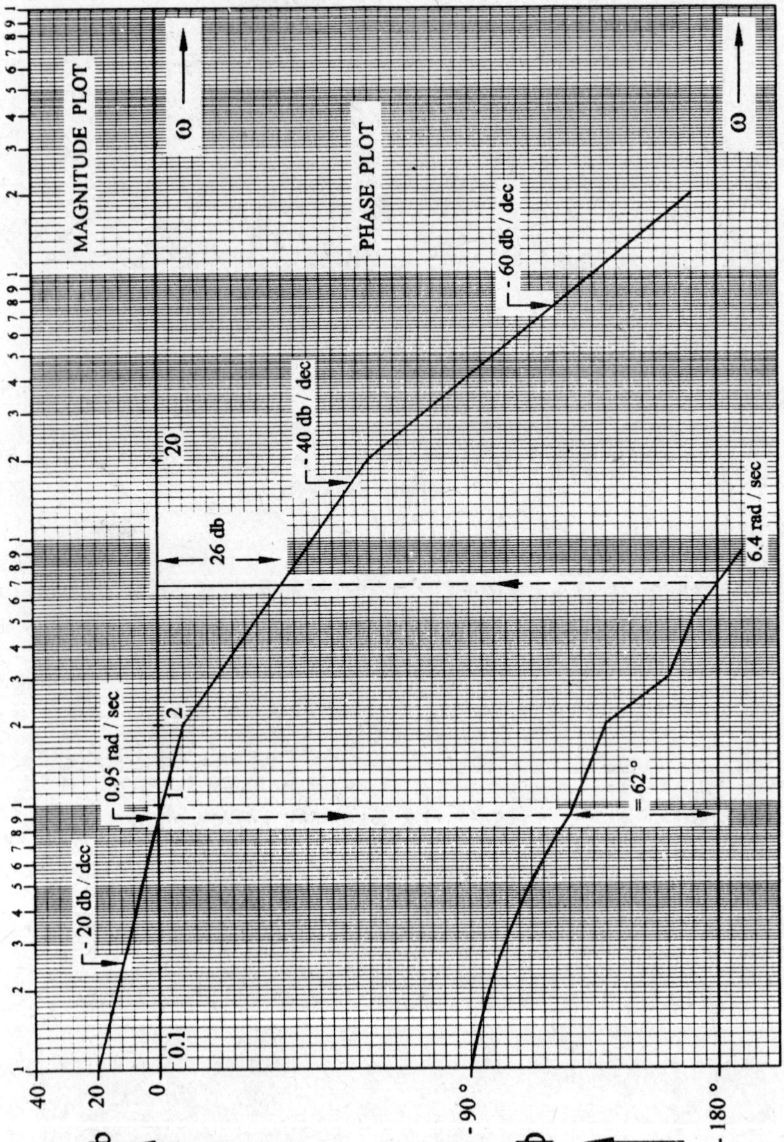

Fig. 8.5

(a) From the curves, the gain margin is 26 db

$$\therefore \qquad 20 \log K_1 = 26$$

$$\text{or} \qquad K_1 = 19.95$$

or $\qquad 0.025\ K = 19.95$

or $\qquad K = 798$ **Ans.**

(b) For the gain margin to be 10 db, the graph has to be lifted up by 26 – 10 = 16 db

$\therefore \qquad 20 \log K_1 = 16$

or $\qquad K_1 = 6.3$

or $\qquad 0.025\ K = 6.3$

or $\qquad K = 252$ **Ans.**

(c) For the phase margin to be 50°, the value of ω at –180° + 50° = –130° is 1.9 rad/sec. Gain Margin at 1.9 rad/sec is 5.5 db. Therefore, to have phase margin of 50°, magnitude plot has to be lifted up by 5.5 db, so that gain cross over frequency is 1.9 rad/sec

$\therefore \qquad 20 \log K_1 = 5.5$

or $\qquad K_1 = 1.88$

or $\qquad 0.025\ K = 1.88 \qquad$ or $\quad K = \dfrac{1.88}{0.025} = 75.35$ **Ans.**

Problem 8.5 Construct the Bode plots on a semilog graph-paper for a unity feedback system whose open-loop transfer function is given by

$$G(s) = \frac{10}{s(1+s)(1+0.02s)}$$

From the Bode-plot determine

(a) Gain and phase crossover frequencies

(b) Gain and Phase Margin, and

(c) Stability of the closed loop system. **(Pune University)**

Solution

Magnitude Plot

Ser. No.	*Factor*	*Corner frequency rad/sec*	*Asymptotic log-magnitude Characteristic*
1	10	None	Straight line of 0 db/dec from point 20 log 10 = 20 db
2	$\dfrac{1}{s}$	None	Straight line of – 20 db/dec passing through ω = 1
3	$\dfrac{1}{(1+s)}$	1	Straight line of – 20 db/dec originating from ω = 1
4	$\dfrac{1}{(1+0.02s)}$	50	Straight line of – 20 db/dec originating from ω = 50

Phase Plot $\phi = -90° - \tan^{-1}\omega - \tan^{-1} 0.02\,\omega$

Ser. No.	*ω rad/sec*	φ
1	0	–90°
2	0.2	–101.5°
3	0.5	–117°
4	1	–136°
5	10	–185.6°
6	50	–224°

Bode Plots are shown in Fig. 8.6

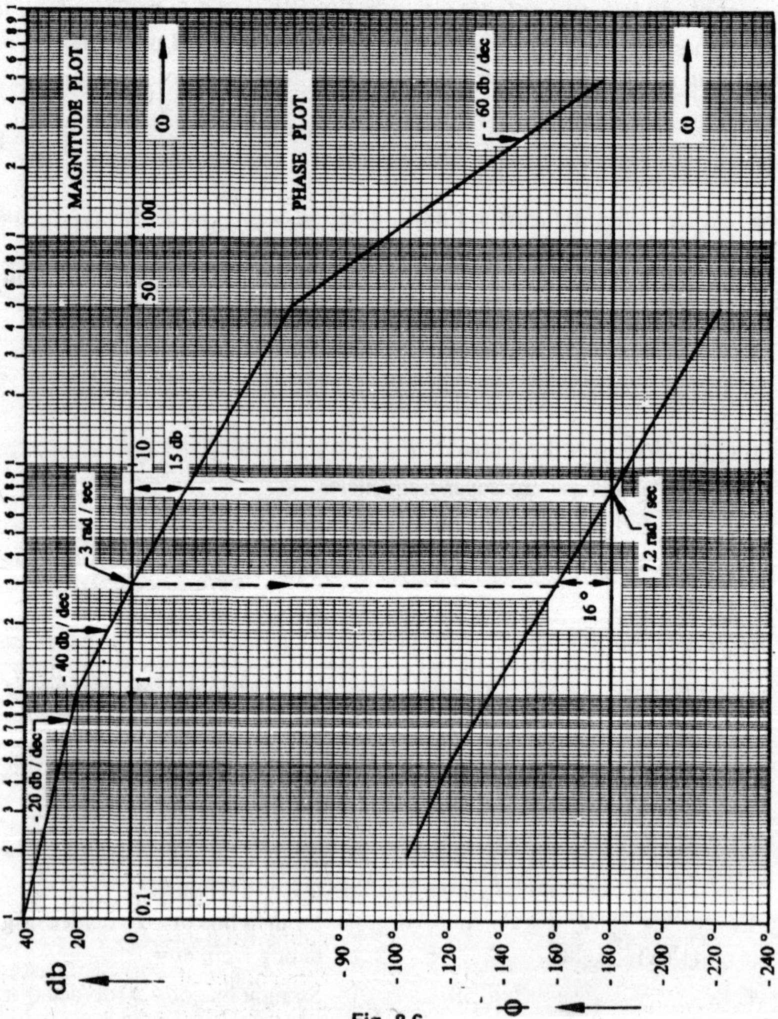

Fig. 8.6

Result

1. Gain Crossover frequency = 3 rad/sec
2. Phase Crossover frequency = 7.2 rad/sec
3. Gain Margin = 15 db
4. Phase Margin = 16°

Stability

Since phase margin and gain margin are positive, the system is absolutely stable.

Problem 8.6 Draw the Bode plot for the system having

$$G(s) = \frac{10}{s(1+0.01s)(1+0.1s)}, \; H(s) = 1$$

Determine

(a) The gain crossover frequency and corresponding phase margin

(b) the phase crossover frequency and corresponding gain margin

(Pune University)

Solution

Magnitude Plot

Ser. No.	*Factor*	*Corner frequency rad/sec*	*Asymptotic log-magnitude Curve*
1	10	None	Straight line of 0 db/dec starting from point 20 log 10 = 20 db
2	$\frac{1}{s}$	None	Straight line of – 20 db/dec passing through ω = 1
3	$\frac{1}{(1+0.1s)}$	10	Straight line of –20 db/dec originating from ω = 10
4	$\frac{1}{(1+0.01s)}$	100	Straight line of –20 db/dec originating from ω = 100

Phase Plot

$$\phi = -90^\circ - \tan^{-1} 0.1\,\omega - \tan^{-1} 0.01\,\omega$$

Ser. No.	ω *rad/sec*	φ	*Ser. No.*	ω *rad/sec*	φ
1	0	–90°	6	10	–141°
2	0.1	–90.6°	7	30	–178.3°
3	0.5	–93.15°	8	50	–195°
4	1	–96.3°	9	100	–219.3°
5	5	–119°			

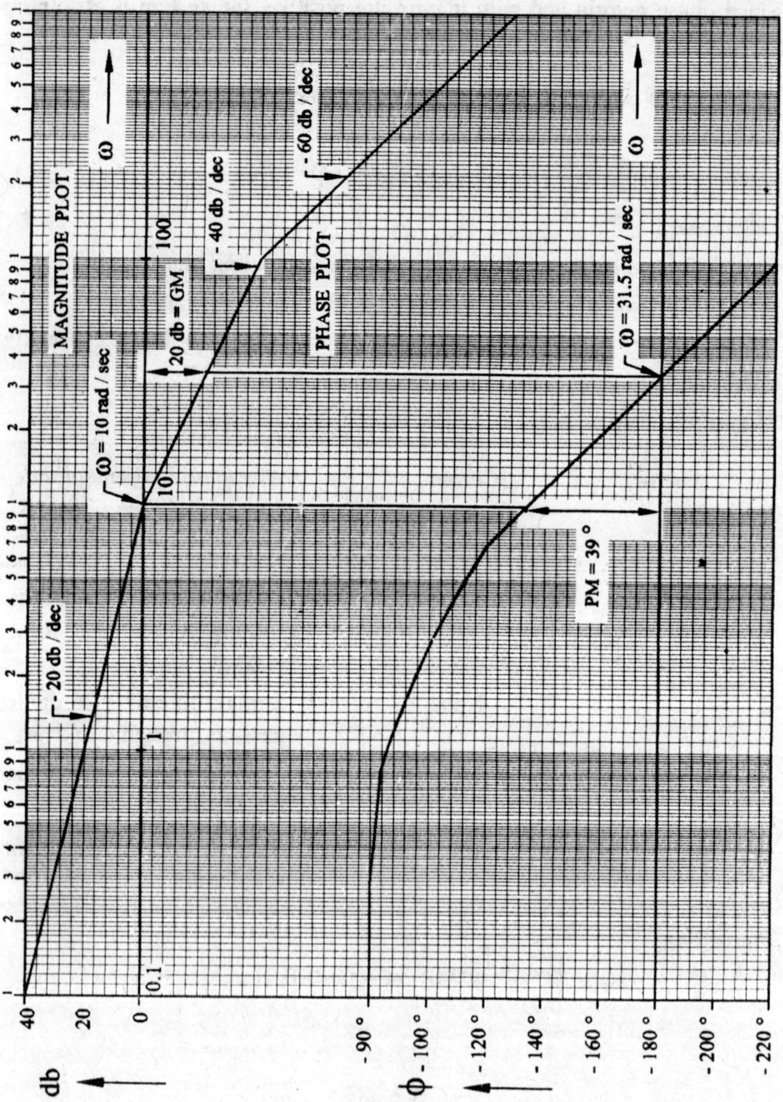

Fig. 8.7

Result

1. Gain Crossover frequency : 10 rad/sec
2. Phase Crossover frequency : 31.5 rad/sec
3. Gain Margin : 20 db
4. Phase Margin : 39°

Problem 8.7 A feedback control system is described by

$$G(s) = \frac{10}{s(1+0.2s)(1+0.01s)},\ H(s) = 1$$

Construct an asymptotic log-magnitude plot and an exact phase plot. From this determine,

1. Gain crossover and phase crossover frequency.
2. Gain Margin and phase Margin.
3. The stability of the closed-loop system. **(Pune University)**

Solution

Magnitude Plot

Ser. No.	*Factor*	*Corner frequency rad/sec*	*Asymptotic log-magnitude Characteristics*
1	10	None	Straight line of 0 db/dec starting from point 20 log 10 = 20 db
2	$\frac{1}{s}$	None	Straight line of – 20 db/dec passing through ω = 1
3	$\frac{1}{(1+0.2s)}$	5	Straight line of –20 db/dec originating from ω = 5
4	$\frac{1}{(1+0.01s)}$	100	Straight line of –20 db/dec originating from ω = 100

Phase Plot

$$\phi = -90° - \tan^{-1} 0.2\,\omega - \tan^{-1} 0.01\,\omega$$

Ser. No.	ω rad/sec	φ	Ser. No.	ω rad/sec	φ
1	0	−90°	7	15	−170°
2	0.1	−91.2°	8	20	−177°
3	0.5	−96°	9	25	−183°
4	1	−102°	10	50	−200.8°
5	5	−138°	11	100	−222°
6	10	−159°			

Bode plots are shown in Fig. 8.8.

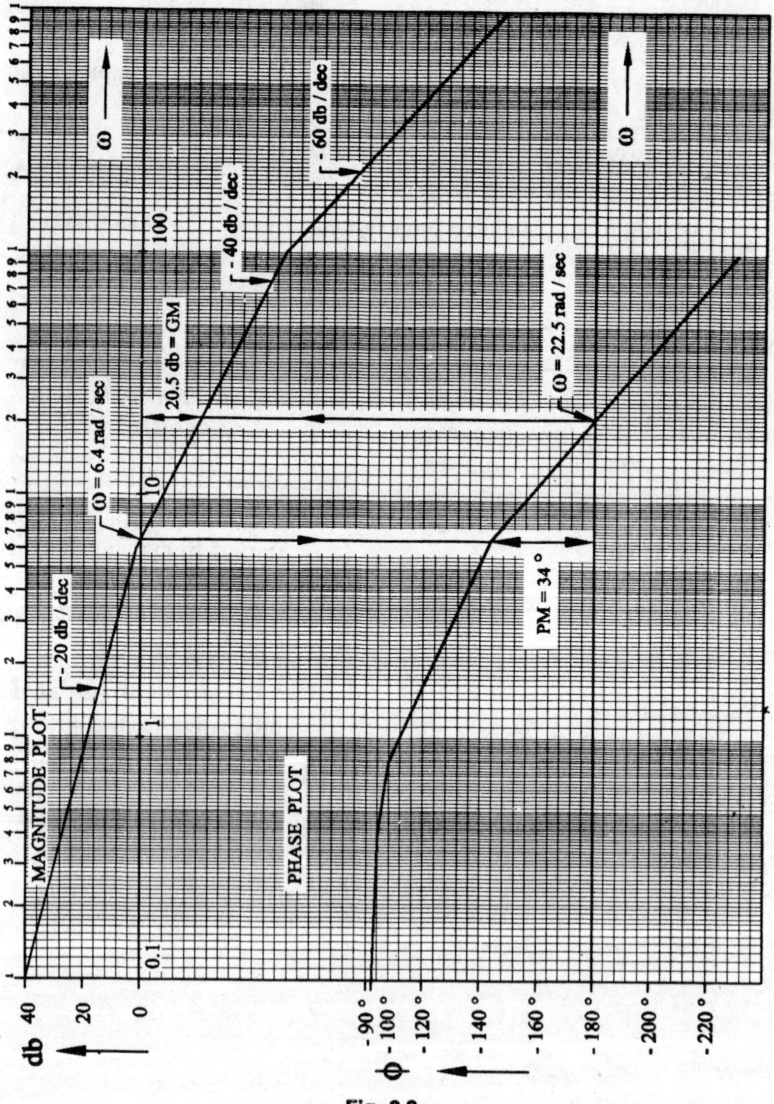

Fig. 8.8

Result

1. Gain Crossover frequency : 6.4 rad/sec
2. Phase Crossover frequency : 22.5 rad/sec
3. Gain Margin : 20.5 db
4. Phase Margin : 34°
5. ***Stability closed loop***

Since gain and phase margin are positive, the system is absolutely stable.

Problem 8.8 Draw the Bode plot for a system having

$$G(s) = \frac{3}{s(1+0.05s)(1+0.2s)}, \; H(s) = 1$$

Determine

(1) Gain Crossover frequency and corresponding phase-margin.
(2) Phase Crossover frequency and corresponding Gain-margin.
(3) Stability of the closed-loop system.

Solution

Magnitude Plot

Ser. No.	*Factor*	*Corner frequency rad/sec*	*Asymptotic log-magnitude Characteristics*
1	3	None	Straight line of 0 db/dec starting from 20 log 3 = 9.5 db
2	$\frac{1}{s}$	None	Straight line of –20 db/dec passing through ω = 1
3	$\frac{1}{(1+0.2s)}$	5	Straight line of –20 db/dec originating from ω = 5
4	$\frac{1}{(1+0.05s)}$	20	Straight line of –20 db/dec originating from ω = 20

Phase Plot $\phi = -90° - \tan^{-1} 0.2\,\omega - \tan^{-1} 0.05\,\omega$

Ser. No.	*ω rad/sec*	ϕ
1	0	–90°
2	0.1	–91.4°
3	0.5	–97°
4	1	–104.2°
5	5	–149°
6	10	–180°
7	20	–211°

Bode plots are shown in Fig. 8.9.

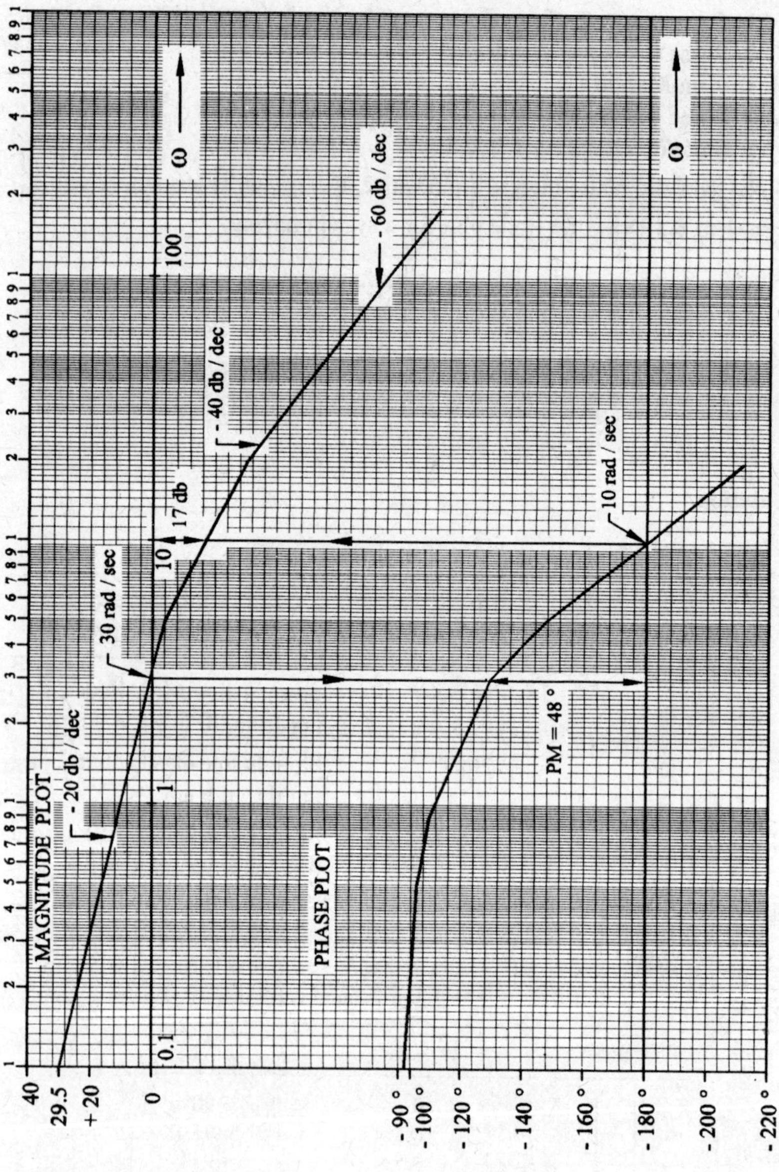

Fig. 8.9

Result

1. Gain Crossover frequency : 3.0 rad/sec
2. Phase Crossover frequency : 10 rad/sec
3. Gain Margin : 17 db
4. Phase Margin : 48°
5. *Stability*

Since gain and phase margin are positive the system is absolutely stable.

Problem 8.9 Construct the Bode plot for a unity feedback control system having

$$G(s)\frac{10(s+10)}{s(s+2)(s+5)}$$

From the plot obtain the gain margin and phase margin. Comment on the stability of the system. **(Pune University)**

Solution

$$G(s)\frac{10(1+0.1s)}{s(1+0.5s)(1+0.2s)}$$

Magnitude Plot

Ser. No.	*Factor*	*Corner frequency rad/sec*	*Asymptotic log-magnitude Characteristics*
1	10	None	Straight line of 0 db/dec starting from 20 log 10 = 20 db point.
2	$\frac{1}{s}$	None	Straight line of –20 db/dec passing through ω = 1.
3	$\frac{1}{(1+0.5s)}$	2	Straight line of –20 db/dec originating from ω = 2.
4	$\frac{1}{(1+0.2s)}$	5	Straight line of –20 db/dec originating from ω = 5.
5	$(1 + 0.1s)$	10	Straight line of +20 db/dec originating from ω = 10.

Phase Plot $\phi = -90^\circ - \tan^{-1} 0.5\,\omega - \tan^{-1} 0.2\omega + \tan^{-1} 0.1\,\omega$

Ser. No.	ω *rad/sec*	φ	*Ser. No.*	ω *rad/sec*	φ
1	0	–90°	6	5	–176.6°
2	0.1	–93°	7	8	–185.3°
3	0.5	–107°	8	10	–187°
4	1	–122°	9	20	–205°
5	2	–145.5°			

Bode plots are shown in Fig. 8.10.

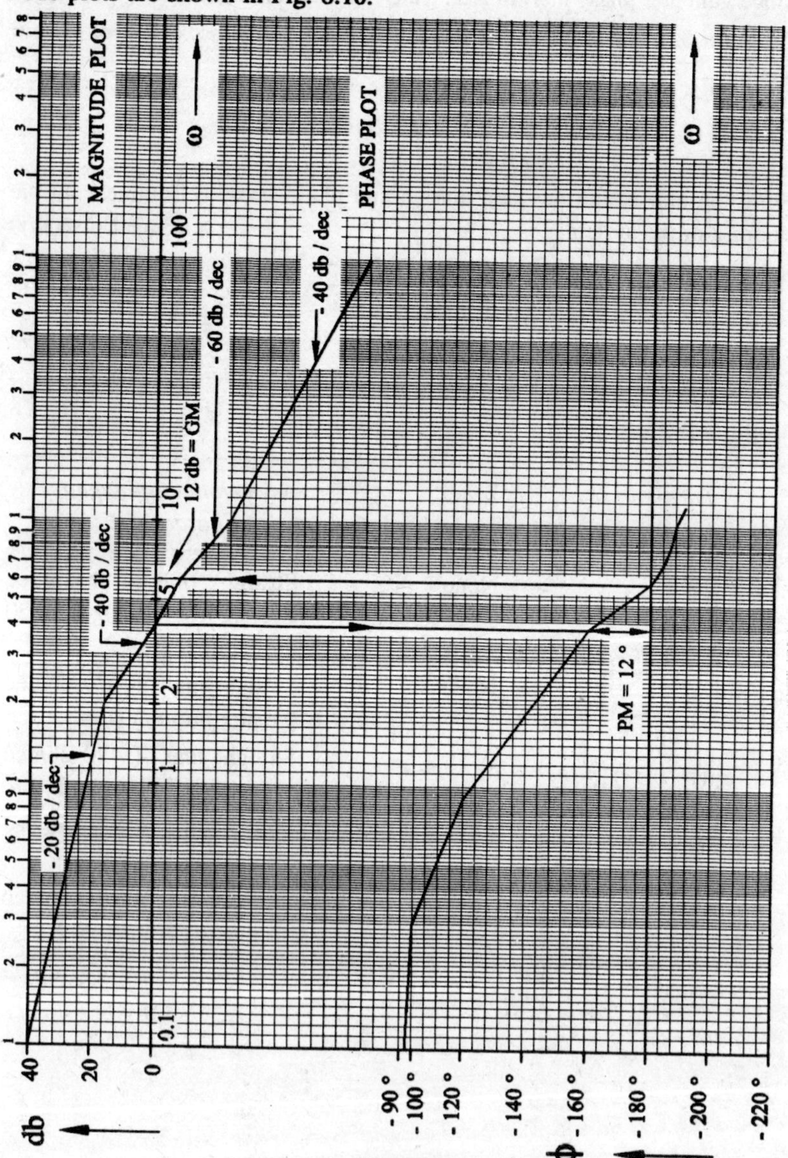

Fig. 8.10

Result

1. Gain Margin : 12 db
2. Phase Margin : 12°
3. ***Stability*** Since gain and phase margin are positive system is absolutely stable.

Problem 8.10 A certain unity feedback control system is given by

$$G(s)\frac{K}{s(1+s)(1+0.1s)}$$

Draw the Bode-plot of the above system. Determine from the plot the value of K so as to have

(a) Gain Margin = 10 db
(b) Phase Margin = 50°. **(Pune University)**

Solution

Let $K = 1$

Magnitude Plot

Ser. No.	*Factor*	*Corner frequency rad/sec*	*Asymptotic log-magnitude Characteristics*
1	$\frac{1}{s}$	None	Straight line of –20 db/dec passing through ω = 1.
2	$\frac{1}{(1+s)}$	1	Straight line of –20 db/dec originating from ω = 1.
3	$\frac{1}{(1+0.1s)}$	10	Straight line of –20 db/dec originating from ω = 10.

Phase Plot

$$\phi = -90° - \tan^{-1}\omega - \tan^{-1} 0.1\,\omega$$

Ser. No.	*ω rad/sec*	*Asymptotic log-magnitude Characteristics*
1	0	–90°
2	0.1	–96.3°
3	0.5	–119.4°
4	1	–140.7°
5	3	–178.3°
6	5	–195°
7	10	–219°

Bode plot is shown in Fig. 8.11.

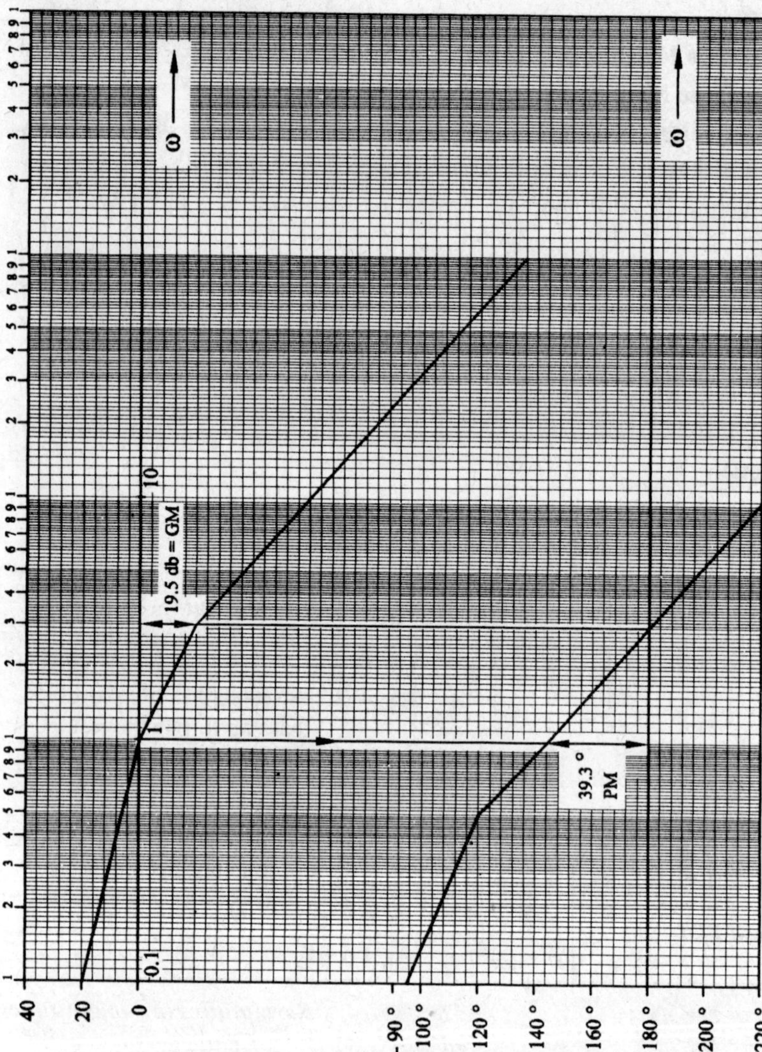

Fig. 8.11

(a) For the gain margin to be 10 db the magnitude plot has to be lifted up by

$$19.5 - 10 = 9.5 \text{ db}$$

$$\therefore \quad 20 \log K = 9.5 \quad \text{or} \quad K = 2.98$$

(b) For the phase margin to be 50° the value of ω at – 180 + 50 = – 130° is 0.73 rad/sec and gain margin is – 5 db. Therefore, to have phase margin of 50° the magnitude plot has to be brought down by 5 db so that gain crossover frequency is 0.73 rad/sec.

$$20 \log K = -5 \qquad \text{or} \qquad K = 0.56$$

Problem 8.11 The open-loop transfer function of a unity feedback system is

$$G(s) = \frac{1}{s(1+0.5s)(1+0.1s)}$$

Find gain and phase margin. If a phase-lag element with transfer function of $\frac{(1+2s)}{(1+5s)}$ is added in the forward path, find by how much the gain must be changed to keep the margin same.

Solution

Magnitude Plot

Ser. No.	*Factor*	*Corner frequency rad/sec*	*Asymptotic log-magnitude Characteristics*
1	$\frac{1}{s}$	None	Straight line of –20 db/dec passing through $\omega = 1$.
2	$\frac{1}{(1+0.5s)}$	2	Straight line of –20 db/dec originating from $\omega = 2$.
3	$\frac{1}{(1+0.1s)}$	10	Straight line of –20 db/dec originating from $\omega = 10$.

With phase lag element

$$G(s) = \frac{(1+2s)}{s(1+0.5s)(s+0.1s)(1+5s)}$$

Ser. No.	*Factor*	*Corner frequency rad/sec*	*Asymptotic log-magnitude Characteristics*
1	$\frac{1}{s}$	None	Similar to previous one.
2	$\frac{1}{(1+5s)}$	0.2	Straight line of –20 db/dec originating from $\omega = 0.2$.
3	$(1 + 2s)$	0.5	Straight line of –20 db/dec originating from $\omega = 0.5$.
4	$\frac{1}{(1+0.5s)}$	2	Similar to previous one.
5	$\frac{1}{(1+0.1s)}$	10	–do–

Phase Plot

(a) ***Without phase lag*** $\phi_1 = -90° - \tan^{-1} 0.5\,\omega - \tan^{-1} 0.1\,\omega$

(b) *With phase lag*

$$\phi_2 = -90^\circ - \tan^{-1} 0.5\,\omega - \tan^{-1} 0.1\,\omega + \tan^{-1} 2\omega - \tan^{-1} 5\,\omega$$

Ser. No.	*ω rad/sec*	ϕ_1	ϕ_2	*Ser. No.*	*ω rad/sec*	ϕ_1	ϕ_2
1	0.1	–93.4°	–105°	5	2	–146.31°	–154.6°
2	0.2	–96.8°	–120°	6	5	–184.76°	–188°
3	0.5	–107°	–130°	7	10	–228.7°	–230.4°
4	1	–122.3°	–137.5°				

Bode plots are shown in Fig. 8.12.

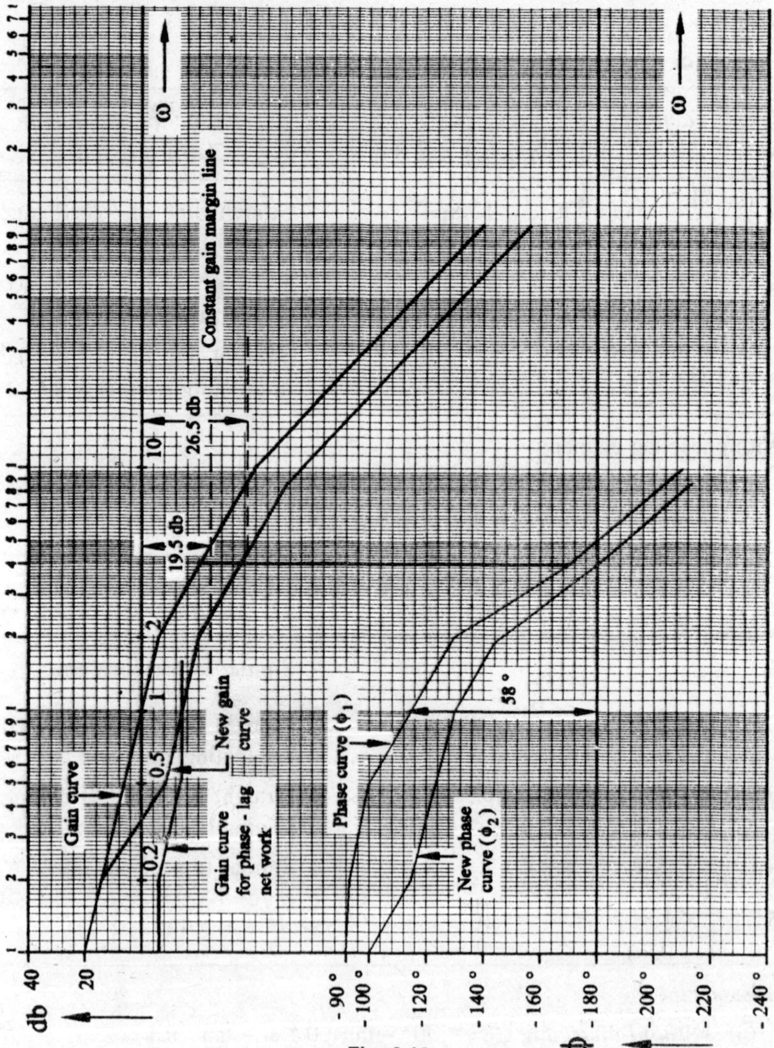

Fig. 8.12

1. Gain Margin : 19.5 db
2. Phase Margin : 58°

With phase-lag network introduced the gain margin is 26.5 db. But since gain margin is required to be kept at constant level of 19.5 db the gain must be changed by

$$26.5 - 19.5 = 7 \text{ db}$$

$$20 \log K = 7 \quad \text{or} \quad K = 2.2$$

Problem 8.12 The open-loop transfer function of a unity feedback system is

$$G(s) = \frac{50\,K}{s(s+10)(s+5)(s+1)}$$

(a) Gain Margin and Phase Margin

(b) The value of steady-state error co-efficient for a gain of 10 db and the value which will make the closed-loop system marginally stable.

Solution

Let $K = 1$, then

$$G(s) = \frac{50}{s(s+10)(s+5)(s+1)}$$

or

$$G(s) = \frac{1}{s(s+1)(1+0.2s)(1+0.1s)}$$

Magnitude Plot

Ser. No.	*Factor*	*Corner frequency rad/sec*	*Asymptotic log-magnitude Characteristics*
1	$\frac{1}{s}$	None	Straight line of –20 db/dec passing through ω = 1
2	$\frac{1}{(1+s)}$	1	Straight line of –20 db/dec originating from ω = 1.
3	$\frac{1}{(1+0.2s)}$	5	Straight line of –20 db/dec originating from ω = 5.
4	$\frac{1}{(1+0.1s)}$	10	Straight line of –20 db/dec originating from ω = 10.

Phase Plot

$$\phi = -90° - \tan^{-1}\omega - \tan^{-1} 0.1\,\omega - \tan^{-1} 0.2\,\omega$$

Ser. No.	ω *rad/sec*	φ	*Ser. No.*	ω *rad/sec*	φ
1	0	−90°	5	1	−152°
2	0.1	−97°	6	2	−187°
3	0.2	−105°	7	3	−209°
4	0.5	−125°	8	4	−227°

Bode plots are shown in Fig. 8.13.

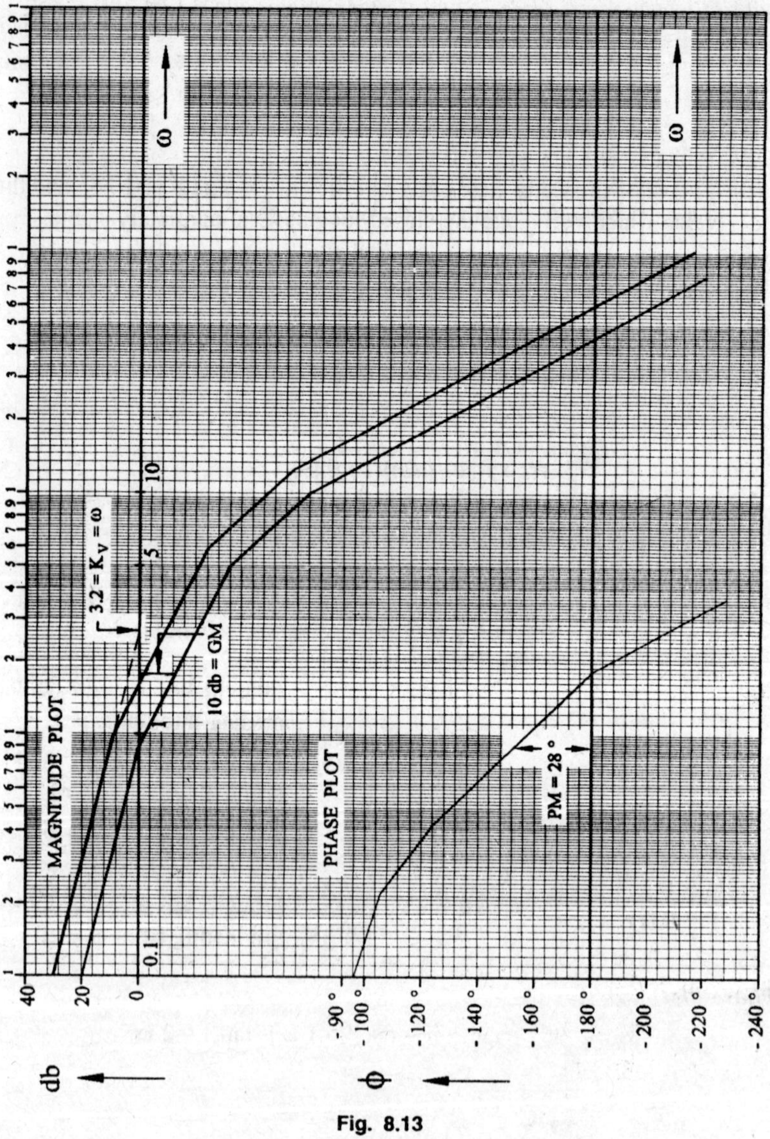

Fig. 8.13

(a) (i) Gain Margin = 10 db
(ii) Phase Margin = 28°

(b) Initial slope of the line is – 20 db/dec and intersects 0 db line at $\omega = 1$ rad/sec.

Therefore, $K_v = 1$

Steady-state error co-efficient for the system to be marginally stable is found by lifting the magnitude curve by 10 db (GM). The new magnitude curve is drawn and initial line having a slope of – 20 db/dec is extended till it intersects 0 db line. Intersection occurs at $\omega = 3.2$ rad/sec.

$\therefore \qquad K_v = 1$

Problem 8.13 Transfer function of a phase advance circuit is $\dfrac{1+0.2s}{1+0.02s}$.

Find the maximum phase lag.

Solution

Ser. No.	*Factor*	*Corner frequency rad/sec*	*Asymptotic log-magnitude Characteristics*
1	$(1 + 0.2s)$	5	Straight line of –20 db/dec originating from $\omega = 5$
2	$\dfrac{1}{(1+0.02s)}$	50	Straight line of –20 db/dec originating from $\omega = 50$

$$\phi = \tan^{-1} 0.2\,\omega - \tan^{-1} 0.02\,\omega$$

Ser. No.	*ω rad/sec*	ϕ
1	0.1	1°
2	0.2	2.1°
3	0.4	4.1°
4	0.5	5.1°
5	1	10.2°
6	5	39.3°
7	10	52°
8	15	54.8°
9	16	54.9°
10	17	54.8°
11	20	54.2°
12	50	39.3°

Bode plots are shown in Fig. 8.14. From the phase-plot.

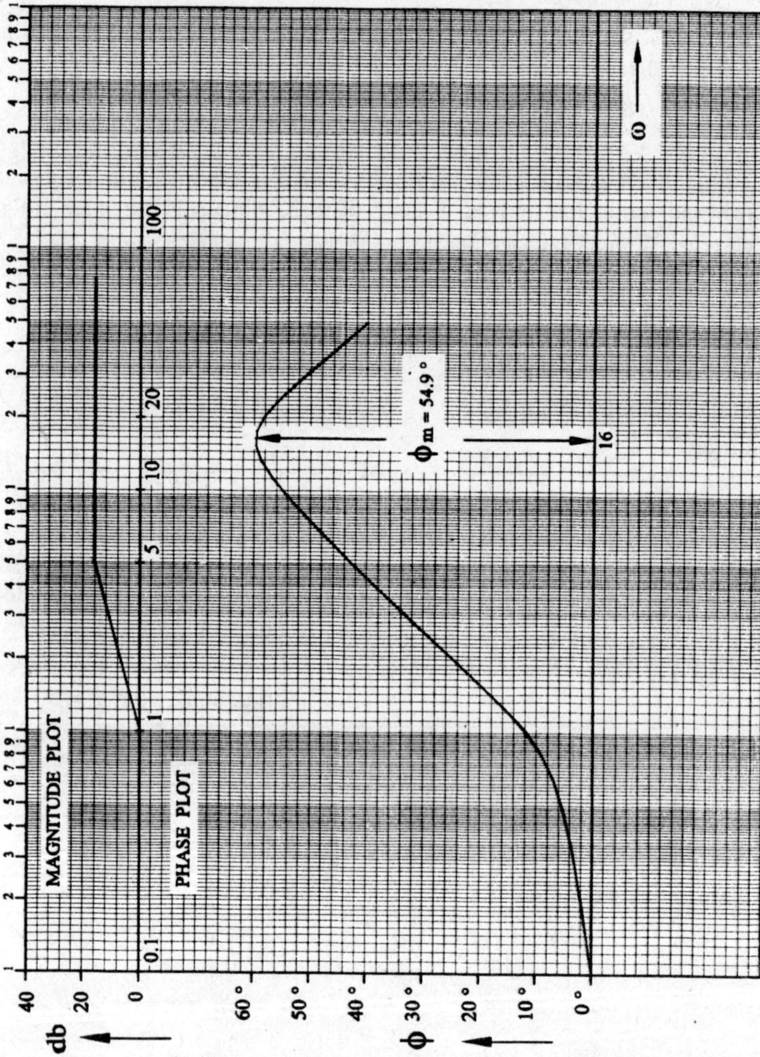

Fig. 8.14

Maximum phase advance = $\phi_m = 54.9°$

Problem 8.14 Sketch the Bode plot for the following transfer function and determine the system gain K for the gain cross-over frequency to be 5 rad/sec

$$G(s) = \frac{K e^{-0.1s}}{s(1+s)(1+0.1s)}$$

Solution

$$M = 20\log|G(s)|_{s=j\omega} = 20\log K + 20\log\left|e^{-0.1j\omega}\right|$$

$$+ 20\log\left|\frac{1}{j\omega}\right| + 20\log\left|\frac{1}{1+0.1j\omega}\right|$$

Now $\left|e^{-0.1j\omega}\right| = |\cos 0.1\omega - j\sin 0.1\omega| = 1$

∵ $20\log\left|e^{-0.1j\omega}\right| = 0$ db

and $\underline{/e^{-0.1j\omega}} = -57.3\times 0.1\omega$

$= -5.73\omega$ (degrees)

Assuming $K = 1$, the magnitude and Phase plots are drawn as given below.

Magnitude Plot

Ser. No.	*Factor*	*Corner frequency rad/sec*	*Asymptotic log-magnitude Characteristics*
1	$\frac{1}{s}$	None	Straight line of –20 db/dec passing through $\omega = 1$
2	$e^{-0.1s}$	None	Coincides with 0 db line
3	$\frac{1}{(1+s)}$	1	Straight line of –20 db/dec originating from $\omega = 1$
4	$\frac{1}{(1+0.1s)}$	10	Straight line of –20 db/dec originating from $\omega = 10$

Phase Plot

$$\phi = -90^\circ - 5.73\,\omega - \tan^{-1}\omega - \tan^{-1}0.1\,\omega.$$

Ser. No.	*ω rad/sec*	ϕ
1	0	–90°
2	0.1	–97°
3	0.3	–110°
4	0.5	–122°
5	1	–146°
6	2.0	–176°

Ser. No.	ω *rad/sec*	φ
7	2.5	−186.5°
8	3	−190°
9	5	−223°
10	10	−276.6°

Bode plot is shown in Fig. 8.15.

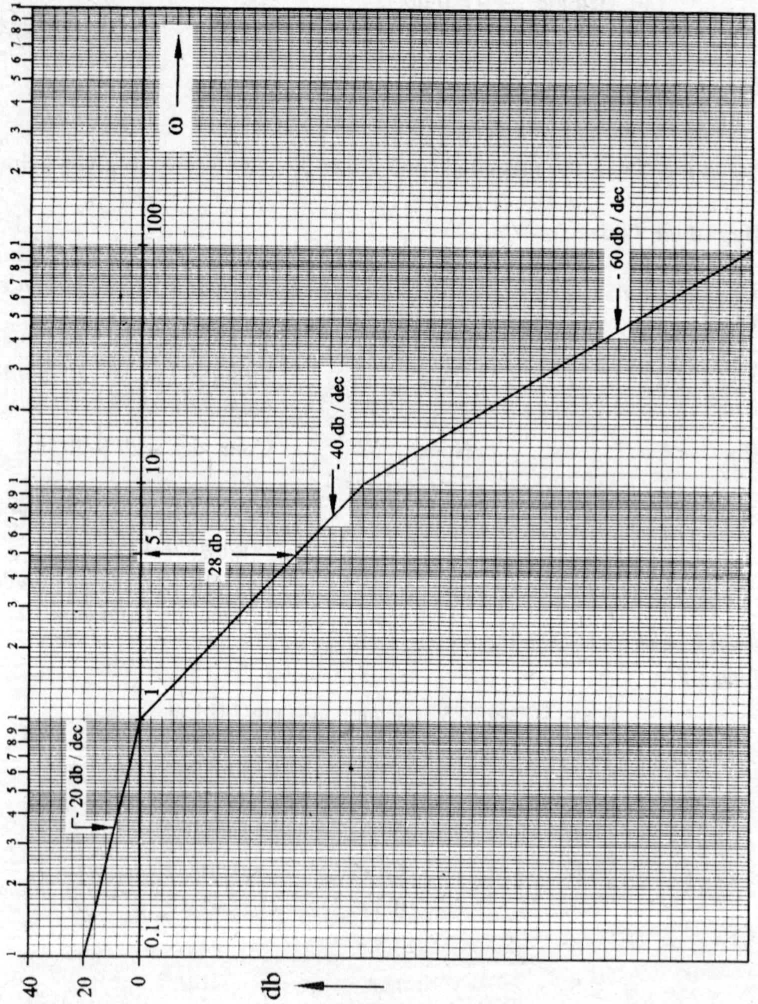

Fig. 8.15

If the gain crossover frequency is required to be 5 rad/sec then the magnitude plot is to be lifted up by 28 db i.e.

$$20 \log K = 28$$

or $$K = 25.1$$

Problem 8.15 Consider the system shown in Fig. 8.16 Design lead compensator of this system to meet the following specification.

Damping ratio $\zeta = 0.7$

Setting time $t_s = 1.4$ secs

Velocity error constant $K_v = 2\ \text{sec}^{-1}$

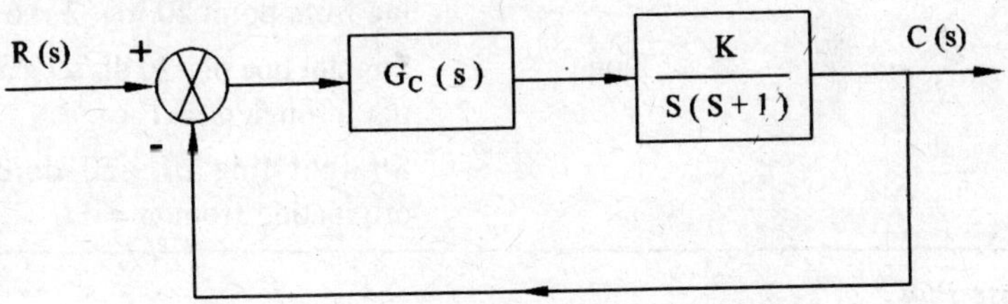

Fig. 8.16

Solution

$$\phi_m = \tan^{-1}\left\{2\zeta/[(1+4\zeta^4)-2\zeta^2]^{1/2}\right\}$$

$$= \tan^{-1}\left\{2\times0.7/[(1+4\times0.7^4)-2\times0.7^2]^{1/2}\right\}$$

$$= 65°$$

For 5% tolerance band

$$t_s = \frac{3}{\zeta\omega_n}$$

or $$\omega_n = \frac{3}{\zeta t_s} = \frac{3}{0.7\times1.4} = 3.06 \text{ rad/sec}$$

$$\omega_b = \omega_n\left[1-2\zeta^2+\sqrt{2-4\zeta^2+4\zeta^4}\right]^{1/2}$$

$$= 3.1 \text{ rad/ sec}$$

$$K_v = 2\ \text{sec}^{-1}$$

Open loop transfer function of a uncompensated system is

$$G(s) = \frac{K}{s(s+1)}$$

$$= \frac{2}{s(1+s)}$$

Magnitude Plot

Ser. No.	*Factor*	*Corner frequency rad/sec*	*Asymptotic log-magnitude Characteristics*
1	2	None	Straight line of 0 db/dec starting from point 20 log 2 = 6 db
2	$\frac{1}{s}$	None	Straight line of –20 db/sec passing through ω = 1
3	$\frac{1}{(1+s)}$	1	Straight line of –20 db/dec originating from ω = 1

Phase Plot

$$\phi = -90° - \tan^{-1}\omega$$

Ser. No.	*ω rad/sec*	φ
1	0.1	–96°
2	0.2	–101°
3	0.3	–107°
4	0.5	–116.5°
5	1	–135°
6	2	–153°
7	3	–161.5°
8	5	–169°
9	10	–174°
11	12	–175°
12	14	–176°
13	20	–177°
14	∞	–180°

Bode plot is shown in Fig. 8.17.

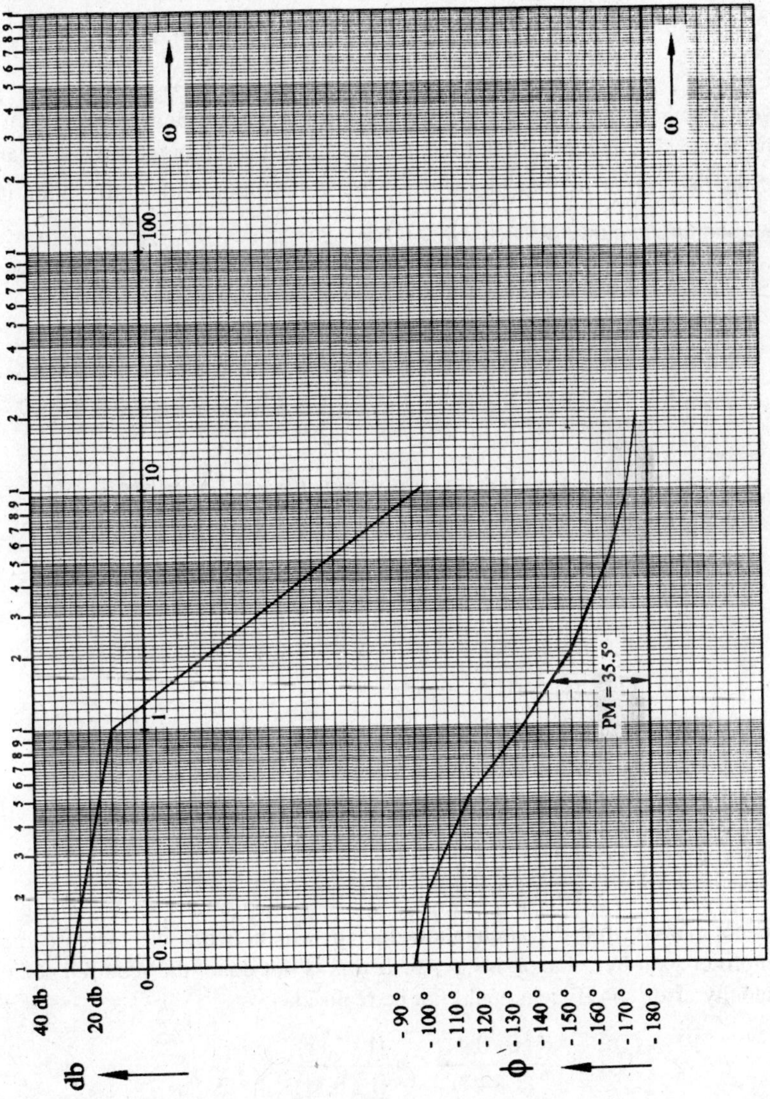

Fig. 8.17

From the plots the phase margin and gain margin are 35.5° and ∞ db respectively. But the phase margin required is 65°.

Additional phase lead necessary to satisfy the relative stability requirement is 65° − 35.5° = 29.5° ≃ 30° (say). Addition of a lead compensator modifies the magnitude curve. Gain crossover frequency will shift. Therefore, we may assume that the maximum phase lead re-

$$30°+10° = 40°$$

$$\alpha = \frac{1-\sin 40°}{1+\sin 40°} \cong 0.2$$

Next step is to determine the corner frequencies $\omega = 1/T$ and $\omega = 1/(\alpha T)$ of the phase lead network. Maximum phase lead occurs at the geometric mean of the two corner frequencies or $\omega = \sqrt{\alpha T}$. The amount of modification in the magnitude curve at $\omega = 1/\left(\sqrt{\alpha T}\right)$ is

$$\left|\frac{1+j\omega T}{1+j\omega\alpha T}\right|_{\omega=1/\sqrt{\alpha T}} = \frac{1}{\sqrt{\alpha}}$$

Now $$\frac{1}{\sqrt{\alpha}} = \frac{1}{\sqrt{0.2}} = \frac{1}{0.45} = 2.22 = 6.9 \text{ db.}$$

$$|G(j\omega)| = -6.9 \text{ db corresponds to } \omega = 2.1 \text{ rad/sec}$$

We shall select $\omega = 2.1$ rad/sec as the new gain crossover frequency.

$$\frac{1}{T} = \sqrt{\alpha}\omega = \sqrt{0.2} \times 2.1 = 0.94$$

$$\frac{1}{\alpha T} = \frac{\omega}{\alpha} = \frac{2.1}{\sqrt{0.2}} = 4.7$$

Lead network is

$$= \frac{s+0.94}{s+4.7}$$

$$= \frac{0.94(1+1.1s)}{4.7(1+0.2s)} = \frac{0.2(1+1.1s)}{(1+0.2s)}$$

To compensate for the attenuation due to lead network we increase the amplifier gain by a factor $1/0.2 = 5$. If this is not done, the required static velocity error co-efficient could not be realised.

$$G_c(s) = \frac{5(s+0.94)}{(s+4.7)} = \frac{(1+1.1s)}{(1+0.2s)}$$

Problem 8.16 Find the open-loop transfer function of a system whose approximate plot is shown in Fig. 8.18

Solution Corner frequencies are

$$\omega_1 = 2.5;\ \omega_2 = 10;\ \omega_3 = 25.$$

Change in magnitude in db

= Slope × (Number of decades between two frequencies)

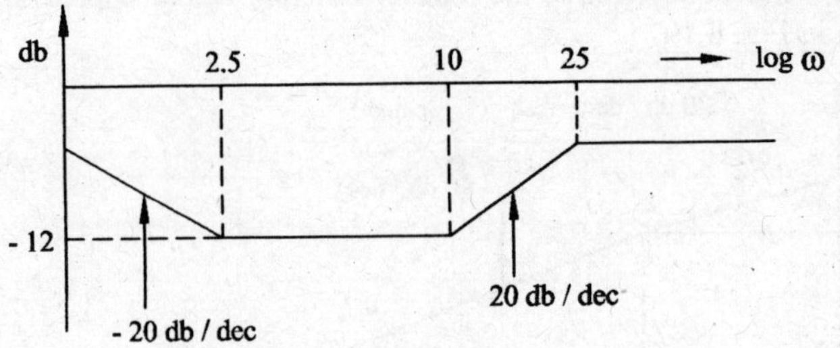

Fig. 8.18

$= -20\ (\log 2.5 - \log 1)$

$= -20 \log 2.5 = -7.95$

Magnitude $= -12 + 7.95$ db $= -4.05$ db

$\therefore\ 20 \log K = -4.05$

or $K = 0.63$

Since first line has a slope of – 20 db/dec and starts from a point – 4.05 db at ω = 1 rad/sec the factor contributing this is

$$= \frac{K}{s} = \frac{0.63}{s}$$

Plot between ω = 2.5 and ω = 10 is having a slope of 0 db/dec. At ω = 2.5 the slope has changed from – = 20 db/dec and this can only happen due to a factor in the numerator and is

$$= \left(\frac{s}{2.5} + 1\right) = (1 + 0.4s)$$

At ω = 10, the slope has changed from 0 db/dec to + 20 db/dec and is due to a factor in the numerator and is

$$= \left(\frac{s}{10} + 1\right) = (1 + 0.1s)$$

At ω = 25, the slope has changed from + 20 db/dec to 0 db/dec and is due to a factor in the denominator and is

$$= \left(\frac{s}{25} + 1\right) = (1 + 0.4s)$$

The open-loop transfer function is this

$$G(s) = \frac{0.63(1 + 0.4s)(1 + 0.1s)}{(1 + 0.04s)}$$

Problem No. 8.17 Determine the transfer function whose approximate plot is shown in Fig. 8.19.

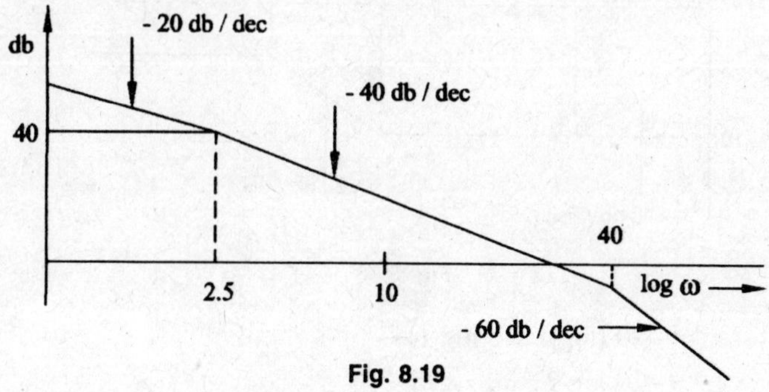

Fig. 8.19

Solution

Corner frequencies are 2.5 and 40 rad/sec

$$20 \log K = 40 + 20 \log 2.5$$
$$= 47.95$$

or
$$K = 250$$

At ω = 2.5 rad/sec slope changes from – 20 db/dec to – 40 db/dec due to a factor $\dfrac{1}{\left(1+\dfrac{s}{2.5}\right)}$. At ω = 40 rad/sec slope changes from – 40 db/dec to – 60 db/dec due to a factor $\dfrac{1}{\left(1+\dfrac{s}{40}\right)}$. Also, since initial slope is – 20 db/dec, it is due to factor. $1/s$. Therefore open-loop transfer function is

$$G(s) = \frac{250}{s\left(1+\dfrac{s}{2.5}\right)\left(1+\dfrac{s}{40}\right)}$$

$$= \frac{250}{s(1+0.4s)(1+0.025s)}$$

Problem 8.18 Determine the open-loop transfer function of a system whose approximate plot is shown in Fig. 8.20.

Solution

First line is having a slope of 12 db/oct (40 db/dec). Therefore, there is a s^2 term in the numerator. At ω = 0.5 rad/sec slope changes to 6 db/oct (20 db/

dec) due to a term in the denominator equal to

$$\left(1+\frac{s}{0.5}\right)$$

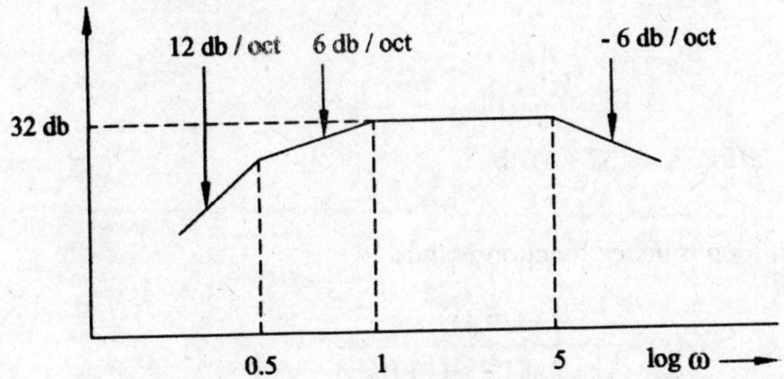

Fig. 8.20

At ω = 1 rad/sec slope becomes 0 db/dec due to a term in the denominator equal to $(1 + s)$.

At ω = 5 rad/sec slope becomes – 6 db/oct (– 20 db/dec) due to a term in the denominator equal to $\left(1+\frac{s}{5}\right)$

$$G(s) = \frac{K s^2}{\left(1+\frac{s}{0.5}\right)(1+s)\left(1+\frac{s}{5}\right)}$$

Calculation of 'K'

Refer Fig. 8.21

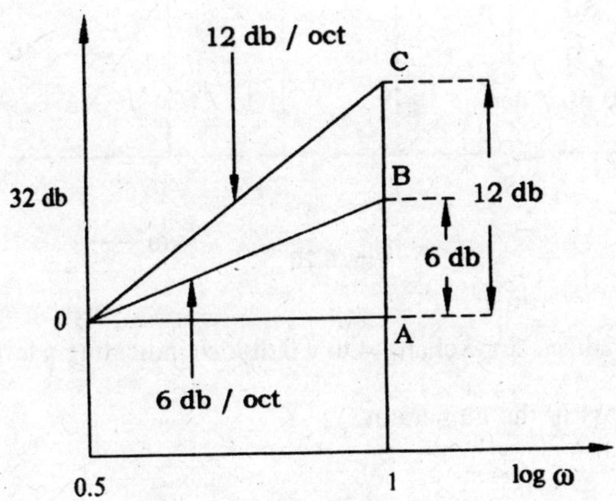

Fig. 8.21

Difference between $\omega = 0.5$ and $\omega = 1$ rad/sec is one octave i.e. $AB = 6$ db since the slope of line OB is 6 db/oct. OC is the extended line having a slope of 12 db/oct.

$$\therefore \quad AC = 12 \text{ db}$$

$$BC = AC - AB$$

$$= 12 - 6 = 6 \text{ db}$$

$$\therefore \quad 20 \log K = (32 + 6) \text{ db}$$

or $$K = 79.4$$

The open loop transfer function is thus

$$G(s) = \frac{79.4s^2}{(1+2s)(1+s)(1+0.2s)}$$ **Ans.**

Problem 8.19 From the asymptotic magnitude (in db) versus frequency (log scale) plot of Fig. 8.22, find the associated transfer function. Assume no right half plane poles or zeros present. **(Pune University)**

Solution

1. Slope of the first line is – 20 db/dec indicating a term $\frac{1}{s}$

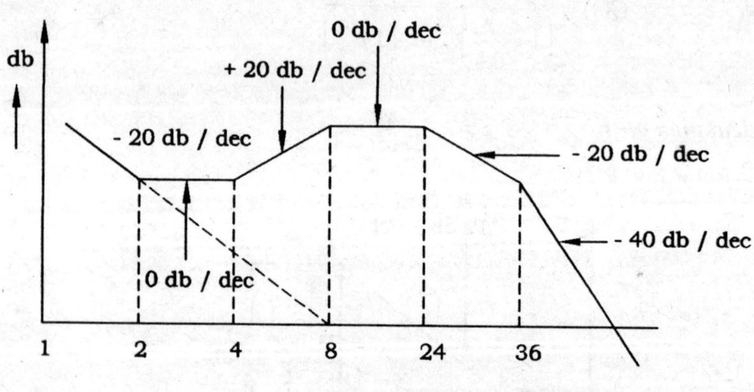

Fig. 8.22

2. At $\omega = 2$ rad/sec slope changes to a 0 db/dec indicating a term $\left(1+\frac{s}{2}\right)$ or $(1 + 0.5s)$ in the numerator.

3. At ω = 4 rad/sec slope changes to + 20 db/dec indicating a term $\left(1+\frac{s}{4}\right)$ or (1 + 0.25s) in the numerator.

4. At ω = 8 rad/sec slope changes to 0 db/dec indicating a term $\left(1+\frac{s}{8}\right)$ or (1 + 0.125s) in the denominator.

5. At ω = 24 rad/sec slope changes to – 20 db/dec indicating a term $\left(1+\frac{s}{24}\right)$ or (1 + 0.042s) in the denominator.

6. At ω = 36 rad/sec slope changes to – 40 db/dec indicating a term $\left(1+\frac{s}{40}\right)$ or (1 + 0.028s) in the denominator.

Transfer function is thus

$$\frac{K(1+0.5s)(1+0.25s)}{s(1+0.125s)(1+0.042s)(1+0.028s)}$$

– ***Calculation of 'K'***

$$20 \log K = 20 \log 8$$

or $$K = 8$$

$$\therefore \quad G(s) = \frac{8(1+0.5s)(1+0.025s)}{s(1+0.125s)(1+0.042s)(1+0.028s)}$$

Problem 8.20 Derive the transfer function of the system from the data given on the Bode diagram shown in Fig. 8.23 below. (AMIE)

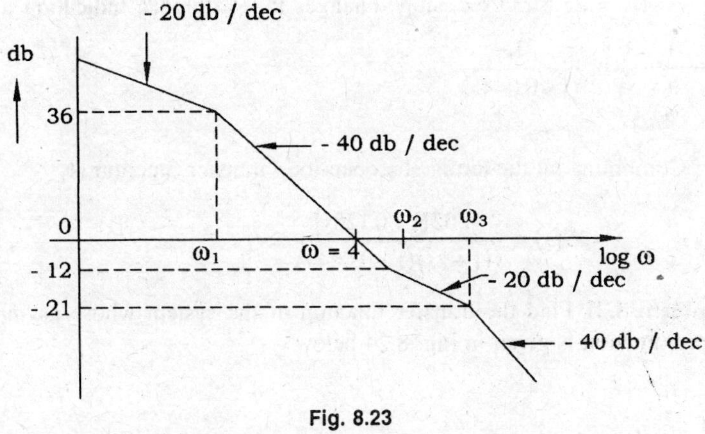

Fig. 8.23

Solution

Between ω_1 and $\omega = 4$ rad/sec there is a decrease of 36 db

$$\therefore \quad -36 = -40(\log 4 - \log \omega_1)$$

or $$\omega_1 = 0.5036 \cong 0.5 \text{ rad/ sec}$$

Calculation of 'K' $$20 \log K = 36 + 20 \log 0.5$$

or $$K = 31.62$$

Calculation of 'ω_2' $$-12 = -40 (\log \omega_2 - \log 4)$$

or $$\omega_2 = 8 \text{ rad/sec}$$

Calculation of 'ω_3' $$-21 + 12 = -20 (\log \omega_3 - \log 8)$$

or $$\omega_3 \doteq 22.5 \text{ rad/sec}$$

First line has a slope of – 20 db/dec indicating a term $\frac{1}{s}$ and since it is not passing through $\omega = 1$ rad/sec, the term is $\frac{K}{s}$ or $\frac{31.62}{s}$.

At $\omega_1 = 0.5$ rad/sec slope changes to –40 db/dec indicating a term $\frac{1}{\left(1+\frac{s}{0.5}\right)}$, or $\frac{1}{(1+2s)}$.

At $\omega_2 = 8$ rad/sec, slope changes to –20 db/dec indicating a term $\left(1+\frac{s}{8}\right)$ or $(1+0.125s)$

At $\omega_3 = 22.5$ rad/sec, slope changes to –40 db/dec indicating a term $\frac{1}{\left(1+\frac{s}{22.5}\right)}$ or $\frac{1}{(1+0.044s)}$

Combining all the terms, the open-loop transfer function is

$$G(s) = \frac{31.62(1+0.125s)}{s(1+2s)(1+0.044s)}$$

Problem 8.21 Find the transfer function of the system whose asymptotic approximation is given in Fig. 8.24 below.

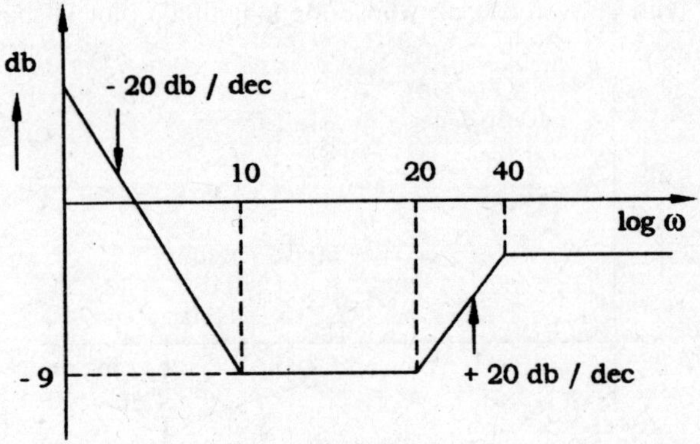

Fig. 8.24

Solution

First line has a slope of −20 db/dec and is not passing through $\omega = 1$ rad/sec. Therefore, it indicates a term $\frac{K}{s}$

$$20 \log K = -9$$

or

$$K = 0.35$$

∴ the term is $\frac{0.35}{s}$

At $\omega = 1$ rad/sec, slope changes to 0 db/dec indicating a term $(1 + s)$.

At $\omega = 20$ rad/sec, slope changes to +20 db/dec indicating a term

$\frac{1}{\left(1+\frac{s}{20}\right)}$ or $(1+0.05s)$.

At $\omega = 40$ rad/sec, slope changes to 0 db/dec indicating a term

$\frac{1}{\left(1+\frac{s}{40}\right)}$ or $\frac{1}{(1+0.025s)}$.

Combining all terms, we get

$$G(s) = \frac{0.35(1+s)(1+0.05s)}{s(1+0.025s)}$$

Problem No. 8.22 Obtain the expression for open-loop transfer function for a system with unity feedback whose log-magnitude plot is shown in Fig. 8.25 below:

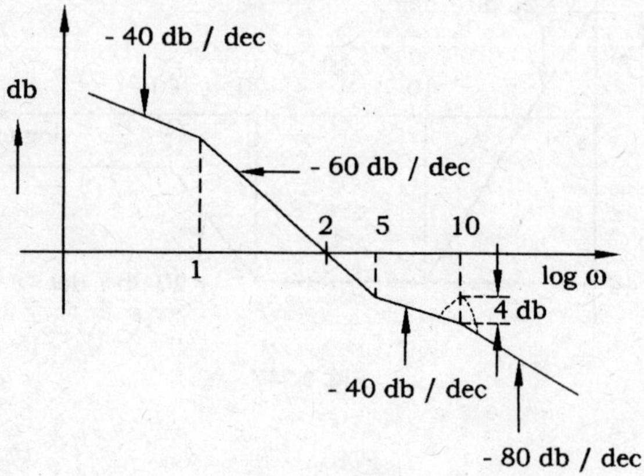

Fig. 8.25

Solution

First line is having a slope of –40 db/dec and since it is not passing through $\omega = 1$ rad/sec it indicates a term K/s^2

$$\text{Magnitude at } \omega = 1 \text{ rad/sec of initial part}$$
$$= 60 \log 2 = 18.06$$

$$\therefore \quad 20 \log K = 18.06 \quad \text{or} \quad K = 8$$

At $\omega = 1$ rad/sec, the slope changes from –40 db/dec to –60 db/dec indicating a term $1/(1 + s)$.

At $\omega = 5$ rad/sec, the slope changes from –60 db/dec to –40 db/dec indicating a term $\left(1+\frac{s}{5}\right)$ or $(1+0.2s)$.

At $\omega = 10$ there is a term of the form

$$\left\{\left(1+\frac{2\zeta s}{\omega_n}+\frac{s^2}{\omega_n^2}\right)\right\}^{-1}$$

because the slope changes from –40 db/dec to –80 db/dec and also a peak of 4 db is shown

$$\omega_n = 10 \text{ rad/sec}$$

$$\text{Value of } \left\{1+\frac{2\zeta s}{\omega_n}+\left(\frac{s}{\omega_n}\right)^2\right\}^{-1} \text{ at } \omega = \omega_n$$

$$= \left\{ \sqrt{\left(1 - \frac{10}{10}\right)^2 + \left(\frac{2 \times \zeta \times 10}{10}\right)^2} \right\}^{-1} = \frac{1}{2\zeta}$$

$\therefore$ log magnitude $= 20 \log \dfrac{1}{2\zeta} = 4$, or $\dfrac{1}{2\zeta} = e^{1/5}$, or $\zeta = 0.316$

$\therefore$ the term is $= \left(1 + \dfrac{2 \times 0.316 s}{10} + \dfrac{s^2}{100}\right)^{-1}$

$$\therefore \quad G(s) = \frac{8(1 + 0.2s)}{s^2 (1 + s)\left(1 + 0.0632 s + \dfrac{s^2}{100}\right)}$$

or

$$G(s) = \frac{800(1 + 0.2s)}{s^2 (1 + s)\left(s^2 + 6.32 s + 100\right)}$$

9

Routh's Criterion

Problem No. 9.1 Apply Routh criterion to the equation

$$P(s) = 2Ts^3 + (2+T)s^2 + (K+1)s + K = 0$$

and obtain the region of stability

Solution The equation is given by

$$P(s) = 2Ts^3 + (2+T)s^2 + (K+1)s + K = 0$$

The Routh's array is

$$\begin{array}{c|ll|} s^3 & 2T & K+1 \\ s^2 & 2+T & K \\ s^1 & (K+1) - \dfrac{2KT}{2+T} & \\ s^0 & K & \end{array}$$

For stability

$2T > 0$ and $2 + T > 0$ $\quad\therefore T > 0$

Also $\quad K > 0$, and

$$K + 1 - \frac{2KT}{2+T} > 0 \text{ gives } K < \frac{2+T}{T-2} \text{ and}$$

$$T < \frac{2(1+K)}{K-1}$$

Therefore, for stability, we have

$K > 0$, $T > 0$ and $K > 1$ and $T > 2$.

The region for stabilit,y is shown on Fig. 9.1

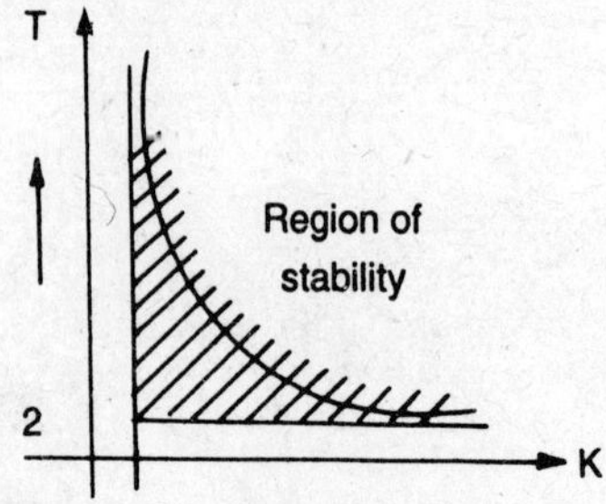

Fig. 9.1

Problem No. 9.2 Using Routh's stability criterion, ascertain stability for each of the following cases.

(a) $s^6 + 2s^5 + 8s^4 + 12s^3 + 20s^2 + 16s + 16 = 0$

(Pune University)

Solution

The Routh's array is

						Remarks
s^6	1	8	20	16		
s^5	2	12	16			
	1	6	8			
s^4	2	12	16			$s^4 + 6s^2 + 8 = 0$
	1	6	8			$4s^3 + 12s = 0$
s^3	0	0	0		or	$s^3 + 3s = 0$
	1	3				
s^2	3	8				
s^1	$\frac{1}{3}$	0				
s^0	8	0				

The s^3 row became a zero row because the elements of s^5 and s^4 rows are same. The zero row was replaced by writing the subsidary equation of the row above the s^3 row and then the first derivative of the subsidary equation was taken as shown under "Remarks". The s^3 row was then formed by the co-efficients.

The first column of the Routh's array shows no sign change. Therefore, no roots are on the R.H.S. of the s-plane. However, presence of a zero row indicates presence of symmetrically located roots in the s-plane. In order to find the roots, we proceed as given below

$$s^6 + 2s^5 + 8s^4 + 12s^3 + 20s^2 + 16s + 16$$
$$= (s^4 + 6s^2 + 8)(s^2 + 2s + 2) = 0$$

Therefore, $\quad (s^4 + 6s^2 + 8) = 0 = (s^2 + 4)(s^2 + 2) = 0$

gives $\quad s = \pm 2j; \quad s = \pm\sqrt{2}j$

and $\quad s^2 + 2s + 2 = 0,$ gives $\quad s = -1 \pm j1$

These roots are shown in Fig. 9.2.

Therefore, there are two pairs of imaginary roots and one pair of complex conjugate root having negative real part. Therefore, the system is limitedly stable.

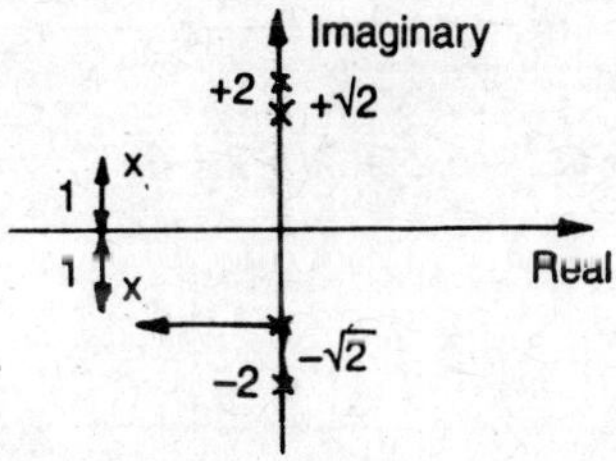

Fig. 9.2

(b) $\quad \boldsymbol{s^6 + s^5 - 2s^4 - 3s^3 - 7s^2 - 4s - 4 = 0}$

The Routh's array is

					Remarks
s^6	1	−2	−7	−4	
s^5	1	−3	−4		*Subsidary equation*
s^4	1	−3	−4		$s^4 - 3s^2 - 4 = 0$
s^3	0	0	0		It's derivative is
	2	−3			$4s^3 - 6s^2 = 0$
s^2	−1.5	−4			or $2s^3 - 3s^2 = 0$
s^1	$-\frac{25}{3}$	0			
s^0	−4				

There is one sign change in the first column of the Routh's array. Therefore, one root is lying on the R.H.S. of the s-plane. The system is unstable. Roots can be ascertained as given below.

$$s^6 + s^5 - 2s^4 - 3s^3 - 7s^2 - 4s - 4 = 0$$

or $$(s^4 - 3s^2 - 4)(s^2 + s + 1) = 0$$

which means $$s^4 - 3s^2 - 4 = 0$$

or $$(s^2 + 1)(s^2 - 4) = 0$$

$\therefore$ $$s = \pm j \text{ and}$$
$$s = \pm 2$$

and $$(s^2 + s + 1) = 0 \text{ gives, } s = \frac{-1}{2} \pm j\frac{\sqrt{3}}{2}$$

Therefore the roots are $+j, -j, +2, -2, \frac{-1}{2} \pm j\frac{\sqrt{3}}{2}$

(c) $s^6 + 3s^5 + 4s^4 + 6s^3 + 5s^2 + 3s + 2 = 0$

The Routh's array is

s^6	1	4	5	2	
s^5	1	2	1		$s^4 + 2s^2 + 1 = 0$
s^4	1	2	1		$4s^3 + 4s = 0$
s^3	0	0	0		$s^3 + s = 0$
	1	1			
s^2	1	1			$s^2 + 1 = 0$
s^1	0	0			$2s = 0$
	2				
s^0	1				

There are two rows which become zero and there is no sign change in the first column of the Routh's array. The original characteristics equation can be written as

$$(s^4 + 2s^2 + 1)(s^2 + 3s + 2) = 0$$

$$(s^4 + 2s^2 + 1) = 0 = (s^2 + 1)^2 \text{ gives } s = \pm j \text{ and}$$
$$s = \pm j$$

and $$(s^2+3s+2)=(s+1)(s+2)$$

gives $$s=-1 \text{ and } -2$$

Because of the multiplicity of the pair of the imaginary roots, the system is unstable.

Problem No. 9.3 The equation of the servo system is given by

$$b_0s^4+b_1s^3+b_2s^2+b_3s+b_4=0$$

Determine the conditions for stability

Solution

The Routh's array is

s^4	b_0	b_2	b_4
s^3	b_1	b_3	
s^2	$\dfrac{b_1b_2-b_0b_3}{b_1}$	$\dfrac{b_1b_4}{b_1}$	
s^1	$\dfrac{\dfrac{b_3(b_1b_2-b_0b_3)-b_1^2b_4}{b_1}}{\dfrac{b_1b_2-b_0b_3}{b_1}}$		
s^0	b_4		

The conditions for stability are

$$b_1>0,\quad b_0>0,\quad (b_1b_2-b_0b_3)>0$$

$$(b_1b_2b_3-b_0b_3^2-b_1^2b_4)>0 \text{ and } b_4>0$$

Problem 9.4 Determine the stability of the following cases, which represent characteristic equations of different systems

(a) $\boldsymbol{3s^4+10s^3+5s^2+5s+2=0}$

Solution

The Routh's array is

s^4	3	5	2
s^3	10	5	
s^2	3.5	2	
s^1	$\dfrac{-2.5}{3.5}$	0	
s^0	2		

There are two sign changes in the first column of the Routh's array. Hence two roots lie on the R.H.S. of *s*-plane. System is unstable.

(b) $s^5 + s^4 + 2s^3 + 2s^2 + 3s + 5 = 0$

Solution

Routh's array is

s^5	1	2	3
s^4	1	2	5
s^3	d	2	
s^2	$\frac{2d+2}{d}$	5	
s^1	$\frac{-4d-4-5d^2}{2d+2}$		
s^0	5		

The first term of the fifth row has a value of -2 as $d \to 0$. Thus there are sign changes in the first column making the system unstable. Alternatively,

Put $s = \frac{1}{z}$ in the characteristic equation which gives

$$5z^5 + 3z^4 + 2z^3 + 2z^2 + z + 1 = 0$$

Routh's array is

z^5	5	2	1
z^4	3	2	1
z^3	−1.33	−0.67	
z^2	0.5	1	
z^1	2		
z^0	1		

There are two sign changes in the first column making the system unstable.

Problem No. 9.5 The equation of a feedback control system is $s^3 + 3Ks^2 + (K+2)s + 4 = 0$. Determine the range K for system is stable.

Solution

Routh's array is

$$\begin{array}{c|cc|} s^3 & 1 & K+2 \\ s^2 & 3K & 4 \\ s^1 & \dfrac{3K^2+6K-1}{3K} & \\ s^0 & 4 & \end{array}$$

For stability, $3K > 0$ i.e. $K > 0$

and $3K^2 + 6K - 4 > 0$ i.e. $K > -1 \pm 1.53$

i.e. $K > 0.53$

The range of K is thus $\infty > K > 0.53$

Problem No. 9.6 The equation of a feedback control system is

$$s^4 + 20Ks^3 + 5s^2 + 10s + 15 = 0$$

Find the range of K for which system is stable.

Solution

Routh's array is

$$\begin{array}{c|ccc|} s^4 & 1 & 5 & 15 \\ s^3 & 20K & 10 & \\ s^2 & 5-\dfrac{1}{2K} & 15 & \\ s^1 & 10-\dfrac{600K^2}{10K-1} & 0 & \\ s^0 & 15 & & \end{array}$$

The system will be stable, if

$$20K > 0 \text{ i.e., } K > 0$$

$$5-\frac{1}{2K} > 0 \quad \text{or} \quad 5 > \frac{1}{2K} \quad \text{or} \quad K > \frac{1}{10}$$

Also, $$10-\frac{600K^2}{10K-1} > 0 \text{ which means}$$

$$-600K^2 + 100K - 10 > 0$$

gives complex roots, hence, system is unstable.

Problem No. 9.7 Find out the conditions for stability for the systems whose characteristics equations are given below. The case where stability is suggested for real values of K, determine the values of K which will cause sustained oscillations. Find the frequency of oscillations.

(a) $s^4 + 20s^3 + 224s^2 + 1240s + 2400 + K = 0$

Solution

The Routh's array is

$$\begin{array}{c|ccc|} s^4 & 1 & 224 & 2400+K \\ s^3 & 20 & 1240 & 0 \\ s^2 & 162 & 2400+K & \\ s^1 & 1240-\dfrac{(2400+K)20}{162} & 0 & \\ s^0 & 2400+K & & \end{array}$$

For stability

$$2400 + K > 0 \text{ or } K > -2400$$

Also, $$1240-\frac{(2400+K)20}{162} > 0 \quad \text{or} \quad K < 7644$$

The range for K for stability is

$$-2400 < K < 7644$$

If $K = 7644$, then the first column is

$$(1, 20, 162, 0, 10044)$$

Thus row No. four becomes a zero row. The subsidary equation for the row above this zero row i.e., row No. 3 is

$$162s^2 + 10044 = 0$$

$$\therefore \qquad s = \pm j7.9$$

The presence of an imaginary pair of roots gives oscillatory response with frequency of oscillations 7.9 rad/sec.

(b) $s^3 + (K + 0.5)s^2 + 4Ks + 0.5 = 0$

Solution The Routh's array is

$$\begin{array}{c|cc} s^3 & 1 & 4K \\ s^2 & (K+0.5) & 50 \\ s^1 & 4K-\dfrac{50}{K+0.5} & \\ s^0 & 50 & \end{array}$$

for stability

$$K + 0.5 > 0 \text{ or } K > -0.5, \text{ and}$$

$$4K - \frac{50}{K+0.5} > 0$$

or $$4K(K+0.5) > 50$$

or $$K(K+0.5) > 12.5$$

or $$K^2 + \frac{K}{2} - 12.5 > 0$$

or $$K > -3.8 \quad \text{and} \quad K > 3.3$$

Therefore, the condition for stability is

$$K > -3.8, \quad K > -0.5 \quad \text{and} \quad K > 3.3$$

Thus, if $K > 3.3$, then all conditions are satisfied.

If we put $K = 3.3$, then we have s^1 row as zero. The subsidary equation for s^2 row for $K = 3.3$ is

$$3.8s^2 + 50 = 0$$

$\therefore$ $$s = \pm j3.63$$

Thus frequency of oscillation is 3.63 rad/sec.

Problem No. 9.8 The open loop transfer function of a unity feedback control system is given as

$$G(s)H(s) = \frac{K}{s(1+Ts)}$$

It is desired that all the roots of the characterstics equation must lie in the region to the left of the lime $s = -a$. Determine the values of K and T required so that there are no roots to right of the line $s = -a$.

Solution

The characteristic equation is

$$1+G(s)\,H(s)=0$$

or $$1+\frac{K}{s(1+Ts)}=0$$

or $$s(1+Ts)+K=0$$

or $$Ts^2+s+K=0$$

or $$s^2+\frac{1}{T}s+\frac{K}{T}=0$$

Replace s by $(s-a)$

Thus $$(s-a)^2+\frac{1}{T}(s-a)+\frac{K}{T}=0$$

or $$s^2+\left(-2a+\frac{1}{T}\right)s+\left(a^2+\frac{K}{T}-\frac{a}{T}\right)=0$$

The Routh's array is

$$\begin{array}{c|cc|} s^2 & 1 & a^2+\dfrac{K}{T}-\dfrac{a}{T} \\ s^1 & \left(-2a+\dfrac{1}{T}\right) & 0 \\ s^0 & a^2+\dfrac{K}{T}-\dfrac{a}{T} & \end{array}$$

If no root is to lie on the line $s=-a$, then the first column of the Routh's array should have no sign change.

i.e. $$\left(-2a+\frac{1}{T}\right)>0 \quad \text{or} \quad \frac{1}{T}>2a$$

or $$T<\frac{1}{2a} \quad \text{or} \quad aT<\frac{1}{2}$$

Also, $$a^2+\frac{K}{T}-\frac{a}{T}>0$$

or $$\frac{K}{T}>\left(\frac{a}{T}-a^2\right)$$

or $$K>(a-a^2T)$$

or $$K > a(1 - aT)$$

$\therefore$ if $$aT = 0 \quad K > 0$$

and if $$aT = \frac{1}{2} \quad K > \frac{a}{2}$$

Problem No. 9.9 The openloop transfer function of a unity feedback system is given by

$$G(s) = \frac{K}{s(s+3)(s^2+s+1)}$$

Determine the values of *K* that will cause sustained oscillations in the closed loop system. Also find the oscillation frequency.

Solution

The characteristic equation is

$$1 + G(s)\frac{K}{s(s+3)(s^2+s+1)} = 0$$

or $$s(s+3)(s^2+s+1) + K = 0$$

or $$s^4 + 4s^3 + 4s^2 + 3s + K = 0$$

The Routh's array is

$$\begin{array}{c|ccc|}
s^4 & 1 & 4 & K \\
s^3 & 4 & 3 & 0 \\
s^2 & \frac{13}{4} & K & \\
s^1 & \frac{\left(\frac{39}{4} - 4K\right)}{\frac{13}{4}} & & \\
s^0 & K & &
\end{array}$$

The condition for system stability is

$$K > 0$$

$$\left(\frac{39}{4} - 4K\right) > 0$$

Therefore, for stability, K should lie in the range

$$\frac{39}{16} > K > 0$$

When $K = \frac{39}{16}$, thre will be a zero at the first entry in the fourth row. This will indicate presence of symmetrical roots, which as shown below, will be pure imaginary.

$$\therefore \qquad K = \frac{39}{16} \text{ will cause sustained oscillations.}$$

The subsidary equation of third row for

$$K = \frac{39}{16}, \text{ is}$$

$$\frac{13}{4}s^2 + \frac{39}{16} = 0$$

$$\therefore \qquad s = \pm j0.75$$

Thus, the frequency of sustained oscillations is 0.75 rad/sec.

Problem No. 9.10 Determine the values of K and b, so that the system whose equation is

$$s^3 + bs^2 + 3s + 1 + K(s+1) = 0$$

oscillates at a frequency of oscillations of 2 rad/sec.

Solution

The characteristic is

$$s^3 + bs^2 + 3s + 1 + K(s+1) = 0$$

or
$$s^3 + bs^2 + (K+3)s + K(s+1) = 0$$

The Routh's array is

$$\begin{array}{c|ll|} s^3 & 1 & K+3 \\ s^2 & b & K+1 \\ s^1 & (K+1) - \dfrac{(K+1)}{b} & \\ s^0 & K+1 & \end{array}$$

The system will have a sustained oscillations if row No. 3 is zero i.e.

$$\frac{b(K+3)-(K+1)}{b}=0$$

i.e. $$b=\frac{K+1}{K+3} \qquad ...(1)$$

The subsidary equation of row No. 2 is

$$bs^2+(K+1)=0$$

$$\therefore \quad s^2=-\frac{(K+1)}{b}=(j\omega)^2=(j2)^2=-4$$

(because ω = 2 rad/sec)

Putting $b=\frac{K+1}{4}$, in equation (1) we get

$$\frac{K+1}{4}=\frac{K+1}{K+3}$$

or $K+3=4$

or $K=1$ **Ans.**

or $b=\frac{K+1}{4}=\frac{1+1}{4}=\frac{1}{2}=0.5$ **Ans.**

Problem No. 9.11 The characteristic equation of a control system is

$$P(s)=(s+1)(s^2+7s+16)=0$$

If K = 8, check, all the roots of the characteristics equation of the system have damping factor greater than 0.5.

Solution

The oscillation will be caused due to the quadratic factor $s^2+7s+16$. $(s+1)$ will cause the response e^{-t} to die out. Therefore considering $(s^2+7s+16)$ and comparing with $s^2+2\xi\omega_n s+\omega_n{}^2=0$, we get

$$\omega_n^2=16 \quad \therefore \omega_n=4$$

and $2\xi\omega_n=7$

$$\therefore \quad \xi=\frac{7}{2\omega_n}=\frac{7}{2\times 4}=\frac{7}{8}=0.874$$

This is greater than 0.5.

Problem No. 9.12 Determine whether the largest time constant of the characteristic equation given below is less than, greater than or equal to 1.0 sec.

$$s^3 + 5s^2 + 8s + 6 = 0$$

Solution

$$s^3 + 5s^2 + 8s + 6 = 0$$

Its roots are

$$(s+2)(s^2 + 2s + 2) = 0$$

The roots are

$$s = -3 \quad \text{and} \quad s = -1 \pm j1$$

for $\quad s = -3$

$$T = \frac{1}{3} = 0.33 \text{ secs.} \qquad (\text{as } \xi\omega_n = 1)$$

for $\quad s = -1 \pm j1$

Refer Fig. 9.3 where the roots have been plotted

$$\xi = \cos\beta = \frac{1}{\sqrt{1+1}} = \frac{1}{\sqrt{2}}$$

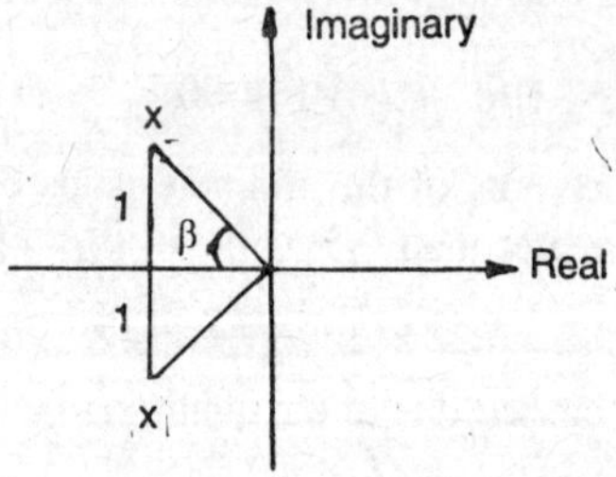

Fig. 9.3

Also,

$$\omega_n^{\,2} = 2$$

$\therefore$

$$\omega_n = \sqrt{2}$$

$$\xi\omega_n = \frac{1}{\sqrt{2}} \times \sqrt{2} = 1$$

$$T = \frac{1}{\xi\omega_n} = \frac{1}{1} = 1\,\text{secs}$$

$\therefore$ Largest time constant is equal to 1 secs.

Problem No. 9.13 A feedback control system has an open loop transfer function of

$$G(s)\,H(s) = \frac{Ke^{-s}}{s(s^2+2s+1)}$$

Determine the maximum value of K for the closed loop stability.

Solution

For low frequencies

$$e^{-s} = (1-s)$$

$$G(s)\,H(s) = \frac{K(1-s)}{s(s^2+2s+1)}$$

$$1 + G(s)\,H(s) = 1 + \frac{K(1-s)}{s(s^2+2s+1)} = 0$$

$\therefore$

$$s(s^2+2s+1) + K(1-s) = 0$$

$$s^3 + 2s^2 + s + K - Ks) = 0$$

$$s^3 + 2s^2 + s(1-K) + K = 0$$

The Routh's array is

$$\begin{array}{c|cc|} s^3 & 1 & (1-K) \\ s^2 & 2 & K \\ s^1 & \dfrac{2(1-K)-K}{2} & \\ s^0 & K & \end{array}$$

For stability $K > 0$ and

$$2(1-K) - K > 0$$

or

$$2 - 3K > 0$$

$$K < \frac{3}{2}$$

Hence, the restriction of K is

$$0 < K < \frac{3}{2}$$ **Ans.**

Problem No. 9.14 Decide the stability of the system, whose characteristic equation is given by

$$s^5 + 2s^4 + 5s^3 + 10s^2 + 4s + 8 = 0$$

(Pune University)

Solution

s^5	1	5	4
s^4	2	10	8
s^4	1	5	4
s^3	0	0	
s^3	2	5	
s^2	$\frac{5}{2}$	4	
s^1	$\frac{9}{5}$	–	
s^0	4	–	

$$2s^4 + 10s^2 + 8 = 0$$

$$8s^3 + 20s = 0$$

$$\text{or} \quad 2s^3 + 5s = 0$$

s^3 row was a zero row and hence subsidary equation was formed by writing the subsidary function of s^4 and then taking its derivative There is no sign change in the first column of the new array so formed. However, there is a zero row.

The subsidary equation formed of s^4 row is one of the factors of the characteristic equation. The other factor is $(s + 2)$

$$\therefore \quad s^5 + 2s^4 + 5s^3 + 10s^2 + 4s + 8$$

$$= (s^4 + 5s^2 + 4)(s + 2)$$

Roots of

$$s^4 + 5s^2 + 4 \Rightarrow s = \pm j2.14, \pm j0.66$$

$$s + 2 \Rightarrow s = -2$$

Imaginary roots make system unstable and response is continuous oscillatory. Therefore, system is unstable.

Problem No. 9.15 Determine the range of K such that feedfack system having characteristic equation

$$s(s^2+s+1)(s+2)+K=0, \text{ will be stable.}$$

(*Pune University*)

Solution

The characteristic equation is

$$s^4+5s^3+5s^2+4s+K=0$$

Routh's Array

s^4	1	5	K
s^3	5	4	
s^2	4.2	K	
s^1	$4-1.2K$	–	
s^0	K	–	

For stability

$$K>0$$

and $$4-1.2K>0$$

i.e. $K<3.33$

Therefore range of K for stability is

$$0<K<3.33$$

Problem No. 9.16 A system is given by equation

$$s^3+2.5s^2+20s+10K=0$$

Find the restriction on K so that the system is absolutely stable.

(*Pune University*)

Solution

The equation is

$$s^3+2.5s^2+20s+10K=0$$

Routh's array is

$$\begin{array}{c|cc|} s^3 & 1 & 20 \\ s^2 & 2.5 & 10K \\ s^1 & \dfrac{50-10K}{2.5} & - \\ s^0 & 10K & - \end{array}$$

For absolute stability

$$10\,K > 0 \text{ or } K > 0$$

and $\quad 50-10K > 0 \text{ or } K < 5$

Therefore, the system will be absolutely stable if the range of K is

$$0 < K < 5$$

Ans.

Problem No. 9.17 A system is given by

$$Ts^3 + (1+2T)s^2 + 2s + 100 = 0$$

Find the value of T such that the system is driven on the verge of instability and the resulting frequency of oscillation.

Solution

The characteristic equation is

$$Ts^3 + (1+2T)s^2 + 2s + 100 = 0$$

The Routh's array is

$$\begin{array}{c|lc|} s^3 & T & 2 \\ s^2 & 1+2T & 100 \\ s^1 & \dfrac{2(1+2T)-100T}{1+2T} & 0 \\ s^0 & 100 & - \end{array}$$

The restriction on T is

$$T > 0$$

$$T > -\frac{1}{2}$$

and $\quad 2-96T > 0 \quad \text{or} \quad T < \dfrac{1}{48}$

The system will become unstable if $T = \frac{1}{48}$

If we substitute $T = \frac{1}{48}$, we will get a zero row in Routh's array and since the zero row occurs on the odd power of s, the s^1 row will be a zero row. The subsidary equation of s^2 row is

$$1.042s^2 + 100 = 0$$

$$s = \pm j9.8$$

Therefore, the angular frequency of oscillation is 9.8 rad/sec.

Problem No. 9.18 The transfer of a closed loop system with unity feedback system is

$\frac{K(s+2)(s+1)}{(s+0.1)(s-1)}$. Comment on stability.

Solution

The characteristic equation is

$$1 + G(s) = \frac{1 + K(s+2)(s+1)}{(s+0.1)(s-1)} = 0$$

or $$(K+1)s^2 + (3K-0.9)s + (2K-0.1) = 0$$

Routh's array is

$$\begin{array}{c|cc|} s^2 & K+1 & 2K-0.1 \\ s^1 & 3K-0.9 & - \\ s^0 & 2K-0.1 & - \end{array}$$

Comments

1. For stability, first column should be greater than unity i.e.

(a) $K > -1$

(b) $K > \frac{0.9}{3}$ i.e. $K > 0.3$

(c) $K > \frac{0.1}{2}$ i.e. $K > 0.05$

There will be no sign change if the condition $K > 0.3$ is satisfied and this condition satisfies all the remaining conditions.

Problem No. 9.19 Applying Routh's criterion, find range of K for stability for a system whose characteristic equation is given by

$$s^3 + 3Ks^2 + (K+2)s + 4 = 0$$

Solution

Routh's array is

$$\begin{array}{c|cc|} s^3 & 1 & K+2 \\ s^2 & 3K & 4 \\ s^1 & \dfrac{3K(K+2)-4}{3K} & - \\ s^0 & 4 & - \end{array}$$

Condition for stability

(a) $3K > 0$ or $K > 0$

(b) $\dfrac{3K(K+2)-4}{3K} > 0$ or $K > 0.527$

Therefore, range of K

$$\infty > K > 0.527$$

Ans.

Problem No. 9.20 Test the stability of this control system characterised by the characteristic equation

$$s^4 + 2s^3 + 3s^2 + 4s + 5$$

Solution

Routh's array is

$$\begin{array}{c|ccc|} s^4 & 1 & 3 & 5 \\ s^3 & 2 & 4 & \\ s^2 & 1 & 5 & \\ s^1 & -6 & - & \\ s^0 & 5 & - & \end{array}$$

System is unstable as there are two sign changes in the first column of Routh's array. It indicates two roots lie on the R.H.S of s-plane.

Problem No. 9.21 A unity feedback control system is described by charac-ter-istic equation

$$s^4+4s^3+7s^2+16s+12=0$$

Test its stability and find the roots on the imaginary axis. *(AMIE)*

Solution

$$\begin{array}{c|ccc|} s^4 & 1 & 7 & 12 \\ s^3 & 4 & 16 & \\ s^2 & 3 & 12 & \\ s^1 & 0 & 0 & \\ & 2 & & \\ s^0 & 12 & & \end{array} \qquad \begin{array}{l} 3s^2+12=0 \\ \text{or} \quad s^2+4=0 \\ 2s=0 \end{array}$$

The s^1 row becomes a zero row. s^1 row is therefore formed by the procedure explained earlier. (Problem 9.2). First column of the Routh's array shows no sign change. But, presence of zero row indicates, presence of symmetrically located roots in the s-plane. Roots are found as given below

$$s^4+4s^3+7s^2+16s+12$$

$$=(s^2+4)(s^2+4s+3)$$

s^2+4 gives $\pm j2$

s^2+4s+3 gives $-3, -1$

Problem No. 9.22 Determine the range of values of K $(K>0)$ such that the characteristics equation

$$s^3+3(K+1)s^2+(7K+5)s+(4K+7)=0$$

has roots more negative than $s=-1$.

Solution

Substitute $s=z-1$

The characteristic equation becomes

$$z^3+3Kz^2+(K+2)z+4=0$$

$$\begin{array}{c|cc|} z^3 & 1 & K+2 \\ z^2 & 3K & 4 \\ z^1 & \dfrac{3K(K+2)-4}{3K} & - \\ z^0 & 4 & - \end{array}$$

For stability,

$$3K > 0 \Rightarrow K > 0, \text{ and}$$

$$\frac{3K(K+2)-4}{3K} > 0$$

or $$3K^2 + 6K - 4 > 0$$

or $$K = -2.53,\ 0.53$$

Hence the range is $\infty > k > 0.53$

Problem No. 9.23 Comment on the stability of the following system using Routh's criteria

$$s^4 + 6s^3 + 21s^2 + 36s + 20 = 0$$

Solution

s^4	1	21	20
s^3	6	36	
	1	6	
s^2	15	20	
	14	4	
s^1	$\frac{4}{3}$		
s^0	4		

There is no sign change in the first column of Routh's array and hence system under consideration is stable.

Problem No. 9.24 Using Routh's criterion, check the stability of the systems

(a) $2s^4 + s^3 + 2s^2 + s + 20 = 0$

(b) $s^5 + s^4 + 2s^3 + s^2 + 4s + 2 = 0$

Comment on the location of roots.

(a)

s^4	2	2	20
s^3	1	1	–
s^2	$0(K)$	20	
s^1	$\frac{K-20}{K}\left(1-\frac{20}{K}\right)$		
s^0	20		

There is a zero in s^2 row. It is replaced by $K \to 0$. First column of s^1 row has the term $1-\frac{20}{K}$ which is a negative term as $K \to 0$. It means there are two signs changes which implies that two roots lie on RHS of s-plane. System is unstable.

(b)

$$\begin{array}{c|ccc} s^5 & 1 & 2 & 4 \\ s^4 & 1 & 1 & 2 \\ s^3 & 1 & 2 & \\ s^2 & -1 & 2 & \\ s^1 & 4 & & \\ s^0 & 2 & & \end{array}$$

There are two sign changes and hence two roots lie on the RHS of s-plane. System is unstable.

Problem No. 9.25 The characteristic equation for a certain feedback control system is given by $s^4 + a_1 s^3 + a_2 s^2 + a_3 s + K = 0$. If the numerical values of $a_1 = 22$, $a_2 = 10$ and $a_3 = 2$, find K which corresponds to the stable system.

Solution

The characteristic equation is

$$s^4 + 22s^3 + 10s^2 + 2s + K = 0$$

$$\begin{array}{c|ccc} s^4 & 1 & 10 & K \\ s^3 & 22 & 2 & \\ s^2 & \frac{218}{22} & K & \\ s^1 & 2-\frac{484K}{218} & & \\ s^0 & K & & \end{array}$$

For stability

$$2-\frac{484K}{218} > 0$$

or

$$\frac{484K}{218} < 2$$

or $$K < \frac{2 \times 218}{484}$$

or $$K < 0.9$$

and $$K > 0$$

$\therefore$ the range of K for stability is

$$0 < K < 0.9$$

$K = 0.9$ will cause sustained oscillations in the closed-loop system.

Problem No. 9.26 Using Routh-Hurwitz criterion check whether systems represented by the following characteristics equations are stable or not. Comment on the location of roots. Determine the frequency of sustained oscillations, if any

(a) $s^3 + 20s^2 + 9s + 100 = 0$

s^3	1	9
s^2	20	100
s^1	4	
s^0	100	

Since there is no sign change in the first column of Routh's array, no roots of characteristic equations are located on the RHS of s-plane. System is stable.

(b) $s^4 + 2s^3 + 6s^2 + 8s + 8 = 0$ *Remarks*

				Remarks
s^4	1	6	8	
s^3	2(1)	8(4)		
s^2	2(1)	8(4)		$s^2 + 4 = 0$
s^1	0(2)			$2s = 0$
s^0	4			

s^1 row becomes a zero. Problem was overcome by forming the auxillary polynomial from the co-efficients of s^2 row. Derivative of the polynomial was taken which was $2s = 0$ and s row was replaced by the co-efficient of s which is 2.

There is no sign change in the first column of Routh's array.

Now

$$s^4 + 2s^3 + 6s^2 + 8s + 8 = (s^2 + 4)(s^2 + 2s + 2)$$

Roots are

$$s^2+4=0 \Rightarrow s=\pm j2$$

and $$s^2+2s+2=0 \Rightarrow s=-1\pm j1$$

No roots lie on the RHS of s-plane.

Frequency of sustained oscillation is 2rad/sec. System is limitedly stable.

Problem No. 9.27 Apply Routh criterion to the following equation and determine the number of roots

(a) with positive real parts

(b) zero real parts

(c) with negative seal parts

$$P(s)=s^6+11s^5+42s^4+7s^2+71s^2+61s+30$$

Solution Routh array is

S^6	1	42	71	30
S^5	11	72	61	
S^4	390/11	720/11	30	
S^3	672/13	672/13		
S^2	30	30		
S^1	0	0		
S^0				

s^1 row has become a zero row. The subsidary equation of row above S^1 row i.e., s^2 row is formed and is

$$30s^2+30=30(s^2+1)$$

Therefore

$$P(s)=(s^2+1)(s^4+11s^3+41s^2+61s+30)=0$$

or $$P(s)=(s^2+1)(s+1)(s+2)(s+3)(s+5)$$

Roots are

$$S=\pm j,-1,-2,-3,-5$$

Therefore, there are

(a) no roots with positive real parts

(b) two roots with zero real part

i.e., $s=+j$ and $s=-j$

(c) four roots with negative real parts i.e. $s=-1,-2,-3$ and -5

Reference Tables

Table 1

Laplace Transforms

$f(t)$	$F(s)$
$f(t)$	$\frac{1}{s}$
$u(t)$	$\frac{1}{s^2}$
t^n (n is integer)	$\frac{n!}{s^{n+1}}$
e^{at}	$\frac{1}{s-a}$
te^{at}	$\frac{1}{(s-a)^2}$
$t^n e^{at}$	$\frac{n!}{(s-a)^{n+1}}$
e^{-at}	$\frac{1}{s+a}$
$\delta(t)$	1
$\frac{1}{(n-1)!} t^{n-1} e^{at}$	$\frac{1}{(s-a)^n}$
$\frac{e^{at} - e^{bt}}{a-b}$	$\frac{1}{(s-a)(s-b)}$
$\frac{e^{-at}}{(b-a)(c-a)} + \frac{e^{-bt}}{(a-b)(c-b)} + \frac{e^{-ct}}{(a-c)(b-c)}$	$\frac{1}{(s+a)(s+b)(s+c)}$
$\left(1 - e^{at}\right)$	$\frac{-a}{s(s-a)}$

Table 2

Some More Laplace Transforms

$f(t)$	$F(s)$
$\sin\omega t$	$\dfrac{\omega}{s^2+\omega^2}$
$\dfrac{1}{\omega}\sin\omega t$	$\dfrac{1}{s^2+\omega^2}$
$\cos\omega t$	$\dfrac{s}{s^2+\omega^2}$
$e^{-at}\sin\omega t$	$\dfrac{\omega}{(s+a)^2+\omega^2}$
$e^{-at}\cos\omega t$	$\dfrac{s+a}{(s+a)^2+\omega^2}$
$t\sin\omega t$	$\dfrac{2\omega s}{(s^2+\omega^2)^2}$
$t\cos\omega t$	$\dfrac{s^2-\omega^2}{s^2+\omega^2}$
$\sinh\omega t$	$\dfrac{\omega}{s^2-\omega^2}$
$\cosh\omega t$	$\dfrac{s}{s^2+\omega^2}$
$t\sinh\omega t$	$\dfrac{2\omega s}{\left(s^2-\omega^2\right)^2}$
$t\cosh\omega t$	$\dfrac{s^2+\omega^2}{\left(s^2-\omega^2\right)^2}$
$e^{-at}\sinh\omega t$	$\dfrac{\omega}{(s+a)^2-\omega^2}$
$e^{-at}\cosh\omega t$	$\dfrac{(s+a)}{(s+a)^2-\omega^2}$

Table 3

Two-Port Parameters

Parameter	Abbreviation	Variable Function: Base Expression	Variable Function: Related with	Relation
Open-Circuit Impedance	Z	V_1, V_2	I_1, I_2	$\begin{bmatrix} V_1 \\ V_2 \end{bmatrix} = \begin{bmatrix} Z_{11} & Z_{12} \\ Z_{21} & Z_{22} \end{bmatrix} \begin{bmatrix} I_1 \\ I_2 \end{bmatrix}$
Short-Circuit admittance	Y	I_1, I_2	V_1, V_2	$\begin{bmatrix} I_1 \\ I_2 \end{bmatrix} = \begin{bmatrix} Y_{11} & Y_{12} \\ Y_{21} & Y_{22} \end{bmatrix} \begin{bmatrix} V_1 \\ V_2 \end{bmatrix}$
Transmission	T	V_1, I_1	V_2, I_2	$\begin{bmatrix} V_1 \\ I_1 \end{bmatrix} = \begin{bmatrix} A & B \\ C & D \end{bmatrix} \begin{bmatrix} V_2 \\ -I_2 \end{bmatrix}$
Inverse Transmission	T′	V_2, I_2	V_1, I_1	$\begin{bmatrix} V_2 \\ I_2 \end{bmatrix} = \begin{bmatrix} A' & B' \\ C' & D' \end{bmatrix} \begin{bmatrix} V_1 \\ -I_1 \end{bmatrix}$
Hybrid	h	V_1, I_2	I_1, V_2	$\begin{bmatrix} V_1 \\ I_2 \end{bmatrix} = \begin{bmatrix} h_{11} & h_{12} \\ h_{21} & h_{22} \end{bmatrix} \begin{bmatrix} I_1 \\ V_2 \end{bmatrix}$
Inverse hybrid	g	I_1, V_2	V_1, I_2	$\begin{bmatrix} I_1 \\ V_2 \end{bmatrix} = \begin{bmatrix} g_{11} & g_{12} \\ g_{21} & g_{22} \end{bmatrix} \begin{bmatrix} V_1 \\ I_2 \end{bmatrix}$

Table 4

Conversion of Parameters

Parameter Name	Z	Y	T	T'	h	g
Z	$\begin{matrix} Z_{11} & Z_{12} \\ Z_{21} & Z_{22} \end{matrix}$	$\begin{matrix} \frac{Y_{22}}{\Delta Y} & \frac{-Y_{12}}{\Delta Y} \\ \frac{-Y_{21}}{\Delta Y} & \frac{Y_{22}}{\Delta Y} \end{matrix}$	$\begin{matrix} \frac{A}{C} & \frac{\Delta T}{C} \\ \frac{1}{C} & \frac{D}{C} \end{matrix}$	$\begin{matrix} \frac{D'}{C'} & \frac{1}{C'} \\ \frac{\Delta T'}{C'} & \frac{A'}{C'} \end{matrix}$	$\begin{matrix} \frac{\Delta h}{h_{22}} & \frac{h_{12}}{h_{22}} \\ \frac{-h_{21}}{h_{22}} & \frac{1}{h_{22}} \end{matrix}$	$\begin{matrix} \frac{1}{g_{11}} & \frac{-g_{12}}{g_{11}} \\ \frac{g_{21}}{g_{11}} & \frac{\Delta g}{g_{11}} \end{matrix}$
Y	$\begin{matrix} \frac{Z_{22}}{\Delta Z} & -\frac{Z_{12}}{\Delta Z} \\ \frac{-Z_{22}}{\Delta Z} & \frac{Z_{12}}{\Delta Z} \end{matrix}$	$\begin{matrix} Y_{11} & Y_{12} \\ Y_{21} & Y_{22} \end{matrix}$	$\begin{matrix} \frac{D}{B} & \frac{-\Delta T}{B} \\ \frac{-1}{B} & \frac{A}{B} \end{matrix}$	$\begin{matrix} \frac{A'}{B'} & \frac{-1}{B'} \\ \frac{-\Delta'_T}{B'} & \frac{D'}{B'} \end{matrix}$	$\begin{matrix} \frac{1}{h_{11}} & \frac{-h_{12}}{h_{11}} \\ \frac{h_{21}}{h_{11}} & \frac{\Delta h}{h_{11}} \end{matrix}$	$\begin{matrix} \frac{\Delta g}{g_{22}} & \frac{g_{12}}{g_{22}} \\ \frac{-g_{21}}{g_{22}} & \frac{1}{g_{22}} \end{matrix}$
T	$\begin{matrix} \frac{Z_{11}}{Z_{21}} & \frac{\Delta Z}{Z_{21}} \\ \frac{1}{Z_{21}} & \frac{Z_{22}}{Z_{21}} \end{matrix}$	$\begin{matrix} -\frac{Y_{22}}{Y_{21}} & \frac{-1}{Y_{21}} \\ \frac{-\Delta Y}{Y_{21}} & \frac{-Y_{22}}{Y_{21}} \end{matrix}$	$\begin{matrix} A & B \\ C & D \end{matrix}$	$\begin{matrix} \frac{D'}{\Delta'_T} & \frac{B'}{\Delta'_T} \\ \frac{C'}{\Delta'_T} & \frac{A'}{\Delta'_T} \end{matrix}$	$\begin{matrix} \frac{-\Delta h}{h_{21}} & \frac{-h_{12}}{h_{21}} \\ \frac{-h_{22}}{h_{21}} & \frac{-1}{h_{21}} \end{matrix}$	$\begin{matrix} \frac{1}{g_{21}} & \frac{g_{22}}{g_{21}} \\ \frac{g_{11}}{g_{21}} & \frac{\Delta Z}{g_{21}} \end{matrix}$
T'	$\begin{matrix} \frac{Z_{22}}{Z_{12}} & \frac{\Delta Z}{Z_{12}} \\ \frac{1}{Z_{12}} & \frac{Z_{11}}{Z_{12}} \end{matrix}$	$\begin{matrix} \frac{-Y_{11}}{Y_{12}} & \frac{1}{h_{12}} \\ \frac{-\Delta Y}{Y_{12}} & \frac{-Y_{22}}{Y_{12}} \end{matrix}$	$\begin{matrix} \frac{D}{\Delta T} & \frac{B}{\Delta T} \\ \frac{C}{\Delta T} & \frac{A}{\Delta T} \end{matrix}$	$\begin{matrix} A' & B' \\ C' & D' \end{matrix}$	$\begin{matrix} -\frac{1}{h_{12}} & \frac{h_{11}}{h_{12}} \\ \frac{h_{22}}{h_{12}} & \frac{\Delta h}{h_{12}} \end{matrix}$	$\begin{matrix} \frac{-\Delta g}{g_{12}} & \frac{-g_{22}}{g_{12}} \\ \frac{-g_{11}}{g_{12}} & \frac{-1}{g_{12}} \end{matrix}$
h	$\begin{matrix} \frac{\Delta Z}{Z_{22}} & \frac{Z_{12}}{Z_{22}} \\ \frac{-Z_{21}}{Z_{22}} & \frac{1}{Z_{22}} \end{matrix}$	$\begin{matrix} \frac{1}{Y_{11}} & \frac{-Y_{12}}{Y_{11}} \\ \frac{Y_{21}}{Y_{11}} & \frac{\Delta Y}{Y_{11}} \end{matrix}$	$\begin{matrix} \frac{B}{D} & \frac{\Delta T}{D} \\ \frac{-1}{D} & \frac{C}{D} \end{matrix}$	$\begin{matrix} \frac{B'}{A'} & -\frac{1}{A'} \\ \frac{-\Delta'_T}{A'} & \frac{C'}{A'} \end{matrix}$	$\begin{matrix} h_{11} & h_{12} \\ h_{21} & h_{22} \end{matrix}$	$\begin{matrix} \frac{g_{22}}{\Delta g} & \frac{-g_{12}}{\Delta g} \\ \frac{-g_{21}}{\Delta g} & \frac{g_{11}}{\Delta g} \end{matrix}$
g	$\begin{matrix} \frac{1}{Z_{11}} & \frac{-Z_{12}}{Z_{11}} \\ \frac{Z_{21}}{Z_{11}} & \frac{\Delta Z}{Z_{11}} \end{matrix}$	$\begin{matrix} \frac{\Delta Y}{Y_{22}} & \frac{Y_{12}}{Y_{22}} \\ \frac{-Y_{21}}{Y_{22}} & \frac{1}{Y_{22}} \end{matrix}$	$\begin{matrix} \frac{C}{A} & \frac{-\Delta T}{A} \\ \frac{1}{A} & \frac{B}{A} \end{matrix}$	$\begin{matrix} \frac{C'}{D'} & \frac{-1}{D'} \\ \frac{\Delta'_T}{D'} & \frac{B'}{D'} \end{matrix}$	$\begin{matrix} \frac{h_{22}}{\Delta h} & \frac{-h_{12}}{\Delta h} \\ \frac{-h_{21}}{\Delta_h} & \frac{h_{11}}{\Delta_h} \end{matrix}$	$\begin{matrix} g_{11} & g_{12} \\ g_{21} & g_{22} \end{matrix}$